D0828281

The Enzymes

STRUCTURE AND CONTROL

VOLUME I

Third Edition

CONTRIBUTORS

DANIEL E. ATKINSON
LOUIS A. COHEN
DAVID J. COX
DAVID EISENBERG
D. E. KOSHLAND, JR.
LESTER J. REED
MILTON J. SCHLESINGER
ELLIOTT SHAW
EMIL L. SMITH
E. R. STADTMAN

ADVISORY BOARD

THE ENZYMES

Structure and Control

Edited by PAUL D. BOYER

Molecular Biology Institute
University of California
Los Angeles, California

Volume I

THIRD EDITION

ACADEMIC PRESS New York and London 1970

ACADEMIC PRESS, INC.
111 Fifth Avenue, New York, New York 10003

United Kingdom Edition published by
ACADEMIC PRESS, INC. (LONDON) LTD.
Berkeley Square House, London W1X 6BA

LIBRARY OF CONGRESS CATALOG CARD NUMBER: 75-117107

PRINTED IN THE UNITED STATES OF AMERICA

Contents

8. Mechanisms of Enzyme Regulation in Metabolism

E. R. Stadtman

9. Enzymes as Control Elements in Metabolic Regulation

Daniel E. Atkinson

List of Contributors

Numbers in parentheses indicate the pages on which the authors' contributions begin.

DANIEL E. ATKINSON (461), Department of Chemistry, University of California, Los Angeles, California

LOUIS A. COHEN (147), Laboratory of Chemistry, National Institute of Arthritis and Metabolic Diseases, National Institutes of Health, Bethesda, Maryland

DAVID J. COX (213), Clayton Foundation, Biochemical Institute, and Department of Chemistry, University of Texas at Austin, Austin, Texas

DAVID EISENBERG (1), Molecular Biology Institute, and Department of Chemistry, University of California, Los Angeles, California

D. E. KOSHLAND, JR. (341), Department of Biochemistry, University of California, Berkeley, California

LESTER J. REED (213), Clayton Foundation, Biochemical Institute and Department of Chemistry, University of Texas at Austin, Austin, Texas

MILTON J. SCHLESINGER (241), Department of Microbiology, Washington University, School of Medicine, St. Louis, Missouri

ELLIOTT SHAW (91), Biology Department, Brookhaven National Laboratory, Upton, New York

EMIL L. SMITH (267), Department of Biological Chemistry, University of California, Los Angeles, California

E. R. STADTMAN (397), Laboratory of Biochemistry, National Heart Institute, National Institutes of Health, Bethesda, Maryland

Preface

The remarkable expansion of information on enzymes since 1959 when Volume I of the Second Edition of "The Enzymes" appeared is so widely recognized that justification for a third edition is unnecessary. The growth rate of information is akin to that of a bacterial culture in early log phase; further expansion of knowledge leading to a deeper understanding of enzymes at the molecular level shows promise of continuing at this rate.

During the last decade, the primary sequence of an increasing number of enzymes has been unraveled, and, even more important, the elegant beauty of the three-dimensional structure for some has been determined. Today, a student of enzymology and metabolism has a high base for biocatalytic insight and experimentation. Enzyme subunits and their interplay have been recognized and a prominent beginning has been made into the understanding of mechanisms of control whose very existence was only hazily forecast ten years ago. Correlated with advances in molecular genetics has come a realization of the dictation of the tertiary and quaternary structure by the primary amino acid sequence. These and related achievements have implications for biological development and evolution considerably beyond understanding the function of single protein molecules.

Somewhat overshadowed by these dramatic thrusts into new intellectual regions has been a continued progress in enzyme isolation, physical characterization, and chemical modification, and in further understanding kinetics, specificity, and mechanism. The goal is to acquire sufficient information to allow generalization and to achieve the satisfaction and power of understanding. In these steady advances lies the hope for a rational molecular pharmacology.

For the present edition, it seemed that any editor, or group of editors, would lack the perspective and competence to cover the entire field. Also, Professor Henry Lardy, who contributed so much to the second edition, was not available. Thus, the approach evolved for this edition was to plan each volume with the help of a highly competent Advisory Board composed of researchers in a particular field. Invaluable service has been given in organizing Volumes I and II by Advisory Board members John M. Buchanan, W. W. Cleland, William P. Jencks, Daniel E. Koshland, Jr., and Emil L. Smith.

The objective of "The Enzymes" is to present fully the information available on enzymes and enzyme catalysis at a molecular level. The scope of the third edition is similar to that of the second edition, except that separate volumes are not planned on cofactors. The chapters will stress contributions of the past ten years, depending on previous editions for much of the earlier material. The treatise does not attempt to give information on all enzymes that have been described. Indeed, there are hints that among enzymes it is considered something of an honor to achieve sufficient status to be presented as a separate chapter in "The Enzymes."

The first two volumes cover basic properties of enzymes in general. Volume I contains chapters on structure and control, Volume II on kinetics and mechanism. Subsequent volumes will cover specific areas of catalysis. Those in preparation cover hydrolysis of peptides and other C—N linkages; cleavage of glycoside bonds and of carboxyl, phosphate, and sulfate esters; phosphorolysis; hydration, dehydration, and elimination; carboxylation and decarboxylation; aldol cleavage and condensation and other C—C cleavage; and isomerization. Volumes for which Advisory Boards are still to be chosen will cover group transfer, energy-requiring syntheses, oxidation–reduction, and special processes.

Timeliness is difficult to achieve in multiauthored treatises. I am pleased that the present volumes excel in this respect. Manuscripts included in Volumes I and II were received within a three-month period. Through the splendid cooperation of the authors and the publisher, these volumes are published in about a year after receipt of the manuscripts! The users of the volume owe their appreciation to the authors and to the staff of Academic Press. Any credit to me must be shared by my capable and charming wife, Lyda, for many organizational details, manuscript handling, and editorial assistance.

Paul D. Boyer

Los Angeles, California

August, 1970

Contents of Other Volumes

Volume II: Kinetics and Mechanism

Volume III: Hydrolysis: Peptide Bonds

Volume IV: Hydrolysis: Other C–N Bonds, Phosphate Esters

1

X-Ray Crystallography and Enzyme Structure

DAVID EISENBERG

I. Introduction

A. Scope of the Chapter

One sign of the growing prominence of X-ray crystallography in the study of enzymes is that, whereas the combined indices of the eight volumes of the second edition of "The Enzymes" contain only seven references to crystallography and related topics, the present edition will have at least that number of chapters on these subjects. Most of these chapters will deal mainly with the results of crystallographic studies of particular enzymes; this chapter, in contrast, deals mainly with the methods of enzyme crystallography, though it also includes a survey of recent results. It is addressed primarily to enzyme chemists who wish to assess critically the results of crystallographic studies, to utilize results of these studies in their own work, or to participate to one extent or another in crystallographic studies. For this reason, an attempt has been made to define many of the terms and to introduce many of the concepts that are used in papers on protein crystallography. Emphasis has also been placed on some of the more chemical problems of enzyme crystallography such as crystallization of enzymes and formation of heavy atom derivatives. These problems are, in fact, the most troublesome ones of current research in this area. The survey of results of structural studies (Section IV,A) is presented partly as a guide to the literature, and partly to illustrate the usefulness of X-ray diffraction for learning about the structures, specificities, and mechanisms of enzymes.

The compounds discussed in this chapter are restricted to enzymes and other globular macromolecules such as the oxygen binding and electron transport proteins that can be studied by the same techniques and thus may be termed *honorary enzymes* (*1*). Previous summaries of the results of crystallographic studies of these compounds include those given by Dickerson (*2*), Kraut (*3*), Davies (*4*), Stryer (*5*), and Perutz (*1*), and also the recent monograph and supplement of stereopair photographs prepared by Dickerson and Geis (*6, 7*). Several reviews of the methods used to obtain the results have also been published. The superb discussion of Crick and Kendrew is the most qualitative of these (*8*) but is now somewhat out of date. Other excellent reviews, in increasing order of detail and mathematical content, are those by Dickerson (*2*), Holmes and Blow (*9*), and Phillips (*10*). Each of these refers to the standard texts

1. M. F. Perutz, *European J. Biochem.* **8,** 455 (1969).
2. R. E. Dickerson, *in* "The Proteins" (H. Neurath, ed.), Vol. 2, p. 603. Academic Press, New York, 1964.
3. J. Kraut, *Ann. Rev. Biochem.* **34,** 247 (1965).
4. D. R. Davies, *Ann. Rev. Biochem.* **36,** 321 (1967).
5. L. Stryer, *Ann. Rev. Biochem.* **37,** 25 (1968).
6. R. E. Dickerson and I. Geis, "The Structure and Action of Proteins." Harper, New York, 1969.
7. R. E. Dickerson and I. Geis, "Stereo Supplement to the Structure and Action of Proteins." Harper, New York, 1969.
8. F. H. C. Crick and J. C. Kendrew, *Advan. Protein Chem.* **12,** 133 (1957).
9. K. C. Holmes and D. M. Blow, *Methods Biochem. Anal.* **13,** 113 (1965); also published in monograph form by Wiley (Interscience), New York, 1966.
10. D. C. Phillips, *Advan. Struct. Res. Diffraction Methods* **2,** 75 (1966).

and reference works on X-ray crystallography. One newer text, by Stout and Jensen (*11*), and one review article, by Jacobson and Lipscomb (*11a*), merit special note as being helpful for beginners.

B. Power and Limitations of X-ray Crystallography

X-ray crystallography and her sister technique neutron crystallography (*12*) are the only methods that can presently yield the relative positions of all, or nearly all, the atoms in an enzyme. Even these methods are capable of achieving this end only under favorable circumstances and then only with the aid of additional information such as the amino acid sequence of the enzyme. The end product of these methods is a plot of the electron density of the crystal $\rho(x, y, z)$ as a function of three coordinate axes. This function is usually displayed on a stack of Plexiglas sheets, each showing the contours of electron density in one slice through a unit cell of the crystal. In crystallographic studies of smaller molecules this function is often sufficiently detailed to give a structure with no prior knowledge of the arrangement, or even of the composition, of atoms in the molecule. With enzymes, however, other information is usually indispensable for complete identification of atomic positions in the electron density map. In the absence of such information, the crystallographer can sometimes determine the path of the polypeptide backbone of an enzyme and can identify amino acid residues with distinctive side chains. In many cases he could probably locate peptides of known sequence in his electron density map; thus, X-ray crystallography could probably be used profitably in conjunction with chemical methods in sequence determination. Once the structure of an enzyme is known, crystallographic methods can yield important knowledge about catalysis and binding from studies of crystals containing inhibitors, cofactors, or compounds similar to substrates. These methods involve the *Fourier difference* maps discussed in Section III,D,2.

In less favorable circumstances, or in the preliminary stages of analysis, crystallographic methods can give the overall shape of an enzyme and the regions of the enzyme to which smaller molecules bind. For multisubunit enzymes, even such limited information can be of great interest. The salt difference map method of determining molecular shapes is described in Section III,D,2.

With very little work the crystallographer can often determine the

11. G. H. Stout and L. H. Jensen, "X-ray Structure Determination." Macmillan, New York, 1968.

11a. R. A. Jacobson and W. N. Lipscomb, *in* "Technique of Organic Chemistry" (A. Weissberger, ed.). Wiley (Interscience), New York (in press).

12. B. P. Schoenborn, *Nature* **224,** 143 (1969).

molecular weight and molecular symmetry of an enzyme. These procedures are described in Section II,C.

Some of the major limitations of crystallographic methods are practical ones: It is necessary to crystallize the material of interest in a unit cell having edges shorter than roughly 400 Å. The crystal, moreover, must be of sufficient size (more than about 0.1 mm in each dimension) and must be reasonably well ordered. The degree of order, more than anything else, limits the attainable resolution of the final electron density map. These problems of crystallization, which are by no means trivial, are discussed in Section II,D.

A limitation of crystallography, inherent in the phenomenon of diffraction, is that the resulting electron density function is an average over all unit cells of the crystal and over the time of the experiment

$$\rho(x, y, z)_{\text{observed}} = \langle\langle \rho(x, y, z)\rangle_{N\text{ unit cells}}\rangle_{\text{time}} \qquad (1)$$

This means, of course, that crystallography can yield very little direct information on the dynamics of an enzymic reaction and that the full potential of crystallographic results can be realized only when they are combined with other types of experiments.

One concern in the early days of enzyme crystallography was that the conformation of enzymes in crystals might be different from that in solution and thus crystallographic results might have only marginal relevance to biochemistry. This concern has been largely dispelled by several lines of evidence. The first of these is the close similarity in conformation of enzyme molecules in different crystal forms: The structure of chymotrypsin is similar in the α and γ forms of the crystal (*13, 14*), and the two molecules within one asymmetric unit of the α crystals (which have distinct molecular environments) have the same conformation (*15*). Ribonuclease S and ribonuclease A have closely similar structures even though the former was crystallized from an ammonium sulfate solution (*16*) and the latter from a 55% alcohol solution (*17*). These similarities indicate that crystal forces do not significantly deform enzymes. The same conclusion may be drawn from the similar folding of the globin chain in the different crystal environments of sperm-whale myoglobin, seal myoglobin, erythrocruorin, and the hemoglobins of horse, human, and *Glycera*

13. B. W. Matthews, G. H. Cohen, E. W. Silverton, H. Braxton, and D. R. Davies, *JMB* **36,** 179 (1968).
14. H. T. Wright, J. Kraut, and P. E. Wilcox, *JMB* **37,** 363 (1968).
15. B. W. Matthews, P. B. Sigler, R. Henderson, and D. M. Blow, *Nature* **214,** 652 (1967).
16. H. W. Wyckoff, K. D. Hardman, N. M. Allewell, T. Inagami, L. N. Johnson, and F. M. Richards, *JBC* **242,** 3984 (1967).
17. G. Kartha, J. Bello, and D. Harker, *Nature* **213,** 862 (1967).

(*1*), and also from the similarity of folding of the elastase and chymotrypsin molecules (*18*).

The most direct evidence for similarity of enzyme structures in the dissolved and crystalline states is that many enzymes retain activity when crystallized. This has been shown for ribonuclease S (*19*), ribonuclease A (*20*), carboxypeptidase A (*21, 22*), and α- and γ-chymotrypsin (*23, 24*). Still other crystalline enzymes form complexes with inhibitors, cofactors, and substrates. Alcohol dehydrogenase and L-lactate dehydrogenase, for example, combine with NADH in the crystal (*25, 26*), and cytochrome c peroxidase forms an enzyme–substrate complex (*27*). Numerous other examples of the ability of crystalline enzymes to react with substrates and inhibitors could be cited; some of these are mentioned in Sections III and IV.

It should be noted, nevertheless, that rates of catalysis and binding are often slower in crystals than in solution. Stryer (*5*) has summarized this evidence and has suggested that the reduced rates may be caused by either "limited access to the active site" owing to contacts between neighboring molecules or by a "reduced conformational motility of the protein in the crystalline state."

II. Enzyme Crystals

A. Differences from Other Crystals

Enzyme crystals differ in two important ways from crystals of smaller molecules. First, because enzymes contain amino acids having only the L configuration at α-carbon atoms, the crystals cannot possess the types of symmetry (mirror planes, glide planes, and centers of inversion) that require equal numbers of D and L configurations. As noted below, this

18. D. M. Shotton and H. C. Watson, *Phil. Trans. Roy. Soc. Lond. B* **257,** 111 (1970).
19. M. S. Doscher and F. M. Richards. *JBC* **238,** 2399 (1963).
20. J. Bello and E. F. Nowoswiat, *BBA* **105,** 325 (1965).
21. F. A. Quiocho and F. M. Richards. *Proc. Natl. Acad. Sci. U. S.* **52,** 833 (1964).
22. F. A. Quiocho and F. M. Richards, *Biochemistry* **5,** 4062 (1966).
23. P. B. Sigler, B. A. Jeffery, B. W. Matthews, and D. M. Blow, *JMB* **15,** 175 (1966).
24. P. B. Sigler and H. C. W. Skinner, *BBRC* **13,** 236 (1963).
25. C. I. Branden. *ABB* **112,** 215 (1965).
26. M. J. Adams, D. J. Haas, B. A. Jeffery, A. McPherson, Jr., H. L. Mermall, M. G. Rossmann, R. W. Schevitz, and A. J. Wonacott, *JMB* **41,** 159 (1969).
27. T. Yonetani, B. Chance, and S. Kajiwara. *JBC* **241,** 2981 (1966).

means that the crystal symmetries confronting protein crystallographers are both much simpler and far fewer than those often seen by small molecule crystallographers. Second, enzyme crystals normally contain a large fraction of solvent. The solvent may be nearly pure water, a salt solution, or an aqueous alcohol solution, depending on the procedure used for crystallization (Section II,D). Matthews found in a survey of 116 crystal forms of globular proteins that the fraction of the crystal volume occupied by solvent is most commonly near 43% but has been observed to have values from about 27 to 65% (*28*). In the extraordinary case of tropomyosin the crystals are about 95% solvent (*29*). The solvent fills the spaces between molecules in crystals, and these spaces usually form an interconnecting system of channels. Bishop and Richards, from a study of the rates of diffusion of solutes in cross-linked crystals of β-lactoglobulin, concluded that the major part of the crystal liquid in this system can be considered identical to water in bulk (*30*). They noted, however, that their methods could not detect the 10% or so of water that may be tightly bound to the protein. Evidence for some tightly bound water in crystals comes from at least two sources: The high resolution electron density map of myoglobin which shows water molecules (and perhaps ammonium ions) bound to all polar groups on the molecular surface (*31*) and the densities of hemoglobin crystals in different solvents that indicate some of the water of crystallization is impermeable to ions and is presumably bound tightly to the molecule (*32*).

The forces between protein molecules in crystals are weak and nonspecific. The weakness is well known to anyone who has tried to manipulate protein crystals without breaking or disordering them; it arises from the fact that each molecule touches neighbors at only a few regions [six in sperm-whale myoglobin type A (*31*) and four in horse cytochrome c (*33*)], each region consisting of several amino acids. The forces at these few regions of contact may be hydrogen bonds, salt links, or van der Waals forces between nonpolar side chains. Away from the contact points, the forces between molecules are attenuated by the intervening solvent of high dielectric constant. The lack of specificity is evident from the many different ways a given protein molecule can pack in a crystal. Bovine pancreatic ribonuclease, for example, crystallizes in at least four-

28. B. W. Matthews, *JMB* **33**, 491 (1968).
29. D. L. D. Caspar, C. Cohen, and W. Longley, *JMB* **41**, 87 (1969).
30. W. H. Bishop and F. M. Richards, *JMB* **38**, 315 (1968).
31. J. C. Kendrew, *Brookhaven Symp. Biol.* **15**, 216 (1962).
32. M. F. Perutz, *Trans. Faraday Soc.* **B42**, 187 (1946).
33. R. E. Dickerson, M. L. Kopka, C. L. Borders, J. Varnum, J. E. Weinzierl, and E. Margoliash, *JMB* **29**, 77 (1967).

teen different forms depending on the solvent, dissolved ions, and other conditions of crystallization (*34, 35*), and β-lactoglobulin crystallizes in at least twelve different forms (*36*). As noted by Caspar and Cohen (*37*), globular proteins have been selected for properties other than their ability to crystallize, and their tendency to form crystals under some nonphysiological conditions should be regarded as an "adventitious property of the molecular structure." Protein crystallographers might be tempted to add, "adventitious but useful."

B. Structure and Symmetry of Enzyme Crystals

The structure of any crystal may be thought of as being generated by a *lattice,* which is merely a rule for translation, and a *motif,* which is the translated object. The motif may be a single molecule, which may or may not have some internal symmetry of its own, or it may be several molecules (Fig. 1). The symmetry of the crystal structure is described by a *space group.* Because of the restrictions mentioned above on the possible symmetry of motifs in enzyme crystals, enzymes must crystallize in 1 of 65 space groups, instead of the full range of 230 space groups open to more symmetric molecules. Each space group is designated by a letter, giving the type of lattice, followed by a symbol giving the symmetry of the motif.

A *unit cell* of the crystal is the smallest unit, having the full available symmetry, from which the entire crystal can be generated by translations. In most crystals the unit cell is simply a block having lattice points at its eight corners; the lattice in such crystals is said to be primitive and is denoted by the letter *P*. In these crystals the number of motifs per unit cell is one. Seven of the fourteen lattices which can fill space (Table I) are primitive and differ among themselves in the possible symmetries of their motifs. In some crystals the unit cell does not have the full symmetry of its contents unless a lattice is chosen that has additional points in the center of the cell or at the centers of its faces. These are the *centered lattices,* denoted *I* for the body-centered cell having an additional lattice point in the center of the cell, *C* for the face-centered cell with additional lattice points in the centers of two faces, and *F* for the all face-centered cell having additional points in the centers of all faces.

34. M. V. King, B. S. Magdoff, M. B. Adelman, and D. Harker, *Acta Cryst.* **9**, 460 (1956).
35. M. V. King, J. Bello, E. H. Pignataro, and D. Harker, *Acta Cryst.* **15**, 144 (1962).
36. R. Aschaffenburg, D. W. Green, and R. M. Simmons, *JMB* **13**, 194 (1965).
37. D. L. D. Caspar and C. Cohen, *Nobel Symp.* **11**, 394 (1969).

TABLE I
THE 65 "BIOLOGICAL" SPACE GROUPS

Crystal system	Lattice	Minimum symmetry of unit cell	Unit cell edges and angles[a]	Diffraction pattern symmetry[b]	Space groups[c]
Triclinic	P	None	$a \neq b \neq c$	$\bar{1}$	$P1$
			$\alpha \neq \beta \neq \gamma$		
Monoclinic	P	2-fold axis parallel to **b**	$a \neq b \neq c$	$2/m$	$P2$, $P2_1$
	C		$\alpha = \gamma = 90°$		$C2$
			$\beta \neq 90°$		
Orthorhombic	P	3 mutually perpendicular 2-fold axes	$a \neq b \neq c$	mmm	$P222$, $P2_12_12_1$, $P222_1$, $P2_12_12$
	C		$\alpha = \beta = \gamma = 90°$		$C222$, $C222_1$
	I				[$I222$, $I2_12_12_1$]
	F				$F222$
Tetragonal	P	4-fold axis parallel to **c**	$a = b \neq c$	$4/m$	$P4$, ($P4_1$, $P4_3$), $P4_2$
	I		$\alpha = \beta = \gamma = 90°$		$I4$, $I4_1$
				$4/mmm$	$P422$, ($P4_122$, $P4_322$), $P4_222$
					$P42_12$, ($P4_12_12$, $P4_32_12$), $P4_22_12$
					$I422$, $I4_122$
Trigonal/rhombohedral	R^d	3-fold axis parallel to **c**	$a = b = c$	$\bar{3}$	$R3$
	P^d		$\alpha = \beta = \gamma \neq 90°$		$P3$, ($P3_1$, $P3_2$)
				$\bar{3}m$	$R32$
					[$P321$, $P312$]
					[($P3_121$, $P3_221$), ($P3_112$, $P3_212$)]

Hexagonal	P	6-fold axis parallel to **c**	$a = b \neq c$	$6/m$	$P6$, $(P6_1, P6_5)$
			$\alpha = \beta = 90°$		$P6_3$, $(P6_2, P6_4)$
			$\gamma = 120°$	$6/mmm$	$P622$, $(P6_122, P6_522)$
					$P6_322$, $(P6_222, P6_422)$
Cubic	P	3-fold axes along cube diagonals	$a = b = c$	$m3$	$P23$
			$\alpha = \beta = \gamma = 90°$		$P2_13$
	I				$[I23, I2_13]$
	F				$F23$
				$m3m$	$P432$, $(P4_132, P4_332)$
					$P4_222$
					$I432$, $I4_132$
					$F432$, $F4_132$

[a] a, b, and c are lengths of unit cell edges, α, β, and γ are the angles between b and c, c and a, and a and b, respectively.

[b] Symbols: number with an overbar, a rotary inversion axis; m, a mirror plane; $2/m$, a mirror plane perpendicular to a 2-fold axis; and $6/m$, a mirror plane perpendicular to a 6-fold axis.

[c] Pairs of space groups in parentheses differ from each other only in that they are enantiomorphs. Space groups enclosed in brackets (and also those in parentheses) cannot be distinguished from one another by systematic extinctions of reflections in the diffraction pattern. All other space groups can be assigned on the basis of the diffraction pattern.

[d] The rhombohedral system is often regarded as a subdivision of the hexagonal system, and unit cells in this system may be chosen on either hexagonal or rhombohedral axes.

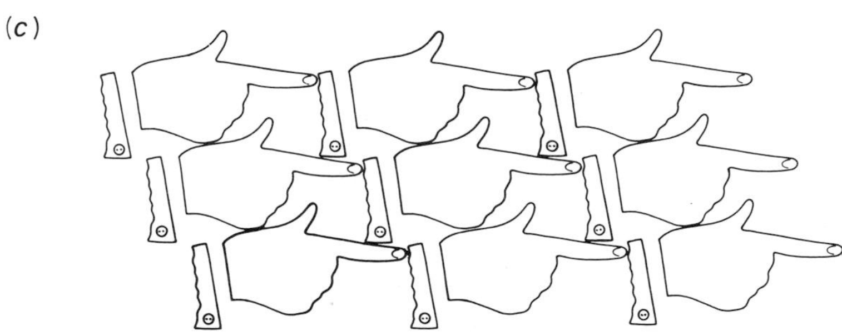

Fig. 1. A lattice (a) specifies the translation of a motif (b) to give a structure (c). Here the lattice is primitive; thus, there is one motif per unit cell. The motif has no internal symmetry in this example.

For the purposes of this chapter, it is sufficient to know that the unit cell in a centered structure contains more than one motif and has a higher symmetry than that of a corresponding primitive cell.

In enzyme crystals there are a limited number of symmetries that the motif may have, owing to the presence of only L amino acids. It may have an m-fold rotation axis, where $m = 1, 2, 3, 4$, or 6. This means that if the motif is rotated by an angle of $2\pi/m$ radians about the axis, it can be superimposed on itself. It may also have a screw rotation axis of order m_p, meaning that if the motif is rotated by $2\pi/m$ radians and then translated along the axis parallel to a unit cell edge by a length equal to the edge divided by p, it can be superimposed on itself. The possible types of screw rotation axes are 2_1, 3_1, 3_2, 4_1, 4_2, 4_3, 6_1, 6_2, 6_3, 6_4, and 6_5. A two-fold rotation axis and a 2_1 screw axis are illustrated in Fig. 2. The motif may also have a symmetry which is a combination of some of the rotation axes and screw rotation axes given here (for example, three mutually perpendicular 2_1 axes).

Table I lists the 65 "biological" space groups, those restricted to the motif symmetries mentioned above. They are grouped in the traditional way into seven *crystal systems*. Each crystal system has one or more of

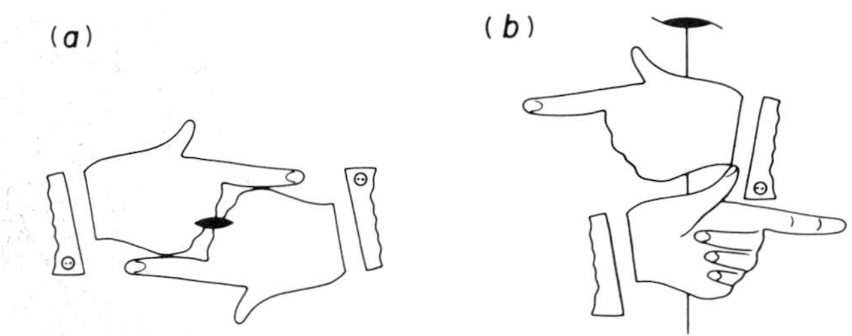

FIG. 2. Two possible motif symmetries: (a) A twofold rotation axis perpendicular to the page, and (b) a 2_1 screw rotation axis in the plane of the page. In both (a) and (b) the motif contains two hands; thus, there would be one pair of hands for each lattice point. The asymmetric unit in both (a) and (b) is a single hand. The motif shown in (a) could represent two molecules, neither having an internal symmetry, or one molecule such as hemoglobin that contains a twofold axis.

the lattices associated with it (given in column two of the table), and each has a minimum symmetry of the unit cell (given in column three) and characteristic cell edges and angles (given in column four). The lengths of the unit cell edges are denoted a, b, and c and the angles between the edges are denoted α, β, and γ. The space groups are determined from the diffraction pattern, first by identifying the symmetry of the pattern (column five), and then by identifying certain systematically extinguished X-ray reflections. This procedure is described by Stout and Jensen (*11*) and Henry and Lonsdale (*38*). One section of the diffraction pattern of diisopropylphosphorofluoridate (DIP) inhibited trypsin shown in Fig. 3 illustrates some of these points. This protein crystallizes in space group $P2_12_12_1$ of the orthorhombic crystal system; thus, according to Table I, the diffraction pattern must display *mmm* symmetry. This means that the symmetry in the three-dimensional diffraction pattern (Section III,A) may be thought of as being reflected across three mutually perpendicular mirrors. Part of this symmetry can be seen in Fig. 3: The pattern is reflected across two mutually perpendicular mirror lines. The figure also shows the systematic extinctions that are characteristic of 2_1 screw axes: every second reflection along the horizontal and vertical lines through the center of the pattern is absent.

One other term is needed in discussing the symmetry of enzyme crystals: the *asymmetric unit* (a.u.) of the crystal is the smallest unit from which the crystal can be generated by the symmetry operations of the space group. The asymmetric unit can be one subunit of an enzyme, it can be one molecule, or it can be several molecules; it cannot be less than

38. N. F. M. Henry and K. Lonsdale, eds., "International Tables for X-ray Crystallography," Vol. I. Kynoch Press, Birmingham, 1952.

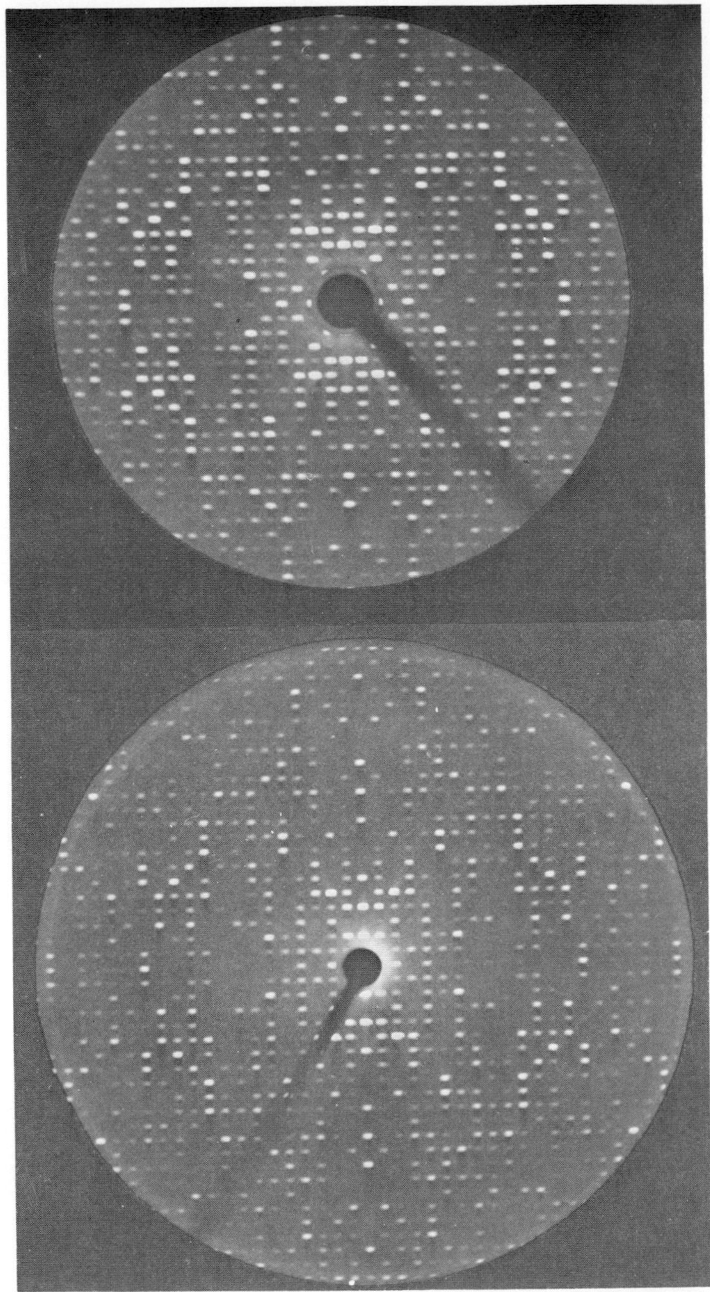

Fig. 3

one polypeptide chain (*38a*). The number of asymmetric units per unit cell n' depends on the symmetry of the motif; n' in the biological space groups can be 1, 2, 3, 4, 6, 12, 24, 48, or 96. One can think of the enzyme crystal being generated from the asymmetric unit in two steps: First the motif is generated from the asymmetric unit by the rotations or screw rotations of the motif symmetry, and then the crystal is generated from the motif by translating it according to the lattice.

C. Determination of Molecular Weight and Molecular Symmetry

1. *Molecular Weight*

X-ray crystallography can provide a value for an integral multiple of the molecular weight of an enzyme, and if the molecular weight is known approximately from some other source, the molecular weight M can be determined accurately. The necessary data for this determination include (1) the volume of a unit cell V, (2) the density of the enzyme crystal ρ, and (3) the weight fraction of enzyme in the crystal X_P. These quantities are related as follows: If the unit cell contained nothing but one protein molecule, its weight would be $NV\rho$ daltons, where N is Avogadro's number. But in fact only the fraction X_P of this weight is protein, the rest being solvent. Moreover, there may be more than one molecule in the unit cell, so that

$$M = \frac{NV\rho X_P}{n} \qquad (2)$$

n being the number of molecules per cell. If an approximate value of M is known, say from a sedimentation study, then Eq. (2) can be solved for an approximate value of n; then the integer nearest to this approximate value can be reinserted into (2) to obtain an accurate value of M.

Of the data needed for this procedure, V is the easiest to obtain. It can be calculated from the unit cell dimensions, which may be determined from just a few X-ray photographs. The density ρ is often measured by

---

Fig. 3. Contact prints of X-ray diffraction photographs of trypsin inhibited by DIP. The upper photograph is the $hk0$ section of the diffraction pattern for the DIP-inhibited enzyme, and the lower photograph is the same section for the thallium acetate derivative. Note the small but definite changes in the relative intensities of reflections that are caused by addition of the thallium ion. The rod-shaped shadow on the films is caused by a trap for the main X-ray beam. The data on these films are sufficient to calculate a projection of the contents of a unit cell of the crystal to a resolution of 2.8 Å. The photographs were taken by Miss L. M. Kay.

38a. Certain cyclic polypeptides are an exception to this rule.

flotation of the crystal in a mixture of liquids such as chlorobenzene and bromobenzene in which the crystal is relatively insoluble. Two more elegant methods, one utilizing a Linderstrøm–Lang gradient column and the other a microbalance, were developed by Low and Richards (*39, 40*).

The weight fraction of protein X_P can be obtained for salt-free crystals by weighing the crystals first when "wet" and then when "vacuum dried" (*41*). But if the crystals contain salt, allowance must be made for the weight of the salt in the dried crystal. This correction is complicated by the fact that the concentration of salt in the solvent within the crystal may be lower than in the mother liquor outside (*32*). Another method of obtaining X_P is based on the relationship (*41*)

$$\frac{1}{\rho} = \bar{v} = \bar{v}_P X_P + \bar{v}_{LC} X_{LC} \tag{3}$$

where $\bar{v}$ is the specific volume of the crystal, $\bar{v}_P$ and $\bar{v}_{LC}$ are the partial specific volumes of the protein in the crystal and of the liquid of crystallization, and X_P and X_{LC} are the weight fractions of the protein and liquid of crystallization in the crystal. Noting that $X_P + X_{LC} = 1$, we get

$$X_P = \frac{\bar{v}_{LC} - (1/\rho)}{\bar{v}_{LC} - \bar{v}_P} \tag{4}$$

The values of $\bar{v}_P$ and $\bar{v}_{LC}$ are usually obtained (e.g., *42*) by assuming that they are equal to the partial specific volumes of, respectively, the protein in dilute solution and the solution used to prepare the crystals. These assumptions were shown to be adequate for crystals of serum albumin (*43*).

It often happens that having measured the volume of the unit cell for a protein of known molecular weight, one would like a preliminary estimate of n without having to resort to measurements of ρ and X_P. Matthews (*28*) pointed out that in many such cases it should be possible to estimate n because the fraction of the crystal occupied by solvent in protein crystals almost always falls between certain limits. He expressed these limits in terms of the quantity V_M, defined as the ratio of the volume of the asymmetric unit to the molecular weight of protein contained in the asymmetric unit (or equivalently, the ratio of V to the molecular weight of protein in the unit cell). Thus

$$V_M = V/(nM) \tag{5}$$

39. B. W. Low and F. M. Richards, *JACS* **74,** 1660 (1952).
40. B. W. Low and F. M. Richards, *Nature* **170,** 412 (1952).
41. B. W. Low, *in* "The Proteins" (H. Neurath and K. Bailey, eds.), 1st ed., Vol. 1, Part A, p. 235. Academic Press, New York, 1953.
42. B. Magdoff-Fairchild, F. M. Lovell, and B. W. Low, *JBC* **244,** 3497 (1969).
43. B. W. Low and F. M. Richards, *JACS* **76,** 2511 (1954).

For the 116 crystal forms considered by Matthews, V_M ranged from 1.68 to 3.53 $Å^3$ $dalton^{-1}$. These limits are often sufficiently restrictive to yield a probable value for n from Eq. (5). As noted by Matthews, it is always possible that the value of V_M for a new protein crystal will be outside this range; thus, the value of n derived by this method should be checked by direct measurement.

2. *Molecular Symmetry*

Enzymes composed of more than one subunit often display some symmetry. The symmetry may be an axis of rotation such as a two-fold axis, or it may be several axes of rotation such as a three-fold axis perpendicular to three twofold axes. The best known example of such molecular symmetry is the hemoglobin molecule, where two pairs of $\alpha\beta$ subunits are related by a twofold axis.

Often a symmetric molecule will crystallize in such a way that its rotation axis is also a rotation axis of the crystal. When this happens, knowing only the space group of the crystal and the number of molecules per unit cell is enough to yield information on the symmetry of the molecule. A simple test for the availability of such information is the following: First determine the space group of the crystal from the diffraction pattern. Each space group is characterized by a certain number of asymmetric units per unit cell n'; this number is given in the "International Tables for X-Ray Crystallography" (*38*). Next determine n, the number of molecules per unit cell, as described in the preceding section. Then if the ratio n/n', the number of molecules per asymmetric unit, is less than unity, some information on the type of rotation axes present can be found simply by consulting the International Tables. Of course, even when $n/n' < 1$, not all the rotation axes of the molecule are necessarily rotation axes of the crystal, and thus this simple procedure may not reveal all of the molecular symmetry. The enzyme aspartate transcarbamylase presents an example of this situation and is discussed below.

This method of studying molecular symmetry in proteins is illustrated in Table II (*26*, *36*, *44–52*). Consider, for example, the three entries for

44. C.-I. Brändén, *ABB* **112,** 215 (1965).
45. D. C. Wiley and W. N. Lipscomb, *Nature* **218,** 1119 (1968).
46. H. C. Watson and L. J. Banaszak, *Nature* **204,** 918 (1964).
47. J. D. Bernal, I. Fankuchen, and M. Perutz, *Nature* **141,** 523 (1938).
48. M. F. Perutz, W. Bolten, R. Diamond, H. Muirhead, and H. C. Watson, *Nature* **203,** 687 (1964).
49. A. C. T. North, *Acta Cryst.* **12,** 512 (1959).
50. W. D. Terry, B. W. Matthews, and D. R. Davies, *Nature* **220,** 239 (1968).
51. D. J. Goldstein, R. L. Humphrey, and R. J. Poljak, *JMB* **35,** 247 (1968).
52. M. G. Rossmann, B. A. Jeffery, P. Main, and S. Warren, *Proc. Natl. Acad. Sci. U. S.* **57,** 515 (1967).

metry that take place when a substrate or cofactor is combined with a crystalline enzyme. For example, L-lactate dehydrogenase from dogfish muscle, an enzyme of four identical subunits, crystallizes in space group *I*422 with one-fourth molecule per asymmetric unit (Table II). This shows that the molecule contains either a fourfold rotation axis or three mutually perpendicular twofold axes; a study of the intensities of X-ray reflections confirmed the latter symmetry. This symmetry can be visualized by thinking of an enlarged methane molecule in which each C—H bond has been replaced by one enzyme subunit. When the cofactor NAD is diffused into the crystals, the space group changes to $P4_22_12$ which necessitates only one twofold axis in the molecule. Adams *et al.* (*26*) believe that this reduction in symmetry is associated with a change in quaternary structure induced by the binding of the NAD. The implications of this hypothesis for enzymic action are discussed elsewhere in this volume (*54*).

In some cases a protein molecule may be suspected to possess a symmetry, but the symmetry may not be expressed in a particular crystal form (i.e., $n/n' \geq 1$). One can always try to recrystallize the material in a more favorable space group by altering the crystallizing conditions, but there are a few devices which may reveal the symmetry with much less work. The first is drying the crystal. Goldstein *et al.* (*51*) found, for example, that when they dried crystals of the Fc fragment of a human immunoglobulin which had been soaked in a solution of $KAuCl_3$, the space group changed from $P2_12_12_1$ to $A2_122$, revealing a suspected twofold axis in the Fc fragment (Table II). In another study, Einstein and Low (*55*) found evidence for a twofold axis in the insulin dimer by controlled drying of crystals. In other cases, just the addition of a reagent can alter the space group to reveal a molecular symmetry. One example is the addition of excess of *p*-chloromercuribenzoate to glyceraldehyde-3-phosphate dehydrogenase which alters the space group from $P2_12_12_1$ to $C222_1$, revealing a molecular twofold axis (Table II).

In other cases where $n/n' \geq 1$, a "pseudo-symmetry" in the diffraction pattern may suggest a symmetry in the molecule that is not directly expressed in the crystal form. The molecular symmetries of both the β-lactoglobulin (*56*) and glyceraldehyde-3-phosphate dehydrogenase (*46*) molecules were indicated by pseudo-symmetries in diffraction patterns before they were found directly from more favorable crystal forms.

54. D. E. Atkinson, this volume, Chapter 9.
55. J. R. Einstein, and B. W. Low, *Acta Cryst.* **15**, 32 (1962).
56. D. W. Green and R. Aschaffenburg, *JMB* **1**, 54 (1959).

D. Growth of Enzyme Crystals

Growing protein crystals is still more of an art than a science. Partly because so many variables are involved, it is sometimes hard to get any crystals whatsoever, and even when small crystals are obtained, it may be difficult or impossible to grow crystals of suitable size, shape, and degree of order for X-ray diffraction work. The optimum crystal size is about 0.3 to 0.5 mm on an edge, though crystals of 0.1 mm or even smaller may be usable in some cases. Much of the recent progress in growing protein crystals has come from development of microchemical methods that enable one to sample a wide range of crystallizing conditions with a limited amount of protein (*57*).

The basic principle of growing large crystals is to suppress the number of crystal nuclei, and at the same time sustain a reasonably rapid rate of growth at those nuclei that are already present. To do this one aims first at eliminating *heterogeneous nucleation*—nucleation on foreign materials such as air bubbles or precipitated protein by removing these potential sources of crystal nuclei. Then one produces a slight supersaturation of the protein by one of several methods:

1. *Traditional batch crystallization.* A protein solution of high concentration is adjusted until a hint of turbidity appears; the adjustment may consist of addition of a salt (often ammonium sulfate), addition of a liquid of lower dielectric constant such as ethanol, or a change in the pH or temperature. The solution is allowed to stand, and the investigator makes further small additions or changes from time to time. According to Zeppezauer *et al.* (*57*), at least 10 mg of protein must be used in each batch.

2. *Crystallization by vapor diffusion.* Slow diffusion of some substance either into or out of the protein solution through the vapor phase can often produce crystals. Evaporation of water, thereby increasing the protein concentration, is a simple example of this method; it works best when the protein is dissolved in just water or a dilute buffer. If the precipitant is an alcohol, it can be diffused into the protein solution through the vapor phase. A convenient experimental arrangement for vapor diffusion with very small samples was described by Hample *et al.* (*58*).

3. *Crystallization by equilibrium dialysis.* In this method, a dialysis

57. This topic has been discussed in detail by M. Zeppezauer, H. Eklund, and E. S. Zeppezauer [*ABB* **126,** 564 (1968)], and the present section follows their presentation to some extent.

58. A. Hample, M. Labanauskas, P. G. Connors, L. Kirkegard, U. L. RajBhandary, P. B. Sigler, and R. M. Bock, *Science* **162,** 1384 (1968).

bag containing the protein solution is immersed in a buffer solution, and periodic additions of salt solution, alcohol, or very dilute buffer are made to the outer solution. The onset of crystallization may be signaled by opalescence in the bag; additions are stopped at this point until any crystals which appear have stopped growing Zeppezauer *et al.* (*57*) described two simple types of microdiffusion cells for work with samples of protein solution smaller than 10 μl. The great advantage of such cells is that they allow one to explore many possible crystallizing conditions with small amounts of protein. The importance of this is illustrated by a number of examples below and also by the finding of Zeppezauer *et al.* that the best crystals of a given protein are often grown only within a range of a few tenths of one pH unit.

Among the variables that may influence crystallization of proteins are the purity of the protein; the nature and concentration of the precipitant; the temperature, pH, and protein concentration; and the nature and concentration of additional electrolytes and nonelectrolytes. Some examples of the wide variety of crystallizing conditions are given in Table III (*33, 59–65*). Many proteins are crystallized by salting out with high concentrations of ammonium sulfate, but others, like carboxypeptidase A, form crystals as the salt concentration is lowered. Still others grow far better from alcohol solutions than from salt solutions; the solvent 2-methyl-2, 4-pentanediol has been found useful for at least three enzymes. As mentioned above, pH is an important variable. Some proteins such as DIP-trypsin and many heavy atom derivatives of hemoglobin crystallize in two forms, one favored at lower pH values and the other at higher values. This phenomenon can be very troublesome until fully understood; indeed, Cullis *et al.* (*60*) noted that it "stopped all progress for several years" on the structure of hemoglobin. Crystals of DIP-trypsin and a number of other enzymes grow more rapidly at room temperature than in the cold, though the reverse is true in many other cases. One example of a requirement of an added compound for crystallization is the apparent need for a

59. W. N. Lipscomb, J. C. Coppola, J. A. Hartsuck, M. L. Ludwig, H. Muirhead, J. Searl, and T. A. Steitz, *JMB* **19,** 423 (1966).

60. A. F. Cullis, H. Muirhead, A. C. T. North, M. F. Perutz, and M. G. Rossmann, *Proc. Roy. Soc.* **A265,** 15 (1961).

61. J. Drenth, J. N. Jansonius, R. Koekoek, J. Marrink, J. Munnik, and B. G. Wolthers, *JMB* **5,** 398 (1962).

62. V. M. Bakulina, V. V. Borisov, V. R. Melik-Adamyan, N. E. Shutskever, and N. S. Andreeva, *Soviet Phys.-Cryst.* (*English Transl.*) **13,** 33 (1968).

63. F. A. Cotton, E. E. Hazen, and D. C. Richardson, *JBC* **241,** 4389 (1966).

64. L. M. Kay, *Acta Cryst.* **16,** 226 (1963).

65. T. Takano, private communication (1970).

TABLE III
SOME CRYSTALLIZING CONDITIONS[a]

Protein	Precipitant	Buffer	Apparent pH	Protein conc. (mg ml^{-1})	Temp. (°C)	Comment	Ref.
Carboxypeptidase A	Water	LiCl tris	7.5	8–20		Protein crystallizes as salt concentration is lowered	*59*
Oxyhemoglobin (horse)	1.25 *M* AS	AP	<7.1	10		At pH > 7.1, *b* axis doubled for heavy atom derivatives	*60*
Papain	60% methanol				0–5	Lower methanol concentrations favor another crystal form (space group *P*2)	*61*
Pepsin	20% ethanol	H_2SO_4	1.5–2	20–25	Room		*62*
Staphylococcus nuclease	34% diol	PB	8.2–8.4	1.5	4	Other crystal forms grow in phosphate buffer	*63*
DIP-trypsin	5% $MgSO_4$		8	10–30	Room	Space group $P2_12_12_1$, but at pH 5 and 11% $MgSO_4$ crystals of space group $P3_121$ also grow	*64*
Ferricytochrome c (horse)	95% saturated AS		6.2	10	4	In 1 *N* NaCl, space group $P4_3$; in 1 *N* NaI, space group $P2_1$	*33*
Ferrocytochrome c (bonito)	AS		8.0	80	Room	Grows from amorphous precipitate	*65*

[a] Abbreviations: AS, ammonium sulfate; diol, 2-methyl-2,4-pentanediol; AP, ammonium phosphate; and PB, phosphate buffer.

sodium halide by horse heart ferricytochrome c. Moreover, the crystal form depends on the anion: When NaCl is added the space group is $P4_3$, but when NaI is substituted for NaCl, the space group changes to $P2_1$. The pattern of growth of crystals of ferrocytochrome c is completely different: They grow from an amorphous precipitate of the protein. These few examples illustrate the problems of finding the optimum conditions for the growth of enzyme crystals.

E. Crystal Mounting and Radiation Damage

Two problems connected with the nature of protein crystals are encountered during collection of X-ray diffraction data. First is the need to keep the crystal in equilibrium with its mother liquor to prevent drying and disordering of the crystal. Bernal and Crowfoot (*66*) were the first to recognize this, and in 1934 they were able to take the first successful X-ray photographs of a single protein crystal by sealing a pepsin crystal in a glass capillary along with some of its mother liquor. From their photographs they inferred that "protein molecules are relatively dense globular bodies, perhaps joined together by valency bridges, but in any event separated by relatively large spaces which contain water." And they concluded, ". . . now that a crystalline protein has been made to give X-ray photographs, it is clear that we have the means of . . . arriving at far more detailed conclusions about protein structure than previous physical or chemical methods have been able to give."

Today protein crystallographers use many variations of this method of mounting crystals. In one procedure, a crystal and a little mother liquor are drawn up by capillary action into a thin-walled glass tube, having an outer diameter of about 1 mm (*67*). To make handling of the fragile capillary easier, rings of sealing wax can first be built up around it near the ends. Then a bristle or small glass fiber is used to push the crystal out of the mother liquor, and if much mother liquor still clings to the crystal it is removed with a wick of thread or filter paper. This step prevents the crystal from slipping when the capillary is moved about on the X-ray camera or diffractometer. Then some extra mother liquor is introduced into both ends of the capillary, taking care not to wet the crystal itself, and the capillary is sealed. The seal can be made by heating the glass or by covering with wax or with silicon stopcock grease capped

66. J. D. Bernal and D. Crowfoot, *Nature* **133,** 794 (1934).

67. Capillaries can be purchased from Unimex-Caine Corp., Room 1817, 3550 N. Lake Shore Drive, Chicago, Illinois 60657.

by a glue. A slightly different mounting procedure was described in some detail by King (*68*).

Another problem encountered during data collection is radiation damage to the crystal. The severity of the problem depends on the substance and crystal form being studied, but it is not uncommon for the intensities of standard reflections to have declined by 10% after 24 hr of radiation at normal levels. In some cases the entire diffraction pattern disappears after several minutes of exposure. Even for proteins only mildly sensitive to radiation, many crystals must be used during the course of data collection, and the data collected from different crystals must be scaled together. A method of decreasing radiation damage would eliminate one of the worst sources of error in protein crystallography.

Two methods have been proposed. Haas and Rossmann (*69*) found that they could greatly impede radiation damage of lactate dehydrogenase by cooling crystals to —50°C, after first having diffused 3 *M* sucrose into the crystals to prevent ice formation. Similarly, Lipscomb and co-workers found that lowering the crystal temperature to 4°C reduced radiation damage by a factor of about three (*69a*). A second way of reducing radiation damage is to use, as the radiation incident on the crystal, a diffracted ray from a crystal of graphite or some other substance. This monochromatized radiation has been found to increase the stability of crystals of cytochrome c peroxidase by a factor of about two (*70*). This suggests that the X-ray wavelengths removed during monochromatization may be especially damaging to protein crystals.

III. X-Ray Diffraction and Principles of Protein Structure Analysis

A. Origin of Diffraction

The X-ray diffraction pattern of a crystal may be thought of as a three-dimensional lattice of spots or "reflections," the position of each reflection on the lattice being given by the indices h, k, and l. Each reflection has a definite intensity $I(hkl)$ that may be measured by either the blackening of a photographic film or the impulses received at the scintillation counter of a "diffractometer" (Fig. 4). Measurement of ac-

68. M. V. King, *Acta Cryst.* **7**, 601 (1954).
69. D. J. Haas and M. G. Rossmann, *Abstr. A.C.A. Meeting, Tucson, Arizona, 1968* p. 60.
69a. W. N. Lipscomb, private communication (1970).
70. L.-O. Hagman, L.-O. Larsson, P. Kierkegaard, and T. Yonetani, private communication (1969); also *Acta Cryst.* **A25**, S185 (1969).

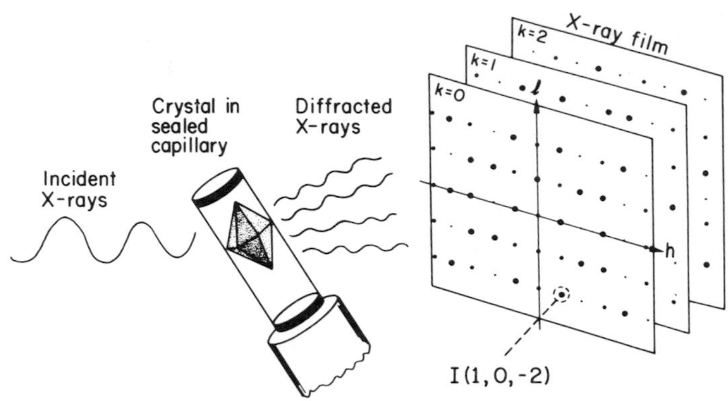

FIG. 4. Photographic collection of data from an enzyme crystal shown schematically. In any given experiment, the film records the blackness of reflections on one section of the three-dimensional X-ray diffraction pattern. The 1, 0, -2 reflection is circled.

curate intensities by either method requires expensive and somewhat complicated equipment, but except for the special problems of macromolecular crystals described in the preceding section, the collection of data may be regarded for present purposes as a straightforward experimental procedure. Whatever method is used, the intensity of each reflection is converted to a structure factor magnitude or structure amplitude $F(hkl)$ by the equation

$$F(hkl) = [kI(hkl)]^{1/2} \tag{6}$$

where k is a known constant that depends on the method and other experimental factors such as the extent of absorption of X-rays by the crystal. These structure factor magnitudes are the primary data of X-ray crystallography.

A crystal diffracts X-rays because it is a periodic object and because its period (the length of the unit cell edge) is comparable in magnitude to the wavelength of the X-rays ($\sim 10^{-8}$ cm). This phenomenon is thus entirely analogous to diffraction of visible light by a grating. Diffraction of light can be observed simply by viewing a candle through a stretched hankerchief, or better, by viewing a pinhole source through a fine mesh wire sieve (Fig. 5). Because the periodic object is two dimensional in this case, the diffraction pattern is two dimensional, but it has two basic characteristics in common with the diffraction pattern of the crystal. The first is that the dimensions of the periodic object and the dimensions of the diffraction pattern bear a reciprocal relationship with one another: A fine mesh screen, for example, will give rise to coarser diffraction pattern (Fig. 5) and vice versa. The term *reciprocal space* for the space of

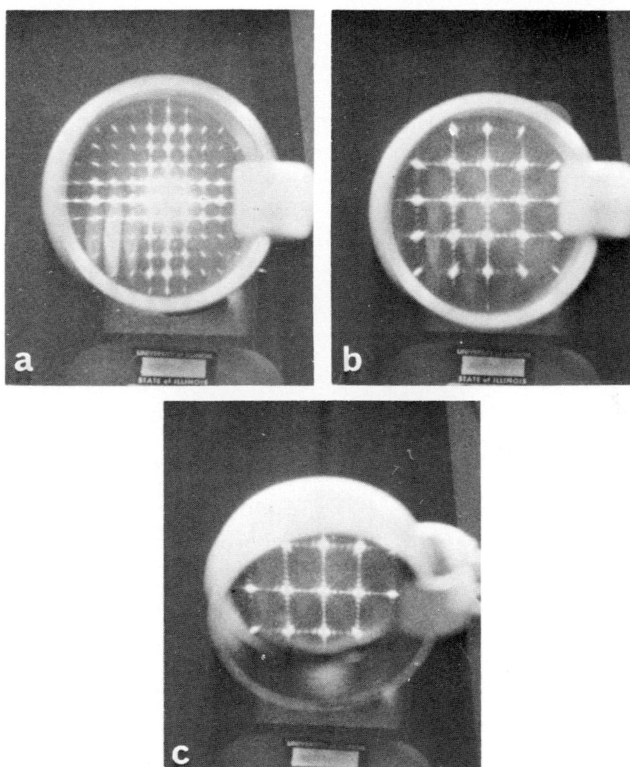

FIG. 5. Diffraction of visible light (from a slide projector) by fine mesh wire sieves. Note that the spacings of the diffraction pattern are reciprocal to those of the wire spacing in the sieves. (a) Sixty μ sieve spacing; (b) 30 μ sieve spacing; and (c) the 30-μ sieve tilted to foreshorten the vertical spacing and thus to increase the vertical spacing in the diffraction pattern (as in an orthorhombic lattice). Reproduced from Dickerson (*2*).

the diffraction pattern is based on this relationship. A second characteristic of diffraction is that, whereas the spacing of the diffraction spots depends only on the periodicity of the diffracting object, the detailed distribution of intensity among the spots depends on the nature of the object. Thus if each intersection of wires in the sieve of Fig. 5 were replaced with, say, a small left hand, the spacings of the spots would not change but their relative intensities would. It is for this reason that one can determine the unit cell volume and the properties related to it (Section II,C) merely by measuring the spacing of spots in the X-ray pattern, but to determine the structure of the cell contents one must measure the intensities of the spots.

B. The Fourier Description of Crystals

1. *The Crystal as a Sum of Waves*

J. B. J. Fourier showed in 1807 that any well-behaved periodic function can be regarded as the superposition of a set of sinusoidal waves. Consider the periodic function $\rho(x)$ shown in Fig. 6 (*71*). It can be represented as a sum of five sinusoidal waves plus a constant; each wave is

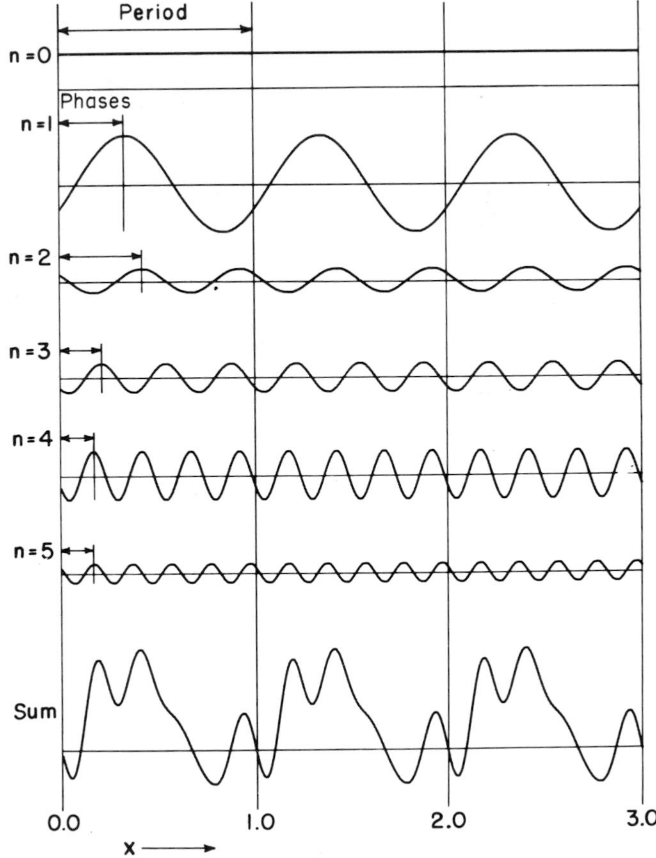

Fig. 6. Superposition of sinusoidal waves to yield a periodic function. The wave numbers and phases of the component waves are shown. Reproduced from Waser (*71*).

71. J. Waser [*J. Chem. Educ.* **45,** 446 (1968)] has presented a superb introduction to Fourier methods in X-ray crystallography.

characterized by a wave number n—the number of oscillations made by the wave within one period of the function, a phase α_n—the horizontal displacement of the wave maximum from a common reference point, and an amplitude F_n—one-half the vertical distance from crest to trough. The mathematical expression for the periodic function as a sum of the component waves is then

$$\rho(x) = \frac{1}{L} \sum_n F_n[\cos(\alpha_n - 2\pi nx) + i \sin(\alpha_n - 2\pi nx)] \tag{7}$$

where L is the length of the period. Making use of Euler's expansion

$$\exp(ix) = \cos x + i \sin x \tag{8}$$

we may write Eq. (7) as

$$\rho(x) = \frac{1}{L} \sum F_n \exp(i\alpha_n) \exp(-2\pi inx) \tag{7a}$$

$$= \frac{1}{L} \sum_n \mathbf{F}_n \exp(-2\pi inx) \tag{7b}$$

In Eq. (7b), $F_n \exp(i\alpha_n)$ is written in the equivalent but more compact form of a complex number $\mathbf{F}_n$.

Since the electron density of a crystal is also a periodic function, it too can be thought of as being built up from component waves, each having the proper amplitude and phase. Two one-dimensional "crystals" are shown in Fig. 7. One contains a "diatomic molecule" and the other contains a "triatomic molecule," yet they are both built up from waves of the same wave number. The phases and the amplitudes of the waves differ, however, for the two crystals. Note that the waves of small wave number define the locations of the molecules within the unit cell, whereas the waves of higher wave number define the molecular details.

To represent a real three-dimensional crystal by a Fourier synthesis it is necessary to include waves running in many directions, and the direction of each wave can be described by three components of the wave number: h, k, and l. The electron density at the point x, y, z can then be written in analogy to Eq. (7a)

$$\rho(xyz) = \frac{1}{V} \sum_h \sum_k \sum_l F(hkl) \exp[i\alpha(hkl)] \exp[-2\pi i(hx + ky + lz)] \tag{9}$$

where V is the volume of the unit cell, $F(hkl)$ is the amplitude of the wave described by the indices hkl, and $\alpha(hkl)$ is the phase of the wave. Note that the amplitude in this equation is the structure factor magni-

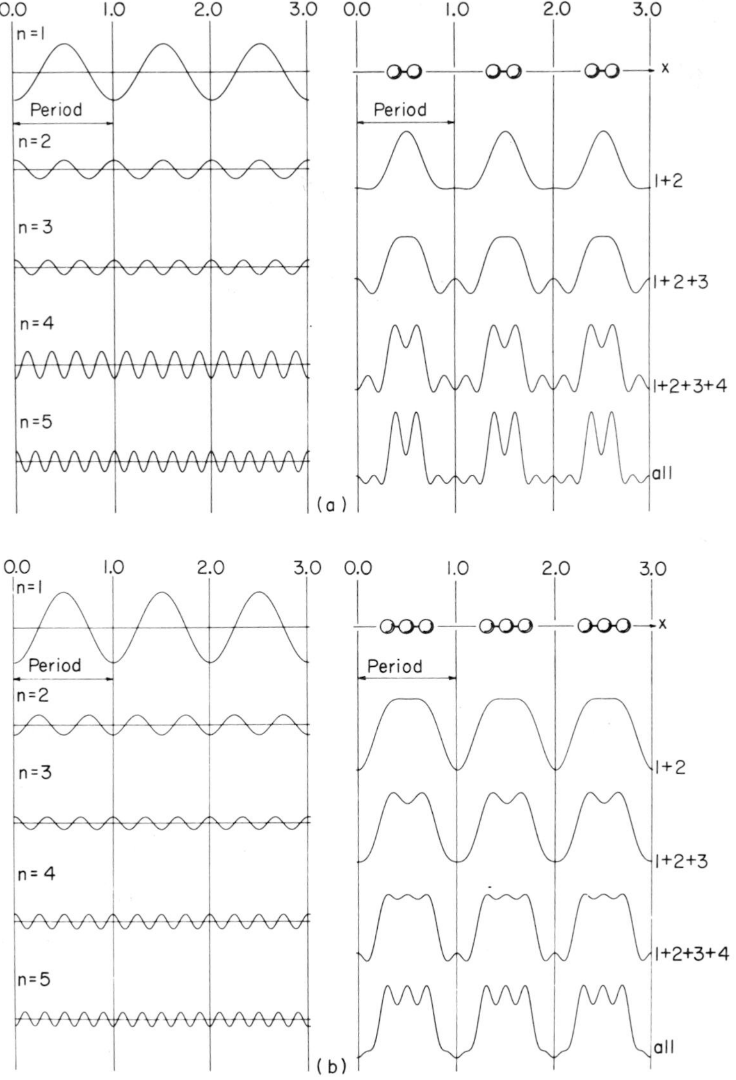

FIG. 7. (a) Representation of the electron density of a one-dimensional "crystal" by a superposition of waves. The crystal is formed by a periodic repetition of a diatomic molecule as shown at the top of the right-hand column. The component waves, each with proper phase and amplitude, are on the left. The curves on the right show the successive superposition of the five waves on the left [from Waser (*71*)]. (b) Representation of another one-dimensional crystal, this one containing a triatomic molecule. Note that this crystal is built up from the same waves as the crystal of (a); only the amplitudes and phases have been changed [from Waser (*71*)].

tude defined in Eq. (6). This is the reason that the Fourier representation of crystals is useful in X-ray diffraction: The amplitude of the wave *hkl* is simply the structure factor magnitude and thus can be derived directly from the intensity of reflection *hkl*. In other words, each spot on an X-ray film corresponds to one wave in the Fourier representation of the crystal, and its blackness is proportional to the square of the amplitude of the wave.

As in the case of the one-dimensional crystal, the waves of higher wave number contribute the fine detail of the molecule; these correspond to reflections far out on the X-ray films (Fig. 3). The "resolution" in angstroms of an electron density map is the reciprocal of the highest wave number used in the Fourier synthesis and represents very roughly the minimum distance between two distinguishable features in the map. To double the resolution takes eight times the number of reflections, and for this reason protein crystallographers usually check their methods with low resolution data (6–4 Å) before going to the trouble of collecting high resolution data (below 3.5 Å). In the case of myoglobin, 400 reflections were used in the 6-Å electron density map, 9600 in the 2-Å map, and more than 25,000 in the 1.4-Å map.

2. *Structure Analysis and the Phase Problem*

Protein structure analysis is based on Eq. (9). The final step in solving a structure is to interpret the electron density map that has been produced by plotting $\rho(xyz)$ on a three-dimensional grid representing the unit cell. To compute $\rho(xyz)$ for any given point xyz, one must measure the intensity of each reflection *hkl*, convert it to a structure factor magnitude $F(hkl)$, arrive somehow at a phase $\alpha(hkl)$ for each reflection, and insert these into Eq. (9), multiplying the term for each reflection by the second exponential (which is a known quantity for any given value of x, y, and z). What keeps this from being a trivial application of the computer once the diffraction pattern of the protein has been measured is that the phase of a reflection cannot be directly observed. This is known as the *phase problem*.

One of the great contributions of Perutz and Kendrew was to demonstrate that the method of *isomorphous replacement* can supply the missing phases for protein molecules. Before describing this method we must say more about the nature of the phase and the structure factor. First, we should note that the phase and amplitude in Eq. (9) can be represented together as a complex number

$$\mathbf{F}(hkl) = F(hkl) \exp [i\alpha(hkl)] \tag{10}$$

called the *structure factor* for the reflection *hkl*. A complex number, moreover, can be represented as a vector in the complex plane (Fig. 8a), the length of the vector being its magnitude [$F(hkl)$ in this case], and the angle measured counterclockwise between the positive real axis and the vector being the phase. Thus the phase for a reflection is an angle between 0 and 2π radians. Knowing the magnitude but not the phase of a structure factor (the situation we are in when we have measured

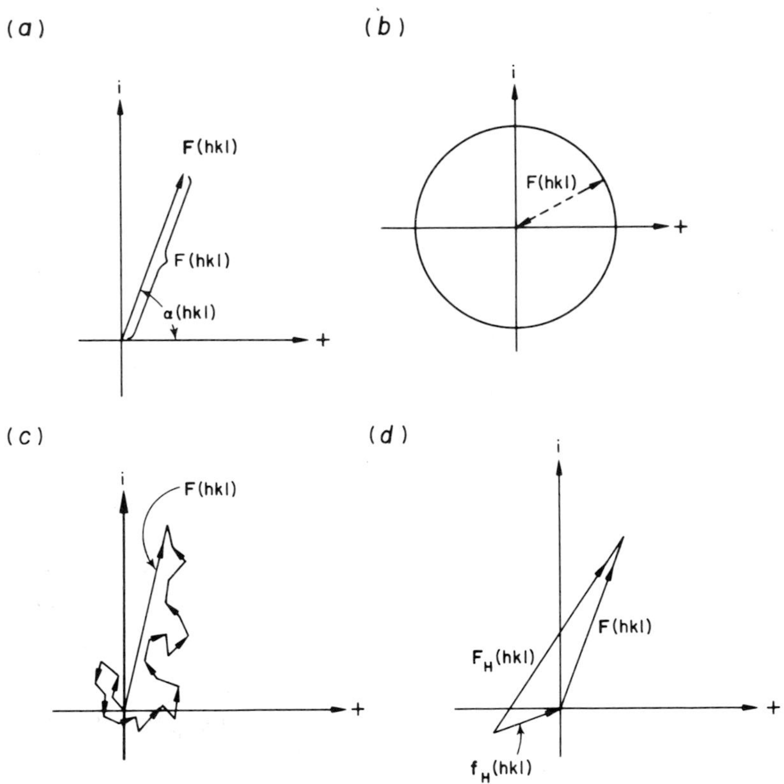

FIG. 8. Representation of structure factors by vectors in the complex plane. (a) The structure factor magnitude $F(hkl)$ is represented by the length of a vector in the complex plane. The phase angle $\alpha(hkl)$ is given by the angle, measured counterclockwise, between the positive real axis and the vector. (b) Knowing only the magnitude of a complex structure factor is equivalent to knowing the radius of a circle in the complex plane. (c) The structure factor for a reflection may be thought of as the vector sum of the X-ray scattering contributions from many atoms. Each of the contributions may be represented as a vector in the complex plane. (d) A special case of (c) in which the structure factor for a given reflection of a heavy atom derivative is represented as the vector sum of the structure factors of the native protein and the heavy atom.

the intensity of a reflection) is equivalent to knowing the length of a vector in the complex plane but not knowing its direction (Fig. 8b).

The complex structure factor $\mathbf{F}(hkl)$ can be regarded in two ways: First, as we have done to this point, as a description of the amplitude and phase of a wave in the Fourier summation of Eq. (9); or, second, as a description of the X-ray scattering of a unit cell in the direction hkl. Taking the second point of view, the scattering of the unit cell is the sum of the scattering of the atoms within it, and thus one can write

$$\mathbf{F}(hkl) = \sum_j f_j \exp\left[2\pi i(hx_j + ky_j + lz_j)\right] \tag{11}$$

$$= \sum_j f_j \exp\left(i\varphi_j\right) = \sum_j \mathbf{f}_j \tag{11a}$$

where x_j, y_j, and z_j give the position of the jth atom in the cell and f_j describes the ability of the jth atom to scatter X-rays and is a known function of the number of its electrons and their distribution around the nucleus. Equation (11) treats the X-ray scattering of the jth atom as a vector f_j in the complex plane and the complex structure factor as a sum of these vectors. This interpretation is illustrated in Fig. 8c. Note that Eq. (11) together with Eq. (6) allows one to compute the diffraction pattern of a substance once one knows the type and positions of atoms within the unit cell. It also allows one to calculate the contribution of any one atom to the total scattering if its coordinates are known.

A special case of Eq. (11) arises when there is one very heavy atom in the unit cell of a protein. Then we can write

$$\mathbf{F}_{\mathrm{H}}(hkl) = \mathbf{F}(hkl) + \mathbf{f}_{\mathrm{H}}(hkl) \tag{12}$$

where $\mathbf{F}_{\mathrm{H}}$ represents the scattering of the protein plus heavy atom, $\mathbf{F}$ represents the scattering of the protein alone, and $\mathbf{f}_{\mathrm{H}}$ represents the scattering of the heavy atom alone; Fig. 8d shows that this is the equation of a closed triangle in the complex plane. It should be noted that if we know the coordinates of the heavy atom within the unit cell (x_{H}, y_{H}, z_{H}) we can calculate $\mathbf{f}_{\mathrm{H}}(hkl)$ from Eq. (11), where the sum extends only over heavy atoms

$$\mathbf{f}_{\mathrm{H}} = \sum_{\text{heavy}} f_{\mathrm{H}} \exp\left[2\pi i(hx_{\mathrm{H}} + ky_{\mathrm{H}} + lz_{\mathrm{H}})\right] \tag{11b}$$

Another special case of Eq. (11) arises when the unit cell has a center of symmetry. Then the imaginary components of $\mathbf{F}$ from symmetry-related atoms cancel, and $\mathbf{F}(hkl)$ is a real number; in other words, the phase must be either 0 or π radians. Protein crystals cannot have a center of symmetry, but in most of the biological space groups at least one of the projections of the unit cell contents has a center of symmetry. In such a case, reflections with indices of the class $0kl$, $h0l$, or $hk0$ have phases restricted to 0 or π, and the vector Eq. (12) becomes a scalar one:

$$F_{\mathrm{H}} = \pm F \pm f_{\mathrm{H}} \tag{12a}$$

This makes phase determination easier for these reflections and is the reason that protein crystallographers often start by locating the positions of heavy atoms in one or more projections of the unit cell.

C. The Method of Multiple Isomorphous Replacement

1. *Outline of the Method*

The determination of an enzyme structure by the method of isomorphous replacement may be thought of as having five stages:

1. Preparation of a set of heavy atom crystals ("derivatives") that are isomorphous with crystals of the parent enzyme. These are crystals of the same space group and cell dimensions, having enzyme molecules packed in the same way as the parent crystals, but having additional heavy atoms bound to the enzyme. The heavy atom may be covalently bound to the enzyme or merely held by noncovalent forces; all that matters is that it sits at the same location (or locations, if there are several sites per cell) in each unit cell of the crystal. At least two derivatives are needed to solve a structure by this method.

2. The intensities of X-ray reflections are measured for the parent and derivative crystals to a desired resolution; often this is done initially for only the class of reflections that corresponds to a centrosymmetric projection of the unit cell. These are converted to structure factor magnitudes $F(hkl)$, $F_{\mathrm{H1}}(hkl)$, $F_{\mathrm{H2}}(hkl)$, etc., where the subscript H1 refers to the first heavy atom derivative.

3. The position of each heavy atom within the unit cell is found by methods described in the following section.

4. The contribution of the heavy atom scattering to each reflection is calculated for each derivative from Eq. (11b). Then the phase $\alpha(hkl)$ for each reflection is computed from the quantities determined in steps 2 and 3 as described in the following section.

5. An electron density map is calculated from Eq. (9) and is then interpreted.

2. *Location of Heavy Atoms*

Green, Ingram, and Perutz discovered that if a sufficiently heavy atom is bound to a protein, it perceptibly changes the intensities of reflections [(*72*); see Fig. 3]. These changes can be used first to determine the coordinates of the heavy atom within the unit cell and eventually to determine the phases of reflections. The coordinates are usually determined by *difference Patterson maps.* These are maps analogous to the electron density maps discussed above but show the relative positions of all pairs of heavy atoms in the cell rather than the heavy atoms themselves; from these relative positions, it is often possible to determine the actual positions. The coefficients of the Fourier summations in these maps involve only structure factor magnitudes so that the phases are not needed to compute them. As an example, suppose that we wish to locate in projection the thallium ion of the thallium acetate derivative of DIP-trypsin for which an X-ray photograph of the $hk0$ section of the diffraction pattern is shown in Fig. 3. The difference Patterson projection $\Delta P(u, v)$ is calculated as follows:

$$\Delta P(u, v) = \frac{1}{A} \sum_u \sum_v |F_{\mathrm{H}}(hk0) - F(hk0)|^2 \exp\left[-2\pi i(hu + kv)\right] \quad (13)$$

where A is the area of the x-y projection, F_{H} is the structure factor magnitude for the thallium acetate derivative, and F is the structure factor magnitude for the parent DIP-trypsin. These structure factor magnitudes can be obtained directly from the intensities of spots on the two films of Fig. 3. Since one unit cell of the heavy atom derivative crystals contains four thallium ions, there are $4^2 = 16$ pairs of heavy ions and hence 16 peaks in the $P(u, v)$ map (some of which overlap); from inspection of these peaks it is possible to determine the positions that the ions must have in the x-y projection of the cell to give rise to the map. Note that even after the positions of the heavy ions have been determined, nothing is known about the structure of the enzyme (except that it presumably touches the ions).

Though difference Patterson maps are straightforward to interpret in principle, troubles often crop up. These stem from several sources. If there are errors in the measured intensities or if the parent and derivative data sets have not been properly scaled together, spurious peaks may appear. If many heavy atoms bind within a unit cell (say, 16) then the number of pairs of atoms, and hence number of peaks, is very large (256), and it may be difficult to unravel the heavy atom positions. Moreover, if some of the binding sites have low occupancies, as often happens when the atoms are not covalently bound, these sites may be missed altogether thus introducing errors into the subsequent phase analysis.

72. D. W. Green, V. M. Ingram, and M. F. Perutz, *Proc. Roy. Soc.* **A225,** 287 (1954).

If the binding of a heavy atom distorts the positions of atoms in the enzyme thereby partially destroying the isomorphism, some of the intensity changes will result from the distortion rather than from the presence of the heavy atoms; though it may be possible to force an interpretation on the difference map in such a situation, the resulting phases may be seriously in error.

Another problem in locating heavy atoms arises from the fact that in most space groups not all coordinates of a heavy atom can be located from centrosymmetric projections and hence difference Pattersons of the form of Eq. (13) are not sufficient for finding all coordinates. Solutions to this problem have been suggested by Blow (*73*) and others and have been reviewed by Phillips (*10*). Still another problem is to place the second, third, and other heavy atoms on the same set of axes with the first. Numerous methods proposed for doing this (*74–78*) have been discussed by Phillips (*10*).

Two newer methods of locating heavy atoms involve anomalous scattering and symbolic addition. Anomalous scattering arises if the heavy atom in a derivative crystal has an absorption edge in the vicinity of the wavelength of the X-rays used in an experiment; its effect is to destroy the equality in intensity of a reflection $I(hkl)$ and its Friedel pair $I(-h - k - l)$. Several authors have shown how measurements of the intensity differences of Friedel pairs can aid in locating heavy atoms (*79–82*). Steitz (*83*) pioneered the application of the symbolic addition method, used frequently to solve structures of small molecules, to the problem of locating heavy atoms in projection. The data required are the same as those for the difference Patterson method, but they are manipulated differently, yielding phases for the heavy atom directly. This method may be especially useful for locating heavy atoms bound to large enzymes, where it seems likely that many atoms will bind to each molecule.

If a study has progressed to the point of locating heavy atoms in a

73. D. M. Blow, *Proc. Roy. Soc.* **A247,** 302 (1958).
74. M. F. Perutz, *Acta Cryst.* **9,** 867 (1956).
75. W. L. Bragg, *Acta Cryst.* **11,** 70 (1958).
76. M. G. Rossmann, *Acta Cryst.* **13,** 221 (1960).
77. G. Bodo, H. M. Dintzis, J. C. Kendrew, and H. W. Wyckoff, *Proc. Roy. Soc.* **A253,** 70 (1959).
78. G. Kartha and R. Parthasarathy, *Acta Cryst.* **18,** 749 (1965).
79. M. G. Rossmann, *Acta Cryst.* **14,** 383 (1961).
80. G. Kartha and R. Parthasarathy, *Acta Cryst.* **18,** 745 (1965).
81. B. W. Matthews, *Acta Cryst.* **20,** 230 (1966).
82. A. K. Singh and S. Ramaseshan, *Acta Cryst.* **21,** 279 (1966).
83. T. A. Steitz, *Acta Cryst.* **B24,** 504 (1968).

second, third, etc., derivative, a Fourier difference map may be useful in finding the positions. This type of map can be calculated from the expression

$$\Delta\rho(xyz) = \frac{1}{V} \sum_h \sum_k \sum [F_{\mathrm{H}}(hkl) - F(hkl)] \exp [i\alpha^*(hkl)] \exp [-2\pi i(hx + ky + lz)] \quad (14)$$

where $\Delta\rho$ is the difference between the electron density of the new heavy atom derivative and of the native protein (and presumably consists of the bound heavy atoms), and α^* is the phase angle determined from all previous heavy atom derivatives. Similar maps can sometimes reveal minor binding sites not detected by other methods. These applications of difference maps offer a trap, however: If the preliminary phases α^* are in error, they will bias the positions determined for the new heavy atoms and can introduce disastrous errors into the final set of phases; thus, caution must be exercised in accepting the indications of this method (*84*).

When preliminary coordinates have been established by one of the above procedures, they should be refined by the method of least squares. Dickerson *et al.* (*85*) have described in detail a program for carrying out these calculations.

3. *Phase Determination and Assessment of Errors*

Once two distinct derivatives have been found and their heavy atoms located, the phases for the reflections of the native protein can be determined. For each reflection, one carries out the following construction, either graphically as done here and as was done in the early days of isomorphous replacement at Cambridge, or analytically on a computer as is usual today. First the complex number $\mathbf{f}_{\mathrm{H1}}$, describing the scattering of the first heavy atom for reflection *hkl*, is calculated from Eq. (11b) and plotted as a vector in the complex plane; as noted by Harker (*86*) it is convenient to take this vector as pointing toward the origin (Fig. 9a). Next a circle of radius F is drawn with its center at the origin, and a second circle of radius F_{H1} is drawn with its center at the tail of the vector $\mathbf{f}_{\mathrm{H1}}$; (it will be recalled from Fig. 8b that knowing the structure factor magnitude of a reflection is equivalent to knowing the radius of a

84. R. E. Dickerson, M. L. Kopka, J. C. Varnum, and J. E. Weinzierl, *Acta Cryst.* **23,** 511 (1967).

85. R. E. Dickerson, J. E. Weinzierl, and R. A. Palmer, *Acta Cryst.* **B24,** 997 (1968).

86. D. Harker, *Acta Cryst.* **9,** 1 (1956).

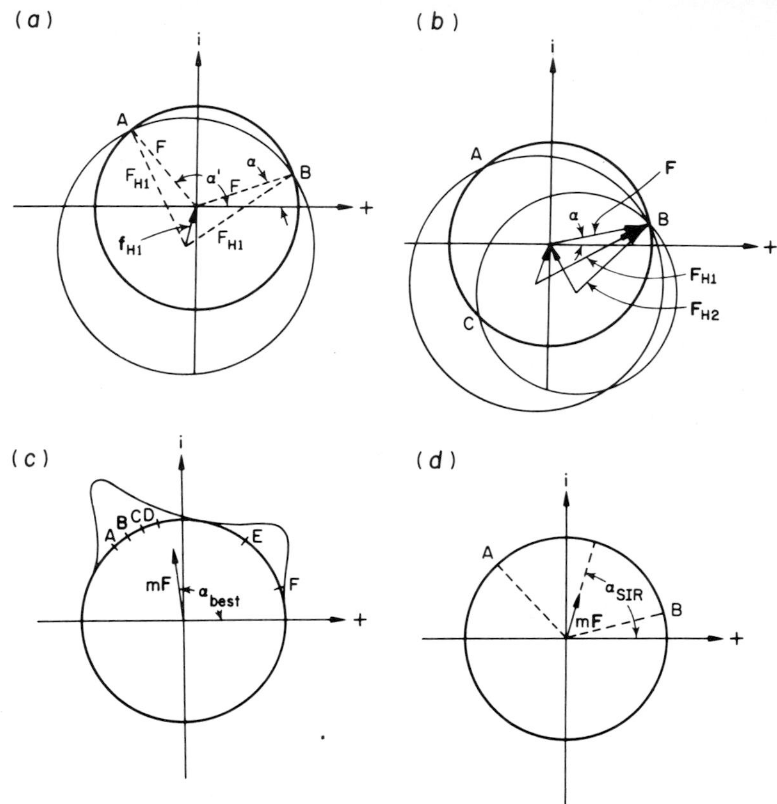

FIG. 9. Phase determination by isomorphous replacement. (a) A single heavy atom derivative gives two possible values (α and α') for the phase of each reflection as explained in the text. The phase circle for the native protein crystal is indicated by a heavy line. (b) Data from a second heavy atom derivative indicate that α is the phase angle for this reflection. (c) Assessment of errors in isomorphous replacement. Suppose that derivative circles intersect the native protein phase circle for a given reflection at points A, B, C, D, E, and F. From the values of F, F_H, and $\mathbf{f}_H$ for these compounds, the probability that any phase angle is the correct angle can be computed and plotted around the phase circle (the double-humped curve). The probability is highest near clusters of intersections. The "best" phase angle is determined by drawing a vector from the origin to the centroid of the probability distribution. The magnitude of this vector is mF, where m is the "figure of merit" for the reflection. (d) Single isomorphous replacement: A single heavy atom derivative gives two possibilities for the phase angle. The average of these two is the single isomorphous replacement phase. The figure of merit varies from unity, when the two possibilities coincide, to zero, when they are 180° apart.

circle in the complex plane). The intersections A and B of the two circles, along with $\mathbf{f}_{H1}$, define two triangles in the complex plane, both of which represent solutions to the vector equation (12). Thus the complex structure factor $\mathbf{F}$ is either the vector from the origin to A or the vector from the origin to B, and the phase for reflection hkl is either α or α'. The question of which of the two is the correct phase can be settled by adding a second derivative to the analysis, as shown in Fig. 9b. The second derivative circle intersects the parent circle at B and C; since B coincides with one intersection of the first derivative, α is the phase.

Up to now we have assumed that the analysis is free of errors. In any real study, however, there will be at least three sources of error: (1) errors in measuring intensities of reflections give errors in F and F_H; (2) errors in locating and describing the scattering of the heavy atom introduce errors in $\mathbf{f}_H$; and (3) any lack of isomorphism between parent and derivative crystals gives further errors. The result of these errors is that the two derivative circles in Fig. 9 usually do not intersect the parent circle at precisely the same point. It is thus desirable to have a third, fourth, etc., derivative to establish the phase more precisely.

Blow and Crick (*87*) and Dickerson *et al.* (*88*) showed how one can assess the probable error in phase determination by isomorphous replacement. They noted that each angle around the parent phase circle can be assigned a probability of being the true phase angle; the probability is high at a given angle if many derivative circles intersect nearby and low if not. Blow and Crick defined the "best" phase angle as the one with the vector drawn to the center of gravity of this probability distribution, and they showed that the electron density calculated from Eq. (9) having the least mean square error is the one in which the "best" phase angles are used and in which each $F(hkl)$ is weighted by a figure of merit, a number ranging from 0 to 1. The figure of merit $m(hkl)$ is the mean cosine of the uncertainty in the phase angle, and it is near unity if the intersections of the circles cluster near a single point. Dickerson *et al.* (*88*) showed that the mean square error in electron density can be calculated directly in terms of these figures of merit

$$\langle \Delta\rho^2 \rangle = \frac{1}{V^2} \sum_h \sum_k \sum_l F(hkl)^2[1 - m(hkl)^2] \tag{15}$$

This number is useful in judging what is probably real detail in the final electron density map and what is merely noise.

4. *How Heavy an Atom?*

How heavy an atom must be used to determine phases by isomorphous replacement? This depends on the molecular weight of protein per heavy atom and also on the accuracy with which intensity differences can be

87. D. M. Blow and F. H. C. Crick, *Acta Cryst.* **12,** 794 (1959).
88. R. E. Dickerson, J. C. Kendrew, and B. E. Strandberg, *Acta Cryst.* **14,** 1188 (1961).

measured. A very rough indication of the fractional change in intensity produced by the binding of a heavy atom can be obtained from an equation derived by Crick and Magdoff (*89*). They showed that the root-mean-square fractional change in intensity produced by the stochiometric binding of N_{H} heavy atoms of scattering power f_{H} to a protein of N_{P} atoms having a mean scattering power f_{P} is roughly

$$\frac{\langle(\Delta I)^2\rangle^{1/2}}{\langle I\rangle} \cong \left(\frac{2N_{\mathrm{H}}}{N_{\mathrm{P}}}\right)^{1/2} \frac{f_{\mathrm{H}}}{f_{\mathrm{P}}} \tag{16}$$

Figure 10 shows an evaluation of this equation for three fractional changes in intensity, assuming that $f_{\mathrm{P}} = 7$ and that the mean atomic weight of a protein atom is 14. A fractional change of intensity of 0.2 is certainly within the limits of measurement using modern methods,

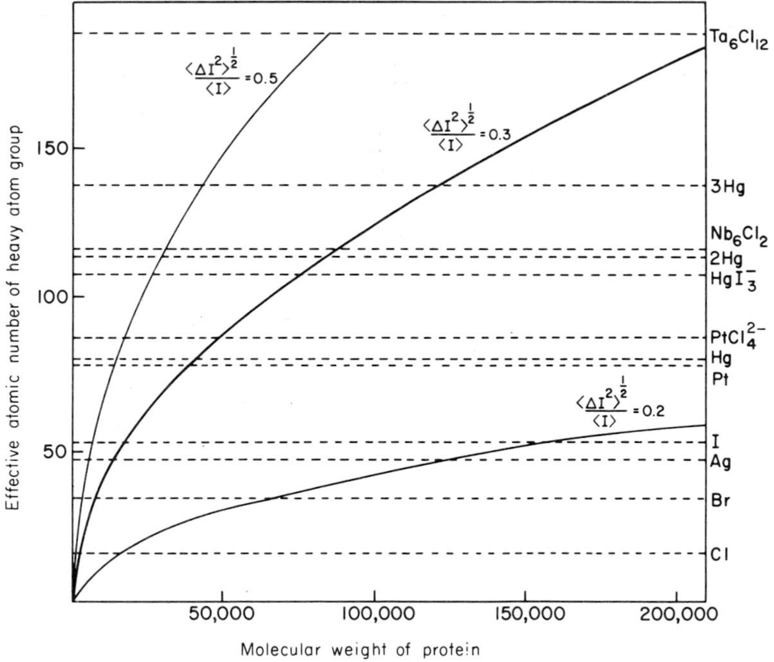

FIG. 10. The approximate size of heavy atom needed to determine the structure of an enzyme of a given molecular weight. The three curves represent evaluations of Eq. (16) for three values of the fractional change in intensity produced by the heavy atom (see text). The curve for a fractional change of 0.3 gives a reasonable estimate of the type of heavy atom group that must be used in most circumstances. This diagram is based in part on an earlier version drawn by Dr. J. E. Weinzierl.

89. F. H. C. Crick and B. S. Magdoff, *Acta Cryst.* **9**, 901 (1956).

but inasmuch as Eq. (16) takes no account of many sources of error in isomorphous replacement, the heavy line for a fractional change of 0.3 probably gives a better estimate of how heavy an atom must be used. Thus, as a rough estimate, an iodine atom would be useful for a protein up to about 15,000 molecular weight and $PtCl_4^{2-}$ for a protein up to about 50,000 molecular weight. In cases where a high degree of isomorphism between derivative and parent is achieved the 0.3 line is probably too pessimistic; for example, the iodine atom of the pipsyl derivative of α-chymotrypsin is useful for an enzyme of 24,000 molecular weight.

For very large enzymes, the complex ions $Nb_6Cl_{12}^{2+}$ and $Ta_6Cl_{12}^{2+}$ may be useful in phase determination, but Stanford and Corey, in the one attempt to date to use these ions, found that their orientations as well as their positions must be known to calculate phases at high resolution, and it was not possible to determine the orientations with certainty (*90*). Another strategy for phase determination with very large molecules has been suggested by Crick and Kendrew (*8*) and others. One first determines the phases at low resolution for a very heavy atom (say, $Ta_6Cl_{12}^{2+}$); then one uses these phases in Eq. (14) to locate a somewhat lighter atom (say, I) substituted at many sites on the molecule. Then the multiple site derivative can be used for phase determination at high resolution.

D. OTHER METHODS OF OBTAINING STRUCTURAL INFORMATION

1. *Single Isomorphous Replacement and Variations*

Because of the time and expense involved in preparing the heavy atom derivatives and in measuring the several sets of intensities needed for multiple isomorphous replacement, any method that reduces the necessary number of derivatives is a welcome development. The method of single isomorphous replacement (SIR) (*91–93*) yields a set of approximate phases, and hence an approximate electron density map, when only one heavy atom derivative is available. Moreover, when the method is used in conjunction with anomalous scattering or with a relationship known as the tangent formula, it can yield a better set of phases, perhaps almost as good as a set obtained from two derivatives. The method is based on the fact that a single heavy atom derivative gives two possible

90. R. H. Stanford, Jr. and R. B. Corey, *in* "Structural Chemistry and Molecular Biology" (A. Rich and N. Davidson, eds.), p. 47. Freeman, San Francisco, California, 1968.

91. D. M. Blow and M. G. Rossmann, *Acta Cryst.* **14**, 1195 (1961).

92. G. Kartha, *Acta Cryst.* **14**, 680 (1961).

93. D. M. Blow and M. G. Rossmann, *Acta Cryst.* **15**, 1060 (1962).

choices for the phase of each reflection (Fig. 9a). Suppose that instead of using a second derivative to choose between these, one simply takes their bisector, α_{SIR}, as the phase. For those reflections having the two choices close to each other, α_{SIR} will be very close to the true phase, whereas for those having the two choices far apart the α_{SIR} will almost certainly be considerably in error. It thus seems reasonable to weight each $F(hkl)$ in the electron density, Eq. (9), by a function such as the cosine of half the angle between the two choices. Using a slightly more elaborate weighting scheme, Blow and Rossmann calculated SIR electron density maps that compare remarkably well to those based on several derivatives (*91*). The SIR method, it should be noted, fails if the heavy atoms in the crystal are related to one another by a center of symmetry.

The question naturally arises if one can select the proper phase from the two choices presented by a single derivative without using the additional derivatives. Several authors (*91, 94*) have shown that the anomalous scattering of the heavy atom compound can in principle resolve the phase ambiguity, even though the anomalous differences [those between $I(hkl)$ and $I(-h-k-l)$] are bound to be small. In at least one case, that of rubredoxin with a HgI_4^{2-} derivative, this method has been used to solve a protein structure, and the structure has been subsequently confirmed by the method of multiple isomorphous replacement (*95*). Anomalous scattering is also helpful for improving phases when several derivatives are available and also for determining the absolute configuration of the enzyme; this is an important aid in the interpretation of maps of a resolution too low to detect the conformations of amino acids about α-carbon atoms (*94, 96, 97*).

A second possible way of resolving the phase ambiguity presented by single isomorphous replacement is to use the tangent formula (*98, 99*). This is a relationship derived by Karle and Hauptman (*100*) that allows one to calculate the phase of one reflection in terms of the phases of other reflections. It can be expressed

$$\alpha_c(h) = \tan^{-1}\left\{\frac{\sum_k E(k)E(h-k)\sin[\alpha(h)+\alpha(h-k)]}{\sum_k E(k)E(h-k)\cos[\alpha(h)+\alpha(h-k)]}\right\} \tag{17}$$

94. A. C. T. North, *Acta Cryst.* **18**, 212 (1965).
95. J. R. Herriott, L. C. Sieker, and L. H. Jensen, *Acta Cryst.* **A25**, S186 (1969).
96. B. W. Matthews, *Acta Cryst.* **20**, 82 (1966).
97. J. Kraut, *JMB* **35**, 511 (1968).
98. C. L. Coulter, *JMB* **12**, 292 (1965).
99. J. E. Weinzierl, D. Eisenberg, and R. E. Dickerson, *Acta Cryst.* **B25**, 380 (1969).
100. J. Karle and H. Hauptman, *Acta Cryst.* **9**, 635 (1956).

where h and k each represent the three indices of a reflection and where the E's are weighting factors that may be easily derived from the measured intensities and a knowledge of the atomic composition of the crystal. If the proper combination of SIR phases is inserted on the right-hand side of Eq. (17), a phase α_c can be calculated for each reflection. This calculated phase will be nearer to either one or the other of the two possible phases for the single derivative, and the correct one can be chosen on this basis. So far this method has been tested for only a limited number of trial cases (*99, 101*) but it seems promising, particularly if high resolution data are available. Figure 11 shows a comparison of electron density maps calculated from phases based on double isomorphous replacement, single isomorphous replacement, and single isomorphous replacement with the tangent formula.

The tangent formula may also prove useful in refining phases obtained from multiple isomorphous replacement (*99, 101*) and in extending phases to reflections for which an intensity has been measured for the parent but not the derivative crystal (*101–103*).

2. *Fourier Difference Maps and Salt Difference Maps*

Once the structure of an enzyme has been solved by the isomorphous replacement method, the Fourier difference method can be used to study the binding of inhibitors, substrate analogs, and other small molecules to the enzyme. The first step in the method is to prepare a crystal isomorphous with the parent enzyme but containing the additional ligand; this can be done either by diffusing the ligand into a crystal of the parent or by reacting the enzyme and ligand in solution and then crystallizing the complex. Next a set of intensities is collected for the crystal of the complex, and these are converted to a set of structure factor magnitudes $F_L(hkl)$ by Eq. (6). Then a Fourier difference map is calculated from Eq. (14), substituting $F_L(hkl)$ for $F_H(hkl)$ and using the phases of the parent enzyme. If the data are sufficiently accurate and if the complex is closely isomorphous with the parent, the difference map will show the position of the ligand within the unit cell, and by comparison with the electron density map of the parent compound, the group on the parent to which the ligand is bound can be identified. The difference map may also show that some group has been displaced in the complex. For example, in the study of the binding of the azide–myoglobin complex, it was discovered that a sulfate ion, normally bound to a imidazole nitrogen atom

101. C. L. Coulter, *Acta Cryst.* **A25**, S190 (1969).
102. M. Zwick, private communication (1969).
103. W. N. Lipscomb, *in* "Structural Chemistry and Molecular Biology" (A. Rich and N. Davidson, eds.), p. 38. Freeman, San Francisco, California, 1968.

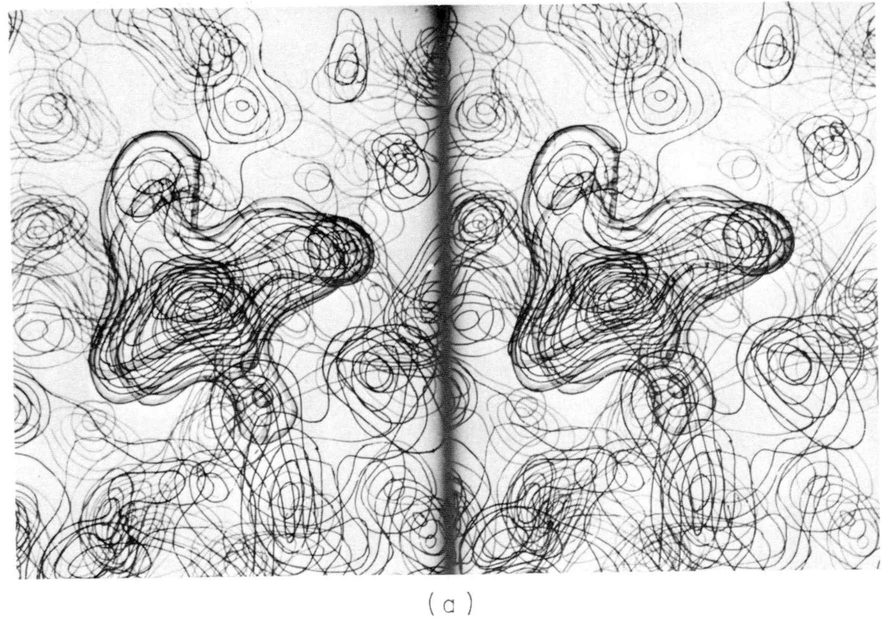

(a)

(b)

Fig. 11

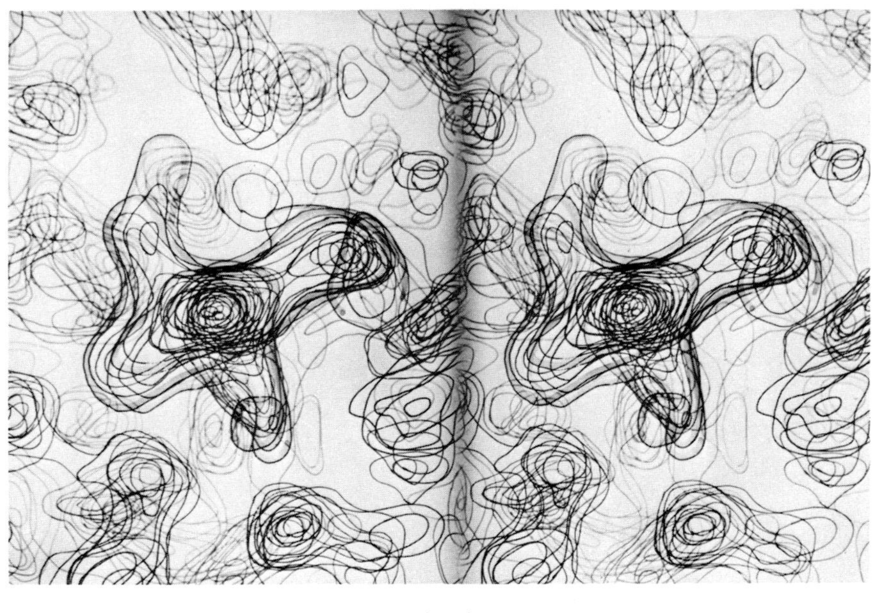

(c)

FIG. 11. Stereopair photographs of the electron density in the region of the heme group in cytochrome c determined from (a) double isomorphous replacement, (b) single isomorphous replacement, and (c) single isomorphous replacement using the tangent formula to choose between the two possible phase angles. All maps are at 4 Å resolution. The most electron dense region in all three maps is the iron atom at the center of the porphyrin ring. Other prominent features are the sulfur atoms above the ring that form the thioether links to the polypeptide backbone and the two propionic acid side chains that extend from the ring toward the bottom of the page. Reproduced from Weinzierl *et al.* (*99*).

of the distal histidine, is ejected when the azide ion binds to the other nitrogen (*104*). The azide–myoglobin study is especially noteworthy in that it demonstrated that the difference Fourier method is capable of locating very light ligands as well ones containing heavy atoms. Stryer *et al.* emphasized that the limiting factor in locating light atoms is the degree of isomorphism between the complex and parent (*104*).

Salt difference maps show the regions of a crystal that are impenetrable to solvent and are thus occupied by protein. These maps are useful for finding the positions of molecules in the unit cell and in obtaining some information about the shape of molecules. Such information is helpful in confirming the interpretation of low resolution studies of proteins (*33*)

104. L. Stryer, J. C. Kendrew, and H. C. Watson, *JMB* **8**, 96 (1964).

and could conceivably reveal the relative positions of subunits in multi-subunit enzymes (see Fig. 12).

To compute a salt difference map one uses an expression similar to Eq. (14), where the two sets of $F(hkl)$'s now refer to crystals soaked in solvents having different electron densities. Since the molecules are the same in both crystals, they will cancel out in the difference map leaving a Swiss cheese-like structure representing the solvent, with regularly spaced holes representing the molecules. Since the positions of molecules are defined by the Fourier waves of low wave number (Section III,B,1),

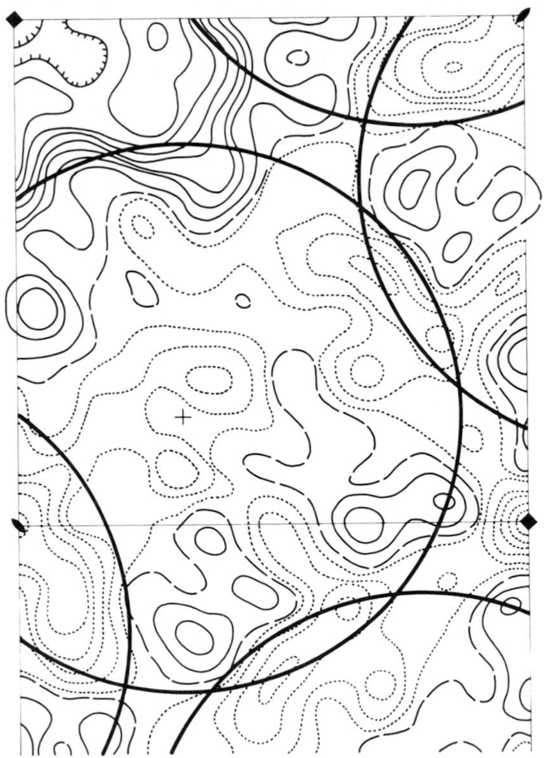

FIG. 12. Salt difference map of cytochrome c showing the regions of the projection of a unit cell that are impenetrable to solvent. The regions shown by solid contours are regions of more than average accessibility to crystallizing medium in projection, and the regions shown by dotted contours are the least accessible to salt solution and hence are occupied by the molecule. The superimposed circles show the approximate positions of molecules deduced from an electron density map. Note that the region most accessible to solvent is at the origin of the cell (upper left corner) where no parts of a molecule are found, and the regions least accessible are along the twofold axes, where molecules overlap. Reproduced from Dickerson *et al.* (*33*).

one needs to include in the summation only reflections having low indices; indeed, these will be the only reflections whose intensities are appreciably affected by the change of solvent. The major difficulties in calculating a salt difference map are finding solvents of different electron densities in which the crystals are stable and in compensating for any change in phase of a reflection as one solvent is replaced by another. The problem of phase changes has been handled differently for each of the four examples of salt maps listed in Table IV (*33, 105–107*).

3. *Methods Involving Noncrystallographic Symmetry*

A completely different method of solving protein structures has been proposed by Rossmann, Blow, Main, and others for the case of crystals containing some noncrystallographic symmetry. Examples of protein crystals having some noncrystallographic symmetry are the form of α-chymotrypsin with two molecules per asymmetric unit and the form of human deoxyhemoglobin with one molecule (and hence two $\alpha\beta$ pairs of subunits) per asymmetric unit (Table V). The method involves three stages. First the relative orientation of the chemically identical units is established by the *rotation function,* which is essentially a systematic

TABLE IV
SALT DIFFERENCE MAPS OF PROTEINS[a]

Protein	Most dense solvent (density in electrons Å^{-3})	Least dense solvent (density in electrons Å^{-3})	Information obtained	Ref.
Horse hemoglobin	4 M $(NH_4)_2SO_4$ + 10% CsCl (0.422)	Water (0.334)	Molecular shape	*105*
Finback whale myoglobin	4 M phosphate buffer (0.46)	75% saturated $(NH_4)_2SO_4$ (0.39)	Molecular position	*106*
Sperm-whale myoglobin	3.1 M $(NH_4)_2SO_4$ + 2 M CsCl (0.43)	2.8 M $(NH_4)_2SO_4$ (0.38)	Molecular position in projection	*107*
Cytochrome c	5.0 M phosphate buffer (0.468)	95% saturated $(NH_4)_2SO_4$ (0.407)	Molecular shape and position in projection	*33*

[a] The resolution of these maps is 4 Å or lower (coarser).

105. W. L. Bragg and M. F. Perutz, *Acta Cryst.* **5,** 277 (1952).
106. J. C. Kendrew and P. J. Pauling, *Proc. Roy. Soc.* **A237,** 255 (1956).
107. M. M. Bluhm, G. Bodo, H. M. Dintzis, and J. C. Kendrew, *Proc. Roy. Soc.* **A246,** 369 (1958).

comparison of the Patterson function for a crystal with a rotated version of itself (*108*). This function has been used successfully to establish noncrystallographic twofold axes of rotation in α-chymotrypsin (*109*) and in insulin (*110*). In the second stage the relative positions of the identical units is established with the translation function (*111*). This function, too, has been applied to chymotrypsin and insulin. In the third and most difficult stage, the phases of the crystal are found using the results of the first two stages (*112–114*). The third stage has been tested on a hypothetical structure containing four molecules related by noncrystallographic symmetry and each composed of ten carbon atoms (*115*); this stage has not yet been used successfully on a protein.

Another use of the rotation function is to determine the relative orientations of identical or closely similar molecules in different crystal forms. Prothero and Rossmann, for example, compared the horse and human hemoglobin molecules in their respective crystals (*116*).

E. Interpretation of Electron Density Maps

The interpretation of the electron density map of a macromolecule is rarely a trivial step in the analysis because these maps do not usually contain enough detail to allow immediate identification of each feature with some chemical group. The ease or difficulty encountered depends on such factors as the resolution of the map, the extent of errors in the phases and intensities, the number of distinctive features of the macromolecule and its derivatives (for example, electron dense prosthetic groups or heavy atoms covalently bound to known amino acids), and the amount of other chemical information such as amino acid sequence. It may be helpful to summarize some experience on these points, and then to mention some newly developed aids for the construction and refinement of a model based on the interpretation of the map.

Because of the many variables involved, there seems to be no definite resolution at which it first becomes possible to trace the path of a poly-

108. M. G. Rossmann and D. M. Blow, *Acta Cryst.* **15,** 24 (1962).
109. D. M. Blow, M. G. Rossmann, and B. A. Jeffery, *JMB* **8,** 65 (1964).
110. E. Dodson, M. M. Harding, D. C. Hodgkin, and M. G. Rossmann, *JMB* **16,** 227 (1966).
111. M. G. Rossmann, D. M. Blow, M. M. Harding, and E. Coller, *Acta Cryst.* **17,** 338 (1964).
112. M. G. Rossmann, and D. M. Blow, *Acta Cryst.* **16,** 39 (1963).
113. M. G. Rossmann and D. M. Blow, *Acta Cryst.* **17,** 1474 (1964).
114. P. Main and M. G. Rossmann, *Acta Cryst.* **21,** 67 (1966).
115. P. Main, *Acta Cryst.* **23,** 50 (1967).
116. J. W. Prothero and M. G. Rossmann, *Acta Cryst.* **17,** 768 (1964).

peptide chain in an electron density map and at which it first becomes possible to identify specific side chains. For at least two enzymes, ribonuclease S and elastase, maps at 3.5 Å were interpreted in terms of known amino acid sequences, whereas in at least one case a resolution approaching 2 Å was required. When the primary structure is not known, interpretation is much more difficult, but as experience with carboxypeptidase and myoglobin demonstrates (*117*, *118*), even then many amino acid residues can be correctly identified in an accurate electron density map of 2 Å resolution. In the 3.5 Å elastase map (*18*), the polypeptide chain is visible as a column of continuous high electron density with prominent peaks marking the branch points of side chains at the α-carbon atoms of amino acids. In contrast, the most prominent peaks along the backbone in maps of higher (finer) resolution are the peptide carbonyl groups. The four cystine disulfide bonds in elastase appear as rods of high density linking two stretches of the main chain, but it should be noted that at lower (coarser) resolution, disulfide bonds are not necessarily the most prominent features of a map (*119*). Other easily recognizable features of the elastase map are the methionine residues with their dense sulfur peaks and the aromatic rings of tryptophan, tyrosine, phenylalanine, and histidine.

An important factor in the ease or difficulty of interpretation is the level of error in the map, and this is intimately related to the accuracy of the phases and intensities. Several quantitative estimates of such errors are available. The first is Eq. (15) for the mean square error in the electron density map in terms of the figures of merit. This equation gives a useful estimate of the overall level of uncertainty, but it probably underestimates errors in the regions of heavy atom binding where there are likely to be greater than average departures from isomorphism. A closely related indicator of the level of error, and one that is often cited in papers on protein structures, is the mean figure of merit for all reflections used in the electron density map. A value approaching unity signifies that the mean uncertainty in phase angle approaches zero, and thus implies a highly accurate electron density map. The mean figure of merit decreases as a map is extended to higher resolution because the principal sources of error in the phase angles such as lack of isomorphism become increasingly serious. Thus the mean figure of merit for the X-ray analysis of carboxypeptidase A dropped from 0.88 at 6 Å to 0.625 at 2 Å resolu-

117. W. N. Lipscomb, J. A. Hartsuck, G. N. Reeke, Jr., F. A. Quiocho, P. H. Bethge, M. L. Ludwig, T. A. Steitz, H. Muirhead, and J. C. Coppola, *Brookhaven Symp. Biol.* **21,** 24 (1968).

118. H. C. Watson. *Progr. Stereochem.* **4,** (in press).

119. G. Kartha, *Nature* **214,** 234 (1967).

tion (though, of course, the information in the map increased enormously because of the greater number of reflections included). Another useful criterion of error, the *centric R factor* (*60*), reflects the accuracy of location and description of the electron density of the heavy atoms as well as the degree of isomorphism and the accuracy of the intensities. It may be defined

$$R_C = \frac{\sum_{hkl} ||F_H(hkl) - F(hkl)| - f_H(hkl)|}{\sum_{hkl} |F_H(hkl) - F(hkl)|} \tag{18}$$

where the sum is over all reflections in a centrosymmetric section of the diffraction pattern. Since the quantity $||F_H - F| - f_H|$ is zero under ideal conditions [see Eq. (12a)], the value of R_C increases from zero as errors enter the analysis. For an exceptionally good derivative (one that is highly isomorphous with the parent and has large intensity changes from the parent, and in which the heavy atom has been precisely located), R_C might be as low as 0.3 at 3 Å resolution, and for a poor derivative R_C might be 0.9. In short, an electron density map that is characterized by a high value of the mean figure of merit and a low value for the centric R factor of each derivative will be easier to interpret at a given resolution than a map with less favorable values of these indicators. Some indication of the level of detail likely to be found in a given map can be obtained by referring to previous studies at similar resolutions and with similar values for the indicators.

The presence of distinctive chemical features makes an electron density map much easier to interpret. In the 4-Å map of horse heart cytochrome c, for example, the heme group was immediately recognizable, and this led to the identification of several amino acids in the environment of the heme (*120*). Regular structure in the polypeptide backbone such as stretches of α helix also helps interpretation. Chemical markers such as added atoms near the active site or heavy atoms attached to known amino acid residues are a great aid. For example, the toluene sulfonyl (tosyl) derivatives of α-chymotrypsin and elastase established unambiguously the location of the active serine residues in the respective electron density maps, and a bound mercury atom showed the location of the active sulfhydryl group in papain. Drenth *et al.* (*121*) used the iodinated form of the Edman reagent, *p*-iodophenylisothiocyanate, to locate the

120. R. E. Dickerson, M. L. Kopka, J. Weinzierl, J. Varnum, D. Eisenberg, and E. Margoliash, *JBC* **242**, 3015 (1967).

121. J. Drenth, J. N. Jansonius, R. Koekoek, H. M. Swen, and B. G. Wolthers, *Nature* **218**, 929 (1968).

N-terminal amino group of papain, and then used the same reagent at a higher pH to label three other positions, all of which are apparently lysine residues. In several studies, including those of papain and subtilisin BPN′, iodine has been used to label tyrosyl residues. The compound $PtCl_4^{2-}$ has been found to bind to methionine when the sulfur atom lies exposed on the surface of a protein, though $PtCl_4^{2-}$ also binds to histidine and to the sulfur atoms of exposed disulfide bridges (*122*). The extent to which chemical markers help in the interpretation, depends, of course, on the other chemical information that is available. For example, if there is only one methionine residue in the sequence, its location may lead to the identification of many neighboring residues.

Even after the electron density map has been interpreted, the construction and refinement of an atomic model of the molecule is not a trivial operation. Richards (*123*) devised a simple yet effective way to fit a skeletal model to the map. His method is to set a half-silvered mirror at an angle of exactly 45° to the floor. Behind the mirror the electron density map is displayed as a stack of contoured sections on Plexiglas sheets. The skeletal model is constructed in front of and below the mirror from accurately formed components. Looking into the mirror, one sees an image of the model superimposed on the electron density, and adjustments of the model can be made to achieve the best possible fit. As noted by Richards, stereophotographs of the superposition of model and map are a useful way of displaying crystallographic results (Fig. 13). The skeletal models of the type used in this procedure have the advantage that the back parts of the model can be seen through the front, but the disadvantage that they do not realistically represent steric restrictions as do space-filling models. It has been reported that one brand of space-filling model can be easily adapted to fit over skeletal models, thus giving the additional information when needed (*124*).

The tentative set of atomic coordinates measured from the completed model can often be improved by a process of computer refinement. Diamond has described two sophisticated programs for this purpose (*125*, *126*). The first folds an idealized polypeptide chain so that it approaches a set of guide points derived from the electron density map. In this way

122. R. E. Dickerson, D. Eisenberg, J. Varnum, and M. L. Kopka, *JMB* **45**, 77 (1969).

123. F. M. Richards, *JMB* **37**, 225 (1968).

123a. H. W. Wyckoff, D. Tsernoglou, A. E. Hanson, J. R. Knox, B. Lee, and F. M. Richards, *JBC* **245**, 305 (1970).

124. J. A. Rupley and T. C. Bruice, *JMB* **37**, 521 (1968).

125. R. Diamond, *Acta Cryst.* **21**, 253 (1966).

126. R. Diamond, *Acta Cryst.* **A25**, S189 (1969).

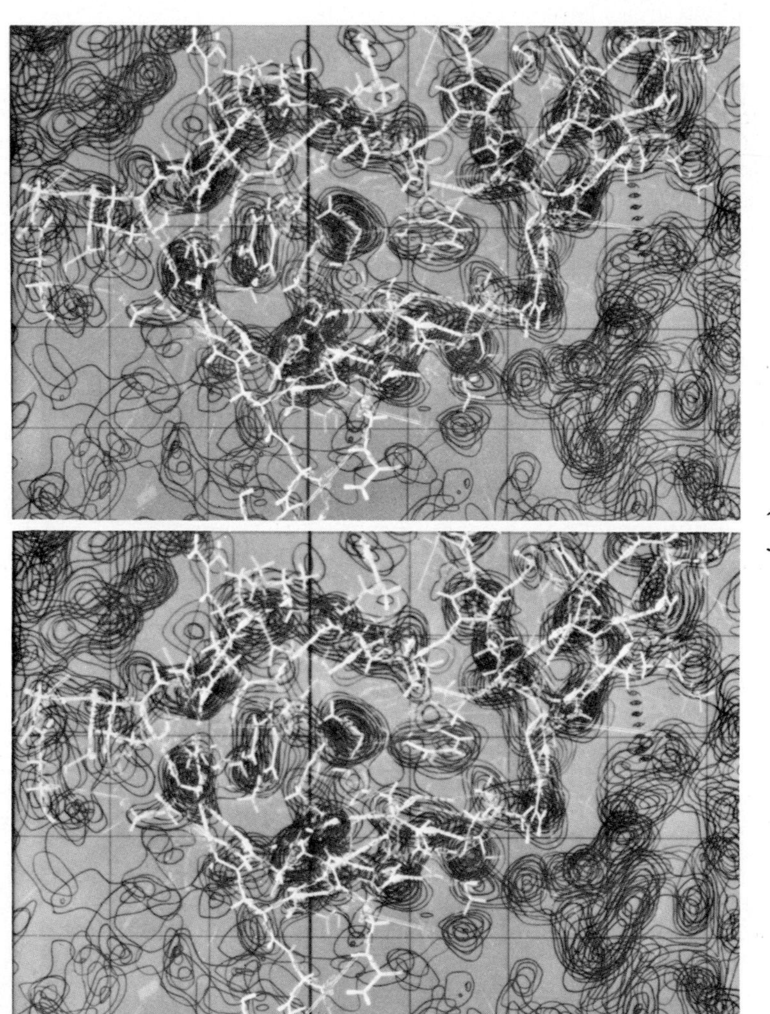

FIG. 13. (a) A stereopair photograph of a portion of the structure of ribonuclease S, shown by optical superposition of eight sections of the 2.0-Å resolution electron density map and the skeletal model derived from the map. The grid squares are 5 Å an a side, and the sections are separated by 7.02 Å.

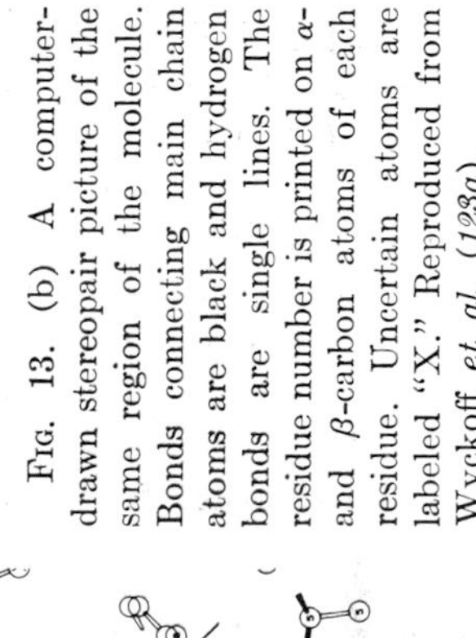

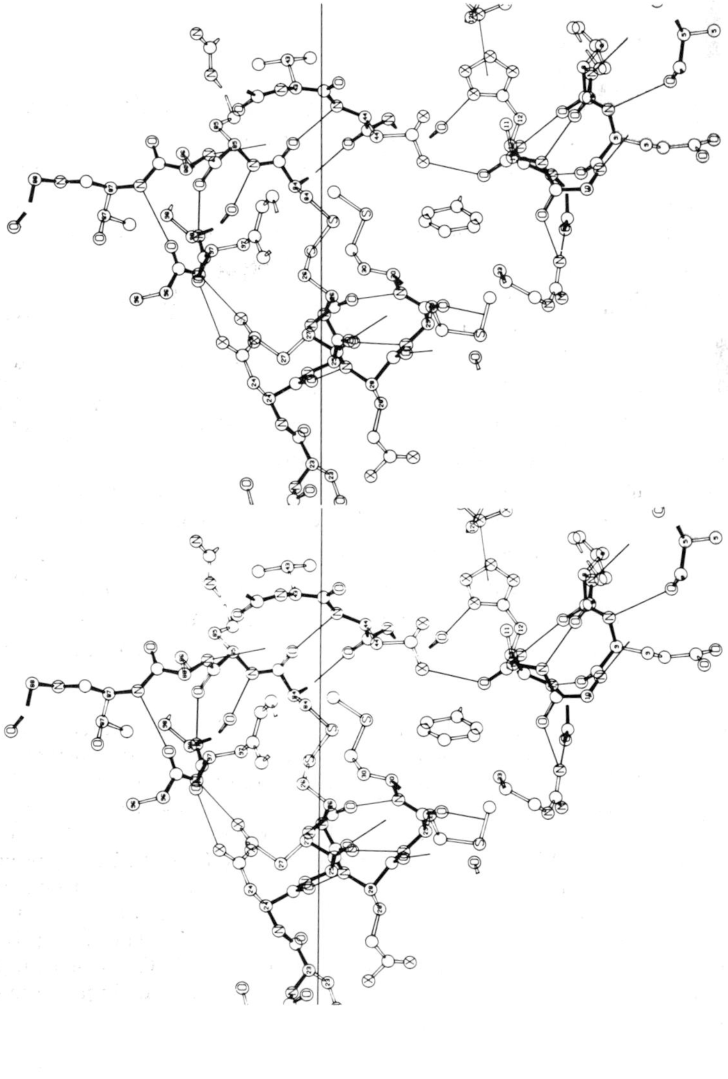

FIG. 13. (b) A computer-drawn stereopair picture of the same region of the molecule. Bonds connecting main chain atoms are black and hydrogen bonds are single lines. The residue number is printed on α- and β-carbon atoms of each residue. Uncertain atoms are labeled "X." Reproduced from Wyckoff *et al.* (*123a*).

an idealized structure can be derived from a selection of approximate coordinates and a knowledge of the sequence, or uncertain regions of the map can be bridged. The second program starts with the idealized coordinates determined by the first and then by treating the model as a flexible chain, fits it to the observed electron density. The calculation is carried out in terms of a "molten zone" of up to ten residues that moves along the polypeptide chain one residue at a time.

IV. X-Ray Diffraction Studies of Globular Macromolecules

A. Molecules Studied

In the decade since the structure of myoglobin was determined to 2 Å resolution, at least fifteen other protein structures have been solved to near-atomic resolution, and electron density maps at lower resolution have been calculated for more than a dozen others. Less detailed structural information such as the presence of some molecular symmetry has been inferred by X-ray methods for at least ten other proteins, and the molecular weights of others have been established partly by diffraction methods. Many other globular macromolecules have been crystallized and characterized by X-ray diffraction, and full structural studies have been initiated on some of them. Table V represents an attempt to summarize some of this information: It catalogs the status of a number of published X-ray diffraction studies and includes a list of many of the heavy atom compounds that have been found to react with or bind to the macromolecules. The symbols for the heavy atom compounds refer to entries in Table VI. The criteria for inclusion of a study in Table V have been somewhat arbitrary; in some cases, such as transfer RNA, the mere report of crystals has been included, while in others, only reports of structural studies or of heavy atom derivatives have been noted (*126a*).

126a. Since Table V was compiled, several important electron density maps of enzymes and other proteins have been computed. These include: erythrocruorin at 2.0 Å resolution (R. Huber, private communication), lamprey hemoglobin at 2.5 Å resolution (W. E. Love, private communication), chymotrypsinogen at 2.5 Å resolution [S. T. Freer, J. Kraut, J. D. Robertus, H. T. Wright, and Ng. H. Xuong, *Biochemistry* **9**, 1997 (1970)], pepsin at 5.5 Å resolution (N. S. Andreeva, private communication), trypsin at 2.8 Å resolution (R. E. Dickerson and R. Stroud, private communication), alcohol dehydrogenase at 5 Å resolution (C. Brändén, private communication), malate dehydrogenase at 5.5 Å resolution (L. Banaszak, private communication), L-lactate dehydrogenase at 2.8 Å resolution (M. G. Rossmann, private communication), and carbonic anhydrase C at 2.0 Å (A. Liljas, private communication).

The structures of those enzymes determined to near-atomic resolution will be described in following volumes of the present series and many of them have been beautifully illustrated by Dickerson and Geis (*6, 7*); moreover the structures of myoglobin and hemoglobin have recently been discussed and depicted in lavish detail by Watson (*118*) and Perutz (*127*). For this reason, the brief discussions of structures given here are intended mainly to accompany Table V as a guide to the literature and to point out applications of techniques described in Section III.

1. *Oxygen Binding Proteins*

Sperm-whale myoglobin consists of a single polypeptide chain of 153 residues and a heme group. All but 21 of the residues participate in one of the eight helical segments that are folded tightly about the heme. The helices are almost entirely of the right-handed α-helical form predicted by Pauling and Corey (*205*), although the final turns of four of them are tightened by one N—H position thus resembling a turn of the 3_{10} helix, and the last turns of two others are loosened by one N—H position (*6, 118*). The heme is entirely buried in polypeptide except for the edge with the two propionic acid side chains; these form hydrogen bonds on the surface of the molecule with His and Arg side chains. The acidic and basic side chains are distributed fairly evenly over the surface, interacting mainly with water molecules or ions in the solvent; apart from these bound waters (or ions) there is no order in the liquid regions. The short stretches of extended chain between the helices serve as corners so that the molecule can fold back on itself and the segment between helices C and D seems also to position additional nonpolar side chains in contact with the heme (*6*). All side chains around the heme are nonpolar, with the important exception of the two His residues that are involved in coordination to the heme and the binding of the oxygen molecule. An imidazole nitrogen atom of one of these is bound to the iron, and the imidazole ring of the other, located on the opposite side of the heme plane, is bridged to the iron through a water molecule. This water molecule in the ferric metmyoglobin takes the place of the oxygen that would be found in the corresponding physiological form, ferrous oxymyoglobin. Kendrew and Perutz believe the function of the cage of nonpolar groups around the heme is to provide an environment in which the iron is safe from oxidation and thus in which its ability to combine reversibly with oxygen is preserved. Deoxymyoglobin was studied by the difference Fourier method at 2.8 Å and found to have the same conformation as metmyoglobin;

127. M. F. Perutz, *Proc. Roy. Soc.* **B173**, 113 (1969).

TABLE V

Class	Molecule	Origin	State and/or prosthetic group	Molecular wt
Oxygen binding proteins	Erythrocruorin	Fly, *Chironomus*	Heme	17,000
	Myoglobin	Sperm whale	Met, heme	17,800
			Deoxy, heme	17,800
	Myoglobin	Seal	Met, heme	17,800
	Myoglobin	Tuna, yellow fin	Met, heme	17,800
	Hemoglobin	Worm, *Glycera*	Heme	18,200
	Hemoglobin	Horse	Oxy, 4 hemes	64,500
	Hemoglobin	Horse	Deoxy, 4 hemes	64,500
	Hemoglobin	Human	Deoxy, 4 hemes	64,500
	Hemoglobin	Ox	Oxy, 4 hemes	64,500
			Oxy, 4 hemes	
Electron transport proteins	Cytochrome b_5	Calf liver	Heme	10,280
	Cytochrome c	Horse heart	Ferri, heme	12,400
	Cytochrome c	Bonito heart	Ferro, heme	12,400
	Cytochrome c_2	*R. rubrum*	Heme	
	Cytochrome c peroxidase	Baker's yeast	Heme	40,000
	Ferredoxin	*M. aerogenes*	8 Fe, 8 S	6,000
	High potential iron protein	*Chromatium*	4 Fe, 4 S	10,000
	Rubredoxin	*C. pasteurianum*	1 Fe	6,000
Proteases	Carboxypeptidase A	Beef	Zn	34,600
	α-Chymotrypsin	Beef		25,000
	γ-Chymotrypsin	Beef		25,000
	δ-Chymotrypsin[d]	Beef		25,000
	π-Chymotrypsin[d]	Beef		25,000

X-RAY DIFFRACTION STUDIES OF GLOBULAR MACROMOLECULES

No. subunits	n′/ a.u.[a]	Space group	Heavy atoms[b]	Stage[c]	Reference
1	1	$P3_1$	Se1, Pt1, Pt7,	5 Å	*128*
			Mersalyl, Hg2	2.8 Å	*129*
1	1	$P2_1$	Ag1, Au1, PCMBS,	HA, 2D	*107*
			PCMA, Hg4, Hg6,	6 Å	*77, 130*
			I1, Xe1	2 Å	*118, 131*
				1.4 Å	*132*
1	1	$P2_1$		2.8 Å	*133*
1	1	$A2$	Ag1, Au1, PCMBS, Hg6	2D	*134*
				5 Å	*135*
1	1	$P2_12_12_1$		C	*136*
1	1	$P2_1$	Au1, Hg2, Hg6, Hg13	5.5 Å	*137*
4	½	$C2$	Ag1, PCMB, DMA,	2D	*138, 139*
			Baker, Hg2, Hg3,	5.5 Å	*60, 140*
			Xe1	2.8 Å	*127, 141, 142*
4	½	$C222_1$	DMA, PCMB	5.5 Å	*143*
4	1	$P2_1$	PCMB, DMA, Hg2, Hg14	5.5 Å	*144*
4	1	$P2_12_12_1$	Ag1, Hg2, Hg3, MMA	2D, HA	*145, 146*
4	½	$P4_132$		C	*49*
1	1	$P2_12_12_1$		C	*147*
1	1	$P4_3$	Pt1, Mersalyl	2D	*33*
				4 Å	*120*
				2.8 Å	*148*
1	1	$P2_12_12_1$	Pt1, U4	6 Å	*65*
1	1	$P2_12_12_1$	Os1, Ir1, Pt1, U4	5 Å	*149*
1	1	$P2_12_12_1$	PCMB	HA	*70*
1	1	$P2_12_12_1$		C	*150*
1	1	$P2_12_12_1$	Pt2, U4	4 Å	*151*
1	1	$R3$	Pt1, Hg6,	2.5 Å	*95*
1	1	$P2_1$	Ag1, Pt1, Hg3, Baker,	2D	*152*
			Pb1	6 Å	*59, 153*
				2.8 Å	*154*
				2.0 Å	*117, 155*
1	2	$P2_1$	Pipsyl, Pt1, Pt2, Pt3,	HA	*23, 156*
			PCMBSF, PMA	2 Å	*15, 157*
1	1	$P4_22_12$	Pipsyl, I1, Pt1, Hg6	HA	*158*
				5.5 Å	*159*
				2.7 Å	*160*
1	1	$P4_22_12$	Pt1, Hg15	5 Å	*161*
1	1	$P4_22_12$		C	*161*

TABLE V (*Continued*)

Class	Molecule	Origin	State and/or prosthetic group	Molecular wt
	Chymotrypsinogen A	Beef		25,000
	Elastase	Pig		25,900
	Extracellular protease	*Arthrobacter*		23,000
	Papain	Papaya latex		23,000
	Pepsin	Pig		35,000
	Subtilisin BPN′	*B. amyloliquefaciens*		27,500
	Trypsin[e]	Beef		24,000
Nucleases	Nuclease	*S. aureus*		16,800
	Ribonuclease	Beef		13,683
	Ribonuclease	Beef		13,683
	Ribonuclease S	Beef		13,700
Oxidoreductases	Alcohol dehydrogenase	Horse liver	Apoenzyme	84,000
			Enzyme + 2NADH	
			Inhibited by pyrazole, etc.	
	Glyceraldehyde-3-phosphate dehydrogenase	Lobster muscle	Enzyme	135,000
			Enzyme + PCMB	
	Malate dehydrogenase	Heart		60,000
	L-Lactate dehydrogenase	Dogfish muscle	Apoenzyme	135,000
			Enzyme + NAD	
Other enzymes	Aspartate transcarbamylase	*E. coli*	Enzyme + CTP	310,000
			Enzyme	
	Carbonic anhydrase	Human C	Zn	34,000

TABLE V (*Continued*)

No. subunits	n'/ a.u.[a]	Space group	Heavy atoms[b]	Stage[c]	Reference
1	1	$P2_12_12_1$	Ir1, Pt2, Hg5, Hg6, U3	5Å 4Å	*162* *161, 163*
1	1	$P2_12_12_1$	PCMBSF, U1	C 3.5 Å	*164* *18*
1	1	$P2_12_12_1$		C	*165*
1	1	$P2_12_12_1$	Ir2, Pt2, PCMB, PCMBS, PCMA, Hg3, Hg6	4.5 Å 2.8 Å	*166* *167*
1	1	$P2_1$	Pt1, Pt9, Hg5, Hg6	HA 2D	*62* *168*
1	1	$C2$	I1, Pt1, Mersalyl, Hg7, Au2, Hg6	2.5 Å	*169*
1	1	$P2_12_12_1$	Mersalyl, Tl1	C HA	*64* *170*
1	1	$P4_1$	I2, Pt2, U2	C 2.8 Å	*63* *171*
1	1	$P2_1$	Ir2, Hg1a, U4	C 5.5	*172* *173*
1	1	$P2_1$	Pt5, Pt6, Pt7, U1	C 2.0 Å	*34, 35* *17*
1	1	$P3_121$	Pt1, Pt8, Pt10, U2	6.0 Å 3.5 Å 2.0 Å	*174* *16* *175, 123a*
2	½	$C222_1$		C, QS	*44*
2	½	$C222_1$		C, QS	*44*
2	1	$P2_1$		C	*44*
4	1	$P2_12_12_1$		HA, QS	*46*
4	½	$C222_1$	PCMB		*46*
2	1	$P2_12_12$		C	*176*
4	¼	$I422$	Pt10, Au1, DMA, Hg1, Baker	C, QS 4 Å	*52* *26*
4	½	$P4_22_12$			*26*
12	⅓	$P321$	Pt1	HA, QS	*177*
12	½	$P422$ or $P4_22_12$		QS	*45*
1	1	$P2_1$	PCMBS, Hg2, Hg8, Hg9, Hg10, Hg11, Hg12	C 2D 5.5 Å	*178* *179* *180*

TABLE V (*Continued*)

Class	Molecule	Origin	State and/or prosthetic group	Molecular wt
	Catalase	Beef liver	Heme	250,000
	Lysozyme-chloride	Hen egg		14,600
	Lysozyme-chloride	Hen egg		14,600
	Lysozyme-nitrate	Hen egg		14,600
Hormones	Glucagon	Pig and sheep		3,482
	Insulin	Pig	Zn	5,780
	Insulin	Pig		5,780
	Insulin	Beef	Wet	5,733
			Dry	5,733
Immunoglobulins	IgG	Human myeloma		155,000
	Fab fragment of IgG	Human myeloma:		
		Patient Hil		55,000
		Patient Smo		55,000
		Patient New		55,000
	Fc fragment of IgG	Human myeloma	Wet	50,000
			Dry with $KAuCl_3$	
Other proteins	Apoferritin	Horse spleen	Protein shell	480,000
	Ferritin B	Horse spleen		747,000
	Ferritin A			747,000
	β-Lactoglobulin	Cow's milk	Cd	36,000
	Tropomyosin	Rabbit muscle		70,000
tRNA	$tRNA_F^{Met}$	*E. coli*		
	$tRNA_F^{Met}$	*E. coli B*		
	$tRNA^{Phe}$	*E. coli*		
	tRNA	Yeast	Unfractionated	

[a] Number of molecules per asymmetric unit.

[b] Symbols refer to the heavy atom compounds listed in Table VI.

[c] Abbreviations: C, preliminary X-ray diffraction study of the crystal; HA, survey for heavy atom derivatives; 2D, a two-dimensional electron density map (a projection); *x* Å,

TABLE V (*Continued*)

No. subunits	n'/ a.u.[a]	Space group	Heavy atoms[b]	Stage[c]	Reference
	1	$P3_121$		QS	*181*
1	1	$P4_32_12$	Pd1, Pt2, PCMBS, MHTS,	6 Å	*182*
			Hg5, Hg6, U1, U4	2 Å	*183–187*
1	1	$P4_32_12$	Nb1, Ta1, Pt4	5 Å	*90, 188*
1	1	$P1$	Pt1, Hg5, Hg6	HA	*189, 190*
1	1	$P2_13$		C	*191*
				3 Å	*192*
1	2	$R3$	Pb1, U2, U4	C, RF	*110, 193*
				2.8 Å	*194*
1	4	$P2_12_12_1$	Hg6	HA	*195*
1	2	$P2_12_12_1$		C, QS	*196*
1	1	$I2_12_12_1$		C, QS	*196*
2 light chains	$\frac{1}{2}$	$C2$		C, QS	*50*
2 heavy chains				HA	*197*
	2	$P2_12_12_1$		C	*198*
	2	$P3_121$		C	*198*
	1	$C2$		C	*199*
2	1	$P2_12_12_1$		QS	*51*
2	$\frac{1}{2}$	$A2_122$		QS	*51*
	1/24	$F432$		QS	*200*
	1/24	$P2_12_12$		QS	*201*
	$\frac{1}{2}$	$F432$		QS	*201*
2	1	$P2_1$		HA	*146*
2	$\frac{1}{2}$	$B22_12$	MMA, PCMBS	QS	*56*
2		$P2_12_12$		QS	*29*
1	1	$P222_1$		C	*202*
1	~6	$P6_2$ or		C	*203*
	3	$P6_222$			
1	2	$P6_222$		C	*58*
				C	*204*

a three-dimensional electron density map to a nominal resolution of x angstroms; RF, use of the Rossmann-Blow rotation function; and QS, study of the quaternary structure.

[d] Inhibited by phenylmethanesulfonyl fluoride.

[e] Inhibited by diisopropylphosphorylfluoridate.

128. R. Huber, H. Formanek, and O. Epp, *Naturwissenschaften* **55,** 75 (1968).
129. R. Huber, O. Epp, and H. Formanek, *Acta Cryst.* **A25,** S186 (1969).
130. J. C. Kendrew, G. Bodo, H. M. Dintzis, R. G. Parrish, H. Wyckoff, and D. C. Phillips, *Nature* **181,** 662 (1958).
131. J. C. Kendrew, R. E. Dickerson, B. E. Strandberg, R. G. Hart, D. R. Davies, D. C. Phillips, and V. C. Shore, *Nature* **185,** 422 (1960).
132. J. C. Kendrew, *Science* **139,** 1259 (1963).
133. C. L. Nobbs, H. C. Watson, and J. C. Kendrew, *Nature* **209,** 339 (1966).
134. H. Scouloudi, *Proc. Roy. Soc.* **A258,** 181 (1960).
135. H. Scouloudi, *JMB* **40,** 353 (1969).
136. R. H. Kretsinger, *JMB* **38,** 141 (1968).
137. E. A. Padlan and W. E. Love, *Nature* **220,** 376 (1968).
138. L. Bragg and M. F. Perutz, *Proc. Roy. Soc.* **A225,** 315 (1954).
139. D. M. Blow, *Proc. Roy. Soc.* **A247,** 302 (1958).
140. A. F. Cullis, H. Muirhead, M. F. Perutz, M. G. Rossmann, and A. C. T. North, *Proc. Roy. Soc.* **A265,** 161 (1962).
141. M. F. Perutz, H. Muirhead, J. M. Cox, L. C. G. Goaman, F. S. Matthews, E. L. McGandy, and L. E. Webb, *Nature* **219,** 29 (1968).
142. M. F. Perutz, H. Muirhead, J. M. Cox, and L. C. G. Goaman, *Nature* **219,** 131 (1968).
143. W. Bolten, J. M. Cox, and M. F. Perutz, *JMB* **33,** 283 (1968).
144. H. Muirhead, J. M. Cox, L. Mazzarella, and M. F. Perutz, *JMB* **28,** 117 (1967).
145. D. W. Green and A. C. T. North, *Acta Cryst.* **A13,** 1055 (1960).
146. P. Dunnill, D. W. Green, and R. M. Simmons, *JMB* **22,** 135 (1966).
147. F. S. Matthews and P. Strittmatter, *JMB* **41,** 295 (1969).
148. D. Eisenberg, O. Battfay, J. C. Varnum, L. Samson, A. Cooper, and R. E. Dickerson, *Acta Cryst.* **A25,** S186 (1969).
149. J. Kraut, S. Singh, and R. A. Alden, *in* "Structure and Function of Cytochromes" (K. Okunuki, M. D. Kamen, and I. Sekuzu, eds.), p. 252. Univ. of Tokyo Press, Tokyo, 1968.
150. L. C. Sieker and L. H. Jensen, *BBRC* **20,** 33 (1965).
151. G. Strahs and J. Kraut, *JMB* **35,** 503 (1968).
152. M. L. Ludwig, I. C. Paul, G. S. Pawley, and W. N. Lipscomb, *Proc. Natl. Acad. Sci. U. S.* **50,** 282 (1963).
153. T. A. Steitz, M. L. Ludwig, F. A. Quiocho, and W. N. Lipscomb, *JBC* **242,** 4662 (1967).
154. M. L. Ludwig, J. A. Hartsuck, T. A. Steitz, H. Muirhead, J. C. Coppola, G. N. Reeke, and W. N. Lipscomb, *Proc. Natl. Acad. Sci. U. S.* **57,** 511 (1967).
155. G. N. Reeke, J. A. Hartsuck, M. L. Ludwig, F. A. Quiocho, T. A. Steitz, and W. N. Lipscomb, *Proc. Natl. Acad. Sci. U. S.* **58,** 2220 (1967).
156. P. B. Sigler and D. M. Blow, *JMB* **14,** 640 (1965).
157. P. B. Sigler, D. M. Blow, B. W. Matthews, and R. Henderson, *JMB* **35,** 143 (1968).
158. P. B. Sigler, H. C. W. Skinner, C. L. Coulter, J. Kallos, H. Braxton, and D. R. Davies, *Proc. Natl. Acad. Sci. U. S.* **51,** 1146 (1964).
159. G. H. Cohen, E. W. Silverton, B. W. Matthews, H. Braxton, and D. R. Davies, *JMB* **44,** 129 (1969).
160. D. R. Davies, G. H. Cohen, E. W. Silverton, H. P. Braxton, and B. W. Matthews, *Acta Cryst.* **A25,** S182 (1969).

161. J. Kraut, H. T. Wright, M. Kellerman, and S. T. Freer, *Proc. Natl. Acad. Sci. U. S.* **58,** 304 (1967).

162. J. Kraut, L. C. Sieker, D. F. High, and S. T. Freer, *Proc. Natl. Acad. Sci. U. S.* **48,** 1417 (1962).

163. J. Kraut, D. F. High, and L. C. Sieker, *Proc. Natl. Acad. Sci. U. S.* **51,** 839 (1964)

164. D. M. Shotton, B. S. Hartley, N. Camerman, T. Hofmann, S. C. Nyberg, and L. Rao, *JMB* **32,** 155 (1968).

165. H. Eklund, M. Zeppezauer, and C.-I. Branden, *JMB* **34,** 193 (1968).

166. J. Drenth, J. N. Jansonius, and B. G. Wolthers, *JMB* **24,** 449 (1967).

167. J. Drenth, J. N. Jansonius, R. Koekoek, H. M. Swen, and B. G. Wolthers, *Nature* **218,** 929 (1968).

168. V. V. Borisov, V. R. Melik-Adamyan, L. E. Sotnikova, N. E. Shutskever, and N. S. Andreeva, *Soviet Phys.-Cryst.* (*English Transl.*) **13,** 329 (1968).

169. C. S. Wright, R. A. Alden, and J. Kraut, *Nature* **221,** 235 (1969).

170. R. M. Stroud, L. Kay, R. H. Stanford, O. Battfay, R. B. Corey, and R. E. Dickerson, *Acta Cryst.* **A25,** S182 (1969).

171. F. A. Cotton, E. E. Hazen, D. C. Richardson, J. S. Richardson, A. Arnone, and C. J. Bier, *Acta Cryst.* **A25,** S188 (1969).

172. C. H. Carlisle and H. Scouloudi, *Proc. Roy. Soc.* **A207,** 496 (1951).

173. H. P. Avey, M. O. Boles, C. H. Carlisle, S. A. Evans, S. J. Morris, R. A. Palmer, B. A. Woolhouse, and S. Shall, *Nature* **213,** 557 (1967).

174. H. W. Wyckoff, K. D. Hardman, N. M. Allewell, T. Inagami, D. Tsernoglou, L. N. Johnson, and F. M. Richards, *JBC* **242,** 3749 (1967).

175. H. W. Wyckoff, N. M. Allewell, K. D. Hardman, D. Tsernoglou, A. Hanson, B. Lee, J. Knox, and F. M. Richards, *Acta Cryst.* **A25,** S189 (1969).

176. L. J. Banaszak, *JMB* **22,** 389 (1966).

177. T. A. Steitz, D. C. Wiley, and W. N. Lipscomb, *Proc. Natl. Acad. Sci. U. S.* **58,** 1859 (1967).

178. R. Strandberg, B. Tilander, K. Fridborg, S. Lindskog, and P. O. Nyman, *JMB* **5,** 583 (1962).

179. B. Tilander, B. Strandberg, and K. Fridborg, *JMB* **12,** 740 (1965).

180. K. Fridborg, K. K. Kannan, A. Liljas, J. Lundin, B. Strandberg, R. Strandberg, B. Tilander, and G. Wirén, *JMB* **25,** 505 (1967).

181. W. Longley, *JMB* **30,** 323 (1967).

182. C. C. F. Blake, R. H. Fenn, A. C. T. North, D. C. Phillips, and R. J. Poljak, *Nature* **196,** 1173 (1962).

183. C. C. F. Blake, D. F. Koenig, G. A. Mair, A. C. T. North, D. C. Phillips, and V. R. Sarma, *Nature* **206,** 757 (1965).

184. D. C. Phillips, *Proc. Natl. Acad. Sci. U. S.* **57,** 483 (1967).

185. C. C. F. Blake, G. A. Mair, A. C. T. North, D. C. Phillips, and V. R. Sarma, *Proc. Roy. Soc.* **B167,** 365 (1967).

186. C. C. F. Blake, L. N. Johnson, G. A. Mair, A. C. T. North, D. C. Phillips, and V. R. Sarma, *Proc. Roy. Soc.* **B167,** 378 (1967).

187. C. C. F. Blake, *Proc. Roy. Soc.* **B167,** 435 (1967).

188. R. H. Stanford, Jr., R. E. Marsh, and R. B. Corey, *Nature* **196,** 1176 (1962).

189. R. E. Dickerson, J. M. Reddy, M. Pinkerton, and L. K. Steinrauf, *Nature* **196,** 1178 (1962).

190. L. K. Steinrauf, J. M. Reddy, and R. E. Dickerson, *Acta Cryst.* **15,** 423 (1962).

however, the space between the iron and second histidine, instead of holding a water molecule, is empty (*133*).

Hemoglobin from mammals consists of two α chains of 141 amino acid residues and two β chains of 146 residues. Perutz and co-workers have found that each of the four chains is folded about a heme in much the same way as myoglobin, and the four are then arranged tetrahedrally. The same "globin fold" exists in erythrocruorin from the fly and in hemoglobin from the worm *Glycera* (*129, 137*). This is a remarkable result when one realizes that only seven of the more than 140 sites in the globin chain are occupied by the same residue in all species studied (*127*). As in myoglobin, polar residues are excluded from the interior of the individual subunits, except for the critical His residues adjacent to the heme and several Ser and Thr residues that are hydrogen bonded to carbonyl groups within the same α helix. Almost all nonpolar side chains are in the interior or at the boundaries between unlike subunits, although a very few, including one Cys and one Leu side chain, protrude into the solvent. The importance of the internal nonpolar groups in maintaining the characteristic globin fold has been emphasized by Perutz and Lehmann, who noted from properties of abnormal human hemoglobins that the molecule is extremely sensitive to small alterations in the pattern of these nonpolar contacts (*206*). The contacts between unlike subunits include some hydrogen bonds but are mainly hydrophobic interactions between nonpolar side chains; contacts between like subunits are not visible in the maps.

191. M. V. King, *JMB* **1,** 375 (1959).
192. W. P. Haugen, and W. N. Lipscomb, *Acta Cryst.* **A25,** S185 (1969).
193. M. M. Harding, D. C. Hodgkin, A. F. Kennedy, A. O'Connor, and P. D. J. Wertzmann, *JMB* **16,** 212 (1966).
194. M. J. Adams, T. L. Blundell, E. J. Dodson, G. G. Dodson, M. Vijayan, E. N. Baker, M. M. Harding, D. C. Hodgkin, B. Rimmer, and S. Sheat, *Nature* **224,** 491 (1969).
195. B. W. Low and C. C. H. Chen, *Acta Cryst.* **19,** 686 (1965).
196. R. J. Einstein and B. W. Low, *Acta Cryst.* **15,** 32 (1962).
197. V. R. Sarma and D. R. Davies, *Acta Cryst.* **A25,** S187 (1969).
198. R. L. Humphrey, H. P. Avey, L. N. Becka, R. J. Poljak, G. Rossi, T. K. Choi, and A. Nisonoff, *JMB* **43,** 223 (1969).
199. H. P. Avey, R. J. Poljak, G. Rossi, and A. Nisonoff, *Nature* **220,** 1248 (1968).
200. P. M. Harrison, *JMB* **6,** 404 (1963).
201. P. M. Harrison, *JMB* **1,** 69 (1959).
202. B. F. C. Clark, B. P. Doctor, K. C. Holmes, A. Klug, K. A. Marcker, S. J. Morris, and H. H. Paradies, *Nature* **219,** 1222 (1968).
203. S.-H. Kim and A. Rich, *Science* **162,** 1381 (1968).
204. J. R. Fresco, R. D. Blake, and R. Langridge, *Nature* **220,** 1285 (1968).
205. L. Pauling and R. B. Corey, *Proc. Natl. Acad. Sci. U. S.* **37,** 205 (1951).
206. M. F. Perutz and H. Lehmann, *Nature* **219,** 902 (1968).

Upon deoxygenation, the hemoglobin molecule undergoes a marked change in structure. The change seems to be mainly one in quaternary structure since the relative positions of atoms within the individual subunits (i.e., the secondary and tertiary structures) are not altered to within the accuracy of the 5.5-Å electron density maps. The heme–heme interaction has its basis in this change of quaternary structure, but the mechanism is not yet understood. Nevertheless, Perutz and his group have recently made an important advance in understanding the physiology of the hemoglobin molecule: The groups responsible for the alkaline Bohr effect have been identified from several Fourier difference maps that show the shifts in electron density that accompany reaction of hemoglobin with *N*-ethylsuccinimide (*207*). They are the imidazole groups of the C-terminal His residues of the β chains and the α-amino groups of the α chains. These groups are free in oxyhemoglobin but are probably bound to carboxyl groups in deoxyhemoglobin, thus having their p*K*'s raised.

2. *Proteases*

Carboxypeptidase A is a zinc metalloenzyme that removes the C-terminal residue of a polypeptide chain and has the highest rate of cleavage for residues with aromatic and branched aliphatic side chains. The study of this molecule by Lipscomb and his group is a striking example of the power of X-ray diffraction for elucidating the structure, the mechanism, and the specificity of an enzyme. In the low (6 Å) resolution analysis, the enzyme was found to be ellipsoidal with dimensions of $52 \times 44 \times 40$ Å and to have a helical content of about 25% (later found to be 30%). The zinc atom was located by means of a difference Fourier map based on data from the native and zinc-free enzymes, and it was found to lie in a depression on the surface. A pocket adjacent to the zinc was surmised to accommodate the bulky side chain of the substrate's C-terminal amino acid (*59*). The 2.8- and 2.0-Å analyses revealed eight vertebra-like folds of parallel and antiparallel β-pleated sheet in the center of the enzyme and eight stretches of α helix packed tightly around them (see Fig. 14). The residues forming the β sheets are mainly nonpolar or hydrogen-bonded Ser and Thr, but a few charged residues are at the edges of the sheet that abut on the solvent. The electron density in the region between molecules is suggestive of partially ordered water in the crystals. The sides of the α helices packed against the β structure consist mainly of nonpolar side chains. The zinc atom was found to be coordi-

207. M. F. Perutz, H. Muirhead, L. Mazzarella, R. A. Crowther, J. Greer, and J. V. Kilmartin, *Nature* **222**, 1240 (1969).

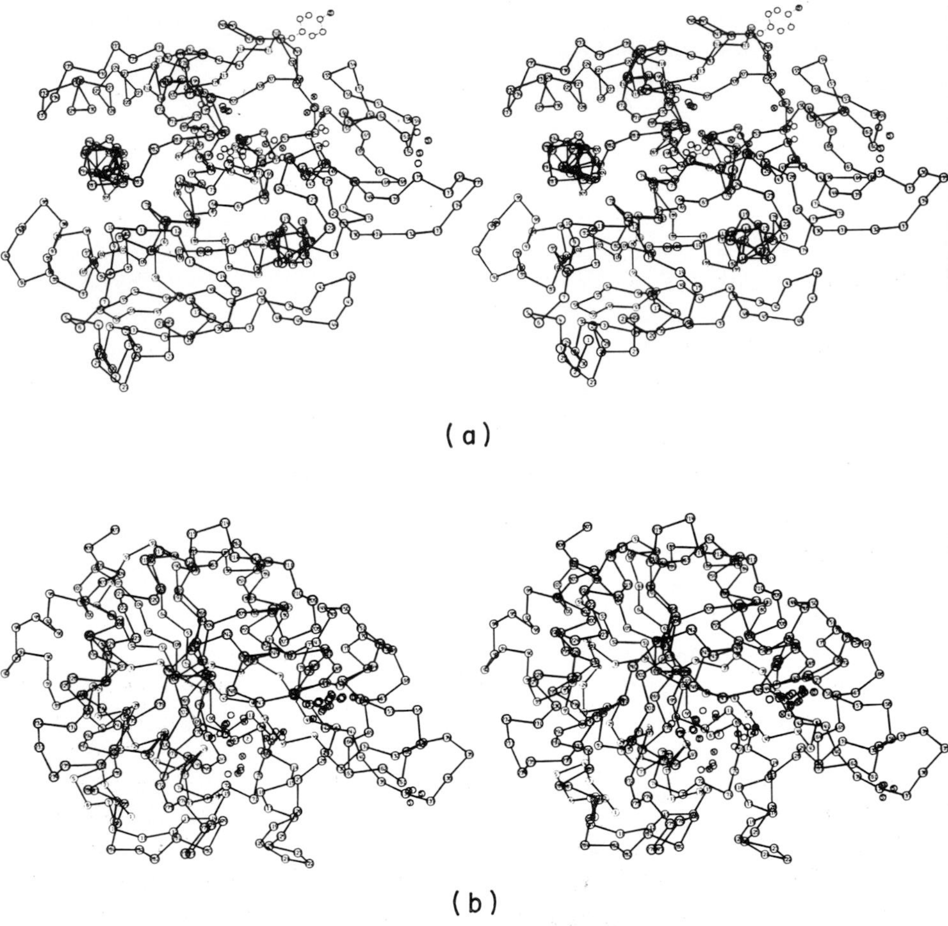

FIG. 14. Computer-drawn stereopair pictures of the polypeptide backbone of carboxypeptidase A. Only α-carbon atoms of amino acid residues and several important side chains are shown. The α-carbon atoms are numbered from the amino-terminal end of the enzyme. (a) A view of the molecule that emphasizes several of the α helices. The side chain of Tyr 248 is shown at the top; the active site lies in the pit below and a bit to the left of it. (b) A view from the top of (a), showing the twisted β sheet near the center of the molecule. Reproduced from Dickerson and Geis (7).

nated to three side chains of the enzyme, none of them being cysteine as had been expected from earlier chemical studies.

Difference Fourier maps of the substrate glycyltyrosine at 2.8 and 2.0 Å resolution provided the key to understanding the mechanism and specificity of the enzyme. The C-terminal carboxylate ion of the substrate forms a salt link with Arg 145 of the enzyme, and the substrate's phenolic side chain extends into the pocket seen on the 6-Å map. Tyrosine 248 of the enzyme swings down some 12 Å so that its hydroxyl group is next to the amide bond of the substrate to be split, and the carbonyl of the amide bond is pressed against the zinc atom. Based on this and a number of chemical observations, Lipscomb *et al.* (*117*) proposed two mechanisms for cleavage, in which Tyr 248 donates a proton to the amide of the susceptible peptide bond and in which Glu 270 attacks or promotes attack on the carbonyl carbon of the same bond. They also noted that the movement of Tyr 248 and other atoms of the active site in response to the binding of substrate is in accord with the induced-fit hypothesis (*208*). The 2.0-Å study also showed how the carboxypeptidase molecule is protected from self-digestion: its C-terminal carboxylate group curls toward the enzyme surface.

α-Chymotrypsin cleaves peptide chains at the C-terminal side of residues having aromatic side chains. The study of this enzyme by Blow and his colleagues has provided information on its specificity and mechanism and has revealed the way in which the zymogen, chymotrypsinogen, is activated. As in carboxypeptidase A, the active site is in a depression in the surface next to a deep cleft that can hold the aromatic side chain of the residue about to be severed. The active site contains Ser 195 and His 57, and buried under His 57 is Asp 102. According to Blow *et al.* (*209*), these three residues act as a charge relay system conducting electrons from the carboxyl group of Asp 102 to hydrogen bonds on the surface of the enzyme: In its nonpolar environment of low dielectric constant, the negatively charged Asp 102 forms a strong link with the positively charged nitrogen on the imidazole ring of His 57; moreover, since the other nitrogen of this ring is hydrogen bonded to the hydroxyl group of Ser 195, the polarizing effect of this link is transmitted to Ser 195. Thus the hydroxyl oxygen of Ser 195 becomes a powerful nucleophile capable of attacking the carbon atom of the amide bond of the substrate. Sigler *et al.* (*157*) found that the activation of the zymogen is associated with the formation of another strong salt link beneath the surface of the enzyme. Activation begins with the cleavage of chymotrypsin by trypsin between residues Arg 15 and Ile 16. The newly created, positively charged α-amino group of Ile 16 pulls the negatively charged side chain of Asp 194 from its previous position in which it covered the active hydroxyl of Ser 195.

The structure of the enzyme is made up of extended chains, often running parallel to each other; a few turns of α helix are found at the C-

208. D. E. Koshland, Jr., *Cold Spring Harbor Symp. Quant. Biol.* **28,** 473 (1963).
209. D. M. Blow, J. J. Birktoft, and B. S. Hartley, *Nature* **221,** 337 (1969).

terminus of the molecule. All charged groups other than those involved in catalysis and activation are on the surface of the molecule; many nonpolar side chains are in the interior.

γ-Chymotrypsin crystals are reversibly interconvertible with crystals of α-chymotrypsin, the two containing molecules of the same covalent structure (*210*). Davies and co-workers recently calculated a 2.7-Å electron density map of γ-chymotrypsin (*160*). An earlier 5.5-Å resolution map had been compared to the 2-Å map of α-chymotrypsin and the molecules were found to have numerous points of similarity (*13*). Kraut and his group obtained 5-Å electron density maps for π-, δ-, and γ-chymotrypsin and a 4-Å map for chymotrypsinogen (*161, 162*). The gross conformations of all four molecules were found to be very similar, the small differences being interpreted in terms of the activation of chymotrypsinogen. Wright *et al.* compared π-, δ-, and γ-chymotrypsin to α-chymotrypsin and proposed a new scheme for the genesis of the various forms of the enzyme (*14*).

Elastase is a serine protease of 240 residues with an amino acid sequence highly homologous with those of chymotrypsin and trypsin. The 3.5-Å electron density map obtained by Shotton and Watson (*18*) confirms earlier suggestions that the three-dimensional structure and catalytic site are closely similar to those of chymotrypsin and also reveals the atomic basis for the different specificities of the enzymes. Elastase cleaves peptide bonds on the C-terminal side of residues lacking charged or aromatic side chains, whereas chymotrypsin requires an aromatic side chain. The polypeptide backbone of elastase winds about in the same pattern as that of chymotrypsin, and even the side chains often have similar orientations, particularly in the interior of the molecule. The points in the chain at which insertions and deletions of residues take place between the two molecules are generally at the surface of the structures. As in chymotrypsin, the α-amino group of the N-terminal residue of elastase is buried in the molecular interior where it can form a salt bridge with the side chain of aspartic acid 194. All residues that form the "charge relay system" at the catalytic site of chymotrypsin are present, although the position of His 57 seems to be distorted by interaction with the toluene sulfonate group in the inhibited form of elastase used in this study. A major difference in the structures is observed at the specificity site, where the chymotrypsin molecule has a deep cleft which holds the aromatic side chain of the residue about to be severed. In elastase the mouth of the cleft is covered by Val 216, and it is also partially filled by Thr 226 thus preventing bulky aromatic groups from entering. Since trypsin has many

210. R. B. Corey, O. Battfay, D. A. Brueckner, and F. G. Mark, *BBA* **94,** 535 (1965).

similarities to both chymotrypsin and elastase but has yet another specificity (cleaving at the C-terminal side of basic residues), the results of the X-ray analysis of the enzyme are awaited with anticipation (*170*).

Subtilisin BPN′ is a serine protease from a soil bacterium; it has a catalytic center similar to those of chymotrypsin and elastase, but an amino acid sequence and overall structure that are completely different. Considered together, chymotrypsin and subtilisin seem to present a remarkable example of convergent evolution, in which molecules have evolved from different starting points to perform a similar function in a similar way. The 2.5-Å electron density map of subtilisin produced by Wright *et al.* (*169*) shows a roughly spherical molecule of diameter 42 Å. The nonpolar side chains extending from a parallel β-pleated sheet of five chains form a "hydrophobic core" in the interior of the molecule. The polypeptide backbone also contains eight segments of right-handed α helix, the longest of which runs roughly through the center of the molecule. Several highly polar or charged side chains are either inside or lie in surface crevices of the molecule, and a sizable number of large nonpolar side chains lie on the surface. The active site, located from a difference Fourier map of the phenylmethanesulfonyl fluoride inhibited enzyme, does not lie in a pronounced depression. However, it contains Ser, His, and Asp residues in an arrangement similar to that found in α-chymotrypsin.

Papain is a sulfhydryl protease of 211 residues. The 2.8-Å electron density map calculated by Drenth and his group shows that the single polypeptide chain is folded into two distinct sections that are divided by a groove. The active center lies in the groove; it consists of a Cys and a His, with the sulfur atom about 4 Å from the imidazole ring. One of the two sections has a large hydrophobic core where some 12–14 nonpolar side chains are clumped together. Other than four short α-helical segments and a small portion of β structure, the path of the chain is irregular.

3. *Nucleases and Lysozyme*

Ribonuclease S is formed from ribonuclease A when subtilisin cleaves the latter between residues 20 and 21. The S peptide (residues 1–20) is held tightly to the rest of the molecule, mainly through hydrophobic interactions, and the enzyme retains its activity. The structures of both modifications have now been solved to 2 Å resolution: Ribonuclease A by Kartha, Bello, and Harker, and ribonuclease S by Richards, Wyckoff, and their associates. The only marked structural difference is a gap of 10–15 Å in ribonuclease S between residues 20 and 21 at the point of cleavage. The molecule is roughly kidney-shaped, the active site being located in a deep depression between the two lobes of the kidney. This site was found in ribonuclease A from a difference Fourier map using data from the native phosphate-ribonuclease A and from an isomorphous derivative in which arsenate replaces the phosphate ion. In ribonuclease S the active site was located from the position of binding of uridine-2′(3′)-phosphate and of 5-iodouridine-2′(3′)-phosphate. Histidine residues 12 and 119 are near the site of nucleotide binding. Structural elements of the molecule include a "hydrophobic core" involving about 15% of the residues, three sections of helix totaling another 15% of the residues, and two chains forming a long and twisted antiparallel β structure.

Nuclease from *Staphylococcus aureus* is an extracellular enzyme consisting of a single polypeptide chain of 149 residues and no disulfide bonds. Its structure has recently been solved to 2.8 Å resolution by Cotton and his group, but details have not yet been published.

Lysozyme from hen egg white was the first enzyme whose structure was determined to near-atomic resolution *(183)*. Phillips, North, and their colleagues who achieved this feat also pioneered the application of difference Fourier maps to study the binding of inhibitors to the enzyme and were able to gain important information on the catalytic mechanism. The lysozyme molecule, like ribonuclease and papain, is divided into two lobes by a deep cleft that contains the catalytic site. The interior of one of the lobes is a "hydrophobic core" formed from many nonpolar side chains. One stretch of the main chain folds back on itself to form an antiparallel β-pleated sheet of 14 residues. There are three stretches of helix, each containing about 10 residues; they are distorted from pure α-helical form toward the form of 3_{10} helix. The active site includes Asp 52 and Glu 35, the former having its pK lowered by the presence of neighboring polar residues and the latter having its pK raised by virtue of sitting on a hydrophobic surface near the bottom of the cleft. These changes in pK enable the enzyme to perform its catalytic role. As in the case of carboxypeptidase, the binding of the substratelike compounds induces changes in the structure of the enzyme. For example, when N-acetylglucosamine is bound to lysozyme, the cleft narrows slightly, some atoms moving as much as 0.75 Å toward the substrate *(184)*.

A small channel running in from the molecular surface seems to contain two water molecules; it is not clear whether they interchange freely with the solvent water or whether they are trapped inside. The intermolecular region shows peaks of electron density that often occur at hydrogen bonding distances from each other or from bonding groups on the enzyme surface; Blake *et al.* *(185)* suggest that these peaks indicate water molecules. Several more prominent peaks on the surface are believed to be bound Cl ions.

4. *Other Proteins*

At least three other protein structures have been solved to near-atomic resolution recently. One of these is rubredoxin, a nonheme iron protein of molecular weight 6000. The 2.5-Å electron density map of Jensen and his colleagues shows four cysteine sulfur atoms coordinated to the single iron atom in a nearly tetrahedral arrangement (*95*). Insulin has been solved to 2.8 Å resolution in Hodgkin's laboratory; the results confirm earlier conclusions that the two molecules in the asymmetric unit are related approximately by a noncrystallographic twofold axis (*194*). Cytochrome c in the oxidized state has been solved to 2.8 Å resolution by Dickerson and co-workers (*148*). The path of the polypeptide backbone is highly irregular, containing no extended helix or β structure. The heme is largely buried in a nonpolar pocket.

Two other enzymes solved to lower resolution deserve special note because electron density maps have revealed important structural information about them. One of these is human carbonic anhydrase C studied by Strandberg and colleagues (*180*). The 5.5-Å map shows an ellipsoidal molecule of dimensions $40 \times 45 \times 55$ Å. The zinc atom, located from a difference Fourier study of the zinc-free apoenzyme, is at the bottom of a large cavity in the enzyme surface. The single sulfhydryl group of the molecule is situated at the surface some 14 Å from the zinc atom. A likely folding of the polypeptide chain was proposed on the basis of this information. Rossmann and co-workers obtained a 4-Å electron density map of L-lactate dehydrogenase (*26*). The tetrameric molecule and the shapes of the individual subunits were visible though it was not possible to trace the polypeptide chain. Each subunit is divided into two parts by a narrower neck. Adenosine, a competitive inhibitor, was found to bind to the surface between adjoining subunits.

B. Heavy Atom Derivatives

In solving protein structure by the method of isomorphous replacement, the preparation of heavy atom derivatives has usually been the most time-consuming step. Fortunately a backlog of successful compounds and methods is now building up, and in each new study one can start by trying the relevant ones. This section summarizes some of the useful compounds and methods but is far from exhaustive; a much more extensive discussion of the derivative problem has recently been given by Blake (*211*).

1. *Preparation and Classification*

Derivatives are usually formed by diffusing the heavy atom into the protein crystals; even quite large and relatively insoluble compounds migrate through the channels occupied by the liquid of crystallization. The compound may be added in solid or solution form to a suspension of protein crystals in their mother liquor, or it can be dialyzed in to prevent osmotic shock. Lipscomb *et al.* described a simple procedure for dialysis (*59*). Wyckoff *et al.* developed a much more elaborate flow-cell method that enables one to flow on, and then wash out, a succession of heavy atom containing solutions, meanwhile monitoring the intensities of several reflections to see if binding has occurred (*212*).

A second method of derivative formation is to react the protein and heavy atom compound in solution, and then to crystallize them together This is the only feasible method for forming some covalently bound derivatives where the conditions for reaction cannot be produced in the crystalline state. It requires more protein and more time for each trial, however, and even if the derivative can be formed in solution, it sometimes cannot be crystallized or it sometimes crystallizes in a space group different from that of the parent compound. But if a covalently bound derivative can be prepared by this method, the effort is rewarded: Chemical studies can often establish the particular amino acid that has been altered, thus providing a heavy atom marker that will be helpful in the interpretation of the electron density map (Section III,E). Moreover, the binding is more likely to be stoichiometric than for noncovalently bound derivatives, and hence the derivative will be a greater aid to finding the phases. Stoichiometric binding also enables the crystallographer to establish the absolute level of the electron density on the final map; this too can be helpful in interpretation.

Blake (*211*) and other authors have classified the various heavy atom derivatives according to the mode of binding. The usual classifications include:

a. Sulfhydryl Reagents. The sulfhydryl side chains of cysteine residues often bind mercurials strongly. Perutz and his colleagues exploited the differential reactivity of different sets of sulfhydryls on the horse and

211. C. C. F. Blake, *Advan. Protein Chem.* **23**, 59 (1968).

212. H. W. Wyckoff, M. Doscher, D. Tsernoglou, T. Inagabi, L. N. Johnson, K. D. Hardman, N. M. Allewell, D. M. Kelly, and F. M. Richards, *JMB* **27**, 563 (1967).

human hemoglobin molecules to obtain numerous derivatives. These are listed in Table VI and have been discussed at length by Blake. The same group also pioneered the use of heavy atom compounds (such as DMA and the Baker mercurial) having mercury atoms in addition to those bonded directly to the sulfhydryl groups. These compounds give relatively large intensity changes, and because they contain heavy atoms in new positions in the unit cell, they contribute new phase information. Reactive sulfhydryl groups in papain and carbonic anhydrase C have also been used to yield heavy atom derivatives. Experience with the reactions of sulfhydryl groups in the β-lactoglobulin molecule has been summarized by Dunnill *et al.* (*146*); their paper lists many sulfhydryl reagents in addition to those given in Table VI.

b. Covalent Binding of Heavy Atoms to Other Side Chains. The usefulness of cysteine residues for forming heavy atom derivatives is based not only on their reactivity toward mercurials but also on their distribution in proteins: often only one or a small number of cysteines are on the surface of an enzyme; thus, it is possible to locate the bound mercury atoms from a difference Patterson map without undue difficulty. Other reactions of heavy atom compounds with side chains have been developed, but their usefulness for any given protein will depend on there being a limited number of reactive side chains (as well as on the usual problem of being able to grow crystals isomorphous with the parent). Some of these reactions were mentioned in Section III,E: iodination of tyrosyl residues and the use of *p*-iodophenylisothiocyanate to label the N-terminal α-amino group and lysine residues. Another heavy atom label for amino groups was proposed by Benisek and Richards (*213*). They reacted lysozyme and guanidinated lysozyme with methyl picolinimidate. The picolinamidine group strongly chelates *d*-orbital transition metal ions, such as $PdCl_4^{2-}$. In guanidinated lysozyme only the free α-amino group seems to react with the picolinimidate. A similar suggestion was made by Sokolovsky *et al.* (*214*). They proposed conversion of tyrosine residues to 3-aminotyrosines, which should chelate some metal ions. Blake lists several other reactions that may prove useful for binding of heavy atoms to side chains (*211*).

c. Heavy Atom Containing Inhibitors. A heavy atom attached to an inhibitor can assist in phase analysis and also marks the site of inhibition in an electron density map. Mathews *et al.* (*15*) and Cohen *et al.* (*159*), for example, used *p*-iodophenyl sulfonyl (pipsyl) fluoride in the phase analysis of chymotrypsin, thus labeling the position of Ser 195 in the

213. W. F. Benisek and F. M. Richards, *JBC* **243**, 4267 (1968).
214. M. Sokolovsky, J. F. Riordan, and B. L. Vallee, *BBRC* **27**, 20 (1967).

electron density map; Tilander *et al.* (*179*) used a number of heavy atom labeled sulfonamide inhibitors in their study of carbonic anhydrase. A possible drawback of this type of derivative is that the presence of the bulky heavy atom is likely to distort the positions of atoms around the active site. This produces a lack of isomorphism in just the region of the enzyme that one wants to learn most about. The lack of isomorphism can be eliminated by using an analogous light atom derivative as the parent compound; for example, *p*-toluyl sulfonyl (tosyl) derivatives were used as the parent compound in the α- and γ-chymotrypsin and elastase studies. The close isomorphism between parent and derivative then gives excellent phases, but the light atom substituent of the parent may still cause some distortion of the native enzyme at the active site so that a somewhat distorted picture of the electron density is obtained. This seems to be the case near His 57 in the map of tosyl-elastase (*18*). The positions of atoms in the native enzyme can, nevertheless, be inferred from a difference Fourier map that shows the electron density change produced by the light atom substituent. Such a study has been made for α-chymotrypsin (*157*).

d. Substitution of a Heavier for a Lighter Metal Atom. When a metal atom is bound to a protein, it is sometimes possible to replace it by a heavier atom. The change in intensity produced by such a replacement is equal to that caused by a virtual atom having an electron density equal to the difference of the two metals. Tilander *et al.* (*179*), for example, were able to remove the zinc atom from crystals of human carbonic anhydrase; then by dialysis of the zinc-free crystals against mercuric acetate they obtained a derivative with mercury atoms both at the sulfhydryl group and at about 0.7 Å from the zinc site. Similarly, Hartsuck *et al.* (*215*) were able to replace the zinc atom of carboxypeptidase A with mercury by dialysis of crystals against $HgCl_2$ for five months. Essentially the same technique was used by Adams *et al.* (*194*) to obtain a derivative of zinc–insulin crystals. Zinc was removed from the parent crystals by EDTA, and lead was inserted in its place by soaking the crystals in lead acetate.

e. Noncovalently Bound Derivatives Found by Trial and Error. Most of the useful heavy atom derivatives have been discovered by diffusing heavy atom containing compounds into crystals of the protein. Past attempts to predict which compounds would bind strongly have not been notably successful, and it is only now that we can draw on a decade of studies by trial and error that it is becoming easier to find derivatives of

215. J. A. Hartsuck, M. L. Ludwig, H. Muirhead, T. A. Steitz, and W. N. Lipscomb, *Proc. Natl. Acad. Sci. U. S.* **53,** 396 (1965).

this type. As more studies of the binding of these compounds to proteins are completed, more rational prediction may become possible.

Sometimes the interaction of a heavy atom with a protein arises through an adventitious juxtaposition of charges on different molecules. This seems to be the case in sperm-whale myoglobin where *p*-chloromercuribenzenesulfonate sits between two molecules and seems to be coordinated to groups on both (*216*). Other heavy ions seem to bind to particular amino acid side chains whenever they are accessible in the crystal. An example is the $PtCl_4^{2-}$ ion which binds to the sulfur of methionine. In a survey of five proteins whose structures are known (*122*) this ion has been found to bind to all Met residues exposed on the molecular surfaces. The mode of binding in cytochrome c suggests that the square-planar Pt(II) is oxidized to octahedral Pt(IV) with the methionine sulfur acting as a fifth ligand by donating one of its lone pairs of electrons to the ion. Another ion that binds to several proteins is UO_2^{2+}. In crystals of lysozyme it is held mainly by carboxylic acid side chains (*211*). The compound mersalyl (see Table VI) also binds to several proteins, and the site of attachment of the mercury atom has been found to be His residues in cytochrome c and subtilisin BPN′. In other cases an ion will bind to different proteins in completely different ways: Mercury iodide is chelated by the guanidinium groups of Arg residues in lysozyme, whereas it binds to the heme in myoglobin.

2. *Some Useful Heavy Atom Compounds*

Table VI lists some heavy atom compounds that have been found to bind to proteins. The list is not comprehensive, and the grounds for inclusion of a given compound have been somewhat arbitrary. The proteins associated with each heavy atom compound are those for which the compound has caused intensity changes in the protein diffraction pattern; in many cases the heavy atom was not used for phase determination. These cases are included because they indicate some form of association between the compound and a protein and suggest that the compound may be useful with some other protein in the future. The binding sites are given when they are known, and the composition of the mother liquor is also noted because it may have bearing on the formation of derivatives (see following section).

3. *Some Additional Factors in the Formation of Heavy Atom Derivatives*

a. Composition of the Mother Liquor. Sigler and Blow reported that the composition of the mother liquor can affect the strength of binding of heavy atom compounds (*156*). They discovered that crystals of α-chymo-

216. H. C. Watson, J. C. Kendrew, and L. Stryer, *JMB* **8**, 166 (1964).

TABLE VI

Heavy atom	Abbreviated name	Structure
Se	Se1	KSeCN
Nb	Nb1	$Nb_6Cl_{12}\cdot Cl_2$
Pd	Pd1	K_2PdCl_4
Ag	Ag1	$AgNO_3$
I	Pipsyl	I—(benzene ring)—SO_2F
	I1	$KI + I_2$
	I2	I—(benzene ring)—NO_2
Xe	Xe1	Xe
Ta	Ta1	$Ta_6Cl_{12}\cdot Cl_2$
Os	Os1	$OsCl_3$
Ir	Ir1	$IrCl_3$
	Ir2	Na_3IrCl_6 or K_3IrCl_6
Pt	Pt1	Na_2PtCl_4 or K_2PtCl_4

HEAVY ATOM DERIVATIVES

Protein	Binding site	Medium[a]	pH	Reference
Erythrocruorin				*128*
Lysozyme-chloride		0.09 *M* AB, 0.023 g ml^{-1} NaCl	4.5	*188*
Lysozyme-chloride	Arg 14, His 15, Asn 93, Lys 96	0.8 *M* NaCl	4.7	*182*
Carboxypeptidase A		0.2 *M* $NaOOCCH_3$	8	*59*
Horse oxyhemoglobin	SH group	AS, AP	<7.1	*72*
Ox hemoglobin	SH group			*145*
Seal metmyoglobin		4 *M* AS	5.9–6.5	*107*
Sperm-whale metmyoglobin		3 *M* AS	6.5–7.3	*134*
α-Chymotrypsin	Active Ser	2.6 *M* AS		*23*
γ-Chymotrypsin		65% satd. AS	5.6	*158*
γ-Chymotrypsin		65% satd. AS	5.6	*159*
Subtilisin BPN′	Tyr residues 104, 217, 167–171, 21	2.1 *M* AS	5.9	*217*
Sperm-whale metmyoglobin	Tyr G4, Tyr HC3	75% satd. AS	5.5	*218*
Staphylococcus nuclease		Diol 40%	8.5	*63*
Sperm-whale metmyoglobin	Between heme and His	80% satd. AS	6.8	*219*
Horse oxyhemoglobin	Near AB and GH corners	2 *M* AS	7.0	*220*
Lysozyme-chloride		0.09 *M* AB, 0.04 g ml^{-1} NaCl	4.5	*188*
Cytochrome c_2		50% satd. AS, PB	6.0	*149*
Chymotrypsinogen				*161*
Cytochrome c_2		AS, PB	6.0	*149*
Papain		Methanol < 60%		*166*
Ribonuclease		Ethanol		*173*
Asparate transcarbamylase–CTP complex		PB	5.9	*177*

TABLE VI (*Continued*)

Heavy atom	Abbreviated name	Structure
Pt	Pt1	Na_2PtCl_4 or K_2PtCl_4
	Pt2	K_2PtCl_6
	Pt3	K_2PtBr_4 or K_2PtI_4
	Pt4	K_2PtI_6
	Pt5	*cis*-$Pt(NH_2CH_2COOH)_2$
	Pt6	$Pt(NH_2CH_2CH_2NH_2)_3Cl_4$
	Pt7	$Pt(NH_3)_2(NO_3)_2$
	Pt8	$K_2Pt(CN)_4$
	Pt9	$Pt(C_2O_4)_2$
	Pt10	$Pt(NH_2CH_2CH_2NH_2)_2Cl_2$
Au	Au1	$NaAuCl_4$ or $KAuCl_4$
	Au2	$KAu(SCN)_4$

TABLE VI (*Continued*)

Protein	Binding site	Medium[a]	pH	Reference
Carboxypeptidase A	Cys 161, Met 103 His 303, N-terminus	LiCl tris buffer	7.5	*59*
α-Chymotrypsin	Disulfide 1–127, N-terminus, Met 192, and other sites	2 *M* AS	4.2	*221*
γ-Chymotrypsin	Disulfide 1–127, Met 192			*13*
δ-Chymotrypsin		50% satd. AS	4.5	*161*
Cytochrome c_2		50% satd. AS, PB	6.0	*149*
Erythrocruorin				*128*
Ferricytochrome c	Met 65, His 33	4.6 *M* PB	6.2	*33*
Ferrocytochrome c		AS	7.4	*65*
Lysozyme nitrate		0.3 *M* $NaNO_3$	4.5	*190*
Pepsin		20% ethanol	1.5	*62*
Ribonuclease S	Met 29	75% satd. AS	5.5	*16*
Rubredoxin				*95*
Subtilisin BPN′	Met 50, His 64	2.1 *M* AS	5.9	*169*
α-Chymotrypsin		3.6 *M* NaH_2PO_4 or 2.2 *M* $MgSO_4$		*156*
Chymotrypsinogen		Ethanol 10%		*161*
High-potential iron protein		80% satd. AS		*151*
Lysozyme-chloride		0.85 *M* NaCl	4.7	*183*
Papain		Methanol < 60%		*167*
Staphylococcus nuclease		Diol 40%	8.5	*63*
α-Chymotrypsin	4 sites	2 *M* AS	4.2	*221*
Lysozyme-chloride		AB, 0.04 g ml^{-1} NaCl	4.5	*90*
Ribonuclease		Diol 55%	5.0	*17*
Ribonuclease		Diol 55%	5.0	*17*
Erythrocruorin				*129*
Ribonuclease		Diol 55%	5.0	*222*
Ribonuclease S		75% satd. AS	5.6	*16*
Pepsin		20% ethanol	1.5	*62*
Ribonuclease S		75% satd. AS	5.5	*16, 174*
L-Lactate dehydrogenase				*26*
Glycera hemoglobin		AS	6.8	*137*
L-Lactate dehydrogenase				*26*
Seal metmyoglobin		4 *M* AS	5.9–6.5	*134*
Sperm-whale metmyoglobin	2 His residues	3 *M* AS	6.5–7.3	*77*
Subtilisin BPN′		2.1 *M* AS	5.9	*169*

TABLE VI (*Continued*)

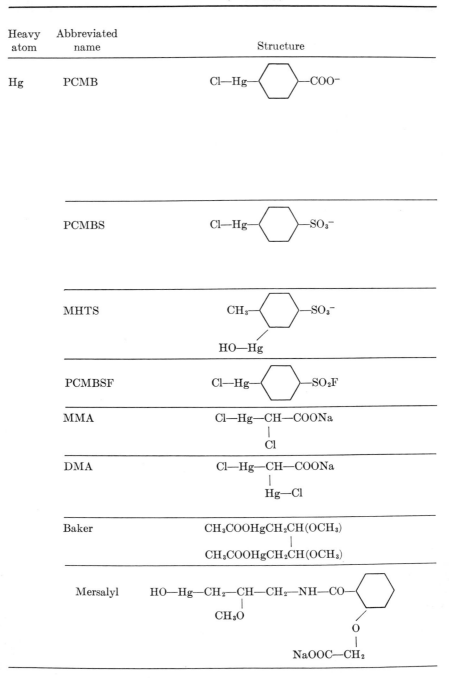

Heavy atom	Abbreviated name	Structure
Hg	PCMB	Cl—Hg—C_6H_4—COO^-
	PCMBS	Cl—Hg—C_6H_4—SO_3^-
	MHTS	CH_3—C_6H_3(Hg—OH)—SO_3^-
	PCMBSF	Cl—Hg—C_6H_4—SO_2F
	MMA	Cl—Hg—CH(Cl)—COONa
	DMA	Cl—Hg—CH(Hg—Cl)—COONa
	Baker	$CH_3COOHgCH_2CH(OCH_3)$—$CH(OCH_3)CH_2HgOOCCH_3$
	Mersalyl	HO—Hg—CH_2—CH(CH_3O)—CH_2—NH—CO—C_6H_4—O—CH_2—COONa

TABLE VI (*Continued*)

Protein	Binding site	Medium[a]	pH	Reference
Cytochrome c peroxidase		Diol, PB		*70*
Horse deoxyhemoglobin	SH group at 93 β	PB	6.4	*143*
Horse oxyhemoglobin	SH groups of β chain	AS, AP	<7.1	*60*
Human deoxyhemoglobin	SH group at 93 β	AS, AP	6.5	*144*
Glyceraldehyde-3-phosphate dehydrogenase		3 *M* AS		*46*
L-Lactate dehydrogenase	1 site/subunit			*26*
Papain	SH group	65% methanol		*167*
Ribonuclease		Ethanol 60%		*223*
Carbonic anhydrase C	SH group	2.5 *M* AS		*179*
β-Lactoglobulin	SH group		5.2	*36*
Lysozyme-chloride	SO_3 group binds	0.85 *M* NaCl	4.7	*182*
Seal myoglobin	3 sites	4 *M* AS	6	*134*
Sperm-whale myoglobin	His G14 and other residues	3 *M* AS	6.5–7.3	*107*
Papain		Methanol 65%		*167*
Lysozyme-chloride		0.85 *M* NaCl	4.7	*182*
α-Chymotrypsin	Active Ser			*156*
Elastase	Active Ser			*18*
β-Lactoglobulin	SH group			*146*
Ox hemoglobin		PB	7	*146*
L-Lactate dehydrogenase	SH groups			*26*
Horse oxyhemoglobin	SH groups of β chains	AS, AP	<7.1	*60*
Horse deoxyhemoglobin	SH group 93 β	PB	6.4	*143*
Human deoxyhemoglobin	β_1 SH groups	AS, AP	6.5	*144*
Carboxypeptidase A		0.2 *M* $NaOOCCH_3$	8.0	*59*
L-Lactate dehydrogenase	SH groups			*26*
Horse oxyhemoglobin	SH groups of β chains	AS, AP	<7.1	*60*
Erythrocruorin[b]				*128*
Ferricytochrome c	His 33	4.6 *M* PB		*33*
Subtilisin BPN′	His 64, NH_3^+ of Ala 1	2.1 *M* AS	5.9	*169*
DIP-trypsin		$MgSO_4$	8.0	*170*

TABLE VI (*Continued*)

Heavy atom	Abbreviated name	Structure
	HgAn	CH_3COOHg—C_6H_4—NH_2
	PCMA	Cl—Hg—C_6H_4—NH_2
	PMA	CH_3COOHg—C_6H_5
	Hg1	HO—Hg—C_6H_4—COO^-
	Hg1a	HO—Hg—C_6H_4—SO_3^-
	Hg2	$Hg(CH_3COO)_2$
	Hg3	$HgCl_2$
	Hg4	HgO
	Hg5	K_2HgBr_4
	Hg6	K_2HgI_4 (binds often as HgI_3^-)

TABLE VI (*Continued*)

Protein	Binding site	Medium[a]	pH	Reference
Carboxypeptidase A		0.2 *M* $NaOOCCH_3$	8.0	*59*
Sperm-whale metmyoglobin		3 *M* AS	6.5–7.3	*107*
Papain		Methanol < 60%		*167*
α-Chymotrypsin	Terminal NH_3^+ group S—S bridge 1–127	3.5 *M* PB		*15*
L-Lactate dehydrogenase	SH group			
Ribonuclease		Ethanol 60%		*173*
Carbonic anhydrase C	SH group			*180*
Erthyrocruorin				*128*
Glycera hemoglobin		AS	6.8	*137*
Horse oxyhemoglobin	SH groups of α chain when β SH groups blocked	AS, AP	<7.1	*60*
Human deoxyhemoglobin	β_2 SH groups after β_1's blocked	AS, AP	6.5	*144*
Ox oxyhemoglobin				*146*
Carboxypeptidase A		0.2 *M* $NaOOCH_3$	8.0	*59*
Horse oxyhemoglobin	SH groups of both α and β chains	AS, AP	<7.1	*60*
Ox hemoglobin	SH groups	PB	7.0	*146*
Papain		Methanol < 60%		*167*
Sperm-whale myoglobin	His GH1	3 *M* AS	6.5–7.3	*10*, *107*
Chymotrypsinogen A				*163*
Lysozyme-chloride		0.85 *M* NaCl	4.7	*224*
Lysozyme-nitrate		0.3 *M* $NaNO_3$	4.5	*189*
Pepsin		20% ethanol	1.5	*62*
γ-Chymotrypsin		65% satd. AS		*159*
Chymotrypsinogen A				*161*

TABLE VI (*Continued*)

Heavy atom	Abbreviated name	Structure
	Hg6	
	Hg7	CH_3—CO—S—Hg—C_6H_4—SO_2F
	Hg8	CH_3COO—Hg—C_6H_3(H_2N)—SO_2NH_2
	Hg9	CH_3—Hg—S—CH_2—CO—NH—SO_2—C_6H_3(SO_2NH_2)—NH_2
	Hg10	CH_3—Hg—S—CH_2—CO—NH—C_6H_4—SO_2NH_2
	Hg11	$CH_3HgOOCH_3$
	Hg12	CH_3—Hg—S—CH_2—COOH
	Hg13	$K_2Hg(CNS)_4$
	Hg14	CH_3HgOH
	Hg15	ClHg—C_6H_4—OH
Tl	Tl1	$TlOOCCH_3$
Pb	Pb1	$PbCl_2$ or $Pb(CH_3COO)_2$

TABLE VI (*Continued*)

Protein	Binding site	Medium[a]	pH	Reference
Glycera hemoglobin		AS	6.8	*137*
Insulin		HCl and KCl	3.0	*195*
Lysozyme-chloride	Arg residues	0.85 *M* NaCl	4.7	*224*
Lysozyme-nitrate		0.3 *M* $NaNO_3$		*189*
Papain	2 sites	Methanol < 60%		*167*
Pepsin		20% ethanol	1.5	*62*
Rubredoxin				*95*
Seal metmyoglobin		4 *M* AS	5.9–6.5	*134*
Sperm-whale metmyoglobin	Lys FG2, between heme and Phe ring	3 *M* AS	6.5–7.3	*107*
Subtilisin BPN′		2.1 *M* AS	5.9	*169*
Subtilisin BPN′	His G4	2.1 *M* AS	5.9	*169*
Carbonic anhydrase C	SH group and "sulfonamide site"	2.5 *M* AS		*179*
Carbonic anhydrase C		2.5 *M* AS		*179, 211*
Carbonic anhydrase C	"Sulfonamide site"	2.5 *M* AS		*179, 211*
Carbonic anhydrase C		2.5 *M* AS		*179*
Carbonic anhydrase C	SH group	2.5 *M* AS		*179*
Glycera hemoglobin				*137*
Human deoxyhemoglobin	112 β and 104 α SH groups	AS, AP	6.5	*144*
δ-Chymotrypsin	3 sites	50% satd. AS	4.5	*161*
DIP-trypsin		$MgSO_4$	8	*170*
Carboxypeptidase A		0.2 *M* LiCl tris	7.5	*117*
Insulin				*194*

TABLE VI (*Continued*)

Heavy atom	Abbreviated name	Structure
U	U1	$UO_2(NO_3)_2$
	U2	$UO_2(OOCCH_3)_2$
	U3	$UO_2(P_2O_7)_n$
	U4	$K_3UO_2F_5$

[a] Abbreviations: PB, phosphate buffer; AS, ammonium sulfate; AP, ammonium phosphate; CB, citrate buffer; diol, 2-methyl-2,4-pentanediol; FCB, ferrous citrate buffer.
[b] Acid form rather than Na salt.

217. B. Weber and J. Kraut, *BBRC* **33,** 280 (1968).
218. R. H. Kretsinger, *JMB* **31,** 315 (1968).
219. B. P. Schoenborn, H. C. Watson, and J. C. Kendrew, *Nature* **207,** 28 (1965).

trypsin can be transferred from the $(NH_4)_2SO_4$ solution in which they are grown to either 3.6 *M* NaH_2PO_4 or 2.2 *M* $MgSO_4$ with no significant change in diffraction pattern. Moreover, they found that three heavy ions that bind weakly or not at all in $(NH_4)_2SO_4$ bind strongly in NaH_2PO_4 or $MgSO_4$. Their interpretation of this result is that in the $(NH_4)_2SO_4$ medium, the nucleophilic side chains of the protein that might bind the metallic ions instead react preferentially with NH_3 derived from the $(NH_4)_2SO_4$. They suggested that since $(NH_4)_2SO_4$ is the salt most frequently used to crystallize proteins this competition of NH_3 could have accounted for difficulty in forming derivatives in other cases. At least one other instance of this has been found: The $PtCl_4^{2-}$ and mersalyl ions bind more strongly to cytochrome c in phosphate buffer than in the $(NH_4)_2SO_4$ medium from which the protein is crystallized.

Mention should be made of one other technique related to the composition of the mother liquor. Dunnill *et al.* (*146*) found a means of increasing the rate at which highly insoluble heavy atoms bind to a protein.

TABLE VI (*Continued*)

Protein	Binding site	Medium[a]	pH	Reference
Lysozyme-chloride	5 sites: COO^- side chains	0.85 *M* NaCl	4.7	*194*
Elastase				*18*
Ribonuclease		Diol 55%	5.0	*17*
Insulin				*194*
Ribonuclease S		3.2 *M* AS	5.6	*16*
Staphylococcus nuclease		Diol 40%	8.5	*63*
Chymotrypsinogen		Ethanol 10%		*161*
Cytochrome c_2		AS, BP	6.0	*149*
Ferrocytochrome c		AS		*65*
High-potential iron protein		80% satd. AS		*151*
Insulin				*194*
Lysozyme-chloride	2 sites: COO^- side chains	0.85 *M* NaCl	4.7	*211*
Ribonuclease		Ethanol		*173*

220. B. P. Schoenborn, *Nature* **208,** 760 (1965).
221. D. M. Blow, M. G. Rossmann, and B. A. Jeffery, *JMB* **8,** 65 (1964).
222. G. Kartha, J. Bello, D. Harker, and F. E. DeJarnette, *in* "Aspects of Protein Structure" (G. N. Ramachandran, ed.), p. 13. Academic Press, New York, 1963.
223. J. D. Bernal, C. H. Carlisle, and M. A. Rosemeyer, *Acta Cryst.* **12,** 227 (1959).
224. R. J. Poljak, *JMB* **6,** 244 (1963).

The method is to use an organic solvent for the initial solution of the reagent; this solution is then finely dispersed in the mother liquor surrounding the protein crystals. This method, of course, increases the rate but not the strength of binding.

b. Restoration of Isomorphism. Sometimes the presence of a heavy atom compound will alter the positions of some atoms in the protein, thus partially destroying the isomorphism of derivative and parent crystals. The lack of isomorphism can be recognized by changes in cell dimensions of the crystal, by an inability to describe the observed changes in intensity by heavy atoms occupying a small number of sites, or by one of the more systematic methods suggested by Crick and Magdoff (*89*). In some such cases one can still carry out a structural analysis if one restores isomorphism by using an analog of the heavy atom derivative as the "parent" crystal. One example of this procedure is the use of tosylchymotrypsin as the parent in studies of the α- and γ-crystal forms of this enzyme. One heavy atom derivative in both these studies is pipsylchymotrypsin, in which the methyl group of the tosyl is replaced by an iodine atom. Another example is the study of lysozyme-chloride by Stanford *et al.* (*90*). It was discovered in this study that crystals of lysozyme containing the complex ions $Ta_6Cl_{12}^{2+}$ and $Nb_6Cl_{12}^{2+}$ are not isomorphous with the native crystals but are highly isomorphous with one another. Lysozyme containing $Nb_6Cl_{12}^{2+}$ was then designated the parent, and the analysis proceeded. Restoration of isomorphism has one obvious drawback: The final electron density map shows not the native enzyme but the enzyme distorted by the restorative group.

c. Several Derivatives at One Binding Site. If a second heavy atom binds at precisely the site occupied by a first, it supplies no new phase information and is thus not truely a new derivative. But if it is separated from the site of the first by means of intervening chemical groups, such as methylene groups or a naphthalene ring, it can be helpful in phase determination. The reaction of both *p*-mercuribenzoate and the Baker mercurial at the sulfhydryl group of the β chain of hemoglobin is an example of forming two derivatives at one binding site. The Baker compound contains a second mercury atom that is several angstroms from the first; the first, like the mercury of *p*-chloromercuribenzoate is bound to the sulfhydryl. Frenselau *et al.* (*225*) suggested that this method could be generalized by preparing a series of compounds based on naphthalene. Each would contain a functional group at one ring position that could bind to an enzyme and heavy atoms at various other positions to help in phase determination.

V. Summary and Conclusions

A. State of the Art of Enzyme Crystallography

The number of enzymes whose structures have been determined to near-atomic resolution is increasing rapidly, and the average time for a structural determination is decreasing. Well over one-half dozen structures were announced in 1969, and one of these, elastase, required less than one man-year of work for its solution (*226*). This acceleration in the pace of protein crystallography is partly a result of automation of data collection and improvement in facilities for data processing, and

225. A. Frenselau, D. Koenig, D. E. Koshland, Jr., H. G. Latham, Jr., and H. Weiner, *Proc. Natl. Acad. Sci. U. S.* **57,** 1670 (1967).

226. H. C. Watson, private communication (1969).

partly a result of accumulated experience in preparing heavy atom derivatives. The growth of well-ordered crystals with good diffraction patterns and the preparation of isomorphous heavy atom derivatives are now more than ever the rate-limiting steps. Further development of the single isomorphous replacement method and its variations or of the methods based on noncrystallographic symmetry could decrease the need for derivatives, but at present multiple isomorphous replacement is the standard method for solving structures.

The size of the structures being solved is increasing. The asymmetric unit of the horse hemoglobin crystal contains 287 amino acid residues, that of the carboxypeptidase A crystal 307 residues, and that of the α-chymotrypsin crystal (holding two molecules) 482 residues. There is no reason to believe that isomorphous replacement should not work for larger molecules, and, in fact, a good start has been made on solving the structure of asparate transcarbamylase with a molecular weight of about 310,000. Even in cases where a high resolution study by isomorphous replacement does not prove possible, salt difference maps and low resolution Fourier difference maps may yield important structural information. These techniques could, in principle, outline the shapes of subunits and show the regions in which various substrates and effectors bind. It is also conceivable that the polypeptide backbone of a large enzyme could be traced in a low resolution electron density map if the amino acid sequence were known and if enough of the residues were labeled with heavy atoms.

B. Common Characteristics of Protein Structures

The structures of globular proteins solved to date have a number of characteristics in common. They are compact, with almost all the highly polar or charged side chains at the molecular surface in contact with the solvent. Those charged or highly polar side chains which are in the interior seem in almost every case to be directly involved in the functioning of the molecule. Examples include the imidazole side chains of the His residues that hold the heme and oxygen molecule in myoglobin and the carboxylate side chain of Asp 102 in chymotrypsin that polarizes the active His 57 residue. By far the majority of side chains in the interior, however, are nonpolar as predicted by Kauzmann in 1959 (*227*). In the globin molecules, and in parts of the lysozyme, papain, ribonuclease, and subtilisin BPN′ molecules, groups of them are clumped together, forming hydrophobic clusters. Even though the nonpolar side chains have a low

227. W. Kauzmann, *Advan. Protein Chem.* **14,** 1 (1959).

affinity for water, there are usually a few on the molecular surface; in the case of subtilisin BPN′ it has been reported that all Trp, Phe, and Tyr side chains are on the surface *(169)*. It is possible, of course, that some nonpolar groups on the surface have special functions, such as binding to other molecules or to membranes, that are presently unknown.

Three regular polypeptide conformations that were first discovered in fibrous proteins have now been found in globular proteins. These are the right-handed α helix and both the parallel and antiparallel β-pleated sheet. Several turns of 3_{10} helix have also been found, usually as the final turn of an α helix. The helices often have a hydrophilic side in contact with the solvent and a hydrophobic side in contact with groups in the interior of the molecule *(6, 228)*. Dickerson and Geis believe that α and β structures in globular proteins play predominantly a structural role analogous to their function in fibrous proteins, in contrast to the stretches of polypeptide chain with no regular order *(6)*.

Ions and water molecules have been found to bind to polar groups on the surfaces of the myoglobin and lysozyme molecules. In crystals of lysozyme and carboxypeptidase A there may be a partial ordering of solvent between enzyme molecules. There has been no report, however, of definite cages or icebergs of water molcules about amino acid side chains.

The active sites of lysozyme, ribonuclease, and papain lie in deep grooves in the surfaces of these molecules; the active site of carboxypeptidase A is in a pitlike depression, and those of elastase and α-chymotrypsin are in shallower depressions. Perutz has suggested that the function of the grooves and depressions is to provide a nonaqueous environment of low dielectric constant in which catalysis can take place *(1)*. For example, the "charge-relay system" at the active sites of chymotrypsin, elastase, and subtilisin BPN′ is particularly effective owing to the nonpolar region in which the participating residues are embedded. The activation of chymotrypsinogen also depends on the strong interaction of charges that is possible only in the nonaqueous interior of the molecule.

Information on the intramolecular movements in proteins induced by substrates, cofactors, and ligands has been inferred from difference Fourier maps and comparisons of electron density maps of different crystal forms. Difference Fourier maps of substrates bound to carboxypeptidase A and lysozyme have confirmed the suggestion *(208)* that the binding of a substrate can induce changes in the structure of the enzyme in the vicinity of the catalytic center. In the case of carboxypeptidase, it is clear that the movements are part of the catalytic process. Much more drastic changes have been seen in hemoglobin when it binds oxygen,

228. M. Schiffer and A. B. Edmundson, *Biophys. J.* **7**, 121 (1967).

and in L-lactate dehydrogenase when it binds NADH. In both cases there is evidence that the changes are mainly in the relative positions and orientations of the subunits, and not in the tertiary structures of the subunits themselves. How can the binding of a small molecule like oxygen transform the structure of a giant molecule like hemoglobin? This is one of the many exciting problems of protein structure that remains to be answered.

Glossary of Symbols

a, b, c	Lengths of unit cell edges
a.u.	Asymmetric unit
$\mathbf{F}(hkl)$	Structure factor for reflection *hkl* of parent compound
$F(hkl) = \lvert\mathbf{F}(\mathrm{hkl})\rvert$	Structure factor magnitude for reflection *hkl* of parent compound
$F_{\mathrm{H}}(hkl)$	Structure factor magnitude for reflection *hkl* of heavy atom derivative
$\mathrm{f_H}(hkl)$	Scattering factor for heavy atom compound
h, k, l	Indices of a reflection (spot) in the diffraction pattern
$I(hkl)$	Intensity of reflection *hkl*
i	$\sqrt{-1}$
M	Molecular weight
N	Avogadro's number
n	Number of molecules per unit cell
n	Wave number
n'	Number of asymmetric units per unit cell
P, I, F, C, R	Symbols for lattices
V	Volume of the unit cell
$\bar{v}$	Specific volume
X_{P}	Weight fraction of protein
x, y, z	Coordinates of unit cell (in fractions of unit cell edges)
$\alpha(hkl)$	Phase for reflection *hkl*
ρ	Macroscopic density (g $\mathrm{cm^{-3}}$)
$\rho(xyz)$	Electron density at point xyz (electrons Å^{-3})
2, 3, 4, 6, m	Symbols for rotation axes (a rotation of $2\pi/m$ radians)
$2_1, 3_1, 3_2, 4_1, 4_2$, etc., m_p	Symbols for screw rotation axes (m_p indicates a rotation by $2\pi/m$ radians and a translation along the axis of p/m of unit cell edge)
$\langle\ \rangle$	Average over variable enclosed in brackets

ACKNOWLEDGMENTS

The author is grateful to Professors R. E. Dickerson and W. N. Lipscomb for discussions on enzyme crystallography, and to Miss L. M. Kay and Professors R. E. Dickerson, J. Waser, and H. W. Wyckoff for permission to reprint figures. Support by U. S. Public Health Service grant GM 16925 for study of the structure of macromolecules and an Alfred P. Sloan Research Fellowship are also gratefully acknowledged.

2

*Chemical Modification by Active-Site-Directed Reagents**

ELLIOTT SHAW

I. Nature of Affinity Labeling and Its Relationship to Chemical Modification in General

The specificity of enzymes results from their ability to form a complex with characteristic substrates and subsequently promote a chemical

* Research carried out at Brookhaven National Laboratory under the auspices of the U. S. Atomic Energy Commission.

change. The region of the enzymic surface in contact with the substrate and therefore containing the structural features responsible both for specificity and catalysis has long been alluded to as the active site (*1*), and this region has been the target of many studies devoted to the understanding of the mechanism of action of enzymes. One method of gaining information about proteins in general is through chemical modification. Group reagents known to alter particular amino acid side chains can be used to study enzymes in the hope of producing a change in properties that can be related to a functional role of a given amino acid. In some circumstances regular group reagents can be used to obtain selective modification of an important group, for example (*2*). However, group reagents generally discriminate chiefly on the basis of accessibility of the side chain, a property not generally related to function, and lead to multiple modifications.

Recently, attention has been given to using the specific binding characteristics of an enzyme in designing a reagent for chemical modification within the active center alone. That is, if the reagent has substratelike structural features adequate to favor complex formation with the enzyme under consideration, then a potential chemical reactivity present in the reagent structure might achieve a covalent modification within the active center during the existence of the complex. The usefulness of this approach for enzyme studies was demonstrated with chymotrypsin in the development of TPCK (see Section II,A,1) (*3, 4*) which led to the iden-

$$\underset{\text{Reagent}}{E + R} \rightleftharpoons \underset{\text{Complex}}{E \cdot R} \rightarrow \underset{\text{Modified enzyme}}{E\text{—}R}$$

tification of an active center histidine (*5*). The potential of this as a general method for enzyme study was appreciated not only at the outset of the above work but independently by Baker (*6*) who chose to exploit it chiefly for its promise in leading to therapeutically useful enzyme inactivators (*7*), as well as by Wofsy *et al.* whose interest has been the characterization of antibody binding sites (*8*). From these investigators has come the descriptively useful terms *active-site-directed reagents* (*7*)

1. D. E. Koshland, Jr., *Advan. Enzymol.* **22,** 45 (1960).
2. E. Shaw, *Physiol. Rev.* **50,** 244 (1970).
3. G. Schoellman and E. Shaw, *BBRC* **7,** 36 (1962).
4. G. Schoellman and E. Shaw, *Biochemistry* **2,** 252 (1963).
5. E. B. Ong, E. Shaw, and G. Schoellman, *JBC* **240,** 694 (1965); *JACS* **86,** 1271 (1964).
6. B. R. Baker, W. W. Lee, E. Tong, and L. O. Ross, *JACS* **83,** 3713 (1961).
7. B. R. Baker, "Design of Active-Site-Directed Irreversible Enzyme Inhibitors," New York, 1967.
8. L. Wofsy, H. Metzger, and S. J. Singer, *Biochemistry* **1,** 1031 (1962).

and *affinity labeling* (*8*, *9*) which will be used interchangeably in this chapter.

In addition to the foregoing, chemical modification studies by active-site-directed approaches were also reported independently by other laboratories (*10–12*) as will be discussed later.

Clarification by Buchanan and his colleagues (*13*, *14*) of the mode of action of the antibiotic azaserine as a glutamine analog which alkylates two enzymes in the purine biosynthetic pathway (cf., for example, formylglycinamide ribotide aminotransferase, Section III,B) was an acknowledged stimulus to the general development of active-site-directed reagents (*7*). In the light of the molecular similarity of azaserine to glutamine, its behavior as a pseudosubstrate, complexing at a glutamine binding site (*13*, *14*) was an easily acceptable idea. The subsequent alkylation with inactivation was chemically rational and was supported by excellent enzymological data. On the other hand, a number of years earlier, diisopropyl fluorophosphate (DFP) and related organophosphorus compounds had already permitted the identification of an active center serine in various hydrolytic enzymes (*15*, *16*). Although recognized as an essentially unique reagent for active center labeling by covalent modification (*17*), DFP did not however suggest any structural resemblance to normal substrates. This possibility was not totally overlooked (*18*); however, interpretation of the selectivity of organophosphorus compounds, in general, emphasized the unique reactivity of the active center serine. Consequently, the capability of such reagents for being active-site-directed did not become established until later (*19*, *20*) when the ability to form complexes with hydrolytic enzymes was examined and even improved by emphasis of substratelike structural features. It must also be admitted that the greater influence of azaserine over DFP as inspiration for a general approach to developing active center reagents for enzymes

9. S. J. Singer, *Advan. Protein Chem.* **22,** 1 (1967).

10. A. Singh, E. R. Thornton, and F. H. Westheimer, *JBC* **237,** PC3006 (1962).

11. G. Gundlach and F. Turba, *Biochem. Z.* **335,** 573 (1962).

12. W. B. Lawson and H. J. Schramm, *JACS* **84,** 2017 (1962).

13. B. Levenberg, I. Melnick, and J.M. Buchanan, *JBC* **225,** 163 (1957).

14. J. M. Buchanan, S. C. Hartman, R. L. Herrmann, and R. A. Day, *J. Cellular Comp. Physiol.* **54,** Suppl. 1, 139 (1959).

15. E. F. Jansen, M. D. F. Nutting, and A. K. Balls, *JBC* **179,** 189 (1949).

16. A. K. Balls and E. F. Jansen, *Advan. Enzymol.* **13,** 321 (1952).

17. M. Dixon and E. C. Webb, "Enzymes," 2nd ed., p. 465. Academic Press, New York, 1964.

18. W. N. Aldridge, *Chem. & Ind.* (*London*) p. 473 (1954).

19. A. R. Main, *Science* **144,** 992 (1964).

20. A. R. Main and F. Iverson, *BJ* **100,** 525 (1966).

results not only from its more reasonable structural basis but also from the more intensive study, in the intervening years, of enzymes from the point of view of primary sequence and mechanism of catalysis.

By now dozens of active center reagents which irreversibly inactivate selected enzymes have been developed for a great variety of enzymes as described below. In the case of the proteolytic enzymes, DFP, until recently considered "the most specific of such reagents" (*17*), has already been superseded by even more specific inhibitors capable, for example, of selective inactivation of either chymotrypsin (*4*) or trypsin (*21*) whereas DFP inactivates both.

In this chapter, reagents have been chosen which apparently meet the criterion of causing progressive inactivation leading to a stoichiometric covalent combination with target enzymes with evidence of prior complex formation. A few examples have been included for chemical interest when there is sufficient reason to suppose these criteria may eventually be met.

II. Modification of Serine Proteases by Active-Site-Directed Reagents

The results obtained with specific reagents designed for active center modification of chymotrypsin and of trypsin have established the usefulness of this approach in enzyme study. Variations in reagent design were examined, the protein chemistry involved in the resultant inhibitions was examined with establishment of the sites of modification in the known sequence, and, finally the results could be compared with the three-dimensional structure of at least one of these enzymes.

It is probable that the conclusions which may now be drawn from the results with these enzymes may apply generally.

A. Chymotrypsin and Neutral Proteases

1. *TPCK, the Chloromethyl Ketone from Tosylphenylalanine*

Chymotrypsin is an endopeptidase selective for bonds formed from carboxyl groups of aromatic amino acids. Fortunately the hydrolysis of simple esters and amides of these acids is also catalyzed. Dissociation constants for simple specific substrates and competitive inhibitors (β-phenylpropionic acid) lie in the range of 10^{-3} to 10^{-4} M. The single

21. E. Shaw, M. Mares-Guia, and W. Cohen, *Biochemistry* **4**, 2219 (1965).

active center of chymotrypsin is responsible for both proteolytic and esterase catalysis.

Toluenesulfonyl-L-phenylalanine (I) was chosen as the basic structure for a reagent designed to label the active center of chymotrypsin since it satisfies the specificity requirements of the enzyme, that is, its esters are substrates. (The choice of this N^{α} derivative of phenylalanine was based on the need to form an acid chloride in the proposed synthetic work. Other blocking groups such as acetyl do not permit this

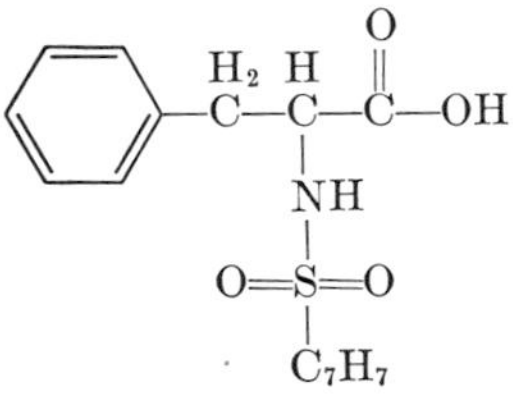

```
                 O                                   O
        H2 H     ‖                          H2 H     ‖   H2
   Ph—C——C——C—OH                       Ph—C——C——C——C—X
            |                                   |
            NH                                  NH
            |                                   |
          O=S=O                               O=S=O
            |                                   |
           C7H7                                C7H7
           (I)                                 (II)
```

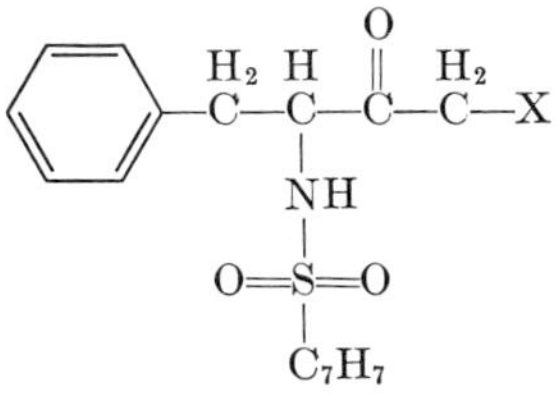

because of azlactone formation.) To this nucleus was added a chemically reactive grouping at the carboxyl end, namely, a bromo- or chloromethyl ketone group as in (II) which, when X = Cl, is known as TPCK (*3, 4*). It was hoped that this structural addition would not result in a loss of ability of the resultant molecule to complex at the active center of chymotrypsin, a hope which did not appear far fetched in view of the fact that substrates of chymotrypsin extend in the same molecular direction. Halomethyl ketones as reagents for the chemical modification of proteins have not been widely studied; however, it was to be expected that in common with other alkyl halides they would react with side chains in proteins by nucleophilic displacement (*22*) and therefore have the potential for modifying any of the normally derivatizable side chains found in proteins.

When the halomethyl ketones derived from tosyl-L-phenylalanine were incubated with chymotrypsin, a progressive loss of activity was observed which ultimately went to completion (*3*). A competitive inhibitor of chymotrypsin, β-phenylpropionic acid, was found to retard the inactivation (*3*). With radioactive TPCK it was established that the combination with chymotrypsin was stoichiometric and covalent (*4*). However, no alkylation took place in 8 M urea, with chymotrypsinogen or with DFP-inhibited chymotrypsin. These observations suggested the need for functioning enzyme for the reaction to proceed. The ability of

22. W. C. J. Ross, "Biological Alkylating Agents." Butterworth, London and Washington, D. C., 1962.

TPCK to form a reversible complex with chymotrypsin with $K_i = 3 \times 10^{-4}\ M$ has been demonstrated (*23*).

The site of attachment of TPCK in chymotrypsin was indicated by a change in amino acid composition of the alkylated enzyme which contained only one histidine residue in contrast to two histidines present in chymotrypsin (*4*). These circumstances were favorable ones for determining the site of chemical modification, that is, the new bond was stable to the strong hydrolytic conditions used for preparing samples for amino acid analysis and the involved residue was not regenerated; its deletion was conspicuous because of the low histidine content. Isolation of the histidine derivative itself from acid hydrolysates is difficult. However, after oxidation with performic acid, 3-carboxymethyl histidine was identified (*24*) which served to confirm the identity of the amino acid and establish the ring position alkylated.

The two histidine residues of chymotrypsin occur at positions 40 and 57 of the primary sequence (*25, 26*). From enzyme inhibited with ^{14}C-labeled TPCK a radioactive peptic peptide was isolated whose composition corresponded to a sequence encompassing His 57 (*5*). This site of alkylation was independently established by dislocation of the peptide in a fingerprint method (*28*) and by its disappearance from a characteristic position in an ion-exchange chromatographic procedure (*29*).

A more extensive study of the reaction of chymotrypsin with TPCK led to observations which confirmed its substratelike behavior (*27*). The reaction rate was dependent on a basic group of p$K = 6.8$ and an acidic group of p$K = 8.9$ already related to the catalytic function of histidine and to the group responsible for binding site integrity. Benzamide blocked the action of TPCK competitively with an inhibition constant equal to that observed when retarding substrate hydrolysis (*27*). TPCK-inhibited chymotrypsin did not react with an organophosphorus compound (*30*).

23. D. Glick, *Biochemistry* **7,** 3391 (1968).
24. K. J. Stevenson and L. B. Smillie, *JMB* **12,** 937 (1965).
25. B. S. Hartley, *Nature* **201,** 1284 (1964); B. S. Hartley and D. L. Kauffman, *BJ* **101,** 229 (1966).
26. B. Keil, Z. Prušik, and F. Šorm, *BBA* **78,** 559 (1963); B. Meloun, I. Kluh, V. Kostka, L. Morávek, Z. Prušik, J. Vanăĉek, B. Keil, and F. Šorm, *ibid.* **130,** 543 (1966).
27. F. J. Kezdy, A. Thomson, and M. L. Bender, *JACS* **89,** 1004 (1967).
28. L. B. Smillie and B. S. Hartley, *JMB* **10,** 183 (1964).
29. B. Meloun and D. Pospíšilová, *BBA* **92,** 152 (1964).
30. H. O. Michel and N. K. Schaffer, *ABB* **117,** 513 (1966).

2. *Other Alkylating Agents. Effect of Structure on Outcome*

In view of the success obtained in labeling the active center histidines in ribonuclease with the simple reagents, iodoacetic acid or bromoacetic acid (*31–34*), the inertness of these reagents and their amides toward chymotrypsin was noteworthy (*11, 35*). It provoked attempts at affinity labeling through the use of substituted derivatives such as iodoacetyl-phenylalanine methyl ester (III) which, however, had an extremely slow inactivating effect, requiring days, and was not stereospecific (*11*). The alkylating potential in this substrate is apparently not in reactive juxtaposition with any side chain of the enzyme.

Some relatively simple halomethyl ketones, rather incomplete as phenylalanine structures owing to lack of α side chains, namely, (IV) (*24*) and (V) (*36, 37*), were subsequently found to be effective inactivators of chymotrypsin and to label His 57 as in the case of TPCK. However, in many other cases, the reaction took another course.

The complete phenylalanine (or tyrosine or tryptophan) structure is not essential for complexing with chymotrypsin: even benzene, toluene, and indole have some measureable affinity (*38*). In consequence, simple, ring-containing alkylating agents such as benzyl bromide (VI), phenacyl bromide (VII), and bromoacetanilide (VIII) can be expected to be somewhat active-site directed, although not necessarily fully oriented. Such compounds react rapidly with chymotrypsin by formation of a sulfonium salt at Met 192 (*35*). The same result was obtained with the epoxide (IX) (*39*). This methionyl side chain is exposed, whereas the second methionine in chymotrypsin (Met 180) is buried. For example, with hydrogen peroxide a selective oxidation is possible, yielding chymotrypsin Met 192 sulfoxide (*40*). Therefore, the formation of a sulfonium salt by benzyl chloride and similar compounds could be interpreted as a nondirected reaction at an exposed site.

31. E. A. Barnard and W. D. Stein, *JMB* **1,** 339 (1959); W. D. Stein and E. A. Barnard, *ibid.* p. 350.
32. H. G. Gundlach, W. H. Stein, and S. Moore, *JBC* **234,** 1754 (1959).
33. A. M. Crestfield, W. H. Stein, and S. Moore, *JBC* **238,** 2413 (1963).
34. A. M. Crestfield, W. H. Stein, and S. Moore, *JBC* **238,** 2421 (1963).
35. H. J. Schramm and W. B. Lawson, *Z. Physiol. Chem.* **332,** 97 (1963).
36. H. J. Schramm, *Biochem. Z.* **342,** 139 (1965).
37. H. J. Schramm, *Z. Physiol. Chem.* **348,** 232 (1967).
38. J. L. Miles, D. A. Robinson, and W. J. Canady, *JBC* **238,** 2932 (1963).
39. J. R. Brown and B. S. Hartley, *BJ* **89,** 59P (1963).
40. D. E. Koshland, Jr., D. H. Strumeyer, and W. J. Ray, Jr., *Brookhaven Symp. Biol.* **15,** 101 (1962).

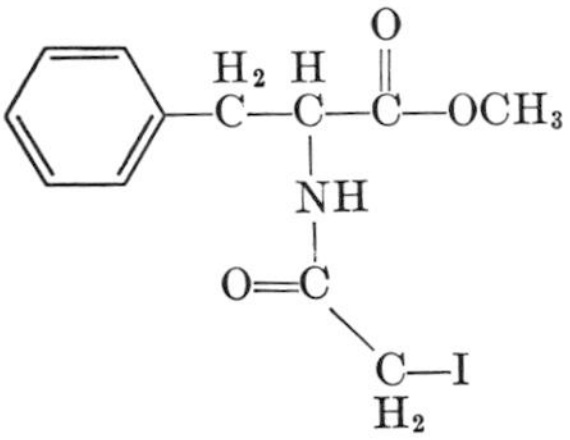

(III) 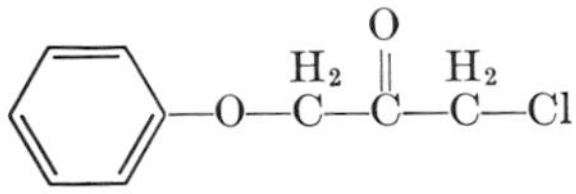 (IV) 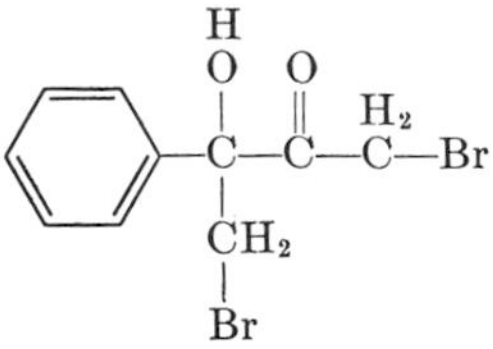 (V)

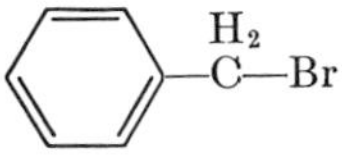

(VI) 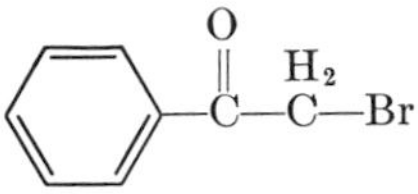 (VII) 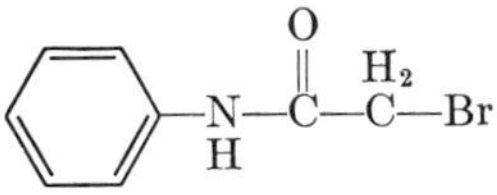 (VIII)

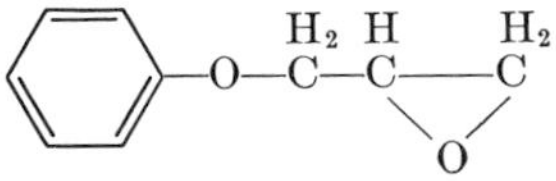

(IX)

On the other hand, the modification of Met 192 by alkylating agents containing an aromatic ring may also be considered affinity labeling of chymotrypsin, in this case of the specificity site. In the study of Schramm and Lawson, benzyl bromide (VI) and phenacyl bromide (VII) were observed to produce a 50% inhibition at a low concentration in a matter of minutes; the bromoacetamido compound (VIII) was somewhat slower (*35*). A number of observations indicated that the presence of the aromatic ring was favoring complexing at the active center: (a) a competitive inhibitor delayed the inactivation; (b) bromoacetamide itself was orders of magnitude slower than (VIII); and (c) no alkylation of methionine took place in chymotrypsinogen or in DIP-chymotrypsin (*35*). Sulfonium salt formation by (VI) and (VII), as well as others, was shown to result in chymotrypsin derivatives catalytically intact but having a diminished affinity for substrate (*35*).

In view of the fact that the hydrophobic binding site of chymotrypsin has no marked geometric requirements (*41*), the adsorption of rather diverse structures such as (VI)–(IX) does not seem improbable in a complex-forming step prior to chemical modification. Methionine 192 is close to, and possibly part of, the specificity site of chymotrypsin according to X-ray structural work (*42*) (cf. Section II,D,1).

41. R. A. Wallace, A. N. Kurtz, and C. Niemann, *Biochemistry* **2**, 824 (1963).
42. D. M. Blow, *BJ* **112**, 261 (1969).

An alkylating agent of a somewhat different type was provided by methyl *p*-nitrobenzenesulfonate (X) which is a methylating agent as a result of cleavage of the carbon–oxygen bond. Its action on chymotrypsin apparently was active-site directed by virtue of the aromatic ring and resulted in methylation of His 57 at N-3 with inactivation (*43*).

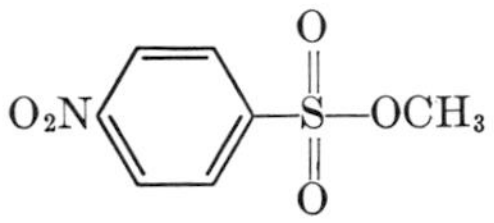

(X) 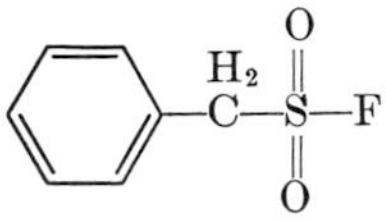 (XI)

Since the rate was greater than the methylation of a model histidine compound, affinity labeling appeared to be involved.

On the other hand, sulfonyl fluorides such as phenylmethanesulfonyl fluoride (XI) and toluenesulfonyl fluoride selectively sulfonate the active center serine. It was anticipated that this might be a substratelike action, but saturation kinetics were not observable (*44*). Eventually, however, it was established that the *p*-toluenesulfonyl group is located in the specificity site in toluenesulfonyl-chymotrypsin by crystallographic studies (*42*), and it seems probable that an intermediate complex is formed and could be detected as eventually proved to be the case with organophosphorus reagents as described in Section III,A,2. The sulfonyl derivatives are very stable with respect to long-term regeneration of active enzyme, a property which had led them to supersede diisopropylphosphoryl derivatives of serine esterases for crystallographic work (*42*).

3. *Relation of Results Obtained with Active-Site-Directed Agents on Chymotrypsin to Other Studies*

A role for histidine in the catalytic mechanism of chymotrypsin had been deduced from the pH dependence of catalytic properties (*45*, *46*), a procedure involving some uncertainty (*47*), as well as from various chemical studies. For example, in the photooxidation of chymotrypsin, although both methionine and histidine are altered, loss of active centers correlates with histidine loss only (*40*). Results with dinitrophenylation were conflicting with respect to the effect of histidine modification on

43. Y. Nakagawa and M. L. Bender, *JACS* **91,** 1566 (1969).
44. D. E. Fahrney and A. M. Gold, *JACS* **85,** 997 (1963).
45. B. R. Hammond and H. Gutfreund, *BJ* **61,** 187 (1955).
46. M. L. Bender and F. J. Kezdy, *JACS* **86,** 3704 (1964).
47. T. C. Bruice and G. Schmir, *JACS* **81,** 4553 (1959).

activity and reflect typical difficulties with reagents that produce modification at multiple sites in proteins (*48, 49*).

On the other hand, TPCK inactivates chymotrypsin by a single chemical modification. The reagent is substratelike in structure and, since it requires functioning enzyme to be effective, the alkylation itself mimics the enzymic process. A covalent change is produced which permits identification of the altered residue. There is nothing inherent in the reagent to favor alkylation of histidine, however; this result must be considered solely due to the proximity of histidine to the alkylating end of the reagent in the intermediate chymotrypsin–TPCK complex. The fact that many other alkylating agents, as described above, are incapable of alkylating His 57 indicates that such an outcome in the case of TPCK cannot be attributed to His 57 being merely a reactive residue in the region of the active center of chymotrypsin but instead results from its proximity to the bond-breaking site and to TPCK being properly oriented. The D isomers of TPCK (*24*) or of the even more reactive ZPCK [(XIII) X = Cl] (*50*) are inactive. These properties indicated the value of active-site-directed reagents in locating enzyme functional groups.

The role favored for histidine in the mechanism of chymotrypsin is that of a general base (*46, 51*). The inactivation caused by TPCK can, of course, be attributed to blocking the entire active site. On the other hand, the reagent methyl *p*-nitrobenzenesulfonate (X) leaves only a methyl group attached to chymotrypsin at N-3 of His 57 (*43*). Since the pK of ^{3}N-methyl histidine (or ^{3}N-carboxymethyl histidine) is close to that of histidine itself (*52*), the altered residue might therefore still be expected to function as a proton acceptor in the usual pH range of chymotryptic action except for an apparent need for this proton transfer to take place in a way blocked by the methyl group or any of the other reagents, including TPCK, which alkylate at N-3 of His 57. In this respect, any size group will cause inactivation. (It would be of interest if the ^{3}N-methyl His 57 derivative of chymotrypsin retains ability to complex with partial substrates.)

When the three-dimensional structure of tosyl-α-chymotrypsin was

48. J. R. Whitaker and B. J. Jandorf, *JBC* **223,** 751 (1956).
49. V. Massey and B. S. Hartley, *BBA* **21,** 361 (1956).
50. E. Shaw, "Methods in Enzymology," Vol. 11, p. 677, 1967.
51. T. C. Bruice and S. Benkovic, "Biorganic Mechanisms," Vol. 1, p. 228. Benjamin, New York, 1966.
52. P. L. Whitney, P. O. Nyman, and B. G. Malmstrom, *JBC* **242,** 4212 (1967).

elucidated, His 57 was found to be very close to Ser 195 (*53*) and, in a later interpretation (*54*), was concluded to be oriented with N-3 at hydrogen bonding distance from the active center oxygen of Ser 195. In fact, these were found to be joined to Asp 102 in what was described as a charge relay system (*54*) in which N-1 of His 57 was hydrogen bonded in turn to the β-carboxyl group of Asp 102 (cf. Fig. 1). The chemical modification studies described above are in excellent agreement with this finding. The unexpected hydrogen bond to N-1 of the imidazole ring of His 57 maintains it oriented toward the interior of the protein (Asp 102 is buried) so that only N-3 is presented to the catalytic region of the enzyme, in accord with the finding that chemical modification at this residue was limited to N-3 of the imidazole ring.

One might wonder why, if the side chain of Ser 195 is the primary nucleophile in chymotryptic action, an active-site-directed agent with a nucleophile-sensitive group complexed in this region of the enzyme leads to alkylation at His 57 instead of to ether formation at Ser 195. In related studies with trypsin, a comparable situation, both alternatives are, in fact, found to take place (*21, 55, 56*). With each reagent the outcome must be determined by the geometry within the reagent–enzyme complex, a situation in which small differences in orientation can be expected to have great influences on relative rates of alkylation and thus on the outcome of the modification. Probably ether formation at Ser 195 will be found to be occurring with other chymotrypsin-directed reagents. It is also possible that the equilibria among the various enzyme–substrate complexes now considered to be intermediates in chymotryptic catalyses (*57, 58*) or of enzyme states in which ligand binding may induce the formation of the histidine serine hydrogen bond (*23*) may also be influenced differently by substratelike alkylating agents. This could result in small structural variations within these complexes leading to a modification in either of the two possible side chains participating in catalysis. It is clear that the affinity labeling reagents may thus not distinguish the primary nucleophile at the catalytic center

53. B. W. Matthews, P. B. Sigler, R. Henderson, and D. M. Blow, *Nature* **214,** 652 (1967); P. B. Sigler, D. M. Blow, B. W. Matthews, and R. Henderson, *JMB* **35,** 143 (1968).

54. D. M. Blow, J. J. Birktoft, and B. S. Hartley, *Nature* **221,** 337 (1969).

55. W. B. Lawson, M. D. Leafer, Jr., A. Tewes, and G. J. S. Rao, *Z. Physiol. Chem.* **349,** 251 (1968).

56. D. Schroeder and E. Shaw, *ABB* (1970) (submitted for publication).

57. T. E. Barman and H. Gutfreund, *BJ* **101,** 411 (1966).

58. G. P. Hess, *Brookhaven Symp. Biol.* **21,** 155 (1968).

of pancreatic proteases (that is serine vs. histidine) but have the ability to label both residues for identification in the primary sequence and thus to complement many other studies that clarify their roles more precisely.

In the case of Met 192, nondirected reagents such as hydrogen peroxide (*40*) and periodate (*59*) lead to selective oxidation to the sulfoxide which was recognized to result in relatively modest affects on its catalytic properties (*59, 60*). Thus, K_m and V_{max} for *N*-acetyl-L-valine ethyl ester remained unchanged whereas for *N*-acetyl-L-tryptophan ethyl ester, K_m, but not V_{max}, roughly doubled (*59*).

Active-site-directed alkylating agents described above (Section II,A,2) which formed sulfonium salts at Met 192 as did the intramolecular alkylation of Lawson and Schramm (Section II,D,1 below) were therefore not essential as a means of modifying this accessible residue but permitted the introduction of a greater variety of groups for establishing more dramatic effects on binding (*35, 61*). By and large, protein chemists were admirably restrained with respect to forceful claims that Met 180 represented a part of the specificity site of chymotrypsin on the basis of these observations; however, this appears to be the case (*42*).

4. *Other Neutral Proteases*

Endopeptidases of a chymotrypticlike specificity are found in a variety of living things possibly because of homology of structure and function. In general, many of these are found to be inactivated by TPCK including enzymes from a sea anemone (*62*), fish (*63*), reptiles (*63*), insect larva (*64*), as well as specialized animal cells not related to digestion (*65*). Some animals have multiple forms of pancreatic chymotrypsin with significant variations in preference for neutral amino acids. For example, porcine chymotrypsin C favors leucine over phenylalanine. The action of TPCK was relatively slow on this enzyme, but the chloromethyl ketone derived from leucine (XII) provided a rapid, stoichiometric alkylation at histidine (*66*).

Bacterial serine proteinases similar to chymotrypsin in specificity have

59. J. R. Knowles, *BJ* **95,** 180 (1965).
60. H. Weiner, C. W. Batt, and D. E. Koshland, Jr., *JBC* **241,** 2687 (1966).
61. F. J. Kezdy, J. Feder, and M. L. Bender, *JACS* **89,** 1009 (1967).
62. D. Gibson and G. H. Dixon, *Nature* **222,** 753 (1969).
63. E. A. Barnard and W. C. Hope, *BBA* **178,** 364 (1969).
64. H. H. Sonneborn, G. Pfleiderer, and J. Ishay, *Z. Physiol. Chem.* **350,** 389 (1969).
65. I. Pastan and S. Almquist, *JBC* **241,** 5090 (1966).
66. T. Tobita and J. E. Folk, *BBA* **147,** 15 (1967).

been encountered which are not affected by TPCK (*67, 68*). In the case of subtilisin this was attributed in part to a low affinity for simple substrates (*69*). An active-site-directed reagent based on phenylalanine but containing an N^{α}-acyl substituent (instead of sulfonamido) and in the form of a bromomethyl ketone, ZPBK (XIII), did provide the desired stoichiometric inactivation (*70*). It was of considerable interest to find that alkylation took place at N^3 of His 64 in both forms of subtilisin (BPN′ and Carlsberg) (*71*). The sequences of these bacterial enzymes (*72*) show no resemblance to mammalian pancreatic chymotrypsin (*73*); for this and other reasons (*74*) an independent origin was indicated for a hydrolytic mechanism involving serine and histidine in common with the mammalian enzymes (see Chapter 6 by Smith, this volume). The participation of histidine had already been surmised from pH activity studies (*69*).

X-ray crystallographic studies on subtilisin BPN′ as the phenylmethanesulfonyl derivative confirmed (*75*) the presence of a hydrolytic mechanism like that of chymotrypsin as had been deduced from chemical and enzymic studies and indicated that His 64 was close to the active center Ser 221. Moreover the histidine was positioned by the side chain of an aspartic acid residue (Asp 32) as had been discovered in chymotrypsin (*54*) (Fig. 1) so that only N-3 was presented to the catalytic site. These features are in accordance with the chemical modification by the site specific reagent ZPBK at N-3 of His 64 as discussed above.

B. Trypsin and Enzymes of Related Specificity

When it had been shown that incorporation of substratelike features in an alkylating agent led to selective chemical modification in the case

67. K. Morihara and H. Tsuzuki, *ABB* **129,** 620 (1969).
68. E. L. Smith, F. S. Markland, C. B. Kasper, R. J. Delange, M. Landon, and W. H. Evans, *JBC* **241,** 5974 (1966).
69. A. N. Glazer, *JBC* **242,** 433 (1967).
70. E. Shaw and J. Ruscica, *JBC* **243,** 6312 (1968).
71. F. S. Markland, E. Shaw, and E. L. Smith, *Proc. Natl. Acad. Sci. U. S.* **61,** 1440 (1968).
72. E. L. Smith, R. J. Delange, W. H. Evans, M. Landon, and F. S. Markland, *JBC* **243,** 2184 (1968).
73. K. Walsh and H. Neurath, *Proc. Natl. Acad. Sci. U. S.* **52,** 884 (1964).
74. E. L. Smith, F. S. Markland, and A. N. Glazer, *in* "Structure-Function Relationships of Proteolytic Enzymes" (P. Desnuelle, H. Neurath, and M. Ottesen, eds.), pp. 160ff. Munksgaard, Copenhagen, 1970.
75. C. S. Wright, R. A. Alden, and J. Kraut, *Nature* **221,** 235 (1969).

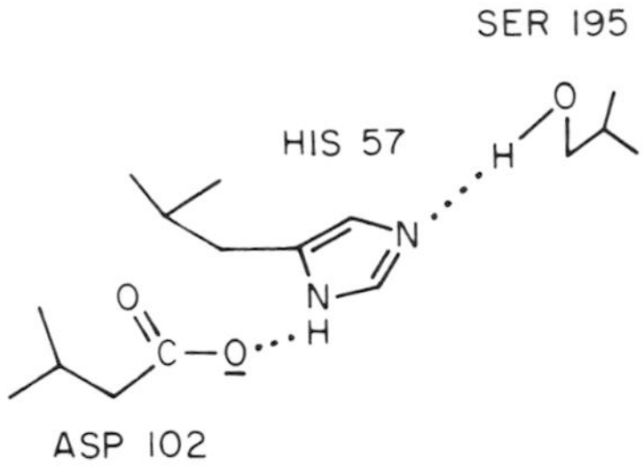

SER 195

HIS 57

ASP 102

FIG. 1. Activation of Ser 195 in chymotrypsin by a hydrogen-bonded system which also includes His 57 and Asp 102 [redrawn from Blow *et al.* (*54*)].

of chymotrypsin as described above, it became of interest to evaluate this approach with an enzyme of different specificity and trypsin was selected. The specificity of trypsin favors hydrolysis at basic amino acids in proteins, that is, at lysine and arginine residues. Here again simple ester derivatives of these amino acids are also readily hydrolyzed, particularly if they possess an N^{α}-acyl substituent. The synthesis of

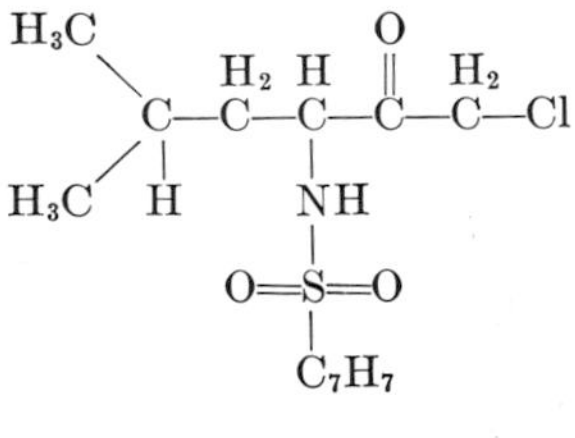

(XII) $(CH_3)_2CH{-}CH_2{-}CH(NH{-}SO_2{-}C_7H_7){-}CO{-}CH_2{-}Cl$

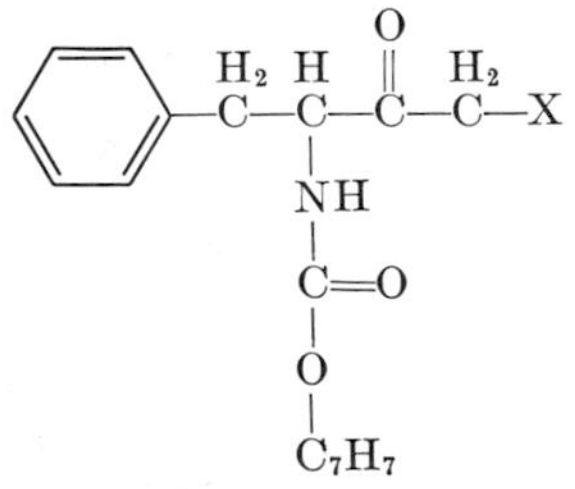

(XIII) $C_6H_5{-}CH_2{-}CH(NH{-}CO{-}O{-}C_7H_7){-}CO{-}CH_2{-}X$

for ZPCK, X = Cl
ZPBK, X = Br

chloromethyl ketones derived from these amino acids is more difficult than in the case of neutral amino acids, and structural variation has consequently been limited. TLCK (XIV), chloromethyl ketone derived from L-lysine (XV) was found to have the anticipated effect on trypsin, that is, inactivation by covalent attachment in an equimolar ratio (*21*) at N-3 of the imidazole ring of a histidine residue (*76*). The D isomer was ineffective (*77*). Given the availability of the amino acid sequence of trypsin (*78, 79*), the location of the altered residue could be determined as His 46 from the composition of a peptide containing radio-

76. P. Petra, W. Cohen, and E. Shaw, *BBRC* **21,** 612 (1965).
77. E. Shaw and G. Glover, *ABB* (1970) (in press).
78. K. A. Walsh, D. L. Kauffman, K. S. V. Sampath Kumar, and H. Neurath, *Proc. Natl. Acad. Sci. U. S.* **51,** 301 (1964).
79. O. Mikeš, V. Tomášek, V. Holeyšovský, and F. Šorm, *BBA* **117,** 281 (1966).

active inhibitor *(80)*. No alkylation took place in 8 *M* urea or with the zymogen.

A striking observation was the reciprocal lack of effect of TLCK on chymotrypsin *(4)* and of TPCK on trypsin *(21)*. Both enzymes utilize histidine and serine as catalytic groups. The agents, when they act, both introduce a substituent on N-3 of the imidazole ring of the histidine residue of the hydrolytic site. Therefore, the selectivity of the agents with respect to the enzymes whose specificity they satisfy is a convincing argument for their being active-site directed. [This selectivity appears to be a quantitative difference as would be expected from the fact that trypsin and chymotrypsin are not "absolutely" specific but have different affinities for charged and uncharged substrates *(81)*.]

A chloromethyl ketone derived from an N^{α}-acyl arginine (XVI) or from lysine with both amino groups free (XVII) was also obtainable and, in either case, inactivated trypsin by alkylation at N-3 of a histidine ring *(77)*, presumably His 46, although this was not established. Kinetic differences were found among these agents as described below.

If the reagent which modifies an enzyme is acting as an active center directed agent, its affinity for the enzyme in reversible complex formation may possibly be established in the usual way as a competitive inhibitor in rate assays. This has been satisfactory with relatively slowly acting inactivators. Kitz and Wilson *(82)* have pointed out that the rate dependence of the inactivation process as a function of inhibitor concentration lends itself to a graphical observation of saturation kinetics. From such plots it may be possible to determine the extent of complex formation K_I as well as the rate of breakdown k_2 of the intermediate complex. These authors were concerned with the inactivation of acetyl-

$$\mathrm{E} + \mathrm{I} \rightleftharpoons \mathrm{E} \cdot \mathrm{I} \xrightarrow{k_2} \mathrm{E}\text{—}\mathrm{I}$$

choline esterase by a series of methanesulfonates and showed that, in fact, some of the agents were substratelike and not inactivating by bimolecular kinetics *(82)* (cf. Section III,A,2 below). This method of analysis was apparently independently evolved by other workers *(19, 44)*. It provides a useful tool in evaluating the affinity of reagents suspected of inactivating by affinity labeling [*(77, 83–85)* among others].

In the case of the alkylation of trypsin by TLCK, the rate of inactiva-

80. E. Shaw and S. Springhorn, *BBRC* **27,** 391 (1967).
81. T. Inagami and J. M. Sturtevant, *JBC* **235,** 1019 (1960).
82. R. Kitz and I. B. Wilson, *JBC* **237,** 3245 (1962).
83. M. Mares-Guia and E. Shaw, *JBC* **242,** 5782 (1967).
84. R. D. O'Brien, *Mol. Pharmacol.* **4,** 121 (1968).
85. H. P. Meloche, *Biochemistry* **6,** 2273 (1967).

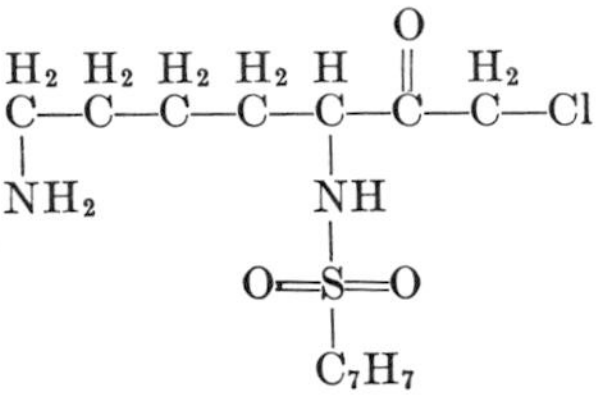

```
                        O
H2 H2 H2 H2 H   ||  H2
C—C—C—C—C—C—C—Cl
|               |
NH2            NH
                |
             O=S=O
                |
               C7H7
             (XIV)
```

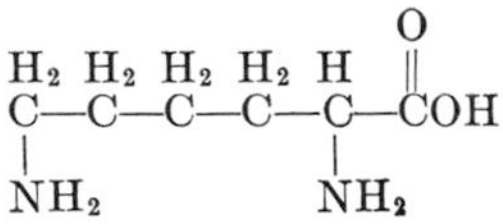

```
                        O
H2 H2 H2 H2 H   ||
C—C—C—C—C—COH
|               |
NH2            NH2
             (XV)
```

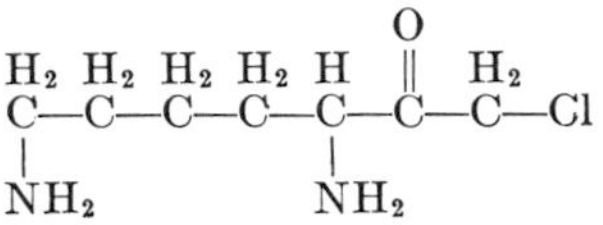

```
                        O
H2 H2 H2 H2 H   ||  H2
C—C—C—C—C—C—C—Cl
|               |
NH2            NH2
             (XVII)
```

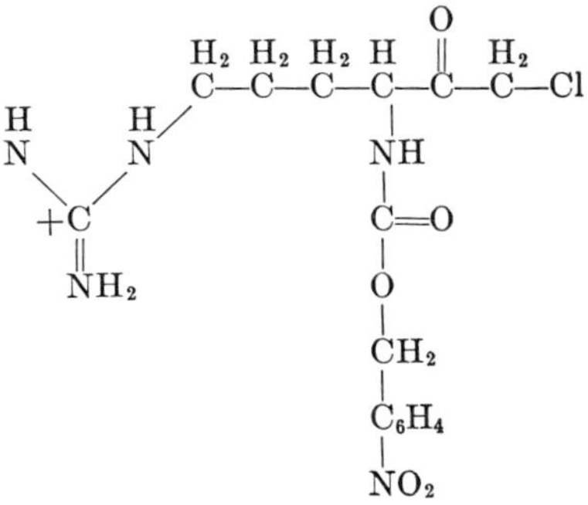

```
                            O
               H2 H2 H2 H   ||  H2
               C—C—C—C—C—C—Cl
H      H  /             |
N      N               NH
  \   /                 |
   +C                  C=O
    ||                  |
   NH2                  O
                        |
                       CH2
                        |
                       C6H4
                        |
                       NO2
                 (XVI)
```

tion was measured at a series of TLCK concentrations each of which was adequate to provide pseudo-first-order kinetics for a major portion of the inactivation (Fig. 2). The rate constants k so obtained were plotted in double reciprocal fashion with inhibitor concentrations as shown in Fig. 3. As discussed by Kitz and Wilson (*82*), since the extrapolation to infinite inhibitor concentration does not pass through the origin, the intercept indicates a limiting rate of inactivation k_2 owing to conversion of all free enzyme to an intermediate complex with the reagent; the slope provides K_I/k_2. The values for TLCK obtained at 25°, pH 7.0, were $K_i = 1.25 \times 10^{-4}\ M$ and $k_2 = 0.16\ \text{min}^{-1}$ (*77*).

Other reagents designed to be active-site directed for trypsin permit comparisons both on the effect of structure on the kinetics of inactivation as well as on the outcome with respect to protein chemistry. These are based on certain simple molecules which may be considered partial structures of normal simple substrates and are known competitive inhibitors of trypsin. Among them are benzamidine (XVIII) and phenylguanidine (XIX) with $K_i = 1.8 \times 10^{-5}\ M$ and $7.3 \times 10^{-5}\ M$, respectively (*86*). The aromatic ring in these molecules is considered to occupy a hydrophobic region of the trypsin active center normally functioning to bind side chain methylene groups (*86*). Consequently, the aromatic

86. M. Mares-Guia and E. Shaw, *JBC* **240,** 1579 (1965).

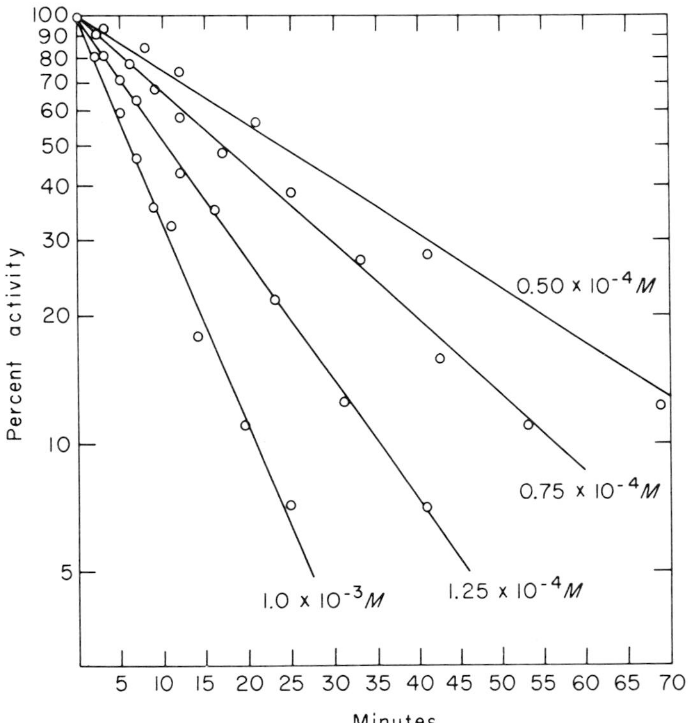

FIG. 2. Inactivation of bovine trypsin ($3 \times 10^{-5}\,M$) by various concentrations of TLCK at pH 7.0, 25°, in 0.05 M tris buffer, 0.02 M in calcium; loss of amidase activity measured (*77*).

ring would extend in the direction of the hydrolytic part of the active center, an idea supported by tryptic cleavage of esters of benzamidine and phenylguanidine extending from the *p* position of these rings and having appropriate overall dimensions (*87*). In addition, aliphatic amines and guanidines appear to be similar to normal side chains, are competi-

$$\overset{+}{}\begin{matrix} H_2N \\ \diagdown \\ \quad C{-}C_6H_5 \\ {/\!/} \\ H_2N \end{matrix} \qquad\qquad \begin{matrix} H_2N \quad \overset{H}{N}{-}C_6H_5 \\ \diagdown \;\; / \\ C \\ +\| \\ NH_2 \end{matrix}$$

(XVIII) (XIX)

87. M. Mares-Guia, E. Shaw, and W. Cohen, *JBC* **242**, 5777 (1967).

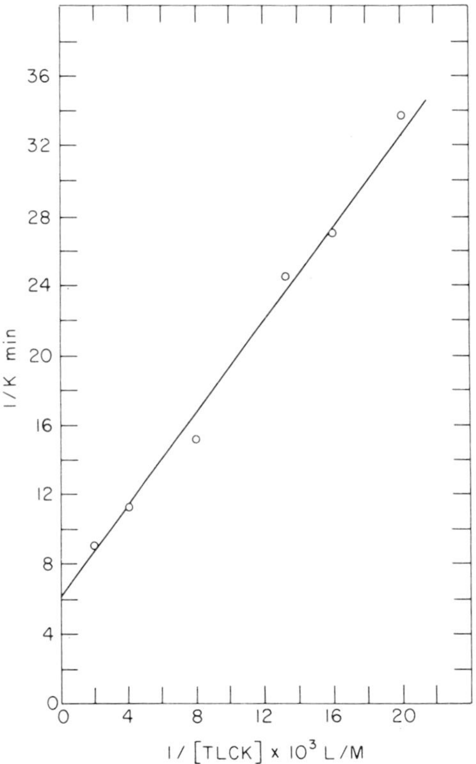

FIG. 3. Saturation kinetics in inactivation of bovine trypsin by TLCK. The K_{app} of inactivation at various TLCK concentrations that provide first-order inactivation kinetics (Fig. 2) replotted in double reciprocal fashion to obtain kinetic properties of intermediate trypsin–TLCK complex (*77*).

tive inhibitors of trypsin, but have less affinity than aromatic derivatives; thus, for *n*-propylamine, $K_i = 8.9$ m*M* (*88*) and for *n*-propylguanidine, $K_i = 0.69$ m*M* (*89*), both at pH 6.6.

Alkylating agents derived from a series of *n*-alkylguanidines of the general structure (XX), where $n = 2$–4, were studied by Lawson *et al.* (*90*). The propyl derivative ($n = 3$) was the most active but quite slow compared to TLCK. However, of greater interest is the fact that all members of this series inactivated trypsin by alkylation of a serine residue (*90*). In rate assays they were shown to be competitive inhib-

88. T. Inagami, *JBC* **239,** 787 (1964).
89. T. Inagami and S. S. York, *Biochemistry* **7,** 4045 (1968).
90. W. B. Lawson, M. D. Leafer, Jr., A. Tewes, and G. J. S. Rao, *Z. Physiol. Chem.* **349,** 251 (1968).

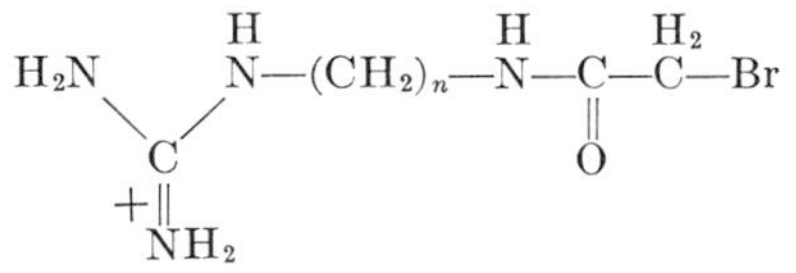

$$\mathrm{H_2N{-}\overset{+}{C}({=}NH_2){-}NH{-}(CH_2)_n{-}NH{-}C({=}O){-}CH_2{-}Br}$$

(XX)

itors. *O*-Carboxymethyl serine was isolated from the inhibited enzymes after hydrolysis. As discussed above, in connection with chymotrypsin-directed reagents, the known presence of at least two nucleophilic side chains in the catalytic region of these hydrolytic enzymes could be expected to lead to alkylation of either side chain by a complexed reagent according to their relative proximity to the reagent carbon at which the displacement occurs. The bromoacetamido guanidines were considered to have a common orientation with respect to Ser 183 in the enzyme–inhibitor complex (*90*). This is reasonable in spite of the variation in length of these reagents because of the known ability of trypsin to hydrolyze synthetic substrates having a comparable spread of chain lengths.

From benzamidine (XVIII) and phenylguanidine (XIX) the corresponding *p*-phenacyl bromides (XXI) and (XXII) emerged as affinity labelers of trypsin (*56*). The latter proved to be quite effective, rapidly alkylating Ser 183.

The value of active-site-directed reagents in labeling the active center histidine and serine of trypsin is demonstrated by these results. It is instructive to observe that $5 \times 10^{-4}\,M$ TLCK inactivates trypsin (10^{-4} to $10^{-5}\,M$) with a half-time of about 6 min at pH 7 as a result of a single modification at a single site, whereas iodoacetamide at a concentration a thousand times higher but with otherwise comparable conditions produces about six chemical modifications per mole of trypsin in 48 hr; however, even this extensive modification has little effect on activity (*91, 92*). The inertness of iodo- or bromoacetic acid and amide has been observed by others (*55, 93*) except in 6 *M* urea in which case the methionines of trypsin are converted to sulfonium salts (*93*) or in the interesting case of the activation of trypsin by alkylguanidines (*92*).

It had been observed that the slow hydrolysis by trypsin of a nonspecific ester substrate, acetyl glycine ethyl ester, was accelerated by alkylamines (*94*) and alkylguanidines (*89*) of limited chain length. The

91. T. Inagami, *JBC* **240,** PC3453 (1965).
92. T. Inagami and H. Hatano, *JBC* **244,** 1176 (1969).
93. V. Holeysovsky and M. Lazdunski, *BBA* **154,** 457 (1968).
94. T. Inagami and T. Murachi, *JBC* **239,** 1395 (1964).

combined size of such molecules did not exceed that of arginine. The effect of the alkylguanidinium ion was interpreted as increasing the pK of the histidine residue of the active center (*92*, *94*) and was shown to involve binding of the ion at the substrate binding site (*89*). In the presence of the methylguanidinium ion, the rate of inactivation by iodoacetamide was increased sixfold (*92*); however, the chemical modification was still complex (*92*) but indicated a correlation of activity loss with histidine modification.

Bromoacetone, on the other hand, although still a small reagent not markedly different from the above in size, revealed more specificity and effectiveness. It modified trypsin at two residues but only the alkylation at His 46 caused loss of enzymic activity (*95*). An unusual observation was the action at this residue even in DIP-trypsin (*95*).

The alkylating agents considered above are listed in Table I. Although they are not numerous, they permit some observations worthy of comment:

(1) With respect to protein chemistry, the active-site-directed reagents for trypsin were found to alkylate the enzyme at either of two residues, namely, the serine and histidine residues of the catalytic region of the active center. As discussed above in the case of chymotrypsin (Section II,A,3) these results are to be expected from substrate-like reagents which, in complexing with these enzymes, localize a nucleophile-sensitive group (alkylating function) in this region. The outcome in the case of any single inhibitor cannot be predicted (prior to having the detailed knowledge we are striving toward).

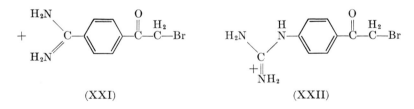

$$(H_2N)_2C^{+}\text{—}C_6H_4\text{—}CO\text{—}CH_2\text{—}Br$$

(XXI)

$$H_2N\text{—}C(=NH_2^{+})\text{—}NH\text{—}C_6H_4\text{—}CO\text{—}CH_2\text{—}Br$$

(XXII)

(2) With respect to inhibitory effectiveness, it can be seen that the rate of trypsin inactivation does not correlate with affinity. Thus *p*-amidinophenacyl bromide (XXI) is more tightly complexed than *p*-guanidinophenacyl bromide (XXII) and yet the latter is by far the more effective irreversible inhibitor (cf. k_2 value in Table I) (*91*). This is not because of a difference in reactivity. The results can be rationalized as owing to a more favorable geometry for alkylation within the

95. J. G. Beeley and H. Neurath, *Biochemistry* 7, 1239 (1968).

TABLE I
TRYPSIN-DIRECTED ALKYLATING AGENTS

Agent	$K_i(M)$	k_2 (min^{-1})[a]	$t_{1/2}$	Conditions[b]	Site of alkylation[c]	Ref.
L-TLCK[d]	1.25×10^{-4}	0.16	4.3[e]	E = 3×10^{-5}, pH 7, 25°	His 46	*77*
D-TLCK				Same		*77*
LCK[d]	1.30×10^{-3}	0.40	1.7[e]	Same	His	*77*
p-Amidinophenacyl bromide (XXI)	1.2×10^{-4}	0.0047		Same	Not histidine	*56*
p-Guanidinophenacyl bromide (XXII)	2.6×10^{-3}	0.17		Same	Ser 183	*56*
Bromacetamidopropylguanidine (XX, n = 3)	1×10^{-2}		4 hr	E = 6×10^{-5}, pH 7, 25° I = 6×10^{-3}	Ser 183	*90*
Bromoacetone			25 min	E = 2×10^{-4}, pH 7, 40° I = 0.28, benzamidine	His 46 (and one nonessential site)	*95*
Methylguanidine and iodoacetamide			10 hr	E = 8.3×10^{-5}, pH 7, 25°; I = 0.5; methylguanidine, 0.4	His and others	*92*
Iodoacetamide			~60 hr	Same, no methylguanidine	His and others	*92*
Iodoacetic acid				E = 1.2×10^{-4}, pH 7, 25° I = 1.2×10^{-4}	Only in 6 M urea; methionines alkylated	*93*

[a] Rate of breakdown of reagent–trypsin complex.
[b] Concentration in molarity.
[c] Lack of residue number indicates position not determined.
[d] Values from double reciprocal plots of Kitz–Wilson type (*82*) (cf. Fig. 3).
[e] At saturation, minutes.

reagent–enzyme complex in the case of guanidinophenacyl bromide. In fact, it is possible that, in attempting to arrive at an active-site-directed reagent, the chemist obtains a highly effective competitive inhibitor which is, however, incapable of achieving alkylation because of absence of suitable side chains in the region of the complexed reagent or because of unfavorable orientation with respect to the ones that are there. Thus affinity is a necessary but possibly not sufficient condition for specific labeling.

It follows from the above considerations that enzymes similar to trypsin in action on simple ester substrates might be expected to give similar, although not identical, information through the use of these affinity labeling reagents. Relatively little detailed information is available for comparison, but TLCK has provided stoichiometric inactivation of plasmin (*96*) and thrombin (*97*) by alkylation of N-3 of a histidine residue in each case. The streptokinase–plasminogen complex is inactivated (*98*) as is an insect form of trypsin, cocoonase (*99*), and a mold serine proteinase (*100*). However, subtilisin (*68*) as well as rat urinary (*101*) and human plasma (*102*) kallikreins were not affected by TLCK. In the latter cases it is not known whether the lack of action is because of poor affinity or poor orientation.

It is of considerable interest that carboxypeptidase B, an exopeptidase which liberates COOH-terminal basic amino acids, was found to be gradually inactivated by 4-bromoacetamidobutylguanidine [(XX) $n = 4$)], an affinity labeler of trypsin at Ser 183. In the case of carboxypeptidase B, tyrosine is the residue alkylated (*103*).

C. Comments on the Results Obtained with TPCK and TLCK

Because TPCK and TLCK result in alkylation of histidine in chymotrypsin and in trypsin, it is sometimes erroneously concluded that they are histidine reagents and may be applied to any enzyme as a test for

96. W. R. Groskopf, B. Hsieh, L. Summaria, and K. C. Robbins, *JBC* **244,** 359 (1969).
97. G. Glover and E. Shaw, *Abstr., 158th ACS Natl. Meeting, New York, 1969* p. B322.
98. F. F. Buck, B. C. W. Hummel, and E. C. DeRenzo, *JBC* **243,** 3648 (1968).
99. F. C. Kafatos, J. H. Law, and A. M. Tartakoff, *JBC* **242,** 1488 (1967).
100. K. Morihara and H. Tsuzuki, *ABB* **126,** 971 (1968).
101. M. Mares-Guia and C. R. Diniz, *ABB* **121,** 750 (1962).
102. R. W. Colman, L. Mattler, and S. Sherry, *J. Clin. Invest.* **48,** 23 (1969).
103. T. H. Plummer, Jr., and W. B. Lawson, *JBC* **241,** 1648 (1966); T. H. Plummer, Jr., *ibid.* **244,** 5246 (1969).

the possible function of histidine. Hopefully, it has been made amply clear in the foregoing sections that active-site-directed reagents are reagents that form a substratelike complex with an enzyme and that if there is any selectivity in the subsequent chemical modification it is determined by the availability of amino acid side chains within the complex. TPCK, TLCK, and other substratelike alkylating agents have by now been found to result in stoichiometric modification of a number of proteolytic enzymes as described above or in Section III,A,1,b and to modify not only histidine but also serine, methionine, half-cysteine, and tyrosine according to reagent structure and the enzyme under study.

The halogen in TPCK or TLCK is not particularly reactive, considerably less reactive, in fact, than that of iodoacetamide (*77, 104, 104a*). The considerably greater rate of displacement by an active-center nucleophile is a clear reflection of a proximity effect produced by complex formation just as in normal catalysis.

Active-site-directed inhibitors of proteolytic enzymes in which the goal is to achieve modification outside the active center region are not within the scope of this review but may be of interest as selective inhibitors (*105, 106*).

D. Consecutive Covalent Modifications Using Acyl-Enzyme Intermediates

The formation of a covalently-bound intermediate during enzymic action is not uncommon, and in the serine proteinases this feature has been used to gain a foothold in the active center to effect a second chemical change. The normal intermediate has a brief existence but acyl analogs can be devised which turn over slowly or not at all providing derivatives of value for physical studies since they permit the observation of a step in the catalytic process. Some of these studies are mentioned below. In this section, another use of this type of intermediate is described, namely, the production of a second covalent change in the active center region that might not otherwise be possible.

1. *Intramolecular Alkylation*

In the esterase action of chymotrypsin on trimethylacetates, the deacylation of the trimethylacetyl enzyme is a slow step presumably

104. B. R. Baker and J. H. Jordaan, *J. Heterocyclic Chem.*, **2**, 21 (1965).
104a. R. Lee and W. D. McElroy, *Biochemistry* **8**, 130 (1969).
105. B. R. Baker and E. H. Erickson, *J. Med. Chem.* **12**, 112 (1969).
106. B. R. Baker and J. A. Hurlburt, *J. Med. Chem.* **12**, 118 (1969).

owing to steric hindrance (*107*). Lawson and Schramm took advantage of this in studying a sterically similar ester containing an alkylating agent (*12, 108*). By use of the nitrophenyl ester for rapid acylation and treatment of chymotrypsin at pH 5 to retard deacylation even further, it was

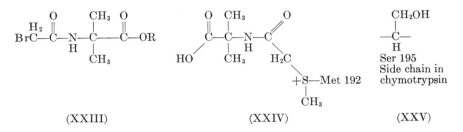

```
       O       CH3  O
   H2  ‖       |    ‖
BrC—C—N—C——C—OR
         H   |
             CH3

           (XXIII)

      O  CH3         O
      ‖   |         //
      C—C—N—C
     /     |  H    \
 HO     CH3      H2C
                     \
                      +S—Met 192
                       |
                       CH3

           (XXIV)

   CH2OH
     |
   —C—
     H
 Ser 195
 Side chain in
 chymotrypsin

           (XXV)
```

possible to prepare the chymotrypsin derivative [(XXIII) R = chymotrypsin]. On incubation, a second covalent change occurred resulting from intramolecular alkylation of Met 192 by sulfonium salt formation. After deacylation at Ser 195 (XXV) took place, the final, stable modification was the sulfonium derivative of chymotrypsin at Met 192 (XXIV). The product was found to be enzymically active but to have reduced affinity for characteristic substrates (*61, 108*). As described above (Section II,A,2) alkyl halides containing an aromatic ring also form sulfonium salts directly at Met 192, and this side chain can also be selectively converted to the monoxide by limited oxidation (*40, 59*). It is generally found that the derivatives have altered specificity (*35, 40, 59*).

The intramolecular alkylation described above is a fine example of enzyme chemistry since it involves a monomodification at the outset and affords the opportunity to utilize controlled reagent geometry.

2. *Photolysis*

Independent recognition of the potential of acyl-enzyme derivatives for active-site modification was made by Westheimer and his colleagues. At the same time attention was given to what is the most difficult aspect of the chemical modification of proteins, that is, the labeling of side chains which have no functional groups. These may be methyl and methylene groups of the hydrophobic amino acid residues (valine, leucine, isoleucine, and alanine), as well as the relatively inert phenyl ring of phenylalanine whose participation in specificity determining regions of an active center is quite likely (*10*).

The possible modification of chemically inert side chains through

107. C. E. McDonald and A. K. Balls, *JBC* **227**, 727 (1957).
108. W. B. Lawson and H. J. Schramm, *Biochemistry* **4**, 377 (1965).

attack by a carbene (XXVII) generated within the active center was studied by the irradiation of diazoacetyl chymotrypsin [(XXVI) R = chymotrypsin, Ser 195 ester] (*10*). Following irradiation at 5° with light of the spectral region near a diazoacetyl absorption band (370 nm), it was established that considerable regeneration of enzymic activity had taken place apparently following conversion of the acyl group to glycollate which was identified. However, modified chymotrypsins were produced resulting from carboxymethylation as hoped. *O*-Carboxymethylserine, 1-carboxymethylhistidine, and *O*-carboxymethyltyrosine were present in the product (*109*). These are derivatives of side chains which can be modified by reagents susceptible to nucleophilic attack (*110*). More remarkable results were obtained in an analogous study involving trypsin in which modification of an aliphatic side chain was observed. In this case, diazomalonyl trypsin [(XXVIII) serine ester] was irradiated and radioactivity was found to appear in glutamic acid after

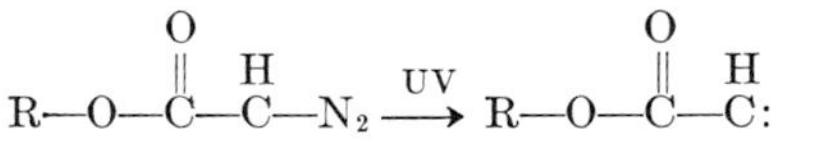

$$\underset{\text{(XXVI)}}{R{-}O{-}\overset{\overset{\displaystyle O}{\|}}{C}{-}\overset{H}{C}{-}N_2} \xrightarrow{\text{UV}} \underset{\text{(XXVII)}}{R{-}O{-}\overset{\overset{\displaystyle O}{\|}}{C}{-}\overset{H}{C}{:}} \qquad \underset{\text{(XXVIII)}}{RO\underset{\underset{\displaystyle O}{\|}}{C}{-}\underset{\underset{\displaystyle N_2}{|}}{C}{-}\underset{\underset{\displaystyle O}{\|}}{C}{-}OC_2H_5}$$

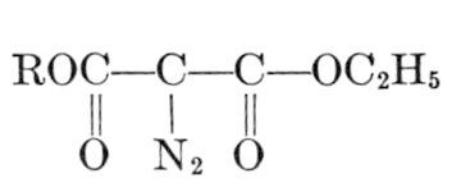

hydrolysis of the photolyzed acyl enzyme (*111*). This indicated the unusual modification of an alanine methyl group by insertion of the carbene, $:\overset{H}{C}{-}COOH$ (*111*). It will be of considerable interest to learn the location of this event.

3. *Elimination Reactions. Anhydrochymotrypsin*

The typical chemical modification adds bulk to an enzyme and the observed effects may solely result from this influence, an inherent danger in interpreting results obtained by this approach. An unusual opportunity to modify the active center serine of chymotrypsin by removal of a functional group was recognized by Koshland and his colleagues (*112, 113*) who made use of tosyl-chymotrypsin (XXIX). This derivative is

109. J. Shafer, P. Baronowsky, R. Lawsen, F. Finn, and F. H. Westheimer, *JBC* **241**, 421 (1966).

110. L. Cohen, *Ann. Rev. Biochem.* **37**, 683 (1968).

111. R. J. Vaughan and F. H. Westheimer, *JACS* **91**, 217 (1969).

112. D. H. Strumeyer, W. N. White, and D. E. Koshland, Jr., *Proc. Natl. Acad. Sci. U. S.* **50**, 931 (1963).

113. H. Weiner, W. N. White, D. G. Hoare, and D. E. Koshland, Jr., *JACS* **88**, 3851 (1966).

readily prepared using toluenesulfonyl fluoride (*44*), which is more selective than the corresponding chloride, and permitted an elimination reaction in dilute alkali at 0°. The result was the formation of "anhydrochymotrypsin" (XXX) which was obtainable in high yield (*112, 113*). In this modification, conversion of Ser 195 to a dehydroalanine residue could not introduce steric hindrance and the loss of catalytic activity that was observed confirmed the essentiality of this side chain without ambiguity of interpretation owing to steric hindrance (*113*). The stability of chymotrypsin to the alkaline conditions required for the elimination was shown. The modified enzyme retained binding capacity for substrates and inhibitors (*114*).

By addition to the double bond in anhydrochymotrypsin as shown with sulfite (XXXI) (*112, 113*) new selective chemical modifications were also possible.

$$\underset{\text{(XXIX)}}{-\overset{\overset{CH_2O-SO_2-C_6H_4-CH_3}{|}}{\underset{H}{C}}-} \xrightarrow{OH^-} \underset{\text{(XXX)}}{-\overset{\overset{CH_2}{\|}}{C}-} \xrightarrow{HSO_3^-} \underset{\text{(XXXI)}}{-\overset{\overset{H_2C-SO_3H}{|}}{\underset{H}{C}}-}$$

4. *Displacement. Thiolsubtilisin*

Subtilisin, a bacterial serine protease, forms an inactive phenylmethanesulfonyl derivative at its active center serine (Ser 221). This derivative underwent a subsequent displacement of the phenylmethanesulfonate group by thiolacetic acid under mild conditions (pH 5.5) leading to an *S*-acetyl substitution on the β-carbon atom of the original serine residue (XXXII) (*115, 116*). This was the only change since subtilisin has no disulfide bonds (these would be susceptible to cleavage by the reagent). The *S*-acetyl derivative spontaneously hydrolyzed to an

$$\underset{\text{Ser 221 of subtilisin}}{\overset{H_2}{\underset{|}{C}}-O-\overset{O}{\overset{\|}{\underset{\underset{O}{\|}}{S}}}-\overset{H_2}{C}-C_6H_5} \xrightarrow{^-S-\overset{O}{\overset{\|}{C}}-CH_3} \underset{\text{(XXXII)}}{\overset{H_2}{\underset{|}{C}}-S-\underset{\underset{O}{\|}}{C}-CH_3} \rightarrow \underset{\text{(XXXIII)}}{\overset{H_2}{\underset{|}{C}}-SH}$$

114. H. Weiner and D. E. Koshland, Jr., *JBC* **240,** PC2764 (1965).

115. K. E. Neet and D. E. Koshland, Jr., *Proc. Natl. Acad. Sci. U. S.* **56,** 1606 (1966).

116. L. Polgar and M. L. Bender, *JACS* **88,** 3153 (1966).

—SH group; thus the overall effect was the remarkable transformation of the active center serine to a half-cysteine residue (XXXIII) which provides the modification with its name, thiolsubtilisin.

It was found that thiolsubtilisin resembled subtilisin in a variety of physical properties measured in solution *(117)* and did not reveal any noteworthy difference in three-dimensional structure examined in crystal structure by X-rays *(75, 118)*. However, thiolsubtilisin retained very little catalytic activity. Nitrophenyl acetate was still a substrate after chromatographic purification that removed any residual subtilisin that may have placed such observations in doubt *(119)*, and this activity was sensitive to *p*-chloromercuribenzoate as would be expected. However, ethyl esters of simple substrates were not split and this simple chemical change can be said to have resulted in almost complete inactivation. This change has been attributed to a loss of ability to promote the acylation step of the catalytic process owing to the small but decisive difference in atomic size of sulfur with respect to oxygen in relation to other groups comprising the active center of subtilisin *(117)*.

III. Active Center Labeling of Other Enzymes by Directed Reagents

The proteolytic enzymes discussed above represent the area of study in which synthetic active-center-directed agents were explored to the level of protein chemistry at a time when the results could be compared with information arriving from other sources. Enzymes of quite different specificity and mechanism are now being subjected to this approach which is becoming a standard procedure for the study of enzymes. We may expect to see some unusual contributions of organic chemistry to the problem of reagent design. It is important, when embarking on this kind of work, to attempt to incorporate the chemically reactive grouping into the reagent in such a way as to bring it into juxtaposition with a side chain of the active region of the enzyme under study. This is quite different from the approach of Baker *(7)* who had in mind the particular goal of selective enzyme inactivation and not functional group detection; reagents with extended side arms may be expected to locate a site for covalent modification somewhere on the enzyme surface which may be

117. K. E. Neet, A. Nanci, and D. E. Koshland, Jr., *JBC* **243**, 6392 (1968).

118. R. Alden, C. S. Wright, F. C. Westall, and J. Kraut, *in* "Structure-Function Relationships of Proteolytic Enzymes" (P. Desnuelle, H. Neurath, and M. Ottesen, eds.), p. 173. Munksgaard, Copenhagen, 1970.

119. L. Polgar and M. L. Bender, *Biochemistry* **8**, 136 (1969).

unrelated to mechanism but useful for inactivation. This is, of course, a danger in any chemical modification study in which the investigator hopes to acquire knowledge about important functional groups and is not necessarily avoided by limiting reagent size and geometry to substrate characteristics.

The examples below appear most likely to involve active center modification and, in general, have been carried out on purified enzyme preparations so that it has been possible to establish the stoichiometry of binding. In a few cases, these criteria are not likely to be met but other considerations have led to their inclusion.

A. Hydrolytic Enzymes

1. *Proteolytic Enzymes*

a. Pepsin. This enzyme is considered representative of a class of acid proteases since its pH optimum ranges from 2 to 4.5, depending on substrate. Its specificity is not easily defined particularly with simple substrates (*120*) probably because of the effect of occupancy of secondary binding sites, in both directions of the susceptible bond, on the rate of catalytic splitting (*121, 122*). However, hydrophobic side chains at the amino acid residues forming the susceptible bond provide a favorable configuration (*120*).

A mechanism involving carboxyl groups as catalytically important groups was considered to be likely based on earlier studies (*120, 123*), but the possibility that this involves an intermediate mixed anhydride between substrate and enzyme (*123*) has been questioned (*124*). A tetrahedral intermediate has also been proposed (*125, 126*). Chemical modification studies have contributed to strengthening the argument for active center carboxyl groups.

Stiochiometric modification of pepsin has been observed by two chemical classes of reagents, and evidence has been provided that carboxyl

120. F. A. Bovey and S. S. Yanari, "The Enzymes," 2nd ed. Vol. 4, p. 63, 1960.

121. R. E. Humphreys and J. S. Fruton, *Proc. Natl. Acad. Sci. U. S.* **59,** 519 (1968).

122. T. R. Hollands and J. S. Fruton, *Biochemistry* **7,** 2045 (1968).

123. M. Bender and F. Kezdy, *Ann. Rev. Biochem.* **34,** 49 (1965).

124. J. L. Denburg, R. Nelson, and M. S. Silver, *JACS* **90,** 479 (1968).

125. G. R. Delpierre and J. S. Fruton, *Proc. Natl. Acad. Sci. U. S.* **54,** 1161 (1965).

126. J. R. Knowles, R. S. Bayliss, A. J. Cornish-Bowden, P. Greenwell, T. M. Kitson, H. C. Sharp, and G. W. Wybrandt, *in* "Structure-Function Relationships of Proteolytic Enzymes" (P. Desnuelle, H. Neurath, and M. Ottesen, eds.), pp. 237ff. Munksgaard, Copenhagen, 1970.

group modification was brought about by members of either class. However, it appears that the two types may differ on the basis of whether they esterify ionized or unionized carboxyl groups. Thus phenacyl halides such as *p*-bromophenacyl bromide (XXXIV) were found to combine irreversibly with pepsin in an equimolar combination (*127*) shown

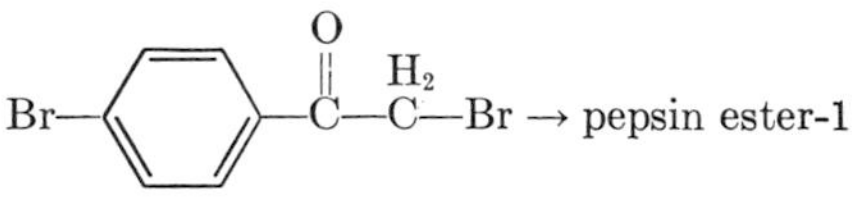

$$Br-C_6H_4-\overset{O}{\overset{\|}{C}}-\overset{H_2}{C}-Br \rightarrow \text{pepsin ester-1}$$

(XXXIV)

$$Br-C_6H_4-\overset{O}{\overset{\|}{C}}-\overset{H}{C}N_2 \rightarrow \text{pepsin ester-2}$$

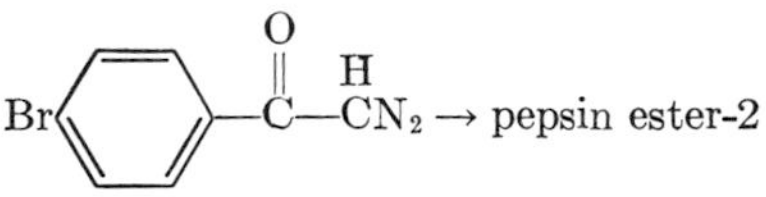

(XXXV)

to be an ester of aspartic acid (*128, 129*). Nucleophilic displacement by a carboxylate group was the probable mechanism (*130*). On the other hand, following the observations of the esterification of pepsin by diazoketones and diazoacetyl derivatives described below, the reaction of pepsin with *p*-bromophenyl diazomethyl ketone (XXXV) was examined and found to lead to a different mono-*p*-bromophenacyl pepsin ester (*130*). It was distinguished from the one formed from the phenacyl bromide by total lack of activity toward hemoglobin as a substrate and by failure to recover activity through the action of thiophenol (*130*). The diazo compound, in common with other members of this class, probably acts on a weakly acidic carboxyl group still protonated at pH 5, possibly by a carbene mechanism (*131*). Phenacyl halides esterify a carboxyl group not essential for activity since considerable (22%) residual hydrolysis of hemoglobin is observable (*127*). Eventually it was shown that, following the esterification of pepsin by bromophenacyl bromide (XXXIV), a second esterification was possible using a diazo reagent (*130, 131*).

A number of laboratories reported independently that pepsin was selectively inactivated by diazomethyl ketones (XXXVI)–(XXXVIII)

127. B. F. Erlanger, S. M. Vratsanos, N. Wasserman, and A. G. Cooper, *JBC* **240,** PC3447 (1965).
128. B. F. Erlanger, S. M. Vratsanos, N. Wasserman, and A. G. Cooper, *BBRC* 23, 243 (1966).
129. E. Gross and J. L. Morell, *JBC* **241,** 3638 (1966).
130. B. F. Erlanger, S. M. Vratsanos, N. Wasserman, and A. G. Cooper, *BBRC* **28,** 203 (1967).
131. R. L. Lunblad and W. H. Stein, *JBC* **244,** 154 (1969).

(*132–134*) or by diazoacetyl derivatives of amino acids (XXXIX) and (XL) (*135, 136*). All of these probably act through esterification of a

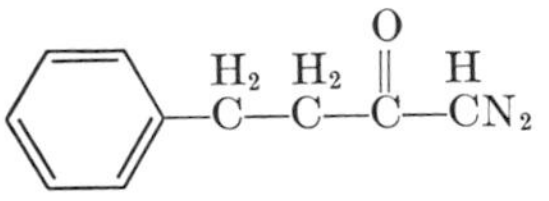

```
            O                          O
     H2 H2  ||  H               H2 H   ||  H
 Ph—C—C—C—CN2               Ph—C—C—C—CN2
                                   |
                                   NH
                                   |
                                   R

   (XXXVI)                 (XXXVII) R = benzyloxycarbonyl-
                           (XXXVIII) R = tosyl-
```

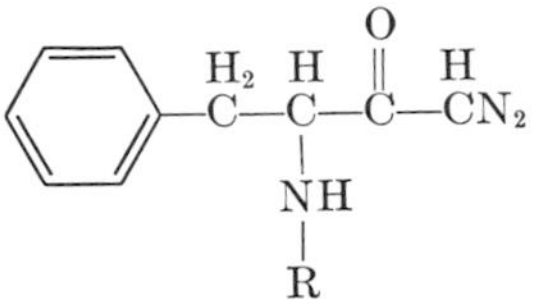

weakly acid carboxyl group (pK near 5) in protonated form. In the case of 1-diazo-4-phenylbutan-2-one (XXXVI) (*133*), the esterification of the β-carboxyl group of an aspartic acid group was demonstrated by isolation of the esterified tetrapeptide Ile—Val—Asp—Thr (*137*). This may be the same aspartic acid residue esterified by the action of diazoacetyl-

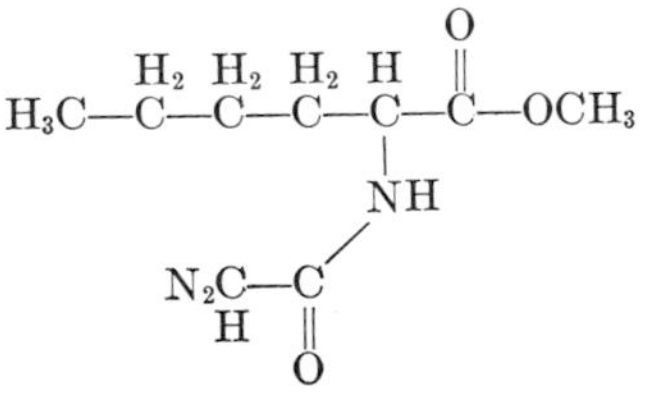

```
                         O                      O
     H2 H2 H2 H   ||               H2 H   ||
H3C—C—C—C—C—C—OCH3          Ph—C—C—C—OCH3
              |                       |
              NH                      NH
             /                        |
      N2C—C                     N2C—C
       H  ||                     H    \\
          O                            O
     (XXXIX)                     (XL)
```

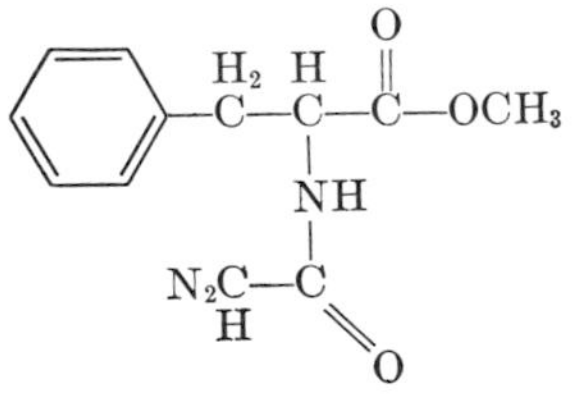

L-phenylalanine methyl ester (XL) on pepsin after which the reagent was located attached to a peptide with a composition (Asp, 2Thr, Ser, Gly, Val, and Ile) which could contain the above tetrapeptide (*136*).

The importance of cupric ion in the stoichiometric inhibition of pepsin by a diazo compound was indicated by the results reported in one of the initial studies of this kind, that is, the action of *N*-diazoacetyl norleucine methyl ester (XXXIX) (*135*). Without copper, the inhibition was slower and less selective. The essentiality of copper may depend on the reagent used. In two cases, the accelerative effect of copper was noted, but the slower action observed in its absence was still selective (*133, 134*).

In further studies with diazoacetyl norleucine methyl ester, preincu-

132. G. R. Delpierre and J. S. Fruton, *Proc. Natl. Acad. Sci. U. S.* **56,** 1817 (1966).
133. G. A. Hamilton, J. Spona, and L. D. Crowell, *BBRC* **26,** 193 (1967).
134. E. B. Ong and G. E. Perlmann, *Nature* **215,** 1492 (1967).
135. T. G. Rajagopalan, W. H. Stein, and S. Moore, *JBC* **241,** 4295 (1966).
136. R. S. Bayliss and J. R. Knowles, *Chem. Commun.* **4,** 196 (1968).
137. K. T. Fry, O. K. Kim, J. Spona, and G. A. Hamilton, *BBRC* **30,** 489 (1968).

bation with copper increased the rate of inactivation and led to the proposal that the active agent was actually a cupric complex attracted by charge interaction. Since the active center region of pepsin contains at least two carboxyl groups as described at the beginning of this section, the ionized carboxyl was visualized as orienting the cupric complex for eventual alkylation of the nonionized carboxyl group (*131*). An extension of this viewpoint led to the conclusion that probably the diazoketone and diazoacetyl esters might not be active-site directed but instead represent a class of copper complexes positioned favorably at the active center of pepsin (*131*). In support of this interesting idea was the observation that the presence of aromatic rings in the foregoing reagents which appear to provide substratelike affinity for pepsin was actually nonessential since the simple diazoacetyl derivative without aromatic ring (XLI) was a selective inactivator in the presence of copper (*131*).

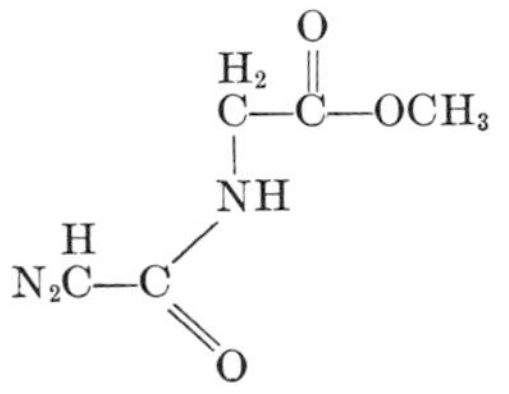

$$N_2CH{-}C(=O){-}NH{-}CH_2{-}C(=O){-}OCH_3$$

(XLI)

As pointed out above, copper has not always been found to be essential for specific modification. There may be two mechanisms available, and substratelike features may be important in both. Thus, the copper-catalyzed inactivation of pepsin by 1-diazo-4-phenyl-3-tosylamido butanone is stereospecific for the L isomer (*132*). In studies without copper, selectivity increased as a substratelike structure was more nearly approached, that is, with 1-diazo-4-phenyl-3-benzyloxycarbonylamido butanone (XXXVII) (*134*), in comparison to diphenyldiazomethane (*125*) or 2-diazocyclohexanone (*133*).

In other respects, the action of these agents with or without copper is substratelike requiring active enzyme and not being observable with pepsinogen (*134, 135*). The pH optimum of the inactivation by the diazoketone (*134*) occurs in the region of 5.3–5.6, well above that for pepsin action, and has been related to the probable esterification of a weakly acidic carboxyl group, but this aspect of the modification is less clear.

In general, evidence for the ester bond has been alkali lability (*136,*

137) or removal of the reagent by hydroxylamine (*129, 130, 137, 138*) leaving the modified carboxyl group of the protein as a hydroxamic acid useful as a label for determining the site of modification (*129, 130*). Finally, in two cases as described above, an ester-containing peptide was isolated.

An acid protease from the mold *Penicillium janthinellum* is inactivated by diazoacetyl-D,L-norleucine methyl ester in the presence of cupric ion; hydroxylamine removes the label (*138*).

b. Thiol Proteases (Papain, Ficin, Stem Bromelain, and Others). A group of very active proteolytic enzymes of vegetable origin are thiol enzymes and typically have broad specificity. Since the thiol group is exposed and often more reactive even than a model SH group such as that in cysteine, chemical modification studies generally result in action at this site. The use of site-directed alkylating agents, therefore, which revealed the function of histidine in serine proteinases as discussed above, or of tyrosine in carboxypeptidase B, would be expected to form thioether in the case of thiol proteinases much as would iodoacetate, although other side chains are undoubtedly also essential. This was particularly to be expected since the SH group functions in these enzymes as the site of thioester formation (*139*) during catalysis, and if TPCK (II) and TLCK (XIV) form substratelike complexes with this class of enzymes, their alkylating potential would be positioned very close to the mercapto group.

As expected, ficin (*140*), stem bromelain, (*141*), and papain (*142–144*) were inactivated by TPCK and TLCK with thioether formation. In the case of TLCK the rate of inactivation of stem bromelain (*141*) or of papain (*142*) was considerably greater than thioether formation with cysteine as a model. In the case of papain, the second-order rate constant was 10^6 higher for the enzyme (*142*) probably owing both to prior complex formation and to the hyper-reactive nature of this SH group relative to that of cysteine. A kinetic study with benzoylarginine ethyl ester as a substrate showed that TLCK did form a complex with papain (*143*); unexpectedly, TPCK was found to inactivate without competition and therefore was not active-site directed.

It is of interest that in crystalline papain, the ϵ-amino group of TLCK

138. J. Sodek and T. Hofmann, *JBC* **243,** 450 (1968).
139. G. Lowe and A. Williams, *BJ* **96,** 199 (1965).
140. M. J. Stein and J. E. Liener, *BBRC* **26,** 376 (1967).
141. T. Murachi and K. Kato, *J. Biochem. (Tokyo)* **62,** 627 (1967).
142. J. R. Whitaker and J. Perez-Villaseñor, *ABB* **124,** 70 (1968).
143. B. C. Wolthers, *FEBS Letters* **2,** 143 (1969).
144. M. L. Bender and L. J. Brubacher, *JACS* **88,** 5880 (1966).

is found near an aspartic acid β-carboxylate group which may therefore represent a specificity site side chain (*145*) when basic substrates are acted upon. The role of a carboxyl group had been deduced from the pH dependence of deacylation (*146–148*), and one interpretation suggested a basic side chain orienting function with a consequent stimulation of deacylation (*146*) for which the primary catalytic group is histidine (*149*).

An active center histidine residue in papain (*150*) was located through the use of a reagent capable of cross-linking after alkylation of the essential thiol group, namely, 1,3-dibromoacetone. In the case of papain, His 159 was identified by this procedure (*150*) and the spatial proximity to Cys 25 also found in the crystallographic studies (*145*). The same approach was successfully applied to ficin and to stem-bromelain (*151*).

Streptococcal proteinase is also a thiol enzyme with a highly reactive SH group (*152*) that presents a problem with respect to obtaining selective chemical modification at other sites. Reversible conversion to the *S*-sulfenylsulfonate derivative (XLII) blocked the SH group and pro-

$$\text{E—SH} \rightarrow \text{E—S—S—SO}_3^-$$

(XLII)

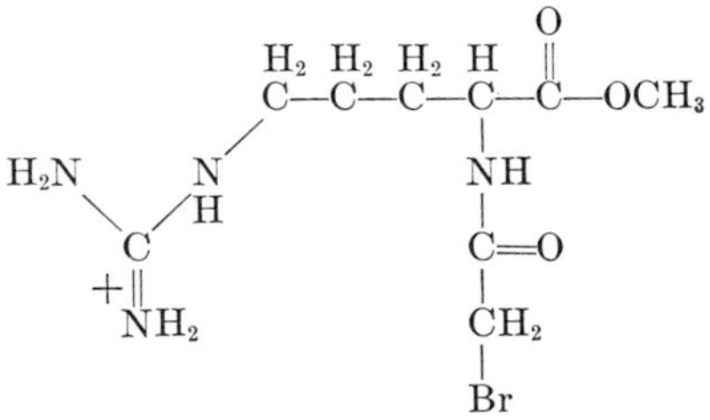

```
                                   O
                    H2 H2 H2 H     ‖
                    C——C——C——C——C——OCH3
                   /         |
      H2N       N            NH
         \     / H           |
          C                  C=O
         +‖                  |
          NH2                CH2
                             |
                             Br
```

(XLIII)

vided a negatively charged group that was considered of value in attracting the positively charged reagent (XLIII) to the active center region. Incubation of the enzyme derivative with the reagent did result in the alkylation of a single histidine residue at N-3 of its imidazole

145. J. Drenth, J. N. Jansonius, R. Koekoek, H. M. Swen, and B. G. Wolthers, *Nature* **218,** 929 (1968); "Structure-Function Relationships of Proteolytic Enzymes" (P. Desnuelle, H. Neurath, and M. Ottesen, eds.), pp. 272ff. Munksgaard, Copenhagen, 1970 (in press).

146. G. Lowe and A. Williams, *BJ* **96,** 194 (1965).

147. J. R. Whitaker and M. L. Bender, *JACS* **87,** 2728 (1965).

148. L. A. Sluyterman, *BBA* **85,** 305 (1964).

149. S. S. Husain and G. Lowe, *Chem. Commun.* p. 345 (1965); *BJ* **108,** 855 (1968).

150. S. S. Husain and G. Lowe, *BJ* **108,** 861 (1968).

151. S. S. Husain and G. Lowe, *BJ* **110,** 53 (1968).

152. B. I. Gerwin, *JBC* **242,** 451 (1967).

ring (*153*). After restoration of the thiol group, the enzyme was found to be inactive.

2. *Esterases*

a. Acetylcholinesterase and Related Esterases. Acetylcholinesterase has been the object of much study with respect to specific inhibitors which covalently combine at its active center serine and which form the basis of many studies relating reagent structure to inhibition. Because of the rarity of pure enzyme, relatively little protein chemistry has been carried out and emphasis has remained on kinetic studies.

Cholinesterase inhibitors which form phosphoryl, phosphonyl, sulfonyl, or carbamoyl enzymes have been widely studied and it has now been established that in many cases a reversible substratelike complex can form prior to covalent attachment. Recognition of this has eventually guided synthetic efforts in attempts to achieve desirable selectivity or to explore the catalytic process.

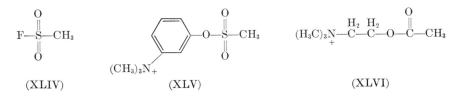

$$F{-}S(=O)_2{-}CH_3 \quad \text{(XLIV)}$$

$$(CH_3)_3\overset{+}{N}{-}C_6H_4{-}O{-}S(=O)_2{-}CH_3 \quad \text{(XLV)}$$

$$(H_3C)_3\overset{+}{N}{-}CH_2{-}CH_2{-}O{-}C(=O){-}CH_3 \quad \text{(XLVI)}$$

Kitz and Wilson considered the possibility of reversible complexes as a step in the formation of methanesulfonyl-acetylcholinesterase by methanesulfonyl fluoride (XLIV) or by methanesulfonate esters and derived a graphical procedure for kinetic analysis of the inactivation process (*82*). As indicated earlier in this chapter, this method should be of general value in evaluating active-site-directed agents for saturation kinetics. In the case of methanesulfonyl fluoride (XLIV) and other sulfonyl fluorides (*44, 82*) a complex could not be detected. Methanesulfonate esters, however, capable of occupying the enzymic binding site for the alcohol portion of acetylcholine (XLVI) such as (XLV) gave evidence of complex formation with $K_i = 4 \times 10^{-4}\,M$ (*82*). Similar evidence was provided for a number of quaternary quinolinium methanesulfonates that irreversibly inactivate acetylcholinesterase (*154*). However, others in spite of excellent affinity (K_i near $10^{-6}\,M$), were incapable of transferring the methanesulfonyl group to the enzyme and remained competitive inhibitors only. It is interesting that the rate of

153. T. Y. Liu, *JBC* **242**, 4029 (1967).
154. J. F. Ryan, S. Ginsburg, and R. J. Kitz, *Biochem. Pharmacol.* **18**, 269 (1969).

modification by methanesulfonyl fluoride was accelerated by tetraethylammonium ion over a thousandfold (*155*).

The organophosphates occupy a unique position historically as inhibitors of acetylcholinesterase (as well as a variety of other enzymes with an active center serine). The significance of this chemical modification with respect to the mechanism of acetylcholinesterase was appreciated by Wilson (*156, 157*). A kinetic examination of the inhibition of human serum cholinesterase by DFP and by malooxon [*O,O*-dimethyl-*S*-(1,2-dicarboethoxyethyl)phosphorothiolate] was made by Main (*19*) and initially only the latter agent provided evidence of complex formation. With improved methods for rapid sampling, it was later shown that DFP combined reversibly with both human serum cholinesterase and bovine erythrocyte acetylcholinesterase; the increased affinity for the former enzyme was then shown to account for the greater effectiveness of DFP for that enzyme, amounting to a second-order rate constant 150-fold higher than for the bovine erythrocyte form (*20*).

The possibility of developing selective inhibitors of hydrolytic enzymes by utilizing substratelike features was explored by Becker with a series of phosphonates (*158, 159*). By suitable alterations he was able to arrive at phosphonates more effective against trypsin than against chymotrypsin or acetylcholinesterase; in addition he was able to diminish anticholinesterase activity relative to other inhibitory effects (proteolytic enzymes).

Finally, it should be noted that certain carbamates, for example, neostigmine (XLVII), are effective inhibitors of acetylcholinesterase that

$(H_3C)_3N^+$–C_6H_4–O–C(=O)–N(CH_3)$_2$

(XLVII)

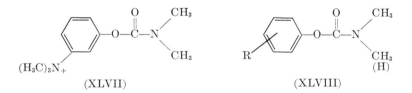

R–C_6H_4–O–C(=O)–N(CH_3)(CH_3 or H)

(XLVIII)

function by carbamylation of the active center serine (*160*) followed by slow hydrolysis. The relation of structure to enzyme carbamylation in a series of mono- and dimethyl carbamates of the general formula

155. R. Kitz and I. B. Wilson, *JBC* **283,** 745 (1963).
156. I. B. Wilson, F. Bergmann, and O. Nachmansohn, *JBC* **186,** 781 (1950).
157. I. B. Wilson, *JBC* **190,** 111 (1951).
158. E. L. Becker, T. R. Fukuto, B. Boone, D. C. Conham, and E. Boger, *Biochemistry* **2,** 72 (1963).
159. E. L. Becker, *BBA* **147,** 289 (1967).
160. I. B. Wilson, M. A. Hatch, and S. Ginsburg, *JBC* **235,** 2312 (1960).

(XLVIII) has been analyzed by O'Brien and his colleagues (*84, 161*) who established the importance of complex formation in the kinetics of the inhibition. Similar data have been provided for physostigmine (eserine) (*162*).

The reagents described above apparently follow a substratelike pathway with covalent modification of the active center serine. Other reagents probably acting elsewhere are beginning to receive attention. *p*-Trimethylammonio-benzenediazonium fluoborate (XLIX) is considered to be an active-site-directed agent for acetylcholinesterase based on the known competitive inhibition by the phenyltrimethylammonium ion

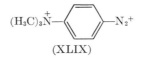

$$(H_3C)_3\overset{+}{N}-C_6H_4-N_2^+$$

(XLIX)

(*163*). A progressive inhibition of the enzyme was observed at pH 6 reaching 86% after 30 min with enzyme and inhibitor concentrations at 10^{-7} and 5×10^{-6} *M*, respectively. The inhibition was considerably less in the presence of a competitive inhibitor (*163*). Although the reagent may complex at the anionic site of the enzyme, the coupling site would have to be some distance away because of geometry and is not likely to shed much light on the anionic site as hoped. The complexity of the modification is indicated by a recovery and enhancement, after initial inhibition, of the ability to hydrolyze a neutral substrate, indophenyl acetate. However, the degree of inhibition toward acetylcholine as a substrate remained unchanged (*164*).

Another reagent with the potential of modifying the anionic site or specificity site of acetylcholinesterase is the ethylenimine quaternary salt (LI) that forms immediately when the chloro compound (L) is dissolved in water (*165*). In this structure, the potential for covalent binding resides on either of the carbons of the three-membered ring, thus giving a more favorable geometry for labeling the specificity site than that of the diazonium reagent above. Acetylcholinesterase (beef erythrocytes) was slowly inactivated by the reagent which behaved as a noncompetitive inhibitor (*165*). The L form was more effective than the D. Evidence for the formation of an ester bond during the inactivation

161. R. D. O'Brien, B. D. Hilton, and L. Gilmour, *Mol. Pharmacol.* **2**, 593 (1966).
162. A. R. Main and F. L. Hastings, *Science* **154**, 400 (1966).
163. L. Wofsy and D. Michaeli, *Proc. Natl. Acad. Sci. U. S.* **58**, 2296 (1967).
164. J. C. Meunier and J. P. Changeux, *FEBS Letters* **2**, 224 (1969).
165. B. Belleau and H. Tani, *Mol. Pharmacol.* **2**, 411 (1966).

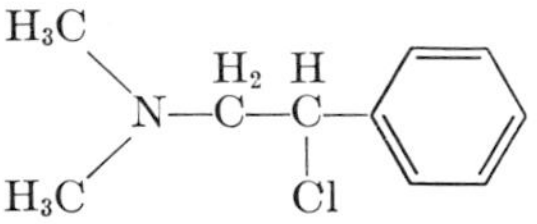

$$(H_3C)_2N\text{—}CH_2\text{—}CH(Cl)\text{—}C_6H_5 \qquad \text{(L)}$$

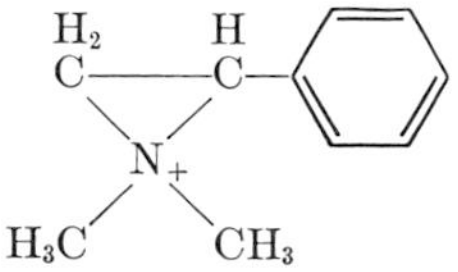

$$\text{(LI)}$$

was not obtainable. It appeared that the specificity of acetylcholinesterase was changed by the modification since esterase action on indophenyl acetate doubled and was sensitive to organophosphate (*166*).

This differential effect of alkylation on the cleavage of various acetylcholinesterase substrates prompted a more extended study (*167*) in which the more reactive ethyleneimine was used (LII). As in the earlier

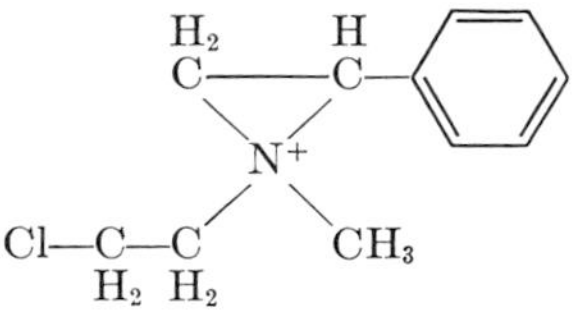

$$\text{(LII)}$$

work, the alkylated enzyme lost hydrolytic activity toward acetylcholine and became more active towards indophenol acetate. It was concluded that $V_{\max}$ was greatly diminished whereas K_m was relatively unaffected in the case of acetylcholine. Examination of the susceptibility of the modified enzyme to acetylcholinesterase inhibitors of the carbamate and organophosphorus classes and other considerations led to the proposal that three binding sites were available in this enzyme, that is, the anionic site and possibly two others (*167*).

3. *Deaminases—Adenosine Deaminase*

The competitive inhibition of calf intestinal adenosine deaminase by 9-substituted adenines in which the ribose of the normal substrate was replaced by phenylalkyl side chains (*168*) provided the basis of attempts to obtain active-site-directed alkylating agents for this enzyme. Incorporation of a bromoacetamido side chain in the *o* or *p* position of 9-benzyladenine [*p* isomer (LIII)] (*169*) or in the *m* position of 9-phenylethyl-

166. J. E. Purdie and R. A. McIvor, *BBA* **128,** 590 (1966).
167. R. D. O'Brien, *BJ* **113,** 713 (1969).
168. H. J. Schaeffer, D. Vogel, and R. Vince, *J. Med. Chem.* **8,** 502 (1965).
169. H. J. Schaeffer, M. A. Schwartz, and E. Odin, *J. Med. Chem.* **10,** 686 (1967).

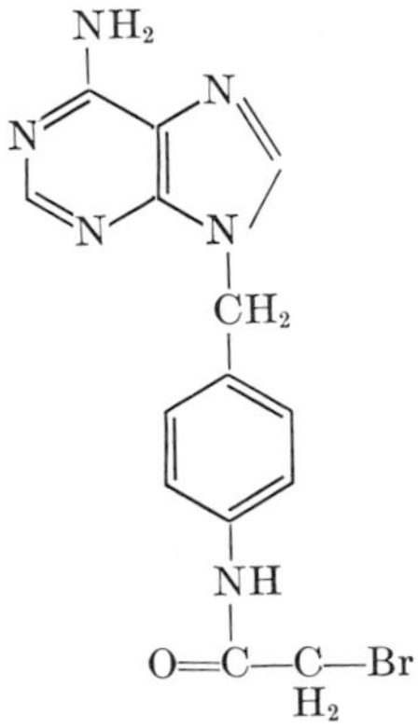

(LIII) 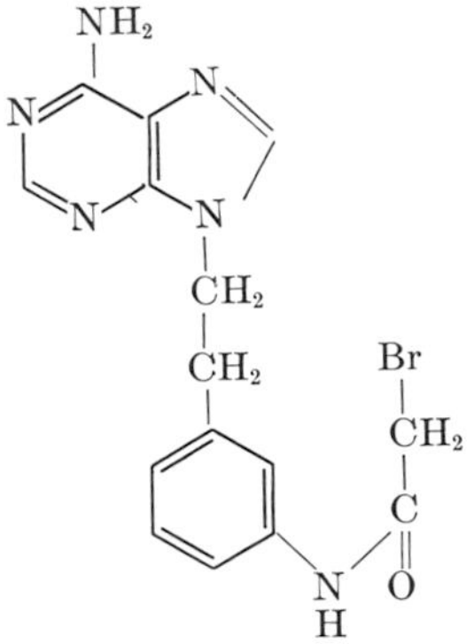 (LIV)

adenine (LIV) (*170*) gave agents which alkylate the enzyme by saturation kinetics with $K_i = 10^{-4}$ to $10^{-5}\,M$, depending on structure. The sites of alkylation appear to be different since adenosine deaminase modified by the meta-substituted compound retains 20% of its original activity in contrast to the *o* and *p* isomers of (*170*). Iodoacetamide was without effect.

4. *Glycosidases*

a. β-Glucosidase A_3. β-Glucosidase A_3, purified from *Aspergillus mentii,* has been reported to be inactivated by an epoxide (LV) related in structure to its substrate (*171*). A stoichiometric incorporation was

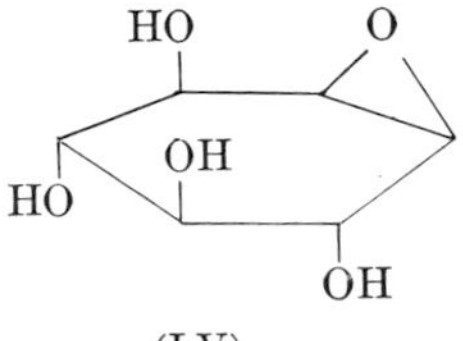

(LV)

established with radioactive inhibitor (one mole of inhibitor per 170,000 g, the molecular weight of the enzyme as established by gel filtration and osmotic measurements). The inhibitor was recovered as (+)-inositol after the action of hydroxylamine suggesting an ester bond with the enzyme (*171*). The inactivation was judged to involve prior complexing

170. H. J. Schaeffer and R. N. Johnson, *J. Med. Chem.* **11,** 21 (1968).
171. G. Legler, *BBA* **151,** 728 (1968).

of the inhibitor, for which K_i near 4 mM was determined, and to depend on an acid group with pK of 6.1 (*172*).

b. Lysozyme. In the hope of obtaining a specific irreversible inhibitor of lysozyme, Thomas *et al.* (*173*) used oligomers of *N*-acetyl-D-glucosamine, with from 1 to 3 hexosamine units, containing a propylene epoxide aglycone (LVI). The epoxide group offers a versatile potential for protein side chain modification. The trimer of acetylglucosamine is not a lysozyme substrate but its position in the enzyme–trimer complex is known relative to the deduced catalytic site (*174*). In the reagent (LVI), the β-epoxypropyl aglycone was expected to reach that region of the active center where catalysis occurs, so-called subsite D (*174*) and possibly combine with carboxyl groups imputed in the mechanism. The mono- and diacetylglucosamine β-glycosides could be expected to complex lysozyme in the same position but with less affinity.

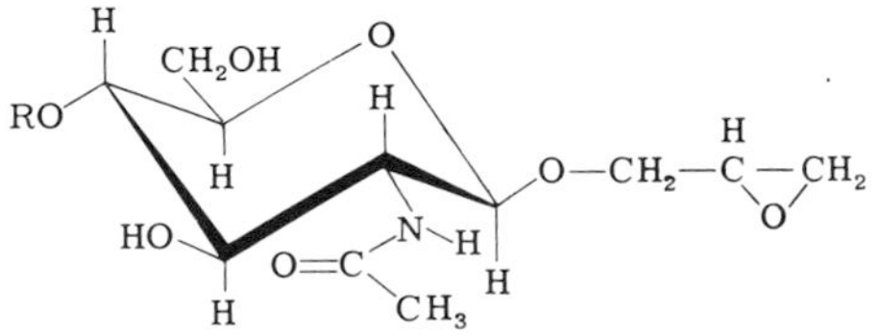

(I) R = H
(II) R = GlcNAc
(III) R = GlcNAc-β-(1→4)-GlcNAc

(LVI)

The epoxypropyl glycoside of the dimer and trimer were found to irreversibly inactivate lysozyme on incubation at pH 5.5. Propylene oxide had no effect even at high concentration. The stoichiometry was established as equimolar using radioactive dimer. The β-methylglycoside of the dimer, a known competitive inhibitor of lysozyme, blocked the action of the epoxypropyl glycoside. The site of attachment is not yet known and is not self-evident from epoxide chemistry.

Many chemical modification studies of lysozyme have been carried out by group specific reagents and are cited elsewhere (*175*).

172. G. Legler, *Z. Physiol. Chem.* **349,** 767 (1968).

173. E. W. Thomas, J. F. McKelvy, and N. Sharon, *Nature* **222,** 485 (1969).

174. C. C. F. Blake, L. N. Johnson, G. A. Mair, A. C. T. North, D. C. Phillips, and V. R. Sarma, *Proc. Roy. Soc.* **B167,** 378 (1967).

175. S. M. Parsons, L. Jao, F. W. Dahlquist, C. L. Borders, Jr., T. Graff, J. Racs, and M. A. Raftery, *Biochemistry* **8,** 700 (1969).

5. *Nucleases*

a. Exoribonuclease (Tumor Nuclei). Several nucleotide derivatives have been described as irreversible inactivators of an exoribonuclease either purified from Ehrlich ascites tumor or observed with isolated nuclei of HeLa cells (*176*). Thymidylic acid 3′-*p*-iodoacetylphenyl ester (LVII) is typical of the group which include the 3′,5′-diester as well as

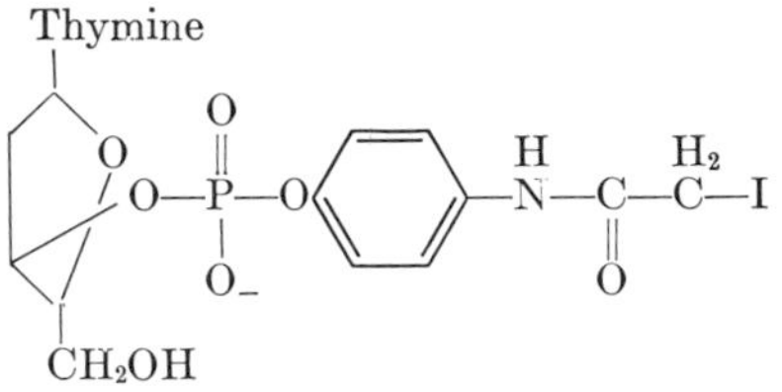

(LVII) 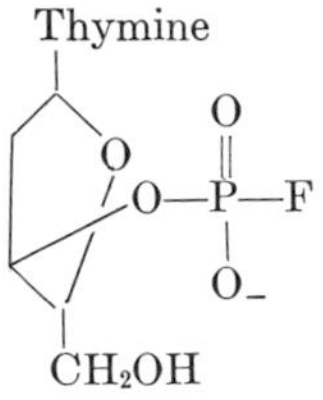 (LVIII)

a dinucleotide diester. The purified exoribonuclease from Ehrlich ascites tumor progressively lost activity on exposure to such derivatives at concentrations (10^{-4} to 10^{-5} *M*) at which iodoacetamide itself was innocuous (*176*).

Thymidine-3′-fluorophosphate (LVIII) and the 5′ isomer were also studied. The former partially inhibited the exoribonuclease of HeLa cell nuclei (at the relatively high concentration of 5 m*M*) without effect on RNA polymerase (*176*). Some fine examples of specific alkylation of staphylococcal nuclease employing bromoacetamidophenyl esters of thymidine 3′,5′-diphosphate have recently been described (*176a*). The 3′-ester 5′-phosphate alkylated Lys 48, Lys 49, and Tyr 115 of the nuclease. The inhibition was stoichiometric and only one of these residues was alkylated in a given nuclease molecule. The residues are considered to be in the substrate binding region.

6. *Amidases*

a. L-*Asparaginase ECII.* L-Asparaginase ECII (*E. coli*) is inactivated by 5-diazo-4-oxo-L-norvaline (DONV) (LIX) an analog of asparagine (LXI) (*177*). Guinea pig serum asparaginase is also affected. Since the bacterial enzyme acts on DONV as a substrate, converting it to the

176. M. B. Sporn, D. M. Berkowitz, R. P. Glinski, A. B. Ash, and C. L. Stevens, *Science* **164,** 1408 (1969).

176a. P. Cuatrecasas, M. Wilchek, and C. B. Anfinsen, *JBC* **244,** 4316 (1969).

177. R. E. Handschumacher, C. J. Bates, P. K. Chang, A. T. Andrews, and G. A. Fischer, *Science* **161,** 62 (1968).

hydroxy acid, the material may be destroyed before it can exert much of its alkylating action which is slow by comparison. Dimethylsulfoxide at 7 M concentration slows down the catalytic action of asparaginase and

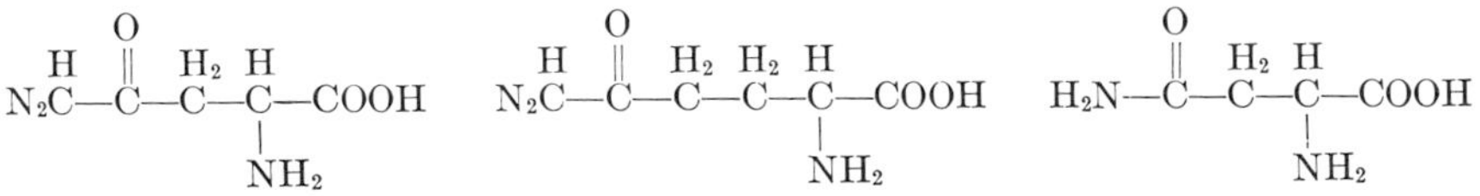

$$N_2\overset{H}{C}\text{—}\overset{\overset{O}{\|}}{C}\text{—}\overset{H_2}{C}\text{—}\underset{NH_2}{\overset{H}{C}}\text{—}COOH$$

DONV (LIX)

$$N_2\overset{H}{C}\text{—}\overset{\overset{O}{\|}}{C}\text{—}\overset{H_2}{C}\text{—}\overset{H_2}{C}\text{—}\underset{NH_2}{\overset{H}{C}}\text{—}COOH$$

DON (LX)

$$H_2N\text{—}\overset{\overset{O}{\|}}{C}\text{—}\overset{H_2}{C}\text{—}\underset{NH_2}{\overset{H}{C}}\text{—}COOH$$

(LXI)

permits inactivation by DONV. The result is not due to unfolding by dimethyl sulfoxide since urea does not promote the labeling. The stereospecificity of the inhibition (only L isomer effective) also suggests affinity labeling as does the fact that DON (LX), the next higher homolog, although chemically similar, is inactive (*178, 179*).

b. Glutaminase. Glutaminase from *E. coli* has been purified and found to act on DON, 6-diazo-5-keto-norleucine with the formation of methanol and glutamic acid, an unusual cleavage reaction (*180*). With adequate amounts of DON, an inactivation can be observed which proceeds to completion, resulting in stoichiometric incorporation of the agent at two catalytic sites per mole of enzyme (MW 110,000). The reaction is inhibited by glutamine and is not observed above pH 6, a range in which substrate binding is also negligible. Ordinary sulfhydryl reagents have no effect on glutaminase, therefore the site of alkylation may be at another amino acid side chain (in contrast to formylglycinamide ribonucleotide aminotransferase (*13, 14*). Glutaminase is not affected by DONV (LIX), the shorter analog (*177*) which inhibits asparaginase as discussed above, showing the importance of close structural resemblance of reagent to substrate in each of the above enzymes.

B. Transferases (Ligases)

1. *Amidotransferases*

a. Formylglycinamide Ribotide Amidotransferase. During the work by Buchanan and his colleagues on the elucidation of the biosynthetic pathway leading to purines (*181*), the inhibitory effect of the antibiotic

178. R. C. Jackson, D. A. Cooney, and R. E. Handschumacher, *Federation Proc.* **28**, 601 (1969).
179. H. A. Campbell and L. T. Mashburn, *Biochemistry* **8**, 3768 (1969).
180. S. C. Hartman, *JBC* **243**, 853 (1968).
181. J. M. Buchanan and S. C. Hartman, *Advan. Enzymol.* **21**, 199 (1959).

azaserine (LXIII) on purine biosynthesis was examined. Two enzymes in the pathway were found to be affected, both of which use glutamine

$$H_2N{-}\overset{\overset{O}{\|}}{C}{-}\overset{H_2}{C}{-}\overset{H_2}{C}{-}\underset{NH_2}{\overset{H}{C}}{-}COOH \qquad\qquad N_2\overset{H}{C}{-}\overset{\overset{O}{\|}}{C}{-}O{-}\overset{H_2}{C}{-}\underset{NH_2}{\overset{H}{C}}{-}COOH$$

(LXII) (LXIII)

(LXII) as an amino donor, namely, phosphoribosyl pyrophosphate amidotransferase and formylglycinamide ribotide amidotransferase. The latter enzyme was the more susceptible and the *in vivo* effect of the antibiotic appears to be due to an inhibition of it. Of unusual interest at that time was the observation that azoserine and the structurally related diazoketone DON (LX), 6-diazo-5-keto-L-norleucine, were irreversibly inhibiting formylglycinamide ribotide amidotransferase, an effect made evident by preincubation of the enzyme with either azaserine or DON (*13*). Irreversible attachment to the enzyme took place. Under normal assay conditions of simultaneous substrate and inhibitor addition, azaserine appeared to be a competitive inhibitor ($K_i = 3.4 \times 10^{-5}\,M$) of glutamine ($K_m$ $6.2 \times 10^{-4}\,M$). Only the L isomer was effective. These antibiotics are clearly naturally occurring active-site-directed reagents and were recognized to be acting as such (*13, 14*).

The site of covalent attachment of azaserine to formylglycinamide ribotide amidotransferase was shown to be the thiol group of a particular half-cysteine residue in the glutamine binding region by isolation and characterization of the inhibitor (R)-bound peptide, Ala—Leu—Gly—Val—$\underset{\text{R}}{\text{Cys}}$—COOH (*182, 183*), in which a thioether bond was demonstrated in the COOH-terminal half-cysteine residue.

Formylglycinamide ribotide (FGAR) amidotransferase catalyzes the formation of formylglycinamidine ribotide (FGAM). By gel filtration,

$$\text{L-Glutamine} + \text{FGAR} + \text{ATP} + H_2O \rightarrow \text{L-glutamate} + \text{FGAM} + \text{ADP} + P_i$$

complexes of the enzyme containing glutamine (*184*) or formylglycinamide ribotide plus ATP (*185*) can be isolated. In agreement with the above discussion of the mechanism of action of azaserine is the observation that it blocks the formation of the first complex not the second

182. T. C. French, I. B. Dawid, R. A. Day, and J. M. Buchanan, *JBC* **283,** 2171 (1963).

183. I. B. Dawid, T. C. French, and J. M. Buchanan, *JBC* **238,** 2178 (1963).

184. K. Mizobuchi and J. M. Buchanan, *JBC* **243,** 4853 (1968).

185. K. Mizobuchi, G. L. Kenyon, and J. M. Buchanan, *JBC* **243,** 4863 (1968).

(*184*). The sulfhydryl group alkylated is considered the site of normal γ-glutamyl enzyme formation during utilization of glutamine as a nitrogen source. The azaserine-labeled enzyme is not inhibited when ammonia itself is used as an amino donor.

This classic work on the specific role of azaserine and DON as glutamine analogs which alkylate the binding site for glutamine has resulted in their useful application with other enzymes cited elsewhere in this chapter and earlier (*186*).

b. Phosphoribosyl Pyrophosphate Amidotransferase. The initial observations of the action of azaserine and DON as glutamine analogs which irreversibly inhibit enzymes using glutamine as amino donors were made with this enzyme and formylglycinamide ribonucleotide amidotransferase (*13, 14*). With purified phosphoribosyl pyrophosphate amidotransferase from chicken liver, the properties of azaserine and DON were reexamined (*187*). Competitive inhibition with glutamine was observed in the usual steady state assay; DON, with a $K_i = 1.9 \pm 0.04 \times 10^{-5}\, M$, was found to be about two orders of magnitude more tightly complexed than azaserine. The norvaline analog (LIX) had no effect. The rate of inactivation of the amidotransferase by DON was greatly accelerated by the cosubstrate, pyrophosphoryl ribose pyrophosphate, and by magnesium in

$$\text{Glutamine} + \text{PP ribose P} + H_2O \xrightarrow{Mg^{2+}} \text{5-phosphoribosylamine} + \text{glutamate} + \text{pyrophosphate}$$

combination. Under these conditions, saturation kinetics in the inactivation process were observable, indicating a binding constant for DON of $2.4 \pm 1 \times 10^{-5}\, M$ in fair agreement with the value measured from competitive inhibition in the rate assay.

The glutamine binding site apparently does not exist in the absence of phosphoribosyl pyrophosphate and magnesium, and under these conditions the action of DON is very slow because of diminished affinity (K_i is at least 100-fold higher) (*187*). All of these properties provide a strong case for the substratelike behavior of DON and increase the likelihood that the site of attachment in the enzyme is at the active center and possibly a catalytically important group.

The stoichiometry of the reaction indicated a single combination of the reagent with a protein of MW 200,000, in agreement with determination by physical methods.

c. DPN, Desamido Amidotransferase. Like the foregoing example,

186. A. Meister, "Enzymes," 2nd ed., Vol. 6, p. 247, 1962.
187. S. C. Hartman, *JBC* **238,** 3036 (1963).

this enzyme uses glutamine as a source of ammonia, in this case to amidify the carboxyl group of desamido-DPN. The transferase, purified from yeast, was found to be inactivated by azaserine as a glutamine analog only in the presence of the nicotinic acid dinucleotide (i.e., desamido-DPN) and ATP (*188*). These results, also as above, suggest the formation of the glutamine binding site after the attachment of the cosubstrates.

d. Carbamyl Phosphate Synthetase. The purified enzyme from *E. coli* catalyzes the reaction

$$\text{L-Glutamine} + CO_2 + 2\text{ATP} \rightarrow \text{Carbamyl phosphate} + \text{glutamate} + 2\text{ADP} + P_i$$

Ammonia at high concentration will serve as an alternate amino donor (*189*). The chloromethyl ketone (LXIV) was found to inactivate the

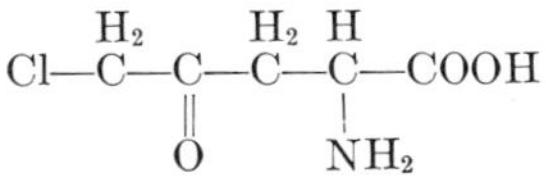

$$Cl{-}\overset{H_2}{C}{-}\underset{\underset{O}{\|}}{C}{-}\overset{H_2}{C}{-}\underset{\underset{NH_2}{|}}{\overset{H}{C}}{-}COOH$$

(LXIV)

$$ClC{-}\overset{H_2}{C}{-}\underset{\underset{O}{\|}}{C}{-}\overset{H_2}{C}{-}\overset{H_2}{C}{-}\underset{\underset{NH_2}{|}}{\overset{H}{C}}{-}COOH$$

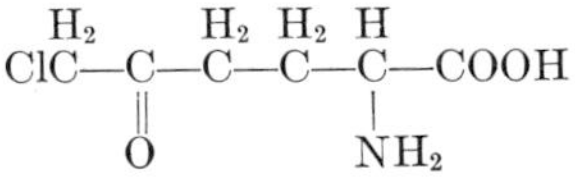

(LXV)

enzyme with respect to glutamine as an amino donor (but not ammonia); at $4 \times 10^{-4}\,M$, the action was complete in 10 min. The corresponding hexanoic acid (LXV) was less effective as were the diazo compounds, azaserine and DON.

2. *Acyltransferases*

a. Choline Acetylase. A preparation of this enzyme from placenta was found to be 95% inhibited by a 5 m*M* solution of the choline analog (LXVI) in 1 hr at 0° (*190*). Activity was not recovered on gel filtration.

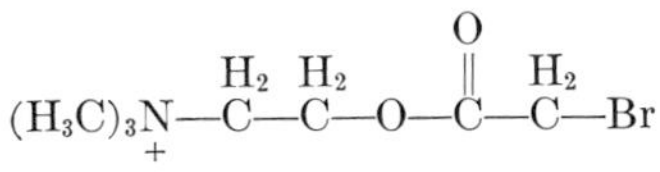

$$(H_3C)_3\overset{+}{N}{-}\overset{H_2}{C}{-}\overset{H_2}{C}{-}O{-}\overset{\overset{O}{\|}}{C}{-}\overset{H_2}{C}{-}Br$$

(LXVI)

Acetylcholinesterase (human erythrocytes) and pseudocholinesterase (human serum) were unaffected.

b. Carnitine, Acetyltransferase. The enzyme from pigeon breast muscle catalyzes the acetylation of the carnitine (LXVII) hydroxyl group using

188. J. Preiss and P. Handler, *JBC* **233,** 493 (1958).
189. E. Khedouri, P. M. Anderson, and A. Meister, *Biochemistry* **5,** 3552 (1966).
190. B. O. Persson, L. Larsson, J. Schumberth, and B. Sorbo, *Acta Chem. Scand.* **21,** 2283 (1967).

acetyl-CoA as a donor. In the hope of revealing the physiological role of the enzyme specific inhibitors were sought and either bromoacetyl carnitine (LXVIII) plus CoA or bromoacetyl-CoA plus carnitine were

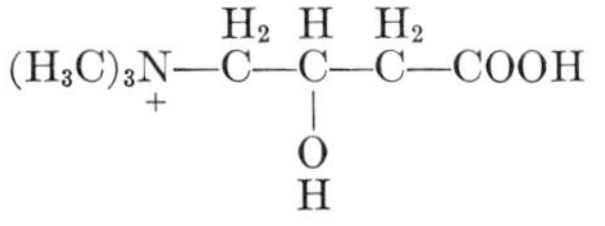

$$(H_3C)_3\overset{+}{N}-\overset{H_2}{C}-\underset{\underset{H}{|}}{\overset{H}{C}}(O)-\overset{H_2}{C}-COOH$$

(LXVII)

$$(H_3C)_3\overset{+}{N}-\overset{H_2}{C}-\overset{H}{C}(O-C(=O)-CH_2Br)-\overset{H_2}{C}-COOH$$

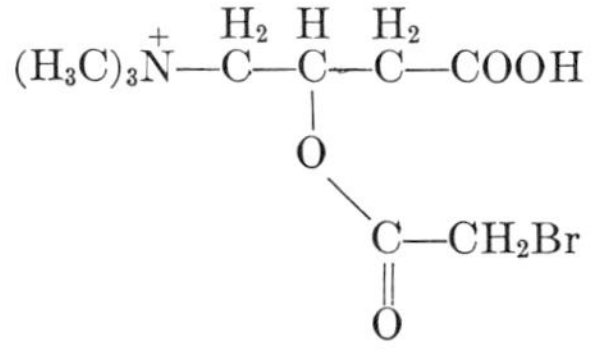

(LXVIII)

effective (*191*). Bromoacetyl carnitine alone slowly inactivates the transferase. However, in the presence of acetyl-CoA, a more rapid inactivation takes place as a result of the formation of a thioether between these two cosubstrates. This is not covalently bound to the enzyme but dissociates so slowly as to provide a moderately stable, inactive state.

3. *Adenylyltransferase*

a. Luciferase (Firefly). Luciferase (firefly) appears to have a hydrophobic binding site for its substrate luciferin (LXIX). The observed inactivation by TPCK (II), which does not obviously resemble luciferin,

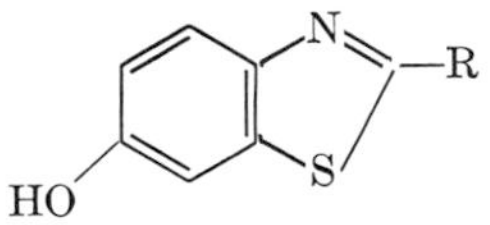

HO—(benzothiazole ring: N, S)—R (R = 4-carboxy-2-thiazolinyl)

(LXIX)

is apparently a result of complex formation with luciferase at this region of the enzyme followed by alkylation of two SH groups per mole of enzyme. For the reversible phase a $K_i = 25 \times 10^{-4}\ M$ was measured (*192*). Iodoacetamide was considerably less effective although more reactive to SH groups in general.

C. Oxidoreductases

NAD-linked dehydrogenases offer at least two sites for affinity labeling: that for the characteristic substrate as well as that for coenzyme binding. In addition there may be allosteric sites.

191. J. F. A. Chase and P. K. Tubbs, *BJ* **111,** 225 (1969).
192. R. Lee and W. D. McElroy, *Biochemistry* **8,** 130 (1969).

1. *Alcohol Dehydrogenase (Yeast)*

A homologous series of nicotinamide quaternary salts containing an alkylating group [(LXX) $n = 2$–4] were prepared as agents possibly capable of attachment to the NAD binding site in dehydrogenases (*193*). For yeast alcohol dehydrogenase, competitive inhibition with NAD was

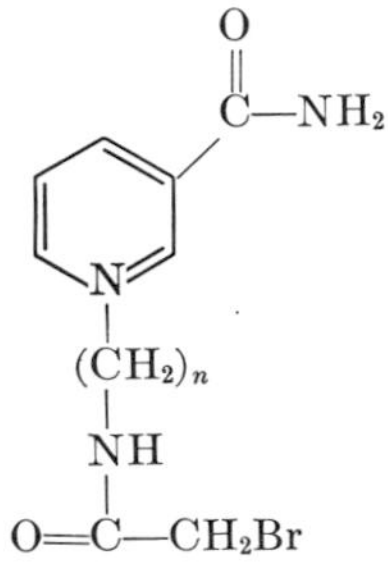

(LXX)

observed for (LXX) $n = 2$, with $K_i = 0.05\ M$. The pyridinium salt was also capable of inactivating the enzyme. ADP increased the affinity of the reagent but slowed the subsequent alkylation step. However, bromoacetamide was more effective than any of the pyridinium compounds. An essential sulfhydryl group is thought to be involved (*193*). Glutamic and lactic dehydrogenases were not affected.

2. *Lactic Dehydrogenase*

The enzyme from pig heart was observed to react with 3-bromoacetyl pyridine (LXXI) at a sulfhydryl group and a histidine residue. The mercuri-enzyme reacted only at histidine, and after removal of mercury, a diminished ability to bind NAD was observed (*194*).

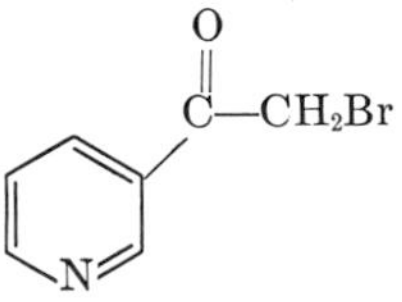

(LXXI) 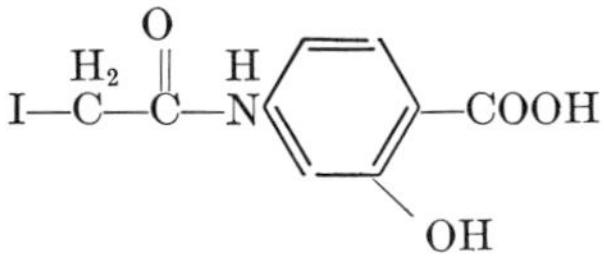 (LXXII)

193. B. V. Plapp, C. Woenckhaus, and G. Pfleiderer, *ABB* **128,** 360 (1968).
194. C. Woenckhaus, J. Berghauser, and G. Pfleiderer, *Z. Physiol. Chem.* **350,** 473 (1969).

3. *Glutamic Dehydrogenase*

In one of the earliest studies on active-site-directed reagents, Baker and his colleagues took advantage of the known competitive inhibition of glutamic dehydrogenase by isophthalic acid to explore benzenoid compounds as the base of active-site-directed inhibitors of glutamic dehydrogenase as well as of lactic dehydrogenase which was also shown to have affinity for aromatic acids [cf. Baker (*7*, Chapter 9)]. 4-Iodoacetamidosalicylic acid (LXXII) was one of the first agents so produced. It inactivated lactic dehydrogenase, but oxamate, a competitive inhibitor of lactate, protected the enzyme (*195*). In a more extended study the beef liver enzyme was found to form a complex with the reagent with K_m of 1.08×10^{-3} M and a rate of breakdown of 2.07×10^{-2} min^{-1}; the active sites reacted independently. Addition of the allosteric inhibitor GTP did not affect the rate of inactivation. Ketoglutaric acid blocked the inactivation and the desensitization of the enzyme to GTP in a way that indicated sequential response of the subunits (*196*).

Another effector site of glutamic dehydrogenase is that for steroids. A mercury derivative of estradiol has been synthesized and suggested for possible use in identifying the sterol binding site of this enzyme on the basis of estradiol protection against reagent uptake (*197*).

4. *Guanosine 5′-Phosphate Reductase*

Guanosine 5′-phosphate reductase forms inosinic acid from GMP using DPNH as the reducing agent. 6-Chloropurine nucleotide (LXXIII) inactivates the enzyme (from *A. aerogenes*) probably by thioether formation with a sulfhydryl group at or near the active center (*198*). The reaction was prevented by substrate but not by DPNH. Since complete inactivation was not demonstrated, the essentiality of the sulfhydryl group cannot be assessed. Similar observations were made with 6-mercaptopurine ribotide and thioguanylic acid whose action could be reversed by reduction and was therefore considered to involve formation of a disulfide bond with the enzyme.

5. *Inosinic Acid Dehydrogenase*

Inosinic acid dehydrogenase leads to the formation of xanthylic acid with NAD as a cofactor. 6-Chloropurine ribotide (LXXIII) and 6-mer-

195. B. R. Baker, W. W. Lee, and E. Tong, *J. Theoret. Biol.* **3**, 459 (1962).
196. A. D. B. Malcolm and C. K. Radda, *Nature* **219**, 947 (1968).
197. C. C. Chin and J. C. Warren, *JBC* **243**, 5056 (1968).
198. L. W. Brox and A. Hampton, *Biochemistry* **7**, 398 (1968).

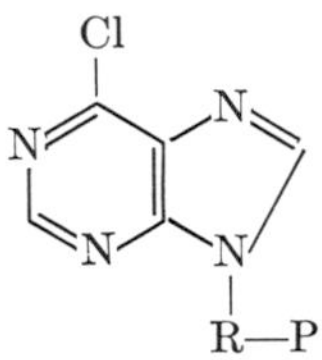

(LXXIII) 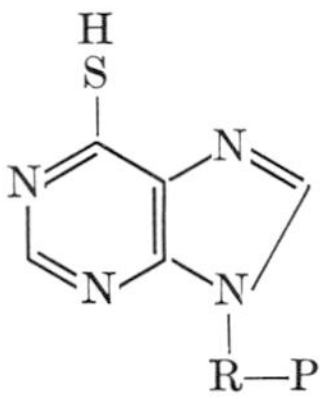 (LXXIV)

captopurine ribotide (LXXIV) are competitive inhibitors of inosinic acid. However, on preincubation with the enzymes they inactivate by thioether and disulfide formation with an SH group of the substrate binding site (*199*). Inosinic acid, but not NAD, protected the enzyme against these actions. The enzyme used was partially purified from *A. aerogenes*.

D. Isomerases—Triosephosphate Isomerase

The enzyme from rabbit muscle was observed to be completely inactivated by the iodo compound (LXXV), an alkylating agent similar in structure to the substrate, dihydroxyacetone phosphate, which pro-

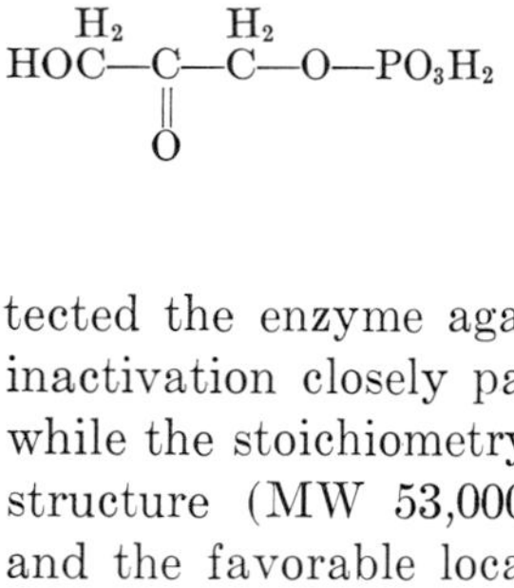

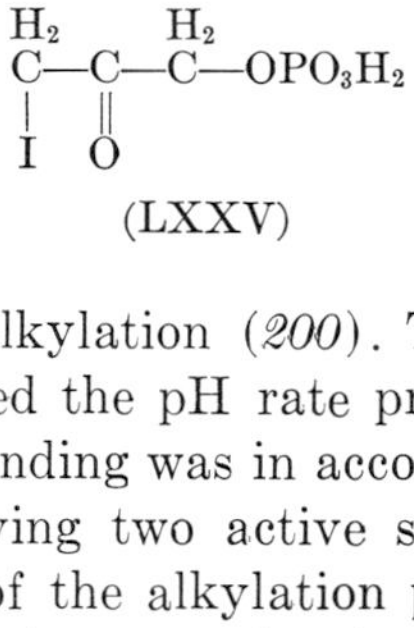

$HOCH_2—C(=O)—CH_2—O—PO_3H_2$, $ICH_2—C(=O)—CH_2—OPO_3H_2$ (LXXV) $CH_2(O)CH—CH_2—OPO_3H_2$ (epoxide) (LXXVI)

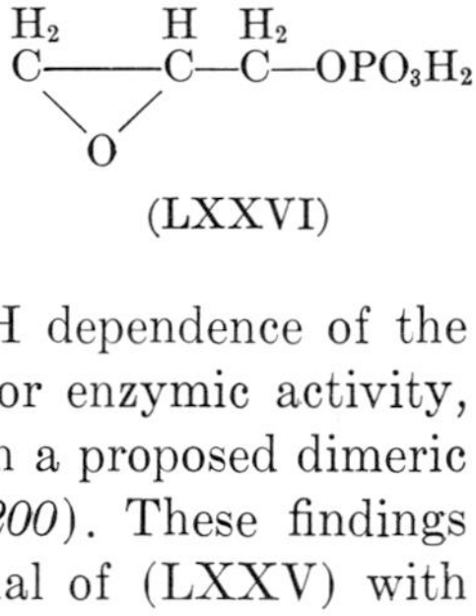

tected the enzyme against alkylation (*200*). The pH dependence of the inactivation closely paralleled the pH rate profile for enzymic activity, while the stoichiometry of binding was in accord with a proposed dimeric structure (MW 53,000) having two active sites (*200*). These findings and the favorable location of the alkylation potential of (LXXV) with respect to enzymic functional groups involved in the isomerase action are suggestive of active-site labeling.

Similar results have been obtained with the propylene oxide phosphate (LXXVI) (*201*). It is of interest that a somewhat similar phosphonic acid derivative (carbon to phosphorus bond) has been found to occur in nature and has antibacterial properties (*202*).

199. A. Hampton and A. Nomura, *Biochemistry* **6,** 679 (1967).
200. F. C. Hartman, *BBRC* **33,** 888 (1968).
201. I. A. Rose and E. L. O'Connell, *JBC* **244,** 6548 (1969).
202. B. G. Christensen, W. J. Leanza, T. R. Beathe, A. A. Patchett, B. H. Arison, R. E. Ormond, F. A. Kuell, Jr., G. Albers-Schonberg, and O. Jardetzky, *Science* **166,** 123 (1969).

E. Lyases. Enzymes Adding to Unsaturated Systems

1. *3-Deoxy-*D*-arabino-heptulosonate 7-Phosphate Synthase*

3-Deoxy-4-arabino-heptulosonate 7-phosphate synthase links phosphoenolpyruvate (LXXVII) and erythrose-4-phosphate together. The enzyme is inactivated by bromopyruvic acid (LXXVIII) acting at the phosphoenolpyruvate binding site (*203*) judging from protection kinetics. On the other hand, erythrose-4-phosphate and phenylalanine, an allosteric effector, have no influence on the bromopyruvic inactivation.

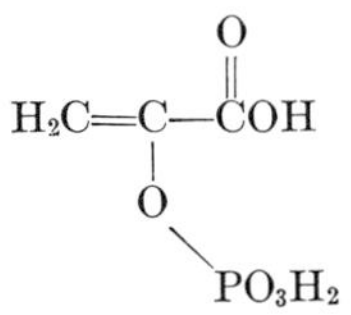

$$H_2C{=}C(OPO_3H_2){-}C({=}O)OH \qquad \text{(LXXVII)}$$

$$Br{-}CH_2{-}C({=}O){-}C({=}O){-}OH \qquad \text{(LXXVIII)}$$

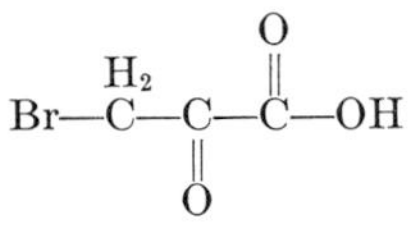

2. *2-Keto-3-deoxy-6-phosphogluconic Aldolase*

2-Keto-3-deoxy-6-phosphogluconic aldolase acts on pyruvic acid and D-glyceraldehyde-3-phosphate to form the gluconic acid derivative. Bromopyruvic acid (LXXVIII) causes inactivation apparently at the pyruvate binding site (*204*) with saturation kinetics (*85*). The inactivation results in the enzyme being alkylated by two moles of bromopyruvate, in agreement with the number of substrate binding sites (*85*).

3. *Carbonic Anhydrase*

A specific alkylation of human carbonic anhydrase B is possible with the normally nonspecific reagents iodoacetamide or bromoacetate (*52*) which form a reversible complex with the enzyme prior to inactivation. It is thus possible to obtain a modification of the enzyme in which alkylation at N-3 of a single histidine residue has occurred to provide a derivative which retains a small amount of activity (*51*, *205*). The susceptible histidine residue has an abnormally low pK (*205*) and has been isolated in a pentapeptide form (*206*). Bovine carbonic anhydrase B,

203. M. Staub and G. Denes, *BBA* **139,** 519 (1967).
204. H. P. Meloche, *BBRC* **18,** 277 (1965).
205. S. L. Bradbury, *JBC* **244,** 2002 (1969).
206. S. L. Bradbury, *JBC* **244,** 2010 (1969).

instead, was reported to be insensitive to bromoacetic acid and iodoacetamide (*207*).

The enzyme from both species was found to be susceptible to inactivation by reagents designed from known sulfonamide inhibitors of carbonic anhydrase, for example, (LXXIX) (*208*) and (LXXX) (*207*) among others. In the studies of Whitney *et al.* a number of compounds were

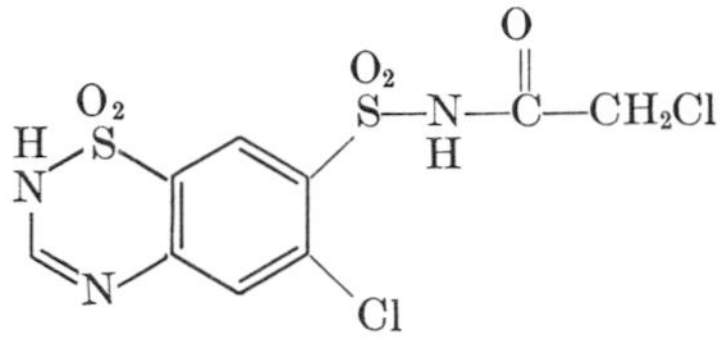

(LXXIX) 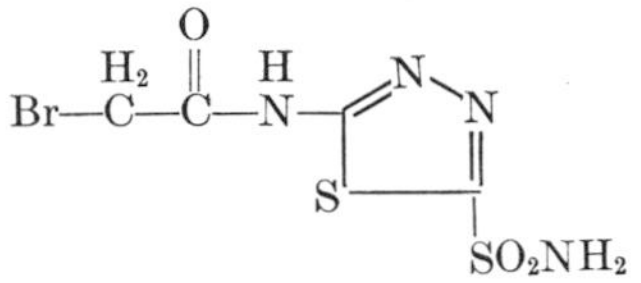 (LXXX)

examined and shown to be reversible inhibitors (*208*). Some of these caused a slow inactivation on incubation with the enzyme, leading eventually to complete inhibition with alkylation of N-3 of a single residue of histidine. The possibility has been considered that the histidine residues alkylated by directed reagents such as (LXXIX) and by iodoacetate may not be the same although both may be at, or near, the active site (*206*, *208*). In fact, considering the small molecular size of carbonic acid, it would indeed be remarkable if a reagent such as (LXXIX) could complex with this enzyme in such a way as to modify a functional group of the enzyme, if a sulfonamide group of the reagent occupies the substrate binding site.

4. *Fumarase*

The purified enzyme from swine heart has been shown to contain thiol groups which are nonreactive near pH 6.5, permitting chemical modification studies by iodoacetate and related compounds without thioether formation (*209*). Of a number of substrate analogs tried, 4-bromocrotonate (LXXXI) and iodoacetate appeared to be effective in causing inactivation without thiol modification and were shown to provide selective modification at methionine and histidine residues in a stoichiometric fashion. Substrate protection and demonstrable competitive inhibition

207. S. I. Kandel, S. C. Wong, M. Kandel, and A. G. Gornall, *JBC* **243**, 2437 (1968).

208. P. L. Whitney, G. Folsch, P. O. Nyman, and B. G. Malmstrom, *JBC* **242**, 4206 (1967).

209. R. A. Bradshaw, G. W. Robinson, G. M. Hass, and R. L. Hill, *JBC* **244**, 1755 (1969).

by bromocrotonate in short-term observations lead to the conclusion that the irreversible modification is active-site directed (*209*). Similar results were reported from the action of bromomesaconate (LXXXII) (*210*). The presence of the free carboxyl group in these agents was considered essential for the selective action.

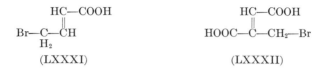

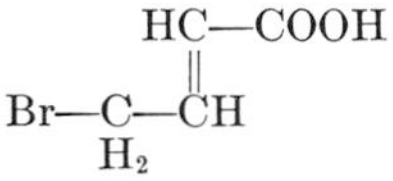

$$\begin{array}{c} \text{HC—COOH} \\ \| \\ \text{Br—}\underset{H_2}{\text{C}}\text{—CH} \end{array}$$

(LXXXI)

$$\begin{array}{c} \text{HC—COOH} \\ \| \\ \text{HOOC—C—CH}_2\text{—Br} \end{array}$$

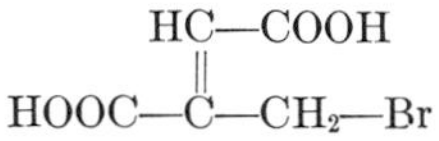

(LXXXII)

5. *β-Hydroxydecanoyl Thioester Dehydrase*

In the characterization of the individual steps in the fatty acid synthetase system of *E. coli*, a single protein has been isolated which catalyzes dehydration, isomerization, and hydration reactions involved in chain elongation. Acetylenic analogs of the substrates were found to be inhibitors (*211*) whose effectiveness varied with chain length in parallel to substrate behavior, suggesting a common binding site. The substrates and inhibitors were studied as thioesters of *N*-acetylcysteamine. One substrate, for example, is the *β,γ*-decenoylthioester [(LXXXIII) R = *N*-acetylaminoethyl] which is isomerized to the *α,β*

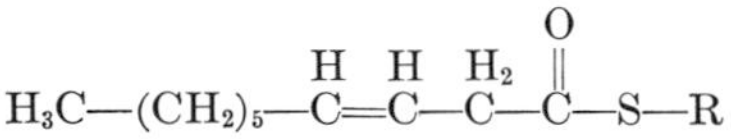

$$H_3C\text{—}(CH_2)_5\text{—}\overset{H}{C}\text{=}\overset{H}{C}\text{—}\overset{H_2}{C}\text{—}\overset{O}{\overset{\|}{C}}\text{—S—R}$$

(LXXXIII)

$$H_3C\text{—}(CH_2)_5C\equiv C\text{—}\overset{H_2}{C}\text{—}\overset{O}{\overset{\|}{C}}\text{—S—R}$$

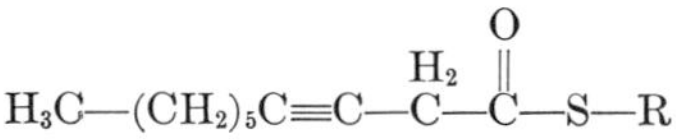

(LXXXIV)

form. The 3-decynoyl analog (LXXXIV) was found to irreversibly inactivate the enzyme essentially immediately even at 10^{-6} *M* concentration (*211*); the 2- and 4-decynoyl derivatives were inactive. Not only the position of the triple bond but also the thioester group were essential for inhibitory action. The very interesting finding that the histidine content of the inhibited enzyme indicates loss of one residue (*212*) suggests an unfamiliar addition of histidine to an isolated triple bond and testifies to the ability of complexed reagents to undergo rate promotion by proximity effects.

210. R. A. Laursen, J. B. Bauman, K. B. Linsley, and W. C. Shen, *ABB* **130,** 688 (1969).

211. G. M. Helmkamp, Jr., R. R. Rando, D. J. H. Brock, and K. Bloch, *JBC* **243,** 3229 (1968).

212. G. M. Helmkamp, Jr. and K. Bloch, *Federation Proc.* **28,** 602 (1969).

F. Heme Proteins

The heme group of myoglobin has two propionic acid side chains which are not covalently bound in myoglobin and in other heme proteins. By conversion to a mixed anhydride with sulfuric acid, these side chains are made reactive acylating agents. The monoanhydride reacted stoichiometrically with apomyoglobin (horse heart) at the side chain of Lys 45, determined by isolation of the heme peptide, as expected from the known three-dimensional structure. Thus verified as a heme binding site-directed label, the application to other heme proteins for this purpose was suggested (*213*).

IV. Particular Usefulness of Active-Site-Directed Reagents

A. Characterization of Functional Groups

This problem of enzyme function has provided the major impetus for the development of specific reagents. As discussed above, it has been possible, in the case of a few enzymes, to compare the results obtained by the approach of selective labeling with those obtained by other methods. This comparison has been favorable. With the many new examples now appearing, it can be expected that, in those cases where the site of covalent change in the reagent appears comparable to that of the substrate in its stereochemical relationship with functional groups of the enzyme, the results will yield information relevant to enzyme function. If, on the other hand, the chemically reactive group is merely appended as a side chain to a structure which occupies the active center, inactivation of the enzyme may occur but the probability is great that, although the active center is occupied, the chemical modification is peripheral and has no functional significance. This result *exoalkylation* (*7*), may be of value in obtaining inhibitors but may not be structurally informative. Some of the reagents described in Section III will probably fall into this category as would a great many others described by Baker (*7*), which have not been reviewed here since they have been developed for another purpose, the achievement of selective irreversible inhibitors, and evidence about their stoichiometry or sites of modification is not available.

213. P. K. Warme and L. P. Hager, *Federation Proc.* **28**, 603 (1969); *Biochemistry* **9**, 1599 (1970).

B. Adjuncts to Physical Methods

The stable derivative formed between an enzyme and an active-site-directed agent may be considered a model of the enzyme–substrate complex for that enzyme and is therefore of particular value for further study.

1. *Enzyme Modifications for Crystallography*

The stability of the sulfonate derivatives of chymotrypsin obtained from toluenesulfonyl fluoride and phenylmethanesulfonyl fluoride (*44*), among others, is greater than that of the DIP derivative and this property rendered them of value as active-center-labeled, inactive forms of the enzyme (*214*). Moreover, the possibility of introducing para-substituents such as iodo- and chloromercuri- provided crystalline forms isomorphous with the *p*-toluenesulfonyl derivative (*215*). These played an important part in solving the three-dimensional structure of α-chymotrypsin (*42*, *53*). The toluenesulfonyl group was found to be occupying the specificity site of chymotrypsin as established by diffusion of formyltryptophan into crystals of unmodified enzyme (*42*). Sulfonyl derivatives of another serine proteinase, subtilisin BPN′, also provided the form of choice for crystallographic studies (*75*).

It has been proposed that the sulfonyl derivatives of serine proteinases offer a way of systematically introducing heavy atoms. Position isomers of chloromercurinaphthalene sulfonyl fluoride were synthesized and used to prepare chymotrypsin derivatives containing a single, covalently bound mercury atom in different fixed positions (*216*). Active-site-directed reagents for any given enzyme may offer similar opportunities.

2. *Spin Labels*

Nitroxide radicals are relatively stable. Their usefulness in proteins as an environmental probe for electron spin resonance studies has been described by Hamilton and McConnell (*217*). Some nitroxide structures

214. P. B. Sigler, H. C. W. Skinner, C. L. Coulter, J. Kallos, H. Braxton, and D. R. Davies, *Proc. Natl. Acad. Sci. U S.* **51,** 1146 (1964).

215. P. B. Sigler, B. A. Jeffery, B. M. Matthews, and R. Henderson, *JMB* **15,** 175 (1966).

216. A. Fenselau, D. Koenig, D. E. Koshland, Jr., H. G. Latham, Jr., and H. Weiner, *Proc. Natl. Acad. Sci. U. S.* **57,** 1670 (1967).

217. C. L. Hamilton and H. M. McConnell, *in* "Structural Chemistry and Molecular Biology" (A. Rich and N. Davidson, eds.), p. 115. Freeman, San Francisco, California, 1968.

have been incorporated into protein group reagents (maleimide, acylating, and alkylating agents) (*217*). To take advantage of acyl enzyme derivatization of chymotrypsin for achieving a monomodification at the active center, the nitrophenyl ester of 2,2,5,5-tetramethyl-3-carboxypyrrolidine-1-oxide (LXXXV) was prepared (*218*). When the acyl chymo-

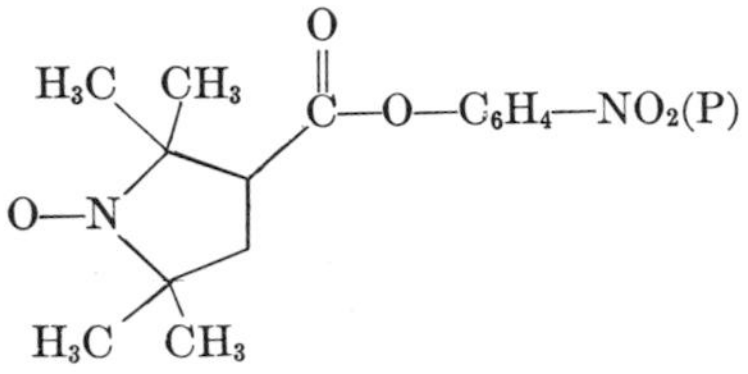

(LXXXV)

trypsin obtained with this reagent was formed at pH 4.5, the ESR spectrum became immobilized. Under conditions allowing deacylation, the release of the spin label was observed by resonance spectroscopy and the rate of change of the signal correlated with known kinetic behavior (*218*). In an extension of this type of study, spin-labeled acyl derivatives of chymotrypsin were prepared containing nitroxide radicals attached to

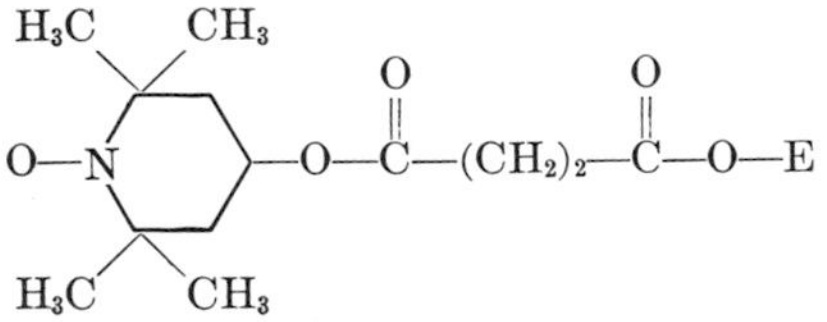

(LXXXVI)

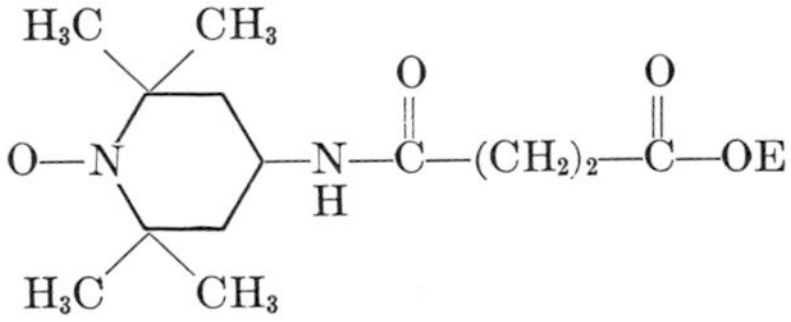

(LXXXVII)

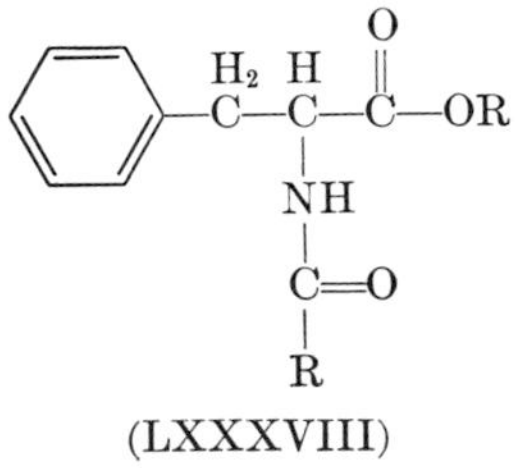

(LXXXVIII)

succinic acid by an ester (LXXXVI) or amide (LXXXVII) bond; the other carboxyl of the succinic acid was used to acylate chymotrypsin [E in (LXXXVI) and (LXXXVII)]. The changes in the ESR spectra were used to measure deacylation rates in the presence and absence of

218. L. J. Berliner and H. M. McConnell, *Proc. Natl. Acad. Sci. U. S.* **55**, 708 (1966).

indole, a known accelerator of deacetylation, and also of acyl enzymes derived from methionine modified chymotrypsins. The observations were interpreted in terms of the possible function of these side chains in the binding sites for the aromatic ring and the α-acylamido group of a normal substrate (LXXXVIII) (*219*). It is hoped that labels can be introduced with a geometry that is more likely to result in interaction with the region of the enzyme of interest.

3. *Other Spectroscopic Methods*

It was demonstrated some time ago by Klotz that the spectral or ionization properties of small molecules changed markedly as a result of binding to proteins and provided information about the local environment (*220*). Selective chemical modification as well as new physical methods have resulted in varied extensions. The ability to modify Met 192 in chymotrypsin by alkylation with retention of enzymic activity led to the introduction of 2-bromoacetamido-4-nitrophenol and 4-bromoacetamido-2-nitrophenol (*221*) to be used as reporter groups to indicate local changes related to chymotryptic function. A structurally similar derivative was used for a different purpose. The Met 192 sulfonium salt formed from 4-nitro- and 2,4-dinitrophenacyl bromide indicated the formation of a new long wavelength band that was interpreted as resulting from charge–transfer complex formation with the indole ring of a nearby tryptophan ring (*222*).

For fluorescence studies it was found that anthraniloyl-chymotrypsin was a highly suitable derivative since, although an acyl-enzyme, it is unusually stable (with respect to deacylation) at neutrality and highly fluorescent (*223*). From fluorescence polarization and nanosecond kinetic measurements it was concluded that the acyl group has no independent rotational mobility within the acyl enzyme which "behaves as a rigid particle." A red shift in emission maximum (as well as spectral broadening) indicated that the environment of the anthraniloyl group is highly polar but nonaqueous (*223*).

Projected uses of modified chymotrypsins in NMR studies are also evident (*224*).

219. D. J. Kosman, J. C. Hsia, and L. H. Piette, *ABB* **133,** 29 (1969).

220. I. M. Klotz and J. Ayers, *JACS* **79,** 4078 (1957).

221. M. Burr and D. E. Koshland, Jr., *Proc. Natl. Acad. Sci. U. S.* **52, 1017** (1964); A. Conway and D. E. Koshland, Jr., *BBA* **133,** 593 (1967).

222. D. S. Sigman and E. R. Blout, *JACS* **89, 1747** (1967).

223. R. P. Haugland and L. Stryer, *in* "Conformation of Biopolymers" (G. N. Ramachandran, ed.), Vol. 1, p. 321. Academic Press, New York, 1967.

224. P. M. Enriquez and J. T. Greig, *Biochemistry* **8,** 3156 (1969).

C. Selective Enzyme Inhibitors

This chapter has dealt with the study of isolated pure enzymes by designed reagents which generally result in inactivation. Success in this effort is adding to the number and variety of enzyme inhibitors available for quite different purposes such as the interpretation of complex metabolic situations by selective inactivation of an enzymic component. Just a step beyond is the possible chemotherapeutic utility of specific enzyme inhibitors (*4, 7*).

V. Conclusions

Affinity labeling is a way of producing chemical modification of proteins in regions with specific binding and catalytic properties. In many cases, ordinary group reagents for modification are either not selective or fail to combine at active center residues which can, however, be covalently modified by substratelike reagents, which localize at the active center. Because the subsequent covalent attachment is accelerated by the proximity effect so produced, the geometry of the reagent is more important than its inherent chemical reactivity. In fact, the resultant chemical reaction may be difficult to reproduce in solution. These properties emphasize the enzymic nature of this type of protein chemistry. It can be expected that chemical ingenuity will result in an ever greater variety of reagent types in an approach that has become a standard method of enzyme study.

3

Chemical Modification as a Probe of Structure and Function

LOUIS A. COHEN

I. Introduction

Any transformation involving the formation or rupture of a covalent, or partially covalent, bond may be treated, in its broadest sense, as a chemical modification of a protein. The subject would, therefore, include such processes as proton transfer, exchange of hydrogen isotopes, metal chelation, enzyme–substrate association, and even hydrogen bonding. The best justification for not including such topics in the present chapter, apart from space limitations, is that the protein chemist would not look for them here. The narrower field of what is commonly considered chemical modification can be disposed of as easily: The encyclopedic coverage of various aspects of the subject, achieved by the publication of ten reviews or books in three years (*1–10*) (Table I), actually requires little of this author but an introduction and conclusion.

Upon closer examination, however, it appears that the initiate in protein chemistry has been provided a vast and bewildering store of information but few principles or guidelines. Since it would be ridiculous to attempt to reproduce the data of references *1–17* in this chapter, the author has tried to complement the cataloging of reagents and reactions with a more general discussion of the principles, methodology, and applications of modification, together with analyses of a few representative

1. G. E. Means and R. E. Feeney, "Chemical Modification of Proteins." Benjamin, New York, 1970.
2. T. F. Spande, B. Witkop, Y. Degani, and A. Patchornik, *Advan. Protein Chem.* **24,** 97 (1970).
3. G. R. Stark, *Advan. Protein Chem.* **24,** 261 (1970).
4. B. L. Vallee and J. F. Riordan, *Ann. Rev. Biochem.* **38,** 733 (1969).
5. L. A. Cohen, *Ann. Rev. Biochem.* **37,** 695 (1968).
6. D. E. Koshland, Jr. and K. E. Neet, *Ann. Rev. Biochem.* **37,** 359 (1968).
7. S. J. Singer, *Advan. Protein Chem.* **22,** 1 (1967).
8. S. N. Timasheff and M. J. Gorbunoff, *Ann. Rev. Biochem.* **36,** 13 (1967).
9. C. H. W. Hirs, Vol. ed., "Methods in Enzymology," Vol. 11, 1967.
10. B. R. Baker, "Design of Active-Site Directed Irreversible Enzyme Inhibitors." Wiley, New York, 1967.
11. J. L. Webb, "Enzyme and Metabolic Inhibitors," Vols. 1–3. Academic Press, New York, 1963–1966.
12. H. Fraenkel-Conrat, *Comp. Biochem.* **7,** 56 (1963).
13. B. Witkop, *Advan. Protein Chem.* **16,** 221 (1961).
14. H. Fraenkel-Conrat, "The Enzymes," 2nd ed., Vol. 1, p. 589, 1959.
15. F. W. Putnam, *in* "The Proteins" (H. Neurath and K. Bailey, eds.), 1st ed., Vol. 1, Part B, p. 893. Academic Press, New York, 1953.
16. R. M. Herriott, *Advan. Protein Chem.* **3,** 169 (1947).
17. H. S. Olcott and H. Fraenkel-Conrat, *Chem. Rev.* **41,** 151 (1947).

TABLE I
BIBLIOGRAPHY OF REVIEWS ON CHEMICAL MODIFICATION AND RELATED TOPICS

Authors	Year	General subject area (mode of classification)
Means and Feeney (*1*)	1970	General (reaction type and residue)
Spande *et al.* (*2*)	1970	Chemical cleavage and some modification
Stark (*3*)	1970	General (reagent and reaction type)
Vallee and Riordan (*4*)	1969	Modification at active sites
Cohen (*5*)	1968	General modification (reaction type)
Koshland and Neet (*6*)	1968	Mechanistic considerations
Singer (*7*)	1967	Affinity labeling
Timasheff and Gorbunoff (*8*)	1967	Accessibility of functional groups
Hirs (*9*)	1967	General modification (methodology)
Baker (*10*)	1967	Affinity labeling
Webb (*11*)	1963–1966	Detailed reviews of selected reagents
Fraenkel-Conrat (*12*)	1963	General modification
Witkop (*13*)	1961	Chemical cleavage
Fraenkel-Conrat (*14*)	1959	General modification
Putnam (*15*)	1953	General modification
Herriott (*16*)	1947	General modification
Olcott and Fraenkel-Conrat (*17*)	1947	General modification

cases. In addition to the principal uses of modification for the study of the three-dimensional structure of proteins and of enzyme function, the compatibility of organic chemistry and X-ray crystallography is examined briefly.

At the considerable risk of turning the reader to other endeavors, certain limitations must be stated at the outset, which are decidedly more realistic than optimistic:

(1) Few, if any, chemical modifiers are absolutely specific for a given type of functional group.

(2) No modification of a protein can be achieved without conformational change, however slight.

(3) Quantitative behavior of a reagent toward amino acids or small peptides provides only a partial basis for its behavior toward proteins.

(4) Behavior of a reagent toward one protein provides only a partial basis for its behavior toward another.

(5) The outcome of modification can be influenced, to some degree, by every type of experimental variable.

(6) Assigning the action of a modifier to a catalytic site, binding site, or conformation-controlling site is often very difficult and sometimes impossible.

(7) Truly meaningful interpretations of the results of chemical modification can be realized only if the investigator pursues the acquisition of analytical data and the exploration of reaction variables with reasonable thoroughness.

The remainder of this chapter will be devoted to demonstrating that, in spite of such limitations, chemical modification has both a respectable past and an inviting future.

II. Principles of Chemical Modification

A. Determinants of Functional Group Reactivity

The X-ray analyses of some ten globular proteins have provided to date clear evidence that the majority of nonpolar (hydrophobic) side chains are located in the core of a very compact, three-dimensional matrix (*18*); conversely, the majority of polar and electrically charged side chains (hydrophilic) prefer the surface and crevices of the matrix. We may assume that the protein matrix is so compact that solvent and other small molecules or ions are effectively excluded from the core. Small molecules may interact with interior residues by penetration of crevices, by slow diffusion, by waiting for random conformational changes to occur (if they ever do), or by effecting a sequence of surface perturbations and modifications which produce a partial or local unfolding of the native structure. Accordingly, the interior of the protein molecule, and certain crevices, will have a diminished polarity (or dielectric constant) approaching that of a hydrocarbon. The surface of the molecule is irregular in contour and inhomogeneous, polarity and charge varying from atom to atom. Thus, a variety of surface polarities may be expected, from the very nonpolar to regions exceeding the polarity of the aqueous medium. The physical and chemical properties of a protein functional group will be strongly influenced by the nature of its local environment. An approaching chemical reagent may also be influenced by surface features and be directed to favorable sites or repelled by hostile sites; furthermore, two identical residues, located in different surface microenvironments, may show significantly different reactivities toward one reagent and similar reactivities toward another. It is pertinent, therefore, to examine in detail the effects of local environment on both the functional group and the reagent. Since the reactivity of a functional group is

18. M. F. Perutz, *European J. Biochem.* **8**, 455 (1969).

measured by its nucleophilicity, which is often (but not always) related to its basicity, the emphasis on pK and its determinants in the following discussion will be apparent.

1. *Polarity of the Microenvironment*

Titration of ionizable groups in a protein often shows a considerable range in apparent pK values, some close to those found for low molecular weight analogs, and some significantly higher (*19*). Values below normal are observed less frequently. In general, a greater range of values will be observed for groups which dissociate by charge separation (carboxyl, phenol, and thiol) than by charge dispersion (ammonium and imidazolium). A striking example is seen in the case of phenols (normal pK_a 10) for which apparent pK values in proteins range from 9.4 to more than 12. Polarity of the microenvironment is one of several determinants of the ease of ionization. The pK of acetic acid (pK_a 4.76) increases to 6.87 in 80% ethanol and to 10.32 in 100% ethanol (*20*). On the other hand, the effect of decreased polarity on the ionization of an ammonium ion is considerably smaller and variable in direction (reflecting ionization by charge dispersion).

Efforts to utilize titration data to estimate surface polarities are hampered by the fact that pK is influenced by additional factors and because functional group ionizations are complex and often overlap. One alternative device, among others, is the incorporation of environmental probes or "reporters." For example, the azomercurial (I) has been coupled to the single sulfhydryl group of bovine serum albumin (*21*).

Me_2N—C_6H_4—N=N—C_6H_4—HgOAc

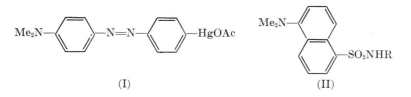

(I)

Me_2N—$C_{10}H_6$—SO_2NHR

(II)

Spectral titration shows a reduction in pK of the anilino nitrogen from 3.3, when coupled to cysteine, to 1.9, when coupled to protein. The pK becomes normalized in the denatured protein. Modification of other proteins with this reagent has provided pK values as low as one (*22*). Reduced polarity of the microenvironment is one of several ways of explain-

19. C. Tanford, *Advan. Protein Chem.* **17,** 70 (1962).
20. R. G. Bates, "Determination of pH," p. 195. Wiley, New York, 1964.
21. I. M. Klotz and J. Ayers, *JACS* **79,** 4078 (1957).
22. I. M. Klotz, E. C. Stellwagen, and V. H. Stryker, *BBA* **86,** 122 (1964).

ing the marked effect. Similarly, the pK values of the anilino nitrogen in dansyl derivatives of proteins (II) are reduced from a normal value of 3.9 to values as low as 1.5 (*22*). A separation of polarity effects from electrostatic influences (Section II,A,3) can sometimes be achieved by the design of reagents which are selective for certain functional groups; are insensitive to normal pH changes; are strongly chromophoric; and have a marked dependence, in their spectra, on solvent polarity. The compound, 2-methoxy-5-nitrobenzyl bromide (III), absorbing at 288 nm

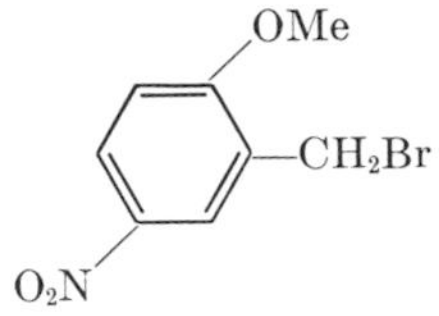

OMe
—CH_2Br
O_2N
(III)

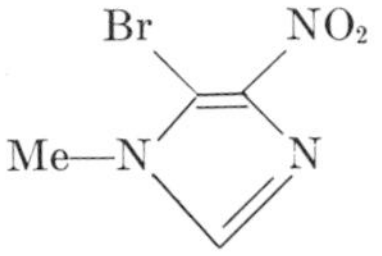

Br NO_2
Me—N N
(IV)

in hexane and at 317 nm in dimethyl sulfoxide, alkylates cysteine, tryptophan, and methionine (*23*). Although the influence of solvent on spectral properties is less marked in the case of (IV), the nitroimidazole derivative appears to be highly selective for thiols (*24*). The application of such environmental probes to proteins awaits exploration.

On the whole, one may expect that variations in local polarity should have the least effect on the reactivities of tryptophan, methionine, and cystine; a greater effect on amino groups and on histidine; and the greatest effect on tyrosine, cysteine, and carboxyl groups. Although a higher pH may be required to generate an active nucleophile (phenolate, carboxylate, etc.) in a low polarity environment, the resultant species will be a more potent nucleophile since it will be solvated to a lesser degree than in water (*25*). The effect of polarity on overall rate depends largely on the type of reaction involved; this aspect will be considered in Section II,B,5.

2. *Hydrogen Bonding Effects*

A second contributing factor to anomalous pK values arises from stabilization of either a neutral species or its ion by hydrogen bonding. Thus, the pK_a of 2-fluorophenol (V) is 0.7 unit higher than the predicted value (*26*). Since the pK_a of 2-bromophenol is normal, the effect is not of steric origin but may be attributed to the considerably greater ability

23. H. R. Horton, H. Kelley, and D. E. Koshland, Jr., *JBC* **240,** 722 (1965).
24. W. Schreiber and L. A. Cohen, unpublished results (1968).
25. J. A. Leary and M. Kahn, *JACS* **81,** 4173 (1959).
26. D. H. McDaniel and H. C. Brown, *JACS* **77,** 3759 (1955).

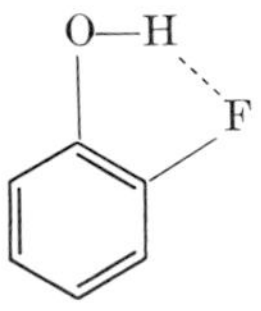

(V) 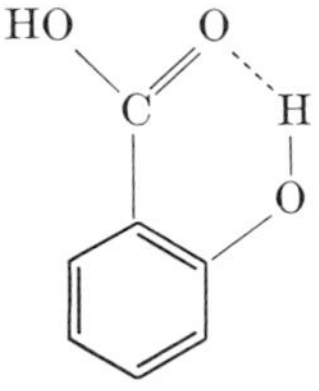 (VI) 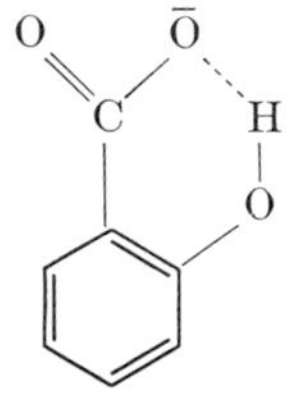 (VII)

of fluorine to form hydrogen bonds. Formation of the phenolate ion from (V) requires the expenditure of additional energy to break the intramolecular hydrogen bond. On the other hand, the carboxyl group of salicyclic acid (VI) is about 1 pK unit more acidic than expected. Although O—H··O interaction exists in both the acid and its anion (VII), the negatively charged oxygen atom of (VII) forms a significantly stronger hydrogen bond. In this case, better stabilization of the anion facilitates ionization. At the same time, the hydrogen bond in (VII) represses phenol ionization and displaces the normal pK of 10 to about 13. Thus, in the case of a phenol–carboxylate interaction in a protein, the pK of the carboxyl should be below normal and that of the phenol above normal. Whereas negative ions make better hydrogen bonds than their parents, bonds to positive ions are weaker than to neutral species, perhaps accounting for the fact that ϵ-amino groups in proteins are slightly weaker bases than normal.

Hydrogen bonding usually depresses reactivity. In β-diketone (VIII)

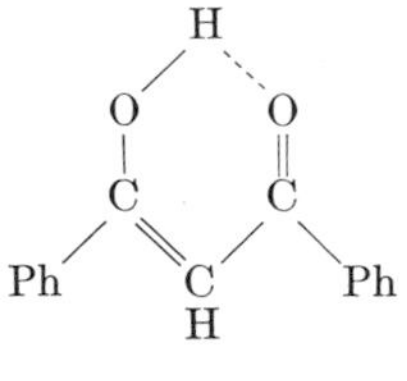

(VIII) 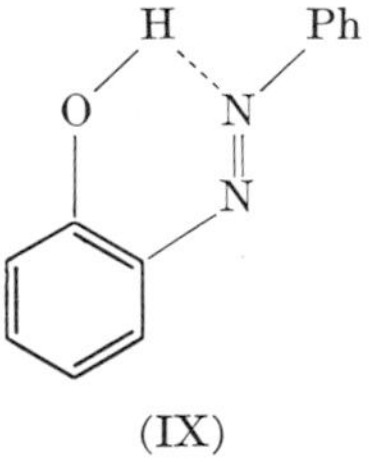 (IX) 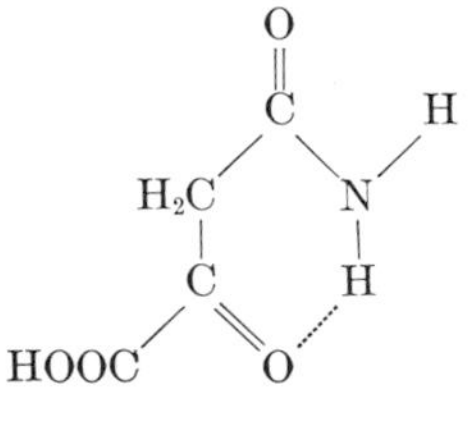 (X)

the enolic hydroxyl fails to react with acetic anhydride or with phenyl isocyanate (*27*). Normal properties of the carbonyl group, such as Schiff base formation, are also blocked. For similar reasons (IX) is insoluble in alkali and resists alkylation or acylaticn (*28*). The keto analog of asparagine (X) fails to give the usual carbonyl derivatives (*29*). Since

27. J. Wislicenus, *Ann. Chem.* **308,** 219 (1899).
28. C. Smith and A. D. Mitchell, *JCS* **93,** 843 (1908).
29. A. Meister, *JBC* **200,** 571 (1953).

most chemical modifications require the proton-poor form (*30*) of the functional group as an effective nucleophile, these examples show that hydrogen bonding hampers the reactivity of one or both partners to the interaction.

3. *Field or Electrostatic Effects*

The presence of a positive charge in the vicinity of an ionizable group will encourage formation of the proton-poor form, either by stabilization of an ion pair or by repressing formation of another positively charged group. The pK_a of the carboxyl group in (XI) is 1.37, but is 3.45 in the meta- or para-isomer. The inductive effect, operating through the aromatic ring, can account for only a small portion of the enhanced acidity of the ortho-isomer. Similarly, the pK_a of (XII) is 7.42, whereas that of

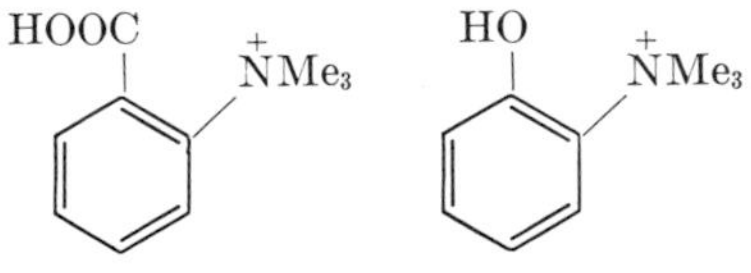

HOOC $\overset{+}{N}Me_3$ HO $\overset{+}{N}Me_3$

(XI) (XII)

the meta-isomer is 8.03 and that of the para-isomer, 8.21. Although anion formation would normally improve the nucleophilicity of a functional group, the formation of an ion-pair bond may serve to negate the increased nucleophilicity (*31*).

4. *Steric Effects*

To date, little information is available regarding the existence of true steric effects on the protein surface. For our purposes, a steric effect could arise by a bulky alkyl group being sufficiently close in space to a functional group to prevent approach of a reagent or formation of a derivative. Until X-ray crystallography can assign atom positions with even greater refinement than at present, it is unlikely that it will be possible to distinguish steric effects from various other causes of depressed reactivity. Steric hindrance may appear in still another way: Since ions owe the ease of their formation, in part, to enclosure in a solvent shell,

30. The term *proton-poor* is used to indicate the species formed by loss of a proton, which may be anionic or uncharged.

31. S. Winstein, L. G. Savedoff, and S. Smith, *Tetrahedron Letters* No. 9, 24 (1960).

a bulky neighbor may provide steric hindrance to solvation and thus repress ionization (*32*, *33*).

5. *Matrix Effects*

A functional group buried within the core or in a deep crevice in the molecule is also sterically inaccessible to reagent. The term *matrix effect* is introduced to differentiate this effect from the surface steric factor described above. In principle, the two types should be capable of differentiation: A group buried within the core would be inaccessible to any reagent, regardless of size; a sterically hindered group on the exterior may be reactive to small but not large molecules, and should respond more readily to stronger reaction conditions than a core group.

6. *Miscellaneous Effects*

Several other factors may be mentioned which are capable of altering functional group reactivity. Because of the limited data available, even for model systems, little discussion can be offered.

a. Charge Transfer. Spectroscopic techniques have provided evidence that the aromatic amino acids can participate in charge–transfer interactions with one another, with cofactors, substrates, and reagents (*34*, *35*). Reactivity of an aromatic ring is generally enhanced by complexing, although deactivation may also occur (*36*).

b. Covalent Bond Formation. The formation of a reversible covalent bond, e.g., a Schiff base or a hemithioacetal, may remove a group temporarily from its normal reactivity class or depress its reactivity.

c. Metal Chelates. The formation of metal complexes or chelates would normally be expected to decrease the reactivity of a functional group since the metal tends to withdraw electron density from the nucleophilic atom (*37*).

d. Freedom of Rotation. The mechanism for interaction with a given reagent often requires a specific orientation of the nucleophilic group or

32. L. A. Cohen and W. M. Jones, *JACS* **85**, 3397 (1963).
33. L. A. Cohen, *J. Org. Chem.* **22**, 1333 (1957).
34. F. J. Bullock, *Comp. Biochem.* **22**, 81 (1967).
35. J. Mauchamp and M. Shinitzky, *Biochemistry* **8**, 1554 (1969).
36. L. J. Andrews and R. M. Kieffer, "Molecular Complexes in Organic Chemistry." Holden-Day, San Francisco, California, 1964.
37. R. C. Voter, C. V. Banks, V. A. Fassel, and P. W. Kehres, *Anal. Chem.* **23**, 1730 (1951).

of its unshared electron pairs. Should the microenvironment, by hydrogen bonding, electrostatic or steric hindrance, hamper free rotation of the nucleophile, decreased reactivity may result. Conversely, restricted rotation, by excluding a number of unprofitable orientations, may enhance reactivity enough to produce a superreactive site (Section II,C,3) (*6, 38*); for example, the reactivity of a disulfide bond may be correlated with the dihedral angle formed by adjacent unshared electron pairs (*39, 40*).

B. Determinants of Reagent Reactivity

Just as the protein conformation and surface features can influence the reactivity of an amino acid side chain, so can they influence the approaching reagent, both favorably and unfavorably. An understanding of the possible modes of interaction is of value in the design of pseudosubstrates and affinity labels.

1. *Selective Adsorption*

Prior to chemical modification, the reagent may be adsorbed selectively in a region of low or high polarity according to its own characteristics. The formation of a protein–reagent complex can sometimes be detected by a saturation effect on rate. Following complex formation, the rate enhancement of an intramolecular process, as in an enzyme–substrate complex, may be observed. Selective adsorption is, in part, responsible for the rate enhancements observed in micellar aggregates. Very few modification reactions have been examined for possible saturation kinetics: Reaction of the sulfhydryl group of D-amino acid oxidase with a series of *N*-alkylmaleimides shows a 15-fold rate enhancement in going from *N*-ethyl to *N*-octyl, suggesting both a nonpolar environment for the thiol and space available for large molecules; saturation kinetics suggests that hydrophobic binding precedes the alkylation step (*41*). The hydrophobic binding of fluorodinitrobenzene to bovine serum albumin, prior to modification, has also been shown by means of the saturation effect (*42*). Undoubtedly, additional studies would reveal the phenomenon to be of more common occurrence.

38. M. R. Holloway, A. P. Mathias, and B. R. Rabin, *BBA* **92,** 111 (1964).
39. O. Foss, *in* "Organic Sulfur Compounds" (N. Kharasch, ed.), Vol. I, p. 75 Pergamon Press, Oxford, 1961.
40. J. A. Barltrop, P. M. Hayes, and M. Calvin, *JACS* **76,** 4348 (1954).
41. M. L. Fonda and B. M. Anderson, *JBC* **244,** 666 (1969).
42. N. M. Green, *BBA* **74,** 542 (1963).

2. *Electrostatic Interaction*

A charged reagent will be attracted selectively to sites of opposite charge on the protein surface. Obviously, such charge may be provided by neighbors of the group undergoing modification, as well as by a substrate, cofactor, or inhibitor. Electrostatic interaction may orient the reagent toward one residue in a polyfunctional site or toward one side of a bifunctional residue. The differences in both rates and sites of alkylation by iodoacetate and iodoacetamide are often determined by electrostatic influences, as are cases of selective alkylation of imidazole at N-1 or N-3. Coupling of diazonium ions to tyrosine is blocked if the phenolic group is acylated. Whereas *p*-carbethoxyphenyldiazonium ion fails to react with acylated tyrosine in bovine serum albumin, *p*-carboxyphenyldiazonium ion does react (*43*). Thus, electrostatic binding of the reagent carrying a carboxylate ion enhances reactivity by converting an intermolecular into an intramolecular reaction; electrostatic orientation of the reagent may also be a contributing factor. Azaserine, which is unreactive toward simple thiols, alkylates the sulfhydryl group of formyl-GAR amidotransferase, probably because of electrostatic binding (*44*). In addition to increased reactivity resulting from electrostatic attraction, repulsive forces may inhibit modification. Reduction of disulfides by neutral, but not by charged, reducing agents has been observed (*45–47*). Other interesting examples have been found for ribonuclease (*48*), yeast alcohol dehydrogenase (*49, 50*) and papain (*51*).

3. *Steric Factors*

As already noted, steric factors inherent in the protein surface (or introduced by substrate, cofactor, or inhibitor) may prevent a reagent from reacting with a site which would normally be preferred. In addition, variation in the size of the reagent may provide anomalous selectivities as well as useful information about the protein surface. Alkylation of His 12 in ribonuclease proceeds readily with α-bromoacids through C_4; alkylation is more difficult with α-bromovalerate and fails with α-bromocapro-

43. A. V. Luisada-Opper and H. Sobotka, *JBC* **238,** 143 (1963).
44. T. C. French, I. B. Dawid, and J. M. Buchanan, *JBC* **238,** 2186 (1963).
45. G. W. Robinson, R. A. Bradshaw, L. Kanarek, and R. L. Hill, *JBC* **242,** 2709 (1967).
46. S. Shall and A. Waheed, *BJ* **111,** 33P (1969).
47. R. Pitt-Rivers and H. L. Schwartz, *BJ* **105,** 28C (1967).
48. R. P. Carty and C. H. W. Hirs, *JBC* **243,** 5244 (1968).
49. B. Eisele and K. Wallenfels, *European J. Biochem.* **6,** 29 (1968).
50. N. Rashed and B. R. Rabin, *European J. Biochem.* **5,** 147 (1968).
51. K. Wallenfels and B. Eisele, *European J. Biochem.* **3,** 267 (1968).

ate. Since the decrease in extent of hydrophilic binding from C_4 to C_6 would be relatively small, the effect may be ascribed to having exceeded the size of the site at which α-bromoacids are bound prior to alkylation (*52*).

4. *Catalytic Factors*

Other functional groups, in the vicinity of the modification site, may assist the reaction by acting as general acids or bases. Since the sensitivity to acid or base catalysis depends not only on the mechanism of the reaction in question but also on the reactivity of the reagent, significant differences in both rate and site of reaction may be observed with different reagents. For example, although *p*-nitrophenyl acetate and phenyl acetate may acetylate the reactive serine of chymotrypsin by the same general mechanism, the latter reagent, being less reactive, will show a greater dependence on acid-base catalysis by the enzyme. The exceptionally high reactivity of fluorophosphates, in contrast to chlorophosphates, toward serine enzymes, probably involves general acid-catalyzed removal of the fluorine atom (*53*).

5. *Polarity of the Local Environment*

The rates of many organic reactions are strongly dependent on solvent polarity while others are almost insensitive. Although solvent effects can be quite complex, several generalizations can be made, for which examples are given in Table II. A hydrophobic region will discourage reactions involving charge separation in the products (1) and accelerate reactions involving charge neutralization (4). Where charge separation is not involved (2) or is simply transferred from one ion to another (3), the

TABLE II
EFFECT OF SOLVENT POLARITY ON REACTION RATE

Reaction type	Effect of decreased polarity on rate
(1) $RSR + ICH_2CONH_2 \rightarrow [R_2\overset{+}{S}CH_2CONH_2]I^-$	Moderate or large decrease
(2) $RSR + H_2O_2 \rightarrow R_2SO + H_2O$	Little effect
(3) $RS^- + ICH_2CONH_2 \rightarrow RSCH_2CONH_2 + I^-$	Little effect
(4) $RNH_3^+ + OCN^- \rightarrow RNHCONH_2$	Moderate or large increase

52. R. L. Heinrikson, *JBC* **241,** 1393 (1966).
53. H. P. Metzger and I. B. Wilson, *Biochemistry* **3,** 926 (1964).

polarity of the medium may be of little importance (*54*). Nevertheless, S_N2 displacement on methyl chloride by chloride ion [a reaction of type (3)] was found to undergo a 10^7-fold enhancement in rate when the medium was changed from water to acetone (*31*).

C. Mechanism and Reactivity

1. *Kinetic Considerations*

If the rate of reaction of *p*-bromophenol with a large excess of chloroacetamide is measured as a function of the pH of the medium (*55*), a plot of the observed rate constants vs. pH will have the general shape of the titration curve for the phenol, rate increasing with increase in pH. This fact alone makes it likely that the phenolate ion is the more, and possibly the only, reactive nucleophile, particularly if the midpoint of the rate curve corresponds in pH to the pK of the phenol. Such data can be condensed into a pH independent or specific rate constant (k_N) by use of Eq. (1), in which α is the fraction of phenolate ion present at any

$$k_{obs} = k_N\alpha = k_N\left(\frac{K_i}{K_i + H^+}\right) \tag{1}$$

given pH and K_i is the ionization constant for the phenol (*56*). It is clear that for any simple, bimolecular reaction involving the proton-poor form of a nucleophile, the observed rate will be 91% of its maximal value at a pH one unit above the pK. Consequently, there is little to be gained, with respect to reaction rate, in conducting the majority of modification reactions at excessively high pH values.

If a series of *m*- and *p*-substituted phenols are studied in the same way, a Brönsted plot of log k_N vs. pK_a will be linear (Fig. 1, line *A*), indicating that the nucleophilicities of the phenolate ions (as measured by log k_N) are proportional to their basicities (as measured by pK). It follows that the higher the pK_a of a phenol, the more reactive will be its anion, although a higher pH may be needed to generate the anion.

Let us now consider a protein containing phenolic groups of pK 8 and 10: From the equation of line *A* in Fig. 1, $k_N = 10\,M^{-1}$ min^{-1} and 100 M^{-1} min^{-1}, respectively. Application of Eq. (1) provides values for k_{obs}, whose ratios for the two types of groups as a function of pH are shown

54. C. K. Ingold, "Structure and Mechanism in Organic Chemistry," p. 457. Cornell Univ. Press, Ithaca, New York, 1969.

55. For simplicity, the absence of buffer catalysis is assumed in this discussion.

56. For a more detailed discussion, see T. C. Bruice and S. Benkovic, "Bioorganic Mechanisms," Vol. I, Chapter 1. Benjamin, New York, 1966.

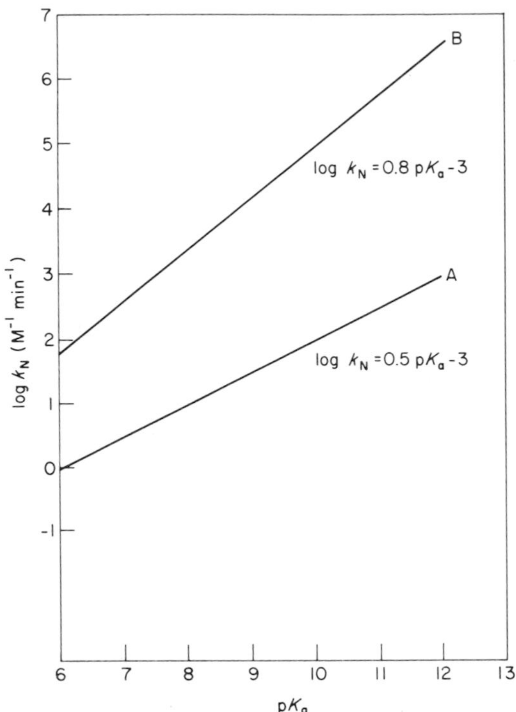

FIG. 1. Hypothetical Brönsted plots (log k_N vs. pK_a) for the alkylation of substituted phenols with chloroacetamide (A) and with iodoacetamide (B).

in Fig. 2, curve *A*. The limiting ratio at low pH is 0.1 and at high pH, 10. Thus, the selective alkylation of the phenol of pK 8 is best attempted at a pH equal to or below its pK and selective alkylation of the phenol of pK 10 at a pH at least one unit above its pK.

Now consider the reaction of the same set of phenols with another alkylating agent, iodoacetamide, k_N following line *B* of Fig. 1. The increased slope indicates a greater sensitivity to the nucleophilicity of the phenolate ion, and the higher value of k_N, a greater reactivity. The ratios of values of k_{obs} vs. pH are shown in curve *B* of Fig. 2. Although the same pH criteria for selective alkylation would apply as previously indicated, it is evident that the opportunity for selective alkylation of the more acidic phenol has decreased 4-fold. Thus, the greater selectivity is achieved by use of the less reactive reagent. This principle has been applied in the use of chlorodinitrobenzene in place of fluorodinitrobenzene, chloroacetate or bromoacetate in place of iodoacetate, and activated esters in place of acid anhydrides (*5*).

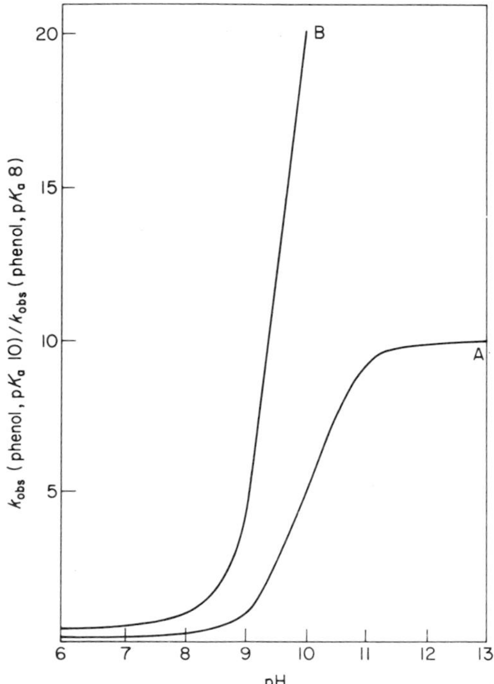

FIG. 2. Ratios of the rates of alkylation of phenols (pK_a 10 and pK_a 8) as a function of pH: (A) with chloroacetamide; (B) with iodoacetamide.

Linear correlations were obtained in the log k_N/pK_a plots of Fig. 1 because all the nucleophiles compared are geometrically similar at the reaction site and possess the same hetero atom. Ortho-substituted phenols would probably not fall on the same slopes, nor would thiophenols or anilines. The steric environments of α- and ϵ-amino groups are sufficiently different to limit correlative comparisons of their reactivities with pK values. The preceding arguments would suggest that some selectivity be observed in modifications of the α-amino group at low pH; however, the ϵ-amino group, because it is significantly less sterically hindered, is a better nucleophile and can compete for reagent at any pH. Here again, the selective modification of an α-amino group in a protein is best attempted by choice of a reagent of relatively low overall reactivity. A comparison of k_N values has been obtained for the reactions of simple amino acids and peptides with trinitrobenzenesulfonic acid (*57*). In this

57. R. B. Freedman and G. K. Radda, *BJ* **108**, 383 (1968).

case, k_N for the ϵ-amino group is approximately ten times that for the α-amino group. On the other hand, the imidazole nitrogen is more nucleophilic than the α- or ϵ-amino nitrogen in alkylation and acylation reactions and has a lower pK. The selective modification of imidazole, often observed at low pH, confirms the expectation drawn from theory.

2. *Nucleophilicities of Protein Functional Groups*

In their proton-poor forms, all the functional groups of amino acids are nucleophilic species (*58*). Unfortunately, the significant differences in the nature of the hetero atom and in the steric environments prevent any serious attempt at correlation of nucleophilic reactivity with pK. At best, one can hope to compile a qualitative order of reactivities, as shown for alkylation reactions in column 1 of Table III. It is readily apparent that the series bears no relationship to pK values of the respective groups. Such a sequence cannot take into account any of the special effects of the protein environment, which can readily move a functional group above or below its normal position. In addition to the difficulties encountered in attempting correlations toward one reagent, even the qualitative order of groups will vary as the nature of the modifying reagent or reaction mechanism changes. Thus, a different sequence is observed in nucleophilic attack at the trigonal carbon atom of a carbonyl group (column 2, Table III) and still another with the trigonal carbon of an activated aryl halide (column 3, Table III) or in photosensitized oxidation (column 4, Table III). Selectivity can be complicated by other mechanistic factors. Picryl chloride arylates amines and phenolate ions; trinitrobenzenesulfonic acid reacts with amines alone (*59*). Seemingly, the difference could be ex-

TABLE III
APPROXIMATE ORDERS OF REACTIVITY IN MODIFICATION REACTIONS[a]

Alkylation	Acylation	Arylation	Photooxidation
Methionine	—	—	Methionine
Cysteine	Cysteine	Cysteine	Cysteine
Imidazole	Phenol	Amine	Indole
Amine	Imidazole	Phenol	Imidazole
Phenol	Amine	Imidazole	Phenol
Carboxyl	Carboxyl	Carboxyl	—

[a] Groups are listed in order of decreasing reactivity.

58. Methionine and tryptophan, although nucleophiles, have nonionizing functional groups.
59. T. Okuyama and K. Satake, *J. Biochem.* (*Tokyo*) **47**, 454 (1960).

plained on the basis of electrostatic repulsion between the sulfonate and phenolate ions; yet, picryl chloride arylates proline but the sulfonic acid does not.

Unexpected or exaggerated selectivities have been observed in a few reactions proceeding by special mechanisms. When the reagent itself requires capture of a proton to become reactive, the general rate equation can be expressed in the kinetically indistinguishable forms shown in Eqs. (2)–(5):

$$\text{Rate} = k_{\text{obs}}[\text{RNH}_3^+][\text{NCO}^-] = k_{\text{obs}}[\text{RNH}_2][\text{HNCO}] \quad (2)$$

$$\text{Rate} = k_{\text{obs}}[\text{PhOH}][\text{AcIm}] = k_{\text{obs}}[\text{PhO}^-][\text{AcImH}^+] \quad (3)$$

$$\text{Rate} = k_{\text{obs}}[\text{RSH}][\text{ethyleneimine}] = k_{\text{obs}}[\text{RS}^-][\text{EIH}^+] \quad (4)$$

$$\text{Rate} = k_{\text{obs}}[\text{RCH{=}NR'}][\text{HBH}_4] = k_{\text{obs}}[\text{RCH{=}NHR'}^+][\text{BH}_4^-] \quad (5)$$

The forms of these equations predict the rate/pH profile to be bell-shaped with optimum rates in the region of neutrality. Carbamylation [Eq. (2)] is one of the very few modifications which select α- over ϵ-amino groups (*60*). Although we have seen that the latter is usually the superior nucleophile, the mechanism of this reaction involves a proton transfer from the ammonium to the cyanate ion, presumably within an ion-pair complex and possibly by a concerted mechanism (*61*). Since the α-amino group has the lower pK, it would be more effective in transferring a proton; apparently, this factor can override nucleophilicity in the ion-pair complex.

Acetylimidazole is unusually selective in the acylation of phenols in preference to amines (*62*). That protonated acetylimidazole is the effective species [Eq. (3)] has been established by careful studies (*63*), suggesting a basis for the selectivity observed. It might be predicted that this reagent should also be selective for α-amino groups; unfortunately, there seems to be no published data relevant to the question.

There is evidence, in model systems, that the ethyleneimmonium ion is significantly more reactive toward nucleophiles than is ethyleneimine [Eq. (4)] (*64*). The high degree of selectivity (and reactivity) observed in the alkylation of thiols with this reagent (*65*) may be the result of

60. G. R. Stark, *Biochemistry* **4**, 1030 (1965).

61. A. A. Frost and R. G. Pearson, "Kinetics and Mechanism," p. 307. Wiley, New York, 1961.

62. J. F. Riordan, W. E. C. Wacker, and B. L. Vallee, *Biochemistry* **4**, 1758 (1965).

63. T. C. Bruice and S. Benkovic, "Bioorganic Mechanisms," Vol. I, p. 65. Benjamin, New York, 1966.

64. L. B. Clapp, *JACS* **73**, 2584 (1951).

65. M. A. Raftery and R. D. Cole, *BBRC* **10**, 467 (1965).

facile proton transfer from the thiol and the high nucleophilicity of the resulting anion.

Finally, the observation that addition of nucleophiles occurs most readily to the protonated form of a Schiff base (*66*) provides a basis for the observation that borohydride reduction of Schiff bases between lysine and carbonyl substrates is an acid-catalyzed process (*67*).

3. *Superreactivity*

The preceding discussion (Section II,A) may suggest that the majority of protein functional groups are less reactive than the same groups in simple amino acids. The fact is inescapable, however, that a lesser number of residues, often no more than one per molecule, show greatly magnified reactivities toward certain reagents and occasionally toward all reagents tested. Although superreactivity is often associated with the enzymic process (such as the active serine of proteases), the majority of modifying reagents have little in common with either substrates or affinity labels. Familiarity with the serine case tends to create the impression that all active sites contain superreactive groups or that they exist only in active sites. Neither of these impressions is correct. Catalytically active groups can be normally reactive or even subreactive to modifying reagents. Conversely, enzymes may possess superreactive groups which have no obvious connection to function or to conformation. For example, only one of the 19 tyrosine residues of papain is reactive toward diisopropyl fluorophosphate, activity being unaltered by the modification (*68*). The catalytically vital sulfhydryl group is not phosphorylated in papain or in other sulfhydryl proteases (*69*), although the thiol is superreactive toward cyanate (*70*). Glutamic-35 of lysozyme has been assigned the catalytic role in activity; yet, it is the only carboxyl group in the enzyme which resists modification (*71*).

The reactivity, or nucleophilicity, of a functional group in a simple molecule is determined by the electron density at the significant atom, by polarizability, and by steric factors which may limit approach of the reagent. The same functional group in a protein has the same features

66. K. Koehler, W. Sandstrom, and E. H. Cordes, *JACS* **86,** 2413 (1964).
67. J. A. Anderson and P.-S. Song, *ABB* **122,** 224 (1967).
68. I. M. Chaiken and E. L. Smith, *JBC* **244,** 4247 (1969).
69. In unrelated studies, the author has observed that whereas thiols react readily with chlorophosphates they are surprisingly unreactive toward fluorophosphates. The impurity in di-isopropyl fluorophosphate which does inhibit sulfhydryl proteases (*68*) may very well be the chloro precursor.
70. L. A. A. Sluyterman, *BBA* **139,** 439 (1967).
71. T.-Y. Lin and D. E. Koshland, Jr., *JBC* **244,** 505 (1969).

to the same degree. To speak of a superreactive group in a protein as one with "enhanced nucleophilicity" is therefore inaccurate. In this case, reactivity has exceeded nucleophilicity, the excess reactivity having been effected by factors external to the functional group. Consider the facile phosphorylation of Tyr 123 in papain, cited above. There is nothing in the electronic or steric structure of this residue which distinguishes it from the remaining tyrosine residues (*71a*) or from simple tyrosine derivatives. One or more of the various reactivity enhancement factors already described, enhanced ionization of the phenolic group, decreased polarity of the microenvironment, and binding of the reagent prior to reaction, would be expected to produce a spread of reactivities within a set of residues (*6*). The all-or-none reactivity observed in the papain case is best explained by invoking general acid catalysis in the removal of the fluorine atom by a neighboring group such as carboxyl or imidazolium (*38*). Similar catalysis must be invoked to explain the rapid and selective reaction of active serine in trypsin or chymotrypsin with diisopropyl fluorophosphate.

In a few instances, quantitative comparisons of reactivity in protein and in model compound have been made. The active thiol of papain is 3,000 times as reactive as cysteine toward cyanate and 15,000 times as reactive toward bromoacetate (*70*). At pH 5, the active-site histidine of ribonuclease is alkylated 500 times as rapidly by bromoacetate as is *N*-acetylhistidine (*72*). Of 17 sulfhydryl groups in rabbit muscle phosphofructokinase, one reacts with 5,5′-dithiobis(2-nitrobenzoic acid) 20,000 times as rapidly as in the denatured enzyme (*73*). Toward the thiol of glutathione, chloroacetamide is almost 20 times as reactive as is chloroacetic acid, as might be anticipated both from electrostatic and inductive effects. Toward the essential thiol of streptococcal proteinase, chloroacetamide is 50–100 times as reactive and chloroacetate 20 times as reactive as toward glutathione (*74*). Of special interest is the fact that the two reagents show considerably different pH/activity relationships toward the enzyme, the acid being 100-fold as reactive as the amide at pH 4.2 but 1/400 as reactive at pH 10.2. It is obvious that superreactivity arises from multiple and complex sources, even involving different binding sites and orientations for only moderately different reagents. An important source of rate enhancement derives from the limitation of conformational populations (restricted freedom of rotation), which may be

71a. J. Drenth, J. N. Jansonius, R. Koekoek, H. M. Swen, and B. G. Wolthers, *Nature* **218**, 932 (1968).

72. H. G. Gundlach, W. H. Stein, and S. Moore, *JBC* **234**, 1754 (1959).

73. R. G. Kemp and P. B. Forest, *Biochemistry* **7**, 2596 (1968).

74. B. I. Gerwin, *JBC* **242**, 451 (1967).

achieved by any of several types of interaction already described. Because so little correlative data are available, it is not yet possible to attempt to evaluate the magnitude of this factor (*6*). One impressive example is found in the comparison of rates of lactonization of the phenolic acids (XIII) and (XIV). The restricted rotation achieved by introduction of the several methyl groups, particularly, the *gem*-dimethyl on the side chain, is sufficient to increase k_N for (XIV) more than 10^8-fold (*75*).

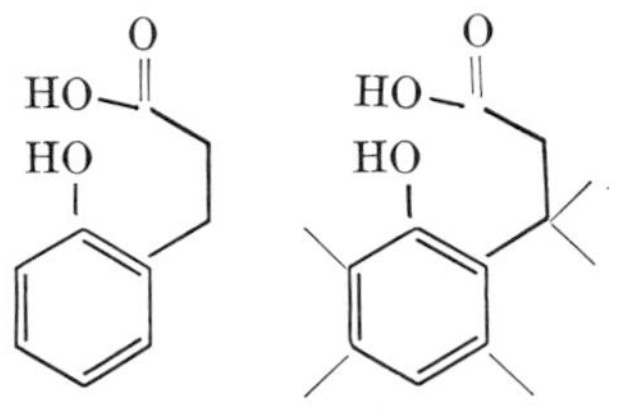

(XIII) (XIV)

III. Methodology

A. General Principles

In preliminary or exploratory studies of chemical modification, there may be little basis or reason for selecting one type of reagent over another. When nothing is known regarding the functional groups involved in catalytic activity, substrate binding, or conformational control, a random, or trial-and-error approach may be the only one usable. The fact that a particular modification appears to have no effect on activity, binding, or conformation is often a more useful observation than the converse. It is also helpful to attempt to eliminate from consideration those functional groups which are present in greatest abundance or which show the least selectivity toward modifying reagents. For example, the demonstration that acylation of the amino groups of soybean trypsin inhibitor has no effect on activity (*76*) permits an immediate concentration on the functional groups which are more difficult to modify.

It is quite difficult, if not foolhardy, to offer general guidelines on the selection of reagents, since each protein presents individual problems. One important generalization can be made, however: Modification of the pro-

75. S. Milstien and L. A. Cohen, *JACS* **91**, 4585 (1969); *Proc. Natl. Acad. Sci. U. S.* **66** (1970).

76. R. Haynes and R. E. Feeney, *BBA* **159**, 209 (1968).

tein and examination of its consequences do not constitute a fully useful study unless a number of the factors listed below are taken into account. The important clues and valuable information, thus obtained, often exceed by far the additional investment in time and effort. Certain of these factors are elaborated in Sections III,B and III,E.

(1) Following exposure of a protein to extremes of pH or to reductive or oxidative modification, a check for the appearance of new amino terminals, arising from unsuspected peptide bond cleavage, should be made.

(2) Total amino acid analyses should be performed, especially for the difficultly analyzable residues such as tryptophan and methionine.

(3) The assumption that a peak on the amino acid chromatogram represents only an intact amino acid may be misleading. Modified amino acids or oxidative degradation products sometimes appear very close to, or coincident with, the positions of normal amino acids.

(4) Extrapolations of selectivity, based on behavior of a reagent toward model compounds or other proteins, may be dangerous. A thorough demonstration of modification or survival is desirable for each residue.

(5) Various reagents may complex (or react) with functional groups without effecting a stable or permanent modification. The usual analytical techniques, including spectroscopy and degradation, often fail to detect such interactions and may require the use of special methods (Section III,E).

(6) Covalent alterations may occur subsequent to the original modification. Transacylation, transthiolation, transhalogenation, and disulfide interchange, may occur spontaneously or in the course of purification or degradation.

(7) The influence of buffer components, or of modification byproducts, on the site and degree of modification, as well as on the activity of the modified enzyme, are sometimes overlooked.

(8) The testing of activity toward one substrate, alone, is inadequate. Numerous examples are known of alterations in activity toward macromolecular substrates but not toward simple ones, toward one type of macromolecule but not toward another, and toward one type of simple substrate but not toward another.

(9) Failure to detect activity at the same pH as the optimum for the native enzyme may be misleading. In a number of cases, the pH optimum for the modified enzyme has been displaced significantly without actual loss of activity. Similarly, metal requirements may be altered in degree or character.

(10) Alterations in activity may result from a change in K_m or V_{max}, or both. Appropriate measurements should be made to permit a differentiation between the two phenomena.

B. Control of Selectivity

1. *Choice in Reagent*

The meaning of selectivity can vary from one experiment to another; for example, blocking of all amino groups and nothing else, blocking of the α-amino group alone, blocking of exposed or highly reactive amino groups, and blocking of catalytically active amino groups. Degree and site of modification may be controlled in three general ways: (a) selection of reagent and reaction conditions, (b) manipulation of protein conformation, and (c) manipulation of functionally significant residues. The two latter methods will be treated in subsequent sections. In choice of reagent, the possibilities are limited not only by the number of reagents and reaction mechanism types available but also by the fact that all protein functional groups are nucleophiles within a rather narrow range of reactivities. Subtle variations in selectivity are sometimes achieved by comparative use of reagents which differ in solubility, pK, charge, size, reactivity, stereochemistry, or mechanistic requirements. Thus, sites or degrees of modification can differ for chlorodinitrobenzene vs. fluorodinitrobenzene (*77*), 4-fluoro-3-nitrobenzenesulfonic acid vs. fluorodinitrobenzene (*78*), *O*-methylisourea vs. *S*-methylisothiourea (*79*), iodoacetamide vs. iodoacetic acid (*80*), *N*-iodosuccinimide vs. iodine (*81*).

2. *Choice of Reaction Conditions*

Selectivity may often be achieved by proper choice of pH, independently of concurrent conformational changes. The many reactions which require the proton-poor forms of nucleophiles may be repressed in acidic media, permitting selective reactions with methionine and tryptophan. Photosensitized oxidation (*82*) and alkylation (*80*) are appropriate examples. As the pH is raised, histidine becomes reactive, followed by tyrosine and ϵ-lysine. Although thiolate ion is more reactive than thiol, the high nucleophilicity of the sulfur species often results in modification, even in acidic media. As already noted (Section II,C,1), high selectivity can be achieved by pH control in neutral or alkaline media primarily with special or with weakly reactive reagents. Electrophilic reagents are often less sensitive to pH control: thus, *N*-bromosuccinimide reacts with

77. J. Kowal, T. Cremona, and B. L. Horecker, *JBC* **240,** 2485 (1965).
78. R. P. Carty and C. H. W. Hirs, *JBC* **243,** 5254 (1968).
79. W. L. Hughes, Jr., H. A. Saroff, and A. L. Carney, *JACS* **71,** 2476 (1949).
80. F. R. N. Gurd, "Methods in Enzymology," Vol. 11, p. 532, 1967.
81. H. Junek, K. L. Kirk, and L. A. Cohen, *Biochemistry* **8,** 1844 (1969).
82. L. A. A. Sluyterman, *BBA* **60,** 557 (1962).

tyrosine, histidine, and tryptophan, as well as with thiol and disulfide, in acidic or neutral media (*13*).

With some reagents, specificity is only apparent because certain derivatives do not survive isolation. Trinitrobenzenesulfonic acid reacts rapidly with thiols, but the product hydrolyzes spontaneously. Similar instability is observed with thiocarbamates, *S*- and *O*-trifluoroacetyl derivatives, and others. Acylation is apparently specific for amino groups because acylthiols and acylimidazoles, which may be formed even more rapidly, hydrolyze spontaneously or during workup. Nevertheless, one should always be prepared for the unique case, in which such a derivative may be stabilized by special features of the protein. Acetylated phenols are deacylated by high concentrations of acetate ion (*83*) or other buffer nucleophiles such as tris. Undoubtedly, buffer-catalyzed deacylation occurs more frequently than has been recognized. Buffer species may also affect modification by altering protein conformation or blocking reactive sites. Since phosphate is a competitive inhibitor of some enzymes, binding of the ion may influence sites of modification. The esterase activity of carbonic anhydrase is inhibited by chloride ion (*84*); similar effects are possible in this case.

Alteration in the reactivity or nature of the reagent by the medium should also be considered. The reactivity of sodium borohydride decreases in media containing organic solvents (*85*) as does that of *N*-iodosuccinimide (*81*). In urea solution, *N*-bromosuccinimide is converted into *N*-bromourea, a milder and more specific reagent (*86*). The slow isomerization of urea itself into ammonium cyanate provided the basis for carbamylation as a modification method (*87*). Loss of reagent, particularly oxidizing and reducing agents, by reaction with buffer components, should be kept in mind. Amines and amino acids are particularly troublesome in this regard. Finally, a number of reagents are stable, or active, only within limited pH ranges.

Selectivity may also be limited by performing the reaction at low temperature, by slow addition of reagent, and by the use of very limited quantities of reagent. Although the limited availability of reagent can result in partial modification of a number of groups, reactivities vary sufficiently, so that all-or-none modification is a common occurrence. Examples may be found in the reaction of lysozyme with triethyloxonium

83. J. F. Riordan and B. L. Vallee, *Biochemistry* **2**, 1460 (1963).
84. Y. Pocker and J. T. Stone, *Biochemistry* **7**, 2936 (1968).
85. H. C. Brown and K. Ichikawa, *JACS* **83**, 4372 (1961).
86. M. Funatsu, N. M. Green, and B. Witkop, *JACS* **86**, 1846 (1964).
87. G. R. Stark, W. H. Stein, and S. Moore, *JBC* **235**, 3177 (1960).

fluoroborate (*88*), muscle fructose diphosphatase with fluorodinitrobenzene (*89*) and staphyloccal nuclease with a cross-linking reagent (*90*).

C. Determination of Degrees and Sites of Modification

1. *Analytical Methods*

Experimental techniques for determination of the nature of groups modified and degree of modification are treated in detail elsewhere (*1, 9*) and need no elaboration here. No analytical technique meets all the requirements demanded of it. Continuous monitoring by spectroscopy is the simplest and most useful technique, also permitting facile calculation of modification rates (Section IV,B,1,b). Unfortunately, few reagents are sufficiently selective and few derivative spectra sufficiently isolated or different from those of their reagents to provide optimal use. Trinitrobenzenesulfonic acid (amine), 2-hydroxy-5-nitrobenzyl bromide (indole), and 5,5′-dithiobis (2-nitrobenzoic acid) (thiol) are examples of reagents approaching the ideal.

Far more often indirect assay methods are used, with or without protein degradation. The degree of incorporation of reagent in the protein may be determined, following isolation and purification, by isotopic label, spectral intensity of a colored modifier, or reversible removal of the modifier and assay of a product (e.g., conversion of acylphenol to acylhydroxamate with hydroxylamine). Identification of the type of site modified requires, in most cases, total degradation of the protein and amino acid analysis. The appearance of one modified amino acid can be determined more accurately, of course, than the disappearance of one of twenty identical residues. Ideally, the modified amino acid would be stable to hydrolytic conditions and appear in an isolated position on the amino acid chromatogram; either of these requirements is rarely achieved (see following section). The instability problem is avoided by use of proteolytic digestion, although some modified residues are unstable even during enzymic hydrolysis and others hamper proteolytic attack at adjacent peptide bonds. In such circumstances, it is common to perform a second modification of residual sites to produce a different and more stable modification, degree of initial change then being obtained by difference. For example, dinitrophenylation of an acylated protein, followed

88. S. M. Parsons, L. Jao, F. W. Dahlquist, C. L. Borders, Jr., T. Groff, J. Racs, and M. A. Raftery, *Biochemistry* **8**, 700 (1969).

89. S. Pontremoli, B. Luppis, W. A. Wodd, S. Traniello, and B. L. Horecker, *JBC* **240**, 3464 (1965).

90. P. Cuatrecasas, S. Fuchs, and C. B. Anfinsen, *JBC* **244**, 406 (1969).

by acid hydrolysis, should give a material balance in the sum of DNP-amino acid and recovered amino acid. Similarly, methionine sulfoxide may be determined by carboxymethylation of the intact methionines, oxidation of sulfoxide to sulfone, and chromatographic assay of the stable sulfone following acid hydrolysis (*91*).

2. *Instability of Modified Residues*

A number of modified residues are unstable to the conditions of acid hydrolysis, to subsequent chemical operations, or even to isolation or purification. As well as being effective for peptide bond hydrolysis, hydrochloric acid is a reducing agent. Under normal hydrolysis conditions, iodotyrosines are reconverted to tyrosine; iodohistidines are partially reduced and partially degraded (*92*). Methionine sulfoxide, cysteine sulfenic and sulfinic acids are also reduced. The action of hydrochloric acid as a reducing agent serves to generate chlorine, which can in turn convert tyrosine to the acid-stable 3-chlorotyrosine (*93*). Chlorine may be effectively trapped by addition of phenol or thioglycollic acid to the hydrolysis mixture. The addition of the latter reagent has been found to protect even tryptophan from oxidative destruction (*94*). The substitution of perfluoropropionic acid for hydrochloric acid may not entirely eliminate the complications described since direct electron transfer between two species can still occur.

Performic acid, when used to cleave disulfide bonds oxidatively, will simultaneously destroy iodotyrosine, aminotyrosine, and iodohistidine, in addition to oxidizing methionine and tryptophan (*95*). Performic acid will also oxidize chloride ion, associated with protein or introduced via buffer, to reactive chlorine.

The instabilities of certain dinitrophenylamino acids and phenylthiohydantoins to acid hydrolysis are well known. Acylated species usually hydrolyze completely. Instances of partial survival of ε-succinyllysine during acid hydrolysis have been reported (*96*). On the other hand, the monoamide of tetramethyleneglutaric acid cyclizes to a fully stable imide useful for chromatographic assay (*97*). Modifications which usually provide derivatives sufficiently stable to hydrolysis and homogeneous for

91. N. P. Neumann, "Methods in Enzymology," Vol. 11, p. 485, 1967.
92. M. E. Koshland, F. M. Englberger, M. J. Erwin, and S. M. Gaddone, *JBC* **238**, 1343 (1963).
93. C. H. W. Hirs, *JBC* **219**, 611 (1956).
94. H. Matsubara and R. M. Sasaki, *BBRC* **35**, 175 (1969).
95. C.-Y. Cha and H. A. Scheraga, *JBC* **238**, 2958 (1963).
96. J. J. Holbrook and R. Jeckel, *BJ* **111**, 689 (1969).
97. M. Z. Atassi, *BJ* **102**, 488 (1967).

column analysis include alkylation, guanidination, nitration, and dansylation. On occasion, instability to hydrolysis is the result of a specific feature in the reagent rather than the residue. Although 4-bromocrotonic acid alkylates and inactivates fumarase in a site-specific manner, the product is degraded by acid hydrolysis, all amino acids being recovered intact (*98*).

In addition to being light and oxygen sensitive, iodotyrosine in proteins is sufficiently unstable to undergo deiodination in the course of chromatography (*99*) or when stored in the presence of iodide ion in mildly acidic media. The reductive cleavage of disulfide bonds in modified proteins requires caution since thiols can reduce methionine sulfoxide or remove aryl groups from phenol or imidazole (*100*).

Whereas some modified amino acids or their degradation products appear in distinct positions on amino acid analyzer chromatograms, and in stoichiometric relationship to their parents, others are partially degraded, produce multiple peaks, or obscure the peaks of normal amino acids. Thus, kynurenine, obtained by ozonolysis of tryptophan, is readily analyzed and provides a useful assay for the total tryptophan content of a protein (*101*). On the other hand, the photooxidation of histidine leads, following acid hydrolysis, to a number of uncharacterized products which overlap the peaks of normal amino acids (*102*).

D. Reversible Modification

Just as an ideal modifier should be highly selective and attachable under mild conditions, it is often desirable that the functional group be capable of regeneration under equally mild conditions. Such reversibility is of value in the control of proteolytic digestion (Section IV,A,1) in demonstrating the absence of unsuspected change by recovery of activity or conformation and in temporarily blocking more reactive residues while modifying those less reactive. Clearly, the demands on a reversible modifier limit the possibilities considerably. Many of the techniques used successfully in peptide synthesis are inapplicable to proteins, either because of the extreme conditions needed for attachment or for removal of the blocking group. Ideally, a reversible modifier would be stable in one

98. R. A. Bradshaw, G. W. Robinson, G. M. Hass, and R. L. Hill, *JBC* **244,** 1755 (1969).

99. K. Hayashi, T. Shimoda, K. Yamada, A. Kumai, and M. Funatsu, *J. Biochem.* (*Tokyo*) **64,** 239 (1968).

100. S. Shaltiel, *BBRC* **29,** 178 (1967).

101. G. Galiazzo, G. Juri, and E. Scoffone, *BBRC* **31,** 158 (1968).

102. U. W. Kenkare and F. M. Richards, *JBC* **241,** 3197 (1966).

mild pH range and removable in another, or it would be sensitive to another nucleophile which has no effect on the protein.

1. *Methionine*

The only effective means for reversible blocking of methionine appears to be oxidation to the sulfoxide. In acidic media, hydrogen peroxide and photosensitized oxygen may be fairly specific for methionine oxidation. The amino acid is regenerated by reduction with mercaptans. Thus, the photosensitized (Rose Bengal) oxidation of ribonuclease in formic acid solution results in the specific modification of the four methionine residues, 13% catalytic activity remaining. Following reduction with thioglycollic acid, full activity is recovered (*103*).

2. *Tryptophan*

Formylation of the indole nitrogen is the only method of blocking tryptophan reversibly at present (*104*). The method appears to be selective and quantitative; however, survival of the modification during subsequent manipulations of a protein remains to be demonstrated.

3. *Histidine*

Surprisingly little attention has been given to the development of reversible blocking agents for histidine. Preliminary studies have shown that sulfonylation of the imidazole ring can be accomplished selectively at pH values below 5. Since the sulfonyl group is easily removed by hydroxylamine or cyanide at pH 7, reversible blocking is feasible and merits further exploration (*105*). Reaction with FDNB is an alternative possibility since the dinitrophenyl group can be removed from imidazole with 2-mercaptoethanol (*100*).

4. *Serine and Threonine*

Aliphatic hydroxyl groups are resistant to most reasonable modifying reagents, the pH necessary for the generation of an alkoxide anion being too high for the survival of most proteins. The acylation of some serine residues of pepsinogen with succinic anhydride has been reported (*106*); hydroxylamine, at neutral pH, is an effective deacylating agent. Since

103. G. Jori, G. Galiazzo, A. Marzotto, and E. Scoffone, *BBA* **154,** 1 (1968).
104. A. Previero, A. Coletti-Previero, and J.-C. Cavadore, *BBA* **147,** 453 (1967).
105. G. L. Schmir, W. M. Jones, and L. A. Cohen, *Biochemistry* **4,** 539 (1965).
106. A. D. Gounaris and G. E. Perlmann, *JBC* **242,** 2739 (1967).

this appears to be the sole observation of serine acylation, the reaction may be far from general.

5. *Tyrosine*

Several good methods are available for the reversible blocking of tyrosine. Acetylation with acetylimidazole is the most specific and the most widely used (removal with mild alkali or neutral hydroxylamine). Alternatively, the phenolic group may be dinitrophenylated (removal with 2-mercaptoethanol), succinylated (deacylation slow at pH 7 or faster with hydroxylamine), glutarylated (*107*) (the monoester formed with glutaric anhydride is significantly more stable than that with succinic anhydride but still removable with hydroxylamine), or carbamylated with cyanate (*108*) (stable at pH 5 but hydrolyzed at pH 8).

6. *Carboxyl Groups*

Reversible blocking of carboxyl groups is surprisingly difficult. Although esters are formed readily by several methods, saponification requires alkaline conditions too severe for most proteins. Benzyl esters can be formed with phenyldiazomethane, but removal by catalytic hydrogenolysis subjects methionine to desulfurization (*109*).

7. *Arginine*

Only recently has the problem of arginine modification received a concentrated effort. By examination of the products of tryptic cleavage, it has been shown that the guanidino groups of ribonuclease are blocked selectively with phenylglyoxal at pH 7, reversal being accomplished by dialysis (*110*).

8. *Sulfhydryl Groups*

The reversible blocking of thiols is readily accomplished by a variety of methods. The reductive cleavage of disulfides with sulfite (*111*), tetrathionate (*112*), phosphorothioate (*113*), or Ellman's reagent (*114*)

107. J. F. Riordan and B. L. Vallee, *Biochemistry* **3,** 1768 (1964).
108. G. R. Stark, "Methods in Enzymology," Vol. 11, p. 590, 1967.
109. C. R. Gunter and L. A. Cohen, unpublished results (1966).
110. K. Takahashi, *JBC* **243,** 6171 (1968).
111. R. D. Cole, "Methods in Enzymology," Vol. 11, p. 206, 1967.
112. T. Y. Liu, *JBC* **242,** 4029 (1967).
113. H. Neumann, R. F. Goldberger, and M. Sela, *JBC* **239,** 1536 (1964).
114. L. F. Kress and L. Noda, *JBC* **242,** 558 (1967).

produce stable derivatives which may be reversed with reagents such as dithiothreitol (*115*). Mercury derivatives of sulfhydryl groups are formed readily and are decomposed by addition of an excess of another thiol. *S*-Dinitrophenyl derivatives can be cleaved with 2-mercaptoethanol. Reaction of a sulfhydryl group with cyanate produces a thiocarbamate, stable at pH 5 but hydrolyzed at pH 8 (*116*). Although thiols react with most acylating agents, the derivatives are usually too unstable to survive manipulation.

9. *Amino Groups*

Interest in reversible blocking has centered largely on amino groups and on the control of tryptic cleavage (Section IV,A,1). Trifluoroacetylation and maleylation have received the greatest attention. Because trifluoroacetic is a strong acid, even its amide derivatives are relatively reactive. The acyl group is best removed by transamidation with aqueous piperidine. The effectiveness of piperidine results not from its basicity but from its nucleophilicity, enhanced by a cyclic structure and the resultant decrease in steric factors. On occasion, trifluoroacetylated derivatives are too insoluble for further work as in the case of phage protein (*117*). Tetrafluorosuccinic anhydride, which reacts rapidly with amino groups at pH 7 and 0°, may be substituted. The adjacent carboxylate ion facilitates hydrolysis by acting intramolecularly as a general base. Negative charge, favorable for solubility, is also introduced by means of the highly reactive maleic anhydride (*118*). In this case, the ease of removal at pH 3.5 results from restricted rotation imposed by the double bond and the intramolecular assistance of an unionized carboxyl group. In general, intramolecular assistance in the hydrolysis of amides is greatest with an adjacent unionized carboxyl group. In the case of the tetrafluorosuccinyl derivative, an undesirably low pH would be required to protonate the carboxylate ion; fortunately, an alternative hydrolytic pathway is available. Of specialized interest is the blocking of the three amino groups of insulin with *N*-*t*-butoxycarbonyl azide in dimethylformamide solution; the blocking groups were removed by exposure of the derivative to trifluoroacetic acid at 25° (*119*).

115. W. W. Cleland, *Biochemistry* **3,** 480 (1964).

116. G. R. Stark, *JBC* **239,** 1411 (1965).

117. G. Braunitzer, K. Beyreuther, H. Fujiki, and B. Schrank, *Z. Physiol. Chem.* **349,** 265 (1968).

118. P. J. G. Butler, J. I. Harris, B. S. Hartley, and R. Leberman, *BJ* **112,** 679 (1969).

119. D. Levy and F. H. Carpenter, *Biochemistry* **6,** 3559 (1967).

E. Undetected and Unsuspected Modifications

1. *Transient Modifications*

Acylation reactions of imidazole, thiol, carboxyl, or even phenol are often undetected because the products are unstable to the reaction conditions or are hydrolyzed during subsequent purification or other manipulation. Innocuous as such transient modifications may seem, their effects on conformation can, undoubtedly, influence the reactivities and accessibilities of other functional groups, producing potentially erroneous conclusions.

Transient formation of a mixed anhydride can occur by interaction of a C-terminal carboxyl group with any reasonably strong acylating agent, with sulfonyl halides, and with cyanate. The adjacent asymmetric center thus becomes liable to racemization, either directly or via an azlactone intermediate. Although the effect of C-terminal racemization on protein conformation would be difficult to predict at present, it would surely present an obstacle to complete removal of C-terminal residues by exopeptidases. Similar activation of nonterminal carboxyl groups, particularly that of aspartic acid, could result in the formation of cyclic imides and β-aspartyl peptide bonds.

Carboxyl groups or reactive esters can react with hydrogen peroxide to produce peracids; the latter are less specific and more potent oxidants than hydrogen peroxide itself, particularly in view of their intramolecular sites. Thus, the addition of *N*-acetyltyrosine ester promotes the inactivation of chymotrypsin by hydrogen peroxide (*120*). Whereas methionine alone is oxidized in the absence of substrate, tryptophan and cystine are also oxidized in its presence. A reasonable explanation is based on the assumption that hydrogen peroxide can deacylate the acylated enzyme to form a peracid.

2. *Unsuspected Modifications*

Some interactions or modifications survive detection simply because they were unanticipated or lack precedent in protein chemistry. Although mercury salts are commonly used to bind free thiols in protein, they are also capable of cleaving disulfide bridges (*121*). The iodine oxidation of

120. H. Schachter, K. A. Halliday, and G. H. Dixon, *JBC* **238**, 3134 (1963).
121. P. R. Brown and J. O. Edwards, *Biochemistry* **8**, 1200 (1969).

thiols may not proceed as far as disulfide but stop at intermediate sulfenyl iodide or sulfenic acid (*122–124*).

Tryptophan can form various complexes or undergo addition reactions not commonly recognized. Because such modifications do not necessarily produce a marked spectral change, or because the tryptophan spectrum is often masked by tyrosine, they may not be observed spectrally. Some tryptophan modifications may escape notice because of the frequent neglect of analyzing for this amino acid. Sulfite can form covalent bonds by addition to the indole ring (*125*); the bisulfite inhibition of papain may involve a reaction of this type (*126*). The possibility of other good nucleophiles (such as nitrite, cyanide, and thiol) adding to the indole ring should be kept in mind. Although the indole nitrogen is too weakly nucleophilic for most reactions, addition to the double bond of acrylonitrile has been observed (*127*). The complexing of mercuric ion to indoles has been demonstrated in model systems (*128*). The inactivation of cellulase by mercuric ion has been attributed to such complexing (*129, 130*).

The time-dependent and irreversible ionization of unreactive phenolic groups may occur in concert with covalent modification, such as the alkali-promoted β elimination of RSS^- from cystine residues (*131, 132*), a postulate which is strengthened by the isolation of lysinoalanine following alkali treatment of proteins (*133–135*).

Acyl shifts at serine and threonine residues may be induced by exposure to strong mineral acid and possibly to formic or trifluoroacetic acid. Despite the fact that such shifts are reversed at neutral pH, nonpermanent covalent and conformational changes have been introduced, which may affect the course of modification of other groups. The addition of amines to *N*-alkylmaleimides has been well documented (*96, 136*). Since the modified amino acid survives acid hydrolysis and can be de-

122. D. J. Parker, and W. S. Allison, *JBC* **244**, 180 (1969).
123. J. H. Freisheim and F. M. Huennekens, *Biochemistry* **8**, 2271 (1969).
124. D. Trundle and L. W. Cunningham, *Biochemistry* **8**, 1919 (1969).
125. A. Hesse, *Chem. Ber.* **32**, 2615 (1899).
126. K. Morihara and K. Nagami, *J. Biochem. (Tokyo)* **65**, 321 (1969).
127. R. C. Blume and H. G. Lindwall, *J. Org. Chem.* **10**, 255 (1945).
128. L. K. Ramachandran and B. Witkop, *Biochemistry* **3**, 1003 (1964).
129. K.-E. Eriksson and G. Pettersson, *ABB* **124**, 160 (1968).
130. See also T. M. Wang, M. Machee, and K. T. Yasunobu, *ABB* **128**, 106 (1968).
131. C. J. Garratt and P. Walson, *BJ* **105**, 51C (1967).
132. J. W. Donovan, *BBRC* **29**, 734 (1967).
133. A. Patchornik and M. Sokolovsky, *JACS* **86**, 1860 (1964).
134. Z. Bohak, *JBC* **239**, 2878 (1964).
135. R. S. Asquith, A. K. Booth, and J. D. Skinner, *BBA* **181**, 164 (1969).
136. C. F. Brewer and J. P. Riehm, *Anal. Biochem.* **18**, 248 (1967).

tected on the amino acid analyzer, a more careful search for its presence is often warranted.

Sodium borohydride is frequently used to reduce disulfide bonds or Schiff bases in proteins. Nonspecific reductive cleavage of peptide bonds may occur simultaneously and has been observed on occasion (*137*). In a recent study of the borohydride reduction of collagen (*138*), the isolation of a number of α-amino alcohols was taken to indicate the presence of α-amino aldehydes in the native protein. Subsequent studies on reduction with borodeuteride (*139*) revealed the principal source of the alcohols to be reductive cleavage. Although the factors favorable to bond cleavage are not yet established, it is clear that the possibility should be taken seriously.

IV. Modification as a Probe of Structure

A. Primary Structure

1. *Modification as an Adjunct to Proteolysis*

By virtue of its high degree of selectivity in the cleavage of peptide bonds following lysine or arginine, trypsin has served as a powerful tool in providing protein segments for sequence determination. Selective modification of lysine or arginine can be used to limit the number of fragments produced. If the fragments are then separated, unblocked, and reexposed to trypsin, several smaller peptides can be obtained from each fragment. Had the unmodified protein been exposed to trypsin directly, the separation of the large number of small peptides would require considerably greater effort. In other cases, it is desirable to expand the sites of tryptic cleavage by conversion of other residues into lysine or arginine analogs. By use of several modifications, together with proteolytic enzymes of different specificities, sufficient overlapping fragments can usually be obtained to elucidate the total primary sequence of a protein.

a. Blocking at Lysine. Many of the reagents available for specific reaction with ε-amino groups have been used to prevent tryptic cleavage adjacent to lysine residues: fluorodinitrobenzene (*140*), carbobenzoxy

137. A. M. Crestfield, S. Moore, and W. H. Stein, *JBC* **238**, 622 (1963).
138. P. M. Gallop, O. O. Blumenfeld, E. Henson, and A. L. Schneider, *Biochemistry* **7**, 2409 (1968).
139. M. A. Paz, E. Henson, R. Rombauer, L. Abrash, O. O. Blumenfeld, and P. Gallop, *Biochemistry,* **9**, 2123 (1970).
140. R. R. Redfield and C. B. Anfinsen, *JBC* **221**, 385 (1956).

chloride (*141*), *O*-methylisourea (*142*), succinic anhydride (*143*), potassium cyanate (*144*), and acetic anhydride (*145*). In these cases, reversible removal of the blocking group is either difficult or impossible. More recently, readily removable blocking groups have been applied in the same way: methyl acetimidate, the amidino group removed with ammonia at pH 11 or hydrazine at pH 9 (*146*); carbon disulfide, the dithiocarbamate decomposing upon acidification (*147*); ethyl thioltrifluoroacetate, the trifluoroacetyl group removed by aqueous piperidine at 0° (*148*); diketene, the acetoacetyl group cleaved with hydroxylamine at pH 7 (*148*); maleic anhydride, the maleyl group removed at pH 3.5 (*118*); citraconic anhydride, reversed at pH 3.5 (*149*); and tetrafluorosuccinic anhydride, reversed at pH 9.5 and 0° (*117*).

In selecting a blocking agent, ease of removal must be weighed against the danger of partial deblocking during proteolysis or subsequent purification. In addition, the conditions for attaching some blocking groups may be unsuitable for certain proteins. To ensure total blocking of lysine, the reaction should be performed under conditions of denaturation or maximum exposure. Even then, it is desirable to demonstrate analytically the absence of free ϵ-amino terminals. Some blocking groups can be removed without eluting the peptide from its chromatographic support. Trifluoroacetyl groups have been removed by exposure of a paper chromatogram to ammonia vapor, permitting the application of a diagonal electrophoresis method (*150*). Both the trifluoroacetyl and maleyl groups were employed in the recent elucidation of the sequence of glyceraldehyde-3-phosphate dehydrogenase (333 residues) (*151*).

b. Blocking at Arginine. A variety of reagents, almost all α-dicarbonyl compounds, have been applied to the modification of the guanidino group with varying degrees of success (*2, 5*). The alkaline conditions needed are often too severe for the survival of the disulfide bonds and other sensitive groups, nor is modification always complete. Nevertheless, encouraging results have been obtained in several cases of tryptic digestion

141. C. B. Anfinsen, M. Sela, and H. Tritch, *ABB* **65,** 156 (1956).
142. G. S. Shields, R. L. Hill, and E. L. Smith, *JBC* **234,** 1747 (1959).
143. C. H. Li and L. Bertsch, *JBC* **235,** 2638 (1960).
144. D. G. Smyth, W. H. Stein, and S. Moore, *JBC* **237,** 1845 (1962).
145. L. Weil and M. Telka, *ABB* **71,** 473 (1957).
146. J. H. Reynolds, *Biochemistry* **7,** 3131 (1968).
147. T. C. Merigan, W. J. Dreyer, and A. Berger, *BBA* **62,** 122 (1962).
148. A. Marzotto, P. Pajetta, L. Galzigna, and E. Scoffone, *BBA* **154,** 450 (1968).
149. H. B. F. Dixon and R. N. Perham, *BJ* **109,** 312 (1968).
150. R. N. Perham and G. M. T. Jones, *European J. Biochem.* **2,** 84 (1967).
151. B. E. Davidson, M. Sajgo, H. F. Noller, and J. I. Harris, *Nature* **216,** 1181 (1967).

following modification. The reaction of ribonuclease with phenylglyoxal has been shown to be reversible at pH 7, expanding significantly the potential application of arginine modification (*110*). Because arginyl–peptide bonds are cleaved more rapidly than lysyl–peptide bonds by trypsin, selective cleavage has also been achieved simply by pH control. At pH 11, arginyl bonds are split preferentially, whereas, at pH 8, the guanidino group is protected by protonation and lysyl bonds are split preferentially (*152–154*).

c. Tryptic Cleavage at Cysteine Residues. The selective alkylation of thiols is effected by use of 2-bromoethylamine or, preferably, ethyleneimine at pH 8.5 (*155*). At this pH, alkylation of other nucleophiles is not observed (*156*). The resulting *S*-(2-aminoethyl)cysteine residues are recognized by trypsin as isosteres of lysine, and adjacent peptide bonds are cleaved, albeit at significantly lower rates and yields. As is the case for lysyl–proline bonds, thialysyl–proline bonds are resistant to attack. Recent applications include ribonuclease (*157*), ferredoxin (*158*), and tryptophan synthetase (*159*). By acetylation of the amino groups of *Ascaris* trypsin inhibitor, followed by conversion of cysteine to thialysine, tryptic attack was directed both to cysteine and arginine (*160*). In order to achieve cleavage solely at cysteine, an immunoglobulin has been consecutively modified by (a) acylation of the ϵ-amino groups with succinic anhydride, (b) reduction of the disulfide bonds with dithiothreitol, (c) alkylation of the newly formed thiols with ethyleneimine, (d) condensation of the guanidino groups with cyclohexan-1,2-dione, and (e) incubation with trypsin (*161*).

d. Tryptic Cleavage at Serine Residues. The anion of 2-mercaptoethylamine has been used to effect a nucleophilic displacement on the sulfonylated or phosphorylated hydroxyl group of serine, generating a residue of thialysine (*162, 163*). Since the introduction of an effective leaving group is a property specific to "active" serine, the technique would have limited applicability to proteins.

152. L. J. Greene and D. C. Bartelt, *JBC* **244,** 2646 (1969).
153. S.-S. Wang and F. H. Carpenter, *JBC* **243,** 3702 (1968).
154. S.-S. Wang and F. H. Carpenter, *Biochemistry* **6,** 215 (1967).
155. R. D. Cole, "Methods in Enzymology," Vol. 11, p. 315, 1967.
156. M. Elzinga and C. H. W. Hirs, *ABB* **123,** 343 (1968).
157. B. V. Plapp, M. A. Raftery, and R. D. Cole, *JBC* **242,** 265 (1967).
158. S. C. Rall, R. E. Bolinger, and R. D. Cole, *Biochemistry* **8,** 2486 (1969).
159. J. R. Guest and C. Yanofsky, *JBC* **241,** 1 (1966).
160. W. Fraefel and R. Acher, *BBA* **154,** 615 (1968).
161. L. I. Slobin and S. J. Singer, *JBC* **243,** 1777 (1968).
162. A. M. Gold, *Biochemistry* **4,** 897 (1965).
163. C. Zioudrou, M. Wilchek, and A. Patchornik, *Biochemistry* **4,** 1811 (1965).

e. Cleavage by Chymotrypsin. Although chymotrypsin is considerably less selective than trypsin as an endopeptidase, preferential attack is usually observed at bonds following phenylalanine, tyrosine, and tryptophan (*164*). Despite the fact that the latter two amino acids are readily modified, little effort has been directed toward limiting chymotryptic cleavage (*165*). The bond adjacent to tyrosine-*O*-sulfate resists cleavage (*166, 167*) although it is split by subtilisin; on the other hand, the diisopropylphosphoryltyrosine residue of a modified papain is released slowly by chymotrypsin (*68*). Nitration of tyrosine does not prevent proteolysis (*168*); presumably, iodinated tyrosine would also be a substrate, although conclusive data seem to be unavailable. On the basis of limited evidence, chemical modification does not appear to block cleavage by nonspecific proteases: Both iodotyrosines, as well as iodohistidines, are released by the action of Pronase (*169*).

2. *Modification as an Adjunct to Chemical Cleavage*

Selective chemical cleavage of peptide bonds may be treated as a technique distinct from modification for the purposes of this chapter. Since the subject has been reviewed in considerable detail (*2, 3, 170*), some general comments will suffice here. Of the various cleavage reactions reported to date, two have been subjected to the demanding tests of general applicability to proteins. Cyanogen bromide cleavage at methionine is remarkable for its specificity and high yields. As evidenced by its widespread use, the method has achieved a status comparable to that of tryptic cleavage. Experience with *N*-bromosuccinimide (NBS) has been less favorable. Modification and cleavage are often competitive pathways with little possibility for control. The reagent finds its greatest utility in the selective oxidation of tryptophan when used in very limited amounts. The fact that NBS is capable of effecting bond cleavage requires a search for new amino terminals as well as total amino acid analysis, even when the reagent is used under supposedly selective conditions (*171*). Selectivity in bond cleavage by NBS can be increased by prior modification of

164. R. L. Hill, *Advan. Protein Chem.* **20,** 37 (1965).
165. E. Scoffone, A. Fontana, F. Marchiori, and C. A. Benassi, *in* "Peptides" (H. C. Beyerman, A. van de Linde, and W. M. van de Brink, eds.), p. 189. Wiley, New York, 1967.
166. A. Anastasi, V. Erspamer, and R. Endean, *ABB* **125,** 57 (1968).
167. J. E. Folk, J. A. Gladner, and K. Laki, *JBC* **234,** 67 (1959).
168. R. A. Kenner, K. A. Walsh, and H. Neurath, *BBRC* **33,** 353 (1968).
169. I. Covelli and J. Wolff, *JBC* **241,** 4444 (1966).
170. B. Witkop, *Science* **162,** 318 (1968).
171. M. J. Kronman, F. M. Robbins, and R. E. Andreotti, *BBA* **143,** 462 (1967).

the protein. Dinitrophenylation (*172*), carbobenzoxylation, or acylation (*173*) prevent attack at tyrosine; modification of tryptophan by alkylation with 2-hydroxy-5-nitrobenzyl bromide (*174*) or by ozonolysis (*175*) directs cleavage to tyrosine and histidine. Sulfonylation of the imidazole ring of histidine failed to block reaction with NBS (*176*).

3. *Location of Sequence Positions of Modified Residues*

In proteins which contain a single residue subject to modification (e.g., cysteine), sequence location poses no problem provided it has been demonstrated that the unique residue is, indeed, the one modified. Spectral change by itself may be a deceptive criterion for specific modification.

In the more general case, specific residues are located by degradation of the modified protein to fragments, either by proteolysis or by chemical cleavage. Assuming that the sequence of the protein has been previously elucidated, the fragments can be related by chromatographic behavior, composition, and sequence to their native counterparts. Modification of residues sensitive to proteolytic or chemical cleavage may decrease the rate of cleavage or prevent it entirely, reducing the total number of peptide fragments obtained. Modification near a sensitive residue may produce the same effect. It is essential that the modifier or blocking group not be lost or transferred to another site during the various manipulations. Since a variety of acylated groups are rather unstable, sulfonylation is a useful alternative.

The location of peptide fragments containing modified residues, on chromatographic columns or papers, is often facilitated by the use of isotopically labeled, colored, or fluorescent reagents (*5*). Some reagents introduce visible color automatically, e.g., tetranitromethane, dansyl chloride, fluorodinitrobenzene, and trinitrobenzenesulfonic acid. Colored analogs have also been employed, e.g., *N*-(4-dimethylamino-3,5-dinitrophenyl)maleimide in place of *N*-ethylmaleimide (*5*). Successive modifications with a reagent containing different isotopic labels at each step permits identification of the order of reactivities of the groups modified, as has been applied to the carboxyl groups of lysozyme (*71*).

172. C. Ressler and V. du Vigneaud, *JBC* **211**, 809 (1954).
173. S. Shaltiel and A. Patchornik, *Israel J. Chem.* **1**, 187 (1963).
174. M. Wilchek and B. Witkop, *BBRC* **26**, 296 (1967).
175. A. Previero, M. A. Coletti-Previero, and C. Axelrud-Cavadore, *ABB* **122**, 434 (1967).
176. G. L. Schmir, W. M. Jones, and L. A. Cohen, *Biochemistry* **4**, 539 (1965).

B. Modification in the Elucidation of Noncovalent Structure

1. *Intramolecular Interactions*

The remarkably specific folding of a polypeptide chain into the three-dimensional structure characterized as "native" or "active" appears to be determined by three groups of factors: (a) steric or conformational interactions between side chain residues; (b) electrostatic, hydrogen bond, charge transfer, and van der Waal's interactions between side chain residues; and (c) attractive or repulsive forces between side chains and solvent or solute molecules. Disulfide bridges were at one time assigned a more important role in structure determination than now appears to be justified since a number of proteins, totally devoid of disulfide bridges, are fully capable of folding uniquely, and since the oxidative re-pairing of the "correct" thiols often seems to be a consequence, rather than a director, of proper folding. Discounting states of denaturation, there would appear to be a small and limited number of three-dimensional structures available from a defined polypeptide sequence. Whether or not the genetically determined sequence of any enzyme produces the conformation optimal for a given catalytic function has yet to be determined.

a. Techniques for Probing Noncovalent Structure. In the determination of general locations of amino acid residues in a three-dimensional structure (surface, crevice, or core), the X-ray crystallographer has a considerable advantage in not being limited to the functionally reactive amino acids, which compose about half of the eighteen common protein constituents. Since relationships between protein conformation in the crystal and in solution are still uncertain (Section IV,B,1,e) and may vary from case to case, the protein chemist is faced with a formidable task.

A variety of physical, chemical, and biochemical techniques are currently in use to probe the solution structure of a native protein or to search for changes in the accessibility of functional groups effected in various ways; each technique, however, has certain inherent limitations. Physical methods such as sedimentation rate, chromatographic behavior, dialysis rate, optical rotatory dispersion, circular dichroism, thermal stability, ease of denaturation, and rate of renaturation examine the molecule as a whole and provide only qualitative data. Results obtained from studies of infrared spectra, electronic spectra, fluorescence spectra, nuclear magnetic resonance spectra, electron spin resonance spectra of spin-labeled derivatives, properties of reporter labels, and solvent perturbation are somewhat quantitative and are dependent on specific functional groups. More quantitative data are obtained from studies of the equilibria of binding of cofactors, metals, inhibitors, modifiers, or substrates. Chemical methods include studies of functional group titration and intrinsic pK values, rates and degrees of modification of reactive groups, and the ease of cleavage of disulfide bonds. Biochemical methods are based on sensitivity to proteolytic cleavage of peptide bonds, catalytic properties, and association with antibodies. Although chemical methods may provide the most quantitative data, an "uncertainty principle" may be an unavoidable consequence *(177)*: Each covalent change produces a new protein, presenting a new conformation to be probed. Some careful studies have shown that several similar functional groups in a protein can be modified at similar rates or that the modification of one group has relatively little effect on the rate of concurrent modification of another. The converse has also been demonstrated by the observation of an induction period in the modification kinetics *(98)*.

b. Measurement of Degrees of Reactivity. Given a set of similar residues, we would like to be able to assign them to subsets according to their locations on the surface, crevices, or core of the protein and, subsequently, to identify the members of each subset with respect to their sequence positions. The degree to which the spectra of chromophoric groups are perturbed, by solvent or solute molecules of varying size, is the most commonly used physical approach to the first problem *(8)*. The chemical approach involves evaluation of the "degrees of exposure" of the residues toward one or more modifying reagents *(8)*. The terms *buried* and *exposed* or *accessible* and *inaccessible* can be misleading since rate or degree of modification is actually a measure of reactivity, not of location. Whereas a functional group which is rapidly modified may be described as exposed or accessible, the converse is not necessarily true. In Section II, a variety of factors were discussed which are capable of decreasing or eliminating the reactivity of a group that is fully exposed in the steric sense. The argument is amply supported by X-ray data in a number of cases (Section VII). It follows that the number of residues per subset, or even the identity of the members of a subset, can vary according to the reagent or reaction conditions used in the study. The more types of modifying reagents and reaction conditions used, the more certain is the conclusion that a particular unreactive residue is truly in the core or in a narrow crevice. The data obtained by

177. This factor is also inherent in some physical and biochemical methods.

studying the "states of residues" in proteins (*8*) can, nevertheless, be of significant value if interpreted with caution and conservatism.

Degrees of functional group reactivities have been studied as a function of concentration or of time. In the first approach, the protein is exposed at fixed pH and temperature to an excess of reagent for a period of time sufficient for all reactive groups to be modified. The degree of modification (preferably of one type of residue) is then determined by spectroscopic or residue analysis (*178*). A series of experiments with steadily increasing concentrations of reagent may or may not result in the modification of a greater number of residues. Residues which are truly buried, or totally unreactive under the experimental conditions, will survive any concentration of reagent, their number being obtained by difference from the total in the protein. In Fig. 3, a hypothetical case is shown of a protein containing eight tyrosine residues. At pH 9, only two residues are modified at any concentration of reagent; at pH 11, three more residues are modified; and, at pH 12, a total of six are reactive. The last two residues do not become reactive unless denaturation is allowed to occur. The dashed line in Fig. 3 represents a case in

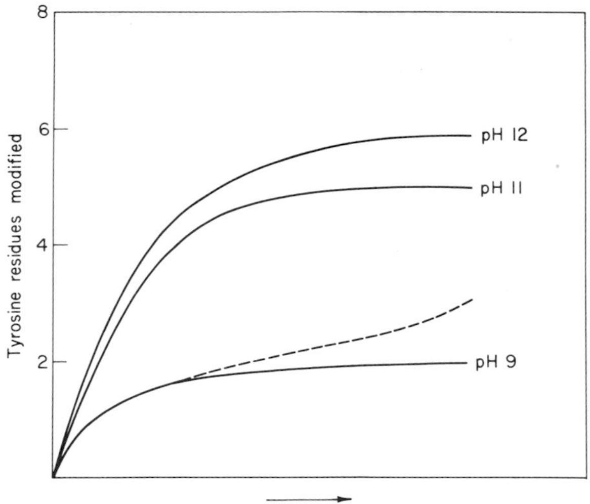

FIG. 3. Hypothetical curves for the modification of tyrosine in a protein with cyanuric fluoride as a function of concentration of reagent and pH. The dashed line illustrates a case of continuous dependence of the degree of modification on reagent concentration.

178. M. J. Gorbunoff, *Biochemistry* **6**, 1606 (1967).

which the degree of modification is concentration dependent. This event could happen if certain residues are weakly reactive or are located in crevices, modification being forced toward completion by use of higher concentrations of reagent. The interpretation of such results presents difficulties, however: Increased reactivity at higher pH may result from unfolding of conformation as is often assumed; it may also result from an increased degree of ionization of hydrogen-bonded residues or from increased reactivity of the reagent.

By following closely or continuously the rate of modification with a fixed amount of reagent, the kinetic curve obtained (Fig. 4) may be

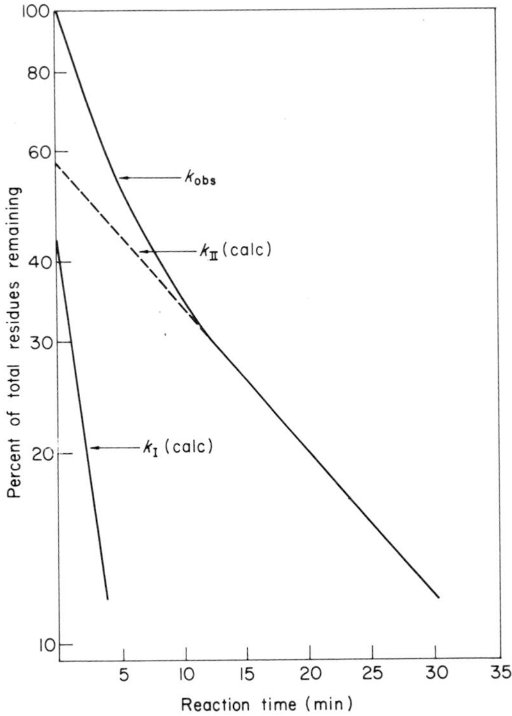

Fig. 4. Kinetic analysis of the modification of a set of residues as a function of time. The observed rate curve is resolved into k_{I} and k_{II} for subsets of differing reactivities.

179. W. J. Ray, Jr. and D. E. Koshland, Jr., *JBC* **236,** 1973 (1961); **237,** 2493 (1962).

180. M. Martinez-Carrion, C. Turano, F. Riva, and P. Fasella, *JBC* **242,** 1426 (1967).

resolved into a series of first-order slopes, each providing a rate constant for modification of different subsets (*179, 180*). Electron spectroscopy is the simplest method for acquisition of kinetic data. Few of the reagents used are truly specific, although modification of only one type of residue may be spectrally apparent. Alternatively, aliquots are removed and subjected to total amino acid analysis.

On the whole, extensive conformational change is most likely to occur with reagents of poor specificity, with reagents which alter charge significantly, with some oxidants and other reagents which are capable of cleaving peptide bonds, and, of course, with reagents which split disulfide bonds. Modification of the tyrosine residues of ribonuclease demonstrates that consistent results are obtainable with various reagents. Three of the six phenolic groups fail to titrate normally, resist solvent perturbation, resist iodination, nitration, acetylation, or reaction with cyanuric fluoride (*8, 181*). On the other hand, none of the four tyrosine residues of cytochrome c are perturbed by solvent, while two titrate in the normal range and two are acetylated by *N*-acetylimidazole (*182*).

c. Detection of Conformational Change by Modification. The demonstration that all tyrosine residues in a protein titrate normally or are equally reactive following denaturation is a common example of the use of modification as a probe of conformational change. The applications of physical, chemical, and biochemical criteria of conformational change are often more reliable and significant than are estimates of geometrical location within the matrix. Conformational changes dependent on pH, temperature, solvent composition, or addition of other species may be examined by changes in rate or extent of modification. If the reactivity or structure of the reagent is itself significantly altered at the same time, interpretation will become less meaningful. Thus, the reactivities of *N*-bromosuccinimide (*183*), hydrogen peroxide (*184*), iodine (*185*), α-haloacids and cyanate (*61*), among other reagents, vary with pH. If effect of pH on conformation is being studied, simultaneous changes in the degree of protonation of the group being modified must be corrected for.

In an acidic medium, horseradish peroxidase is carboxymethylated at methionine with elimination of activity; at pH 7.9, histidine is alkylated

181. H. A. Scheraga, *Federation Proc.* **26**, 1380 (1967).
182. D. D. Ulmer, *Biochemistry* **5**, 1886 (1966).
183. N. M. Green and B. Witkop, *Trans. N. Y. Acad. Sci.* [2] **26**, 659 (1964).
184. Y. Hachimori, H. Horinishi, K. Kurihara, and K. Shibata, *BBA* **93**, 346 (1964).
185. J. Roche and R. Michel, *Advan. Protein Chem.* **6**, 253 (1951).

in preference to methionine (*186*). These observations alone permit several conclusions:

(1) Histidine is inherently the more reactive of the two residues but requires its proton-poor form for alkylation; the degree of alkylation of histidine should then duplicate its titration curve.

(2) As the pH is raised, a negative charge is introduced near the methionine residue, hampering the approach of iodoacetate; according to this interpretation, some methionine alkylation should be observed by use of a much larger concentration of reagent or by use of the neutral iodoacetamide.

(3) With increase in pH, a conformation change is induced which buries methionine sterically, exposes histidine, or both; similar results with a variety of modifying agents would provide strong support.

A number of alternative interpretations could be offered. The effect of pH change on physical properties such as optical rotatory dispersion may reinforce a particular interpretation. The fact that the alkylation of methionine produces a greater change in optical rotatory dispersion than does the alkylation of histidine (*186*) may indicate that the methionine residue is involved in conformational control or that the modification itself, by introducing both positive and negative charge, is responsible for the conformational change.

The majority of the tryptophan residues of avidin are reactive to *N*-bromosuccinimide at pH 4.6. Biotin forms a very tight complex with avidin, each biotin molecule protecting four tryptophans from modification (*187*). The protective action is most readily explained by considering the tryptophan residues to be located in deep crevices; fitting of biotin into the crevice serves to "plug the well" and render the tryptophan residues inaccessible.

A further elaboration involves the use of a second chemical modification to probe conformational changes induced by the first. For example, the ease of tyrosine iodination in rabbit muscle aldolase is increased following photochemical oxidation of histidine (*188*). Whereas none of the disulfide bonds of lysozyme are reducible with mercaptoethanol, one becomes reducible following nitration of two tyrosine residues (*189*).

d. Intramolecular Cross-Linking as a Structural Probe. Bifunctional or cross-linking reagents were introduced to protein chemistry at an early

186. I. Weinryb, *ABB* **124,** 285 (1968).
187. N. M. Green, *BJ* **89,** 599 (1963).
188. P. Hoffee, C. Y. Lai, E. L. Pugh, and B. L. Horecker, *Proc. Natl. Acad. Sci. U. S.* **57,** 107 (1967).
189. M. Z. Atassi and A. F. S. A. Habeeb, *Biochemistry* **8,** 1385 (1969).

data because of the practical benefits to be derived from increased thermal stability, resistance to proteolysis, and insolubilization. More subtle applications in estimating interfunctional distances and identifying spatial neighbors have provided valuable information on the noncovalent structures of proteins (*190*). An early criticism of the latter use was based on the argument that two groups with sufficient flexibility would eventually come within linking distances of each other and therefore fail to reflect their thermodynamically stable loci. Although the argument is still valid, the inferences drawn from cross-linking data have, in numerous cases, been amply supported by X-ray evidence. Unfortunately, X-ray data have also failed to confirm conclusions derived from cross-linking experiments. It would appear that, except for certain terminal portions, polypeptide sequences in proteins do not have the extensive freedom of movement previously supposed. Although X-ray analysis has revealed small movements of a few functional groups upon binding of a substrate or inhibitor, the active center (or crevice) may be even more conformationally rigid than the remainder of the molecule. Interresidue distances estimated with bifunctional affinity labels (*191, 192*) may therefore be accepted with a reasonable degree of confidence.

One of the difficulties in the use of bifunctional reagents is the competitive formation of intermolecular links leading to dimers or polymers. Several ways have been utilized for keeping intermolecular reaction at a minimum: (a) the use of very dilute solutions of protein; (b) the use of a reagent whose functional ends have different specificities or reactivities; and (c) the use of a reagent with one end temporarily protected, activation being effected only after all protein molecules have been modified by the initially reactive end of the reagent. Identification of the specific residues joined by cross-linking requires the isolation and analysis of a proteolytic fragment containing nonneighboring sequences. The problem is facilitated if subsequent to isolation, the fragment can be cleaved chemically at some point along the bridge. Bifunctional reagents with an azo linkage in the center are highly suitable since the bridge can be split by dithionite reduction (*193, 194*).

Alkylation of ficin with 1,3-dibromoacetone-^{14}C results in rapid inactivation and the incorporation of 0.6 mole of reagent; concomitantly, equivalent amounts of histidine and cysteine are lost, indicating that

190. F. Wold, "Methods in Enzymology," Vol. 11, p. 617, 1967.

191. W. B. Lawson and H. J. Schramm, *Biochemistry* **4**, 377 (1965).

192. H. J. Schramm and W. B. Lawson, *Z. Physiol. Chem.* **332**, 97 (1963).

193. H. Fasold, U. Groschel-Stewart, and F. Turba, *Biochem. Z.* **337**, 425 (1963); **339**, 487 (1964).

194. H. Fasold, *Biochem. Z.* **342**, 288 and 295 (1965).

bifunctional alkylation has occurred (*195*). Following performic acid treatment and hydrolysis, 1-carboxymethylhistidine and *S*-carboxylmethylcysteine were identified. Performic acid evidently oxidized the bridge carbonyl (XV) to the two possible esters (*196*), permitting cleavage in both directions. Unless the functional groups were brought together by marked conformational change, formation of the cross-link suggests an interfunctional distance of about 5 Å.

$$R{-}CH_2{-}\overset{\overset{O}{\|}}{C}{-}CH_2{-}R' \xrightarrow{HCO_3H} \begin{cases} R{-}CH_2{-}O{-}\overset{\overset{O}{\|}}{C}{-}CH_2{-}R' \\ R{-}CH_2{-}\overset{\overset{O}{\|}}{C}{-}O{-}CH_2{-}R' \end{cases}$$

(XV)

Under appropriate conditions, staphylococcal nuclease can be mononitrated either at Tyr 85 or Tyr 115 (*197*). Following reduction of the nitro group, the weakly basic aniline was coupled with a bifunctional reagent at pH 5, the other end of the reagent remaining unaltered. At higher pH, the intact end reacted with tyrosine or lysine, according to the reagent used (*90*). The isolation of a fragment cross-linked between Tyr 85 and Tyr 115 demonstrated the stereoproximity of the two functional groups. The formation of intermolecular links in modification of the native enzyme was minimized by use of very dilute solutions of protein and by slow addition of small amounts of reagent.

e. Correlations of Chemical and X-Ray Models. One test of confidence in the reliability of chemical modification as a structural tool depends on its ability to identify residues involved in substrate binding or catalytic function. Assuming that X-ray analysis provides the most reliable assignments presently available, the high level of agreement experienced thus far should serve as an impetus to both disciplines. With respect to conformation, it is unlikely that the protein chemist will ever be able to provide more than a fragmentary description of three-dimensional structure. For noncrystallizable proteins and for proteins in solution, his may be the only source of information. It is indeed gratifying to find many of the chemically based results and inferences made visible in X-ray-derived models. Two kinds of structural assignment are obtainable from chemical data. The first follows directly from experimental fact. Thus,

195. S. S. Husain and G. Lowe, *BJ* **110,** 53 (1968).

196. P. A. S. Smith, *in* "Molecular Rearrangements" (P. de Mayo, ed.), Vol. 1, p. 577. Wiley, New York, 1963.

197. P. Cuatrecasas, S. Fuchs, and C. B. Anfinsen, *JBC* **243,** 4787 (1968).

if a cross-linking reagent is found to bridge Lys 37 and Lys 7 of ribonuclease (*198*), the X-ray model should confirm that the ε-amino groups of these residues exist at, or can be brought to, a distance of 8.6 Å of each other. The degree of conformational change, either of side chains or of backbone, needed to effect such cross-linking may not be readily ascertained by chemical methods. It is of interest that this particular modification of ribonuclease effects a 60% increase in activity toward a model substrate, suggesting that native activity can even be improved by appropriate conformational change.

The second type of conformational assignment to be derived from chemical and physical data requires the use of prediction as well as fact. One of the most impressive examples of conformational prediction comes from studies of tyrosine–carboxylate hydrogen bonds in ribonuclease. By combination of data on partial iodination, partial esterification, titration, electronic spectra and kinetics of heat denaturation, Scheraga (*181*) was able to predict noncovalent interactions of Tyr 25 . . . Asp 14, Tyr 92 . . . Asp 38, and Tyr 97 . . . Asp 83. Examination of X-ray models reveals the first two interactions to be highly probable and the third less so, but not impossible. Considering the presence in the enzyme of six phenolic and eleven carboxyl groups, the three hydrogen-bonded pairs had to be selected from a total of 3300 possible combinations.

2. *Intermolecular Interactions*

The same noncovalent interactions which serve to maintain the three-dimensional integrity of a protein molecule also serve to maintain specific associations between separate molecules. From dimers (insulin) to super-aggregates (virus coat proteins), the identical, or sometimes nonidentical, subunits associate in highly ordered geometrical arrays, depending on specific surface features to provide the attractive forces. Similar forces are responsible for the association of antigen and antibody, enzyme and its polypeptide inhibitor, enzyme and its small or high molecular weight substrate (in most cases), and association of protein with cell wall or membrane. Covalent association via intermolecular disulfide bridges (and possibly other covalent bonds) occurs in certain instances. Except for the latter type, the specific functional groups or surface loci involved in intermolecular association have been identified in very few cases. On occasion, X-ray analysis has revealed the specific residues or regions responsible for intermolecular interaction in the crystal. It is noteworthy that the areas of contact between subunits appear to be relatively small, with only a few groups participating. Similarly, the area of contact be-

198. F. C. Hartman and F. Wold, *JACS* **88**, 3890 (1966).

tween antigen and antibody may be rather confined, the antibody sometimes having multiple combining sites of differing specificities (Section VI,A).

There is considerable variation in the strengths of intermolecular forces, some polymers dissociating to subunits upon dilution of their solutions, increase (or decrease) in temperature, addition of salts or denaturing agents (urea, guanidine, and sodium dodecyl sulfate), or by alteration in pH (*199*). In certain instances, dissociation may occur upon removal of a metal ion (*200*), cofactor (*201, 202*), porphyrin (*203*), or product of reaction (*204*). In other cases, equilibrium is so far in favor of dimer or polymer that chemical blocking of functional groups may be necessary to prevent recombination. The formation of mercury derivatives of thiols is often used for this purpose (*205, 206*). In more extreme cases, the introduction of a countercharge by reaction of amino groups (or thiols) with a cyclic acid anhydride has been utilized. Although the blocking of amino or sulfhydryl groups may result in dissociation, it must be noted that a direct role for the functional groups in the intermolecular interaction is not thereby demonstrated; dissociation may be the indirect consequence of a general change in conformation of the subunit or an alteration in net or local charge.

The use of alkaline media for depolymerization may produce irreversible covalent changes in the protein; subunits may be soluble in media containing strong denaturants but may precipitate upon removal of the dissociating agent. To avoid these and other problems, succinylation was introduced as a method for generating subunits (*207*). Succinylated amino groups are stable, while those of phenolic or sulfhydryl groups deacylate spontaneously, or more rapidly in the presence of hydroxylamine. In addition to their bearing excess negative charge, facilitating the dissociation process, the succinylated subunits may be prepared at neutral pH and are generally soluble over a moderate pH range. A variety of protein aggregates have been dissociated by succinylation, including hemerythrin

199. F. J. Reithel, *Advan. Protein Chem.* **18,** 123 (1963).
200. M. L. Applebury and J. E. Coleman, *JBC* **244,** 308 (1969).
201. D. B. Wilson and D. S. Hogness, *JBC* **244,** 2132 (1969).
202. M.-L. Lee and K. H. Muench, *JBC* **244,** 223 (1969).
203. W. N. Poillon, H. Maeno, K. Koike, and P. Feigelson, *JBC* **244,** 3447 (1969).
204. R. G. H. Cotton and F. Gibson, *BBA* **160,** 188 (1968).
205. J. C. Gerhart and H. K. Schachman, *Biochemistry* **4,** 1054 (1965).
206. S. Ohnishi, T. Maeda, T. Ito, K. J. Hwang, and I. Tyuma, *Biochemistry* **7,** 2662 (1968).
207. I. M. Klotz, "Methods in Enzymology," Vol. 11, p. 576, 1967.

(*208*), aldolase (*209*), α-crystallin (*210*), ceruloplasmin (*211*), aspartic transaminase (*212*), and *p*-diphenol oxidase (*213*). Chemical modification may be readily tolerated when the purpose of dissociation is sequence analysis or identification of intermolecular forces. Clearly, however, dissociation effected by modification may not be acceptable when it leads to marked conformational change or loss of catalytic activity.

Maleic anhydride appears to be more specific than succinic anhydride for amino groups. Such specificity may be more apparent than real since maleylated phenols and thiols simply deacylate much more rapidly (*214*). In fact, the ease of deacylation of maleylated amines (pH 3.5, 37°) makes the method particularly attractive for dissociation. Aldolase (*215*), methionyl tRNA synthetase (*216*), and yeast alcohol dehydrogenase (*217*), among other enzymes, have been dissociated into subunits by maleylation.

Fumarase is dissociated by reaction of its sulfhydryl groups with charged reagents (iodoacetate, cystine, the disulfide of 2-mercaptoethylamine); on the other hand, neutral reagents (*N*-ethylmaleimide, iodoacetamide, the disulfide of 2-mercaptoethanol), although reacting equally well with the sulfhydryl groups, fail to effect dissociation (*45*). Since the sulfhydryl groups are probably embedded in hydrophobic pockets, the neutral modifiers may have less of an effect on overall conformation than charged modifiers. Dissociation via electrostatic repulsion must also be considered.

Rabbit liver fructose-1,6-diphosphatase is considered to be a tetramer composed of two distinct types of subunits, one serving a catalytic (C) and the other a regulatory (R) role. Dissociation by acidification to pH 2 produces unsymmetrical dimers ($R_2C_2 \rightarrow 2RC$), whereas dissociation by means of sodium dodecyl sulfate or maleic anhydride provides symmetrical dimers ($R_2C_2 \rightarrow R_2 + C_2$) (*218*).

208. I. M. Klotz and S. Keresztes-Nagy, *Biochemistry* **2,** 445 (1963).
209. L. F. Hass, *Biochemistry* **3,** 535 (1964).
210. A. Spector and E. Katz, *JBC* **240,** 1979 (1965).
211. W. N. Poillon and A. G. Bearn, *BBA* **127,** 407 (1966).
212. O. L. Polyanovsky, *BBRC* **19,** 364 (1965).
213. J. J. Butzow, *BBA* **168,** 490 (1968).
214. P. J. G. Butler, J. I. Harris, B. S. Hartley, and R. Leberman, *BJ* **112,** 679 (1969).
215. G. Rapoport, L. Davis, and B. L. Horecker, *ABB* **132,** 286 (1969).
216. C. J. Bruton and B. S. Hartley, *BJ* **108,** 281 (1968).
217. P. J. G. Butler, H. Jörnvall, and J. I. Harris, *FEBS Letters* **2,** 239 (1969).
218. C. L. Sia, S. Traniello, S. Pontremoli, and B. L. Horecker, *ABB* **132,** 325 (1969).

The manner in which active sites are related in ribonuclease dimer has been demonstrated by an ingenious application of histidine modification. In the monomer, iodoacetate will alkylate either His 12 or His 119, but not both. Alkylation of the dimer provides a fraction in which His 12 and His 119 of the same subunit have become alkylated. The phenomenon is most readily explained by allowing His 12 of one subunit to associate with His 119 of the other, and vice versa (*219*).

V. Modification as a Probe of Function

A. Modification at the Active Site

A universal definition of the term *active site* has yet to be formulated (*4*). A functional group which participates in the formation of a covalent enzyme–substrate intermediate, such as the active serine of chymotrypsin or a lysine residue of aldolase, is readily identified. Where no such intermediate is formed, and where the precise mechanistic role of the functional group is uncertain, a broader definition may be required. If one residue is known to activate the substrate by proton transfer, another to orient or bind the substrate electrostatically, and a more distant residue to maintain an active conformation, the ranking of the residues by their importance to activity, or by their distances from the focal point of activity, may be largely arbitrary. As difficult as it is to define the limits of what shall be considered the active site, it is even more difficult, experimentally, to define the role of a particular residue with certainty. For example, a metal ion, whose removal results in loss of activity, may participate in the catalytic process as a Lewis acid, may bind or orient the substrate, may have a conformational role near or far from the substrate site, may help bind a regulator or cofactor, or may be required for linking of subunits. The precise role of the metal ion may be as difficult to identify as the residues to which it is linked. The fact that the metal ion may protect a particular functional group from chemical modification reveals little regarding the role of the metal ion or of the functional group. Such dilemmas are encountered in almost every study of enzyme function. Intensive experimentation may help reduce the number of possible interpretations but will rarely provide a unique solution. More exact answers will come from X-ray studies and from techniques yet to be developed; for the present, partial solutions should be considered preferable to none at all.

219. R. G. Fruchter and A. M. Crestfield, *JBC* **240**, 3875 (1965).

Having demonstrated that a particular type of chemical modification results in loss of catalytic activity, subsequent experiments should involve steadily decreasing amounts of reagent or shorter reaction times, the goal being the elimination of activity with the minimum degree of modification. In some cases, inactivation occurs upon modification of the first residue to react; in others, the most reactive residues are not necessarily significant in function, nor is activity always lost by modification of one residue alone. For example, in the reaction of fructose diphosphatase with *N*-acetylimidazole, six tyrosines are acetylated without loss of activity; acetylation of additional tyrosines leads to gradual inactivation with activity almost gone when ten residues have been modified (*220*).

A small residual activity may result from contamination with unmodified or partially modified protein or may be inherent in the total material. Modification of a catalytically active group should result in complete disappearance of activity, whereas even slight residual activity may indicate only a loss of substrate binding potential. Accordingly, the modified material requires purification to a state of maximal homogeneity before any inferences can be made. Whenever possible, it is desirable to demonstrate the recovery of activity by removal of the modifying groups. If a functionally active residue has been modified, disappearance of activity should be demonstrable with different classes of substrate, with large and small substrates, and over a range of pH values. A variety of additional experiments may help determine the role of the group modified: (a) inactivation may be prevented or retarded in the presence of substrate or competitive inhibitor; (b) the pH dependence for activity loss may follow the pH activity curve; (c) loss of activity may occur at approximately the same rate as modification; (d) loss of activity and degree of modification are related stoichiometrically; and (e) the same residue is modified by an appropriate affinity label.

Compliance with any of these criteria is primarily an impetus to seek further. It is only through the accumulated evidence of numerous experiments that a strong, if not absolute, argument can be presented. An interesting example of the pitfalls in active site studies is found in the modification of *E. coli* tryptophan synthetase. Incorporation of just one equivalent of *N*-ethylmaleimide-^{14}C led to total inactivation. Fragmentation of the modified protein, however, revealed the ^{14}C to be equally distributed among the three thiol groups of the enzyme (*221*).

The use of affinity labels (site-specific reagents) is probably the most effective method for identifying functional groups which may be involved

220. S. Pontremoli, E. Grazi, and A. Accorsi, *Biochemistry* **5**, 3568 (1966).
221. J. K. Hardman and C. Yanofsky, *JBC* **240**, 725 (1965).

in catalytic activity (*7, 10*). Even here, the strict criteria listed above must be met to provide a strong case for assignment of active site residues. Final, and still not unequivocal, evidence may be sought in the results of X-ray analysis.

B. Modification at Accessory Sites

Catalytic and binding roles for functional groups are most readily differentiated by modification in cases where activity is not totally lost, where K_m and V_{max} are differently affected for different substrates and where groups known to be involved in catalysis can be shown to be intact. The binding of both small and large substrates is dependent on a variety of noncovalent interactions (and sometimes covalent), ranging from weak van der Waal's to strong electrostatic forces. Unless a variety of substrates are tested for activity, a loss of binding capacity can easily be misinterpreted as a loss of catalytic activity. The ability of lysozyme to degrade some bacterial cell walls depends on an electrostatic association between positive charges on the enzyme and a negatively charged cell wall. If the amino groups of lysozyme are acetylated, activity toward *Micrococcus lysodeikticus* cell walls is lost but is unaltered toward the neutral polysaccharide, glycol chitin (*222*). Gross modifications of this kind can reveal important determinants in the manner and specificity of binding of large molecules. More confined and subtle modifications can do the same for smaller substrates. In such experiments, the goal is selective alteration of binding sites without interference with catalytic sites. Modification in the presence of small inhibitors can often provide information on residues essential to binding.

It is becoming increasingly evident that the regulation of enzymic activity in metabolic complexes is often determined by noncovalent binding, which may take the form of subunit association, binding of metabolic intermediates, or of end products. A temporary decrease in enzymic activity may be effected by a partial or total steric block of the active site, by the acquisition of charge which repels the substrate, or by conformational alterations which diminish substrate binding or activity. Conversely, binding of a regulator may enhance activity by attracting the substrate or by effecting a favorable conformational change. The polypeptide inhibitors of proteases may be viewed as regulators, and it is entirely possible that some hormones function in the same way. As yet, little is known regarding regulator binding sites, their distances from

222. N. Yamasaki, K. Hayashi, and M. Funatsu, *Agr. Biol. Chem.* (*Tokyo*) **32,** 64 (1968).

active sites, and their modes of operation. The design of affinity labels for regulator sites offers a promising research area for the future.

As difficult as it may be to differentiate catalytic from binding sites, the distinction between binding and regulatory sites presents an even more challenging task. The novice can best prepare himself for such investigations by a careful study of published examples. As is true for many other areas of biochemistry, a superhuman degree of conservatism in drawing conclusions is often advisable.

Substrates of D-amino acid oxidase require the presence of a free carboxyl group, benzoic acid acting as a competitive inhibitor. Although benzoic acid failed to protect the enzyme against inhibition by trinitrobenzenesulfonic acid, it did protect against inhibition by glyoxal. Amino acid analysis of the glyoxal-modified enzyme–benzoate complex showed the survival of one arginine residue. Following removal of benzoate, the modified enzyme is partially active. Kinetic studies showed K_m for substrate to be unaltered, reduction having occurred in V_{max} (*223*). A reasonable conclusion is that one guanidino group is needed for the binding of amino acid substrates, but modification of the remaining guanidino groups alters the catalytic site by means of a conformational change.

In the presence of Mn^{2+} or Cu^{2+}, histidine in deoxyribonuclease is alkylated by iodoacetate with inactivation, although iodoacetamide has no effect. Since a metal ion is also required for activity, its function may be to bind a negative charge in the substrate or in iodoacetate near the active site. If the metal is removed with EDTA, alkylation occurs elsewhere in the molecule without affecting activity (*224*). The evidence does not require that the metal bind at the active site; it could have a conformational role in moving some other positively charged group close to the active histidine.

C. Modification of Multifunctional Enzymes

A number of enzymes are now known to have multiple activities involving totally distinct regions of the molecule for different functions or utilizing the same catalytic site and different adjacent binding sites (*4*, *225*). By appropriate chemical modification, it is possible to diminish

223. A. Kotaki, M. Harada, and K. Yagi, *J. Biochem.* (*Tokyo*) **64,** 537 (1968).
224. P. A. Price, S. Moore, and W. H. Stein, *JBC* **244,** 924 (1969).
225. Substrates of isocitrate dehydrogenase may utilize the same binding site and different catalytic sites [R. F. Colman, *JBC* **243,** 2454 (1968)].

or eliminate one activity without affecting (or even enhancing) another. In fact, modification may be the only practical way of detecting the existence of multiple sites. The best known example is that of carboxypeptidase A, which is capable of hydrolyzing both ester and amide bonds. An increase in esterase and a decrease in peptidase activities results from replacement of zinc by other metals, acetylation, succinylation, iodination, nitration, diazonium coupling, and photochemical oxidation (*226*). Thus, a significant role for tyrosine (and possibly histidine) is indicated. Modification, kinetic, and inhibitor studies lead to a model in which a single site for hydrolysis is flanked by separate sites for the binding of esters and amides.

Alkaline phosphatase hydrolyzes *p*-nitrophenyl phosphate or transfers phosphate to other acceptors. Diazonium coupling increases transferase and diminishes hydrolase activity, while *N*-bromosuccinimide enhances both activities (*227*). Four different activities have been demonstrated for 3-phosphoglyceraldehyde dehydrogenase: dehydrogenase (acetaldehyde + $P_i \rightarrow$ acetyl phosphate); transferase (acetyl phosphate + RSH $\rightarrow P_i$ + acetyl-SR); esterase (hydrolysis of *p*-nitrophenyl acetate); and phosphatase (hydrolysis of acetyl phosphate). Alkylation of a thiol with iodoacetate inhibits all but phosphatase activity, whereas addition of cyanide inhibits only phosphatase; DPN is required for all activities except esterase, which is observed only in the absence of the cofactor (*228*). Clearly, identification of the groups involved in each activity presents a considerable challenge.

The availability of X-ray models permits an inspection for possible catalytic sites other than those normally recognized. The exposure of His 15 on the side of lysozyme to the rear of the active cleft, and its proximity to Thr 89, suggested the possibility that the enzyme may have esterase activity. Indeed, hydrolysis of *p*-nitrophenyl acetate was found to occur with a rate dependent on a group of pK 5.2 (*229*). Ester hydrolysis is not inhibited by the binding of sugar derivatives to the normal site but is prevented by iodoacetate alkylation of the histidine residue. The data, therefore, seem to implicate histidine in a second type of catalytic activity, which may be entirely coincidental or may be an evolutionary fossil.

226. B. L. Vallee and J. F. Riordan, *Brookhaven Symp. Biol.* **21**, 91 (1969).
227. I. B. Wilson, J. Dayan, and K. Cyr, *JBC* **239**, 4182 (1964).
228. J. H. Park, B. P. Meriwether, P. Clodfelder, and L. W. Cunningham, *JBC* **236**, 136 (1961).
229. D. Piszkiewicz and T. C. Bruice, *Biochemistry* **7**, 3037 (1968).

VI. Other Uses of Chemical Modification

A. Immunochemistry

Chemical modification was first applied to immunochemistry as a means of demonstrating that antibodies could be generated *in vivo* against haptens of nonbiological nature such as dinitrophenyl groups. The association of antigen and antibody depends on the same types of noncovalent interactions which help determine the conformation of a protein or bind substrate to enzyme. By chemical modification of an antibody and determination of changes in its ability to interact with antigen, information can be obtained on the types of functional groups involved in the bonding (*230–232*). For example, precipitating ability was lost by acetylation of certain antibodies but not by guanidination, suggesting the importance of electrostatic interaction (*233*). Although some modification experiments indicate electrostatic binding at the specific site to be critical (*234*), others suggest overall charge of the antibody molecule to be equally important (*235*).

Because selective modification of tyrosine often leads to inactivation of antibodies, the phenolic group is considered one of the more important sites of binding (*236*, *237*). Just as the active site of an enzyme is often at or near the protein surface, the region of an antibody which is specific for a particular antigen is probably well exposed to permit effective contact. This consideration is supported by the fact that loss of activity often occurs during the initial stages of modification, the contact residues being more reactive than others in the molecule (*238*). The combination of antigen, or small hapten, with antibody often serves to protect the contact sites from modification (as in an enzyme–inhibitor complex),

230. S. J. Singer, *in* "The Proteins" (H. Neurath, ed.), Vol. 3, p. 270. Academic Press, New York, 1965.
231. E. Haber, *Ann. Rev. Biochem.* **37,** 497 (1968).
232. As in all sections of this chapter recent references have been cited as examples; no attempt has been made to provide complete bibliographies.
233. A. F. S. A. Habeeb, H. G. Cassidy, P. Stelos, and S. J. Singer, *BBA* **34,** 439 (1959).
234. A. L. Grossberg and D. Pressman, *Biochemistry* **7,** 272 (1968).
235. E. Rüde, E. Mozes, and M. Sela, *Biochemistry* **7,** 2971 (1968).
236. A. L. Grossberg and D. Pressman, *Biochemistry* **2,** 90 (1963).
237. A. H. Good, P. S. Traylor, and S. J. Singer, *Biochemistry* **6,** 873 (1967).
238. O. A. Roholt and D. Pressman, *BBA* **147,** 1 (1967).

permitting more accurate identification of the residues involved (*237*). Other enzyme techniques, including affinity labeling (*239*, *240*), spin labeling (*241*), and intermolecular cross-linking (*242*), have been adapted to immunochemistry. The uncertainties of protein conformation apply as well to immunoproteins. The binding of antigen to antibody may induce conformational changes in either which affect the reactivities of functional groups unessential to the interaction. Furthermore, chemical modification may affect binding ability not only by direct attack at contact points but also indirectly by producing marked conformational changes from a distance (*243*).

B. Insoluble Enzymes and Antigens

The covalent binding of biologically functional proteins to insoluble supports promises to be one of the most important areas of applied biochemistry. Insoluble enzymes can be used to catalyze chemically inaccessible reactions on a preparative scale (*244*), to perform optical resolutions, and to degrade other proteins. Conversely, polymer-bound substrates, inhibitors, and cofactors can be used in a chromatographic sense to isolate and purify specific enzymes (*245*, *246*). Similarly, the binding of antigens or haptens to insoluble supports (immunoadsorbents) permits the isolation of antibodies in a relatively facile manner (*247*). Areas yet to be explored include *in vivo* generated antibodies to insolubilized viruses, tissue implantation of insoluble enzymes, antigens and antibodies, and purification of body fluids in artificial hearts and kidneys.

The preparation of active insoluble proteins is still a matter of trial and error (*247*). Just as a particular chemical modification in solution inactivates some proteins and not others, and just as certain modifications and not others inactivate a particular protein, the means of covalent coupling to polymer must be chosen with care. Among the more popular techniques have been the acylation and alkylation of protein functional groups by polymeric reagents, diazo coupling to tyrosine and histidine residues, and carboxyl coupling via carbodiimides. The activity of an

239. Y. Weinstein, M. Wilchek, and D. Givol, *BBRC* **35**, 694 (1969).
240. J. Koyama, A. L. Grossberg, and D. Pressman, *Biochemistry* **7**, 1935 (1968).
241. J. C. Hsia and L. H. Piette, *ABB* **129**, 296 (1969).
242. G. E. Frances, *Immunochemistry* **4**, 203 (1967).
243. J. Koyama, A. L. Grossberg, and D. Pressman, *Biochemistry* **7**, 2369 (1968).
244. R. J. H. Wilson, G. Kay, and M. D. Lilly, *BJ* **109**, 137 (1968).
245. P. Cuatrecasas, M. Wilchek, and C. B. Anfinsen, *Proc. Natl. Acad. Sci. U. S.* **61**, 636 (1968).
246. D. M. Jerina, and P. Cuatrecasas, *Proc. 4th Intern. Congr. Pharmacol., Basel, 1969* (in press).
247. I. H. Silman and E. Katchalski, *Ann. Rev. Biochem.* **35**, 873 (1966).

insoluble enzyme depends on the particular functional groups modified, the degree of denaturation or conformational alteration, and the accessibility of active sites to the medium. Loss of activity is sometimes surmounted by first coupling polyamino acid tails to the enzyme or antibody, followed by covalent coupling of polymer to tail residues (*248*). If the active site of an enzyme is too close to the polymer backbone, it may be able to bind small but not large substrates. Introduction of a long flexible bridge between enzyme and polymer may overcome this problem as well.

C. Modification in X-Ray Crystallography

1. *Isomorphous Heavy Atom Replacements*

Determination of the crystal structure of a protein at high resolution requires the preparation of several heavy atom derivatives which are reasonably isomorphous with the parent crystal (*249*). Heavy atoms may be incorporated by selective chemical modification (*250*), affinity labeling (*251*, *252*), metal chelation, noncovalent binding of inhibitors or analogs (*253*, *254*) and nonspecific association with complex ions (*249*). Since it is desirable to limit the incorporation to one or two heavy atoms and to know as accurately as possible the residues with which the heavy atoms are associated, it is obvious that selective chemical modification can make important contributions to X-ray crystallography. The search for an acceptable heavy atom derivative is tedious and is often the result of considerable trial and error. Consequently, the more modification reactions available for this purpose, the better.

Proteins containing a single or small number of sulfhydryl groups are readily amenable to binding of mercury derivatives, isomorphous crystals having been obtained for papain (*255*), carbonic anhydrase (*254*), and hemoglobin (*256*), among others. The dangers in assuming that mercury

248. A. Bar-Eli and E. Katchalski, *JBC* **238,** 1690 (1963).

249. D. R. Davies, *Ann. Rev. Biochem.* **36,** 321 (1967).

250. W. F. Benisek and F. M. Richards, *JBC* **243,** 4267 (1968).

251. C. S. Wright, R. A. Alden, and J. Kraut, *Nature* **221,** 235 (1969).

252. P. B. Sigler, H. C. W. Skinner, C. L Coulter, J. Kallos, H. Braxton, and D. R. Davies, *Proc. Natl. Acad. Sci. U. S.* **51,** 1146 (1964).

253. H. W. Wyckoff, K. D. Hardman, N. M. Allewell, T. Inagami, D. Tsernoglou, L. N. Johnson, and F. M. Richards, *JBC* **242,** 3749 (1967).

254. B. Tilander, B. Strandeberg, and K. Fridborg, *JMB* **12,** 740 (1965).

255. J. Drenth, J. N. Jansonius, R. Koekoek, H. M. Swen, and B. G. Wolthers, *Nature* **218,** 929 (1968).

256. D. W. Green, V. M. Ingram, and M. F. Perutz, *Proc. Roy. Soc.* **A222,** 23 (1954).

compounds bind solely to thiols have already been noted (Section III,E,2). The relative ease of utilizing sulfhydryl groups for metal binding should stimulate the exploration of various means of introducing the thiol group into proteins in a specific and limited manner, as was accomplished in the subtilisin–thiolsubtilisin conversion (*257*, *258*). The reaction of proteins with homocysteine thiolactone offers such a possibility if means could be devised to increase its selectivity (*259*). Although thiols may be generated by reduction of disulfide bonds, the high probability of a resulting conformational change limits the chances of obtaining an isomorphous crystal (*259a*).

Because of the large number of residues of other amino acids in a protein, chemical modification is best applied by taking advantage of super-reactive groups or other residues present in unique environments. Affinity labeling with *p*-iodobenzenesulfonyl fluoride has been successful for α- and γ-chymotrypsins (*254*). Iodination of tyrosine may be of utility in the few cases in which reaction can be limited to one or two residues (*260*). Covalent binding of heavy atoms to tryptophan and reactive histidine is a possibility awaiting exploration.

2. *Comparisons of Proteins in Crystal and Solution*

From the limited data already available, it is clear that the relationships of crystal and solution properties will vary from protein to protein as well as with crystal size and other parameters. A number of physical techniques have been employed in efforts to make such comparisons. Titration studies have been performed on crystals of methemoglobin (*261*) and zinc insulin (*262*); hydrogen–tritium exchange has been examined in insulin (*263*) and lysozyme (*264*); rates of binding of small molecules have been measured for lysozyme (*265*), ferrihemoglobin (*266*), ferrimyoglobin (*267*), and alcohol dehydrogenase (*268*).

257. L. Polgar and M. L. Bender, *Biochemistry* **8**, 136 (1969).
258. K. E. Neet and D. E. Koshland, Jr., *Proc. Natl. Acad. Sci. U. S.* **56**, 1606 (1966)
259. I. M. Klotz, Y. C. Martin, and B. L. McConaughy, *BBA* **100**, 104 (1965).
259a. R. Arnon and E. Shapira, *JBC* **244**, 1033 (1969).
260. P. B. Sigler, D. M. Blow, B. W. Matthews, and R. Henderson, *JMB* **35**, 143 (1968).
261. J. A. Rupley, *Biochemistry* **3**, 1524 (1964).
262. C. Tanford and J. Epstein, *JACS* **76**, 2170 (1954).
263. M. Praissman and J. A. Rupley, *JACS* **86**, 3584 (1964).
264. M. Praissman and J. A. Rupley, *Biochemistry* **7**, 2446 (1968).
265. L. G. Butler and J. A. Rupley, *JBC* **242**, 1077 (1967).
266. B. Chance and A. Ravilly, *JMB* **21**, 195 (1966).
267. B. Chance, A. Ravilly, and N. Rumen, *JMB* **17**, 525 (1966).
268. H. Theorell, B. Chance, and T. Yonetani, *JMB* **17**, 513 (1966).

Enzymic activity toward model substrates has been observed for ribonuclease A (*269*), ribonuclease S (*270*), carboxypeptidase (*271*), papain (*272*), α-chymotrypsin (*273*), and cytochrome c peroxidase (*274*). Although the crystalline enzymes appear qualitatively similar to their solutions in rate of substrate binding and in activity, both are of significantly lower magnitude.

Relatively few comparisons by chemical modification have been made to date. Crystals of γ-chymotrypsin react stoichiometrically and are inhibited by diisopropyl fluorophosphate (*275*); photooxidation of solid ribonuclease S shows the same selectivity for histidine as in solution (*102*); alkylation of histidine in ribonuclease with bromoacetic acid yields the same carboxymethylhistidines as in solution although in a different ratio (*276*); carboxymethylation of myoglobin is qualitatively similar to its solution behavior (*277*). On the other hand, tyrosine-171 is iodinated selectively in α-chymotrypsin crystals, whereas a number of tyrosine residues are iodinated in solution (*260*). Such selectivity is not observed with crystals of γ-chymotrypsin. The data suggest that, whereas conformation of a protein may not be seriously altered by crystallization, reactive groups may become less accessible as a result of aggregation alone.

VII. Discussions of Specific Enzymes

A. Subtilisin BPN′

The catalytically active site of subtilisin BPN′ shows similarities to those of trypsin and chymotrypsin. The reactive Ser 221 is specifically phosphorylated (*278*), sulfonylated, or acylated (*279*) by the same reagents that are effective for other proteinases; the same model substrates (e.g., *N*-acetyl-L-tyrosine ethyl ester) are hydrolyzed and the same sim-

269. J. Bello and E. F. Nowoswiat, *BBA* **105**, 325 (1965).
270. M. S. Doscher and F. M. Richards, *JBC* **238**, 2399 (1963).
271. F. A. Quiocho and F. M. Richards, *Biochemistry* **5**, 4062 (1966).
272. L. A. A. Sluyterman and M. J. M. De Graaf, *BBA* **171**, 277 (1969).
273. J. Kallos, *BBA* **89**, 364 (1964).
274. T. Yonetani, B. Chance, and S. Kajiwara, *JBC* **241**, 2981 (1966).
275. P. B. Sigler and H. C. W. Skinner, *BBRC* **13**, 236 (1963).
276. J. Bello and E. F. Nowoswiat, *Biochemistry* **8**, 628 (1969).
277. F. R. N. Gurd, L. J. Banaszak, A. J. Veros, and J. F. Clark, *in* "Hemes and Hemoproteins" (B. Chance, R. W. Estabrook, and T. Yonetani, eds.), p. 221. Academic Press, New York, 1966.
278. E. F. Jansen, D. Nutting, R. Jang, and A. K. Balls, *JBC* **179**, 189 (1949).
279. L. Polgar and M. L. Bender, *Biochemistry* **6**, 610 (1967).

ple aromatic compounds, phenol or indole, serve as competitive inhibitors (*280*). Furthermore, a histidine residue (His 64) also appears to be associated with catalytic function (*281*). On the other hand, subtilisin is inert to the site-specific inhibitors, *N*-tosyl-L-phenylalanine chloromethyl ketone or *N*-tosyl-L-lysine chloromethyl ketone, but it is slowly inhibited by *N*-carbobenzyloxy-L-phenylalanine chloromethyl ketone (*281*). Recognition of the fact that bromine is a more effective leaving group than chlorine prompted the synthesis of the corresponding bromomethyl ketone, which was found to be at least five times as effective an inhibitor. Use of the tritium-labeled reagent permitted the facile identification of the alkylated histidine on a peptide map (*282*). Despite the absence of disulfide bridges in subtilisin, the molecule shows considerable resistance to denaturation, some proteolytic activity being retained in 6 *M* urea or 50% alcohol (*283*).

The three-dimensional structure of the enzyme has been determined by a 2.5-Å X-ray analysis of crystals of its Ser 221–phenylmethanesulfonyl derivative (*284*). Examination of the perspective drawing (*284*) suggests that ten of the eleven lysine residues are on the surface, and the remaining residue (Lys 94) is directed into a crevice. Furthermore, none of these residues appears to be sufficiently close to the active site or to the binding channel to be involved in either function. These conclusions are in accord with the fact that only ten lysine residues can be succinylated (*283*). Despite the transformation of ten groups from positive to negative charge, no significant changes were observed in activity toward the strongly basic protein, clupein, or toward model substrates, even the pH optimum for activity remaining unaltered. On the basis of stability or optical properties of the modified protein, there was no evidence for major conformational changes.

Carbamylation with potassium cyanate (*285*) effected blocking of all eleven lysine residues; activity was lost but could be restored upon incubation of the modified enzyme with hydroxylamine. Most likely, the active serine was carbamylated, in addition to the ε-amino groups (*286*). The difference in behavior of Lys 94 toward the neutral reagent, succinic anhydride, and toward the anionic species, cyanate, may be the result of

280. A. N. Glazer, *JBC* **242,** 433 (1967).
281. E. Shaw and J. Ruscica, *JBC* **243,** 6312 (1968).
282. F. S. Markland, E. Shaw, and E. L. Smith, *Proc. Natl. Acad. Sci. U. S.* **61,** 1440 (1968).
283. A. Gounaris and M. Ottesen, *Compt. Rend. Trav. Lab. Carlsberg* **35,** 37 (1965).
284. C. S. Wright, R. A. Alden, and J. Kraut, *Nature* **221,** 235 (1969).
285. I. B. Svendson, *Compt. Rend. Trav. Lab. Carlsberg* **36,** 235 (1967).
286. D. S. Shaw, W. H. Stein, and S. Moore, *JBC* **239,** 671 (1964).

charge attraction or of a difference in reaction mechanism (Section II,C,2).

Spectral titration shows that five of the ten tyrosine residues titrate normally (*287*). Similarly, four or five tyrosine residues are nitrated or iodinated preferentially. The nitrotyrosyl residues have been identified as 6, 21, 104, 167 (or 171), and 262 (or 263) (*288*). Examination of the three-dimensional model reveals no obvious steric basis for classification since all the tyrosine residues are on the surface. Since both nitration and iodination require the participation of phenolate ions, the unreactive residues may be hydrogen bonded (Section II,A,2). Interestingly, the iodination of subtilisin crystals seems to occur at a different set of tyrosine sites (*284*).

B. Papain

The functional role of the single cysteine residue of papain (*289*) has been demonstrated by inhibitor studies (*290, 291*) and by the direct observation of an acylenzyme intermediate (*292–294*). Rapid alkylation occurs with α-haloacids (*51, 295, 296*) and with α-haloketones (*290, 291*), that with *N*-tosyl-L-lysine chloromethyl ketone being 10^6 and with bromoacetate 10^4 times as fast as with cysteine (*70, 291*). Since the rate of inactivation increases with pH, it is apparently dependent on the formation of thiolate ion. Even with α-iodopropionic acid, there is a rate enhancement of 10^3, but with α-iodopropionamide only a 10-fold difference is observed (*51*). Alkylation of other residues is not observed, even when the thiol is blocked as its mercury derivative (*297*). These high rates of reaction can result only in part from specific binding or orientation of reagents, since cyanate ion, a much smaller and linear species, forms a thiocarbamate 3000 times as rapidly as with cysteine (*70*). Fur-

287. F. S. Markland, *JBC* **244,** 694 (1969).
288. F. S. Markland, *Federation Proc.* **28,** 877 (1969).
289. I. B. Klein and J. F. Kirsch, *BBRC* **34,** 575 (1969).
290. S. S. Husain and G. Lowe, *BJ* **108,** 855 and 861 (1968).
291. J. R. Whitaker and J. Perez-Villasenor, *ABB* **124,** 70 (1968).
292. G. Lowe and A. Williams, *Proc. Chem. Soc.* p. 140 (1964).
293. M. L. Bender and L. J. Brubacher, *JACS* **86,** 5333 (1964).
294. L. J. Brubacher and M. L. Bender, *JACS* **88,** 5871 (1966).
295. A. Light, R. Frater, J. R. Kimmel, and E. L. Smith, *Proc. Natl. Acad. Sci. U. S.* **52,** 1276 (1964).
296. E. Shapira and R. Arnon, *JBC* **244,** 1026 (1969).
297. A report of histidine alkylation with bromoacetic acid [Y.-K. Sun and C.-L. Tsou, *Sci. Sinica (Peking)* **12,** 879 (1963)] could not be confirmed [A. Light, *BBRC* **17,** 781 (1964)].

ther assistance can be explained by general acid catalysis provided by a neighboring imidazolium ion (*298*). Thiol inactivation is also observed with *N*-ethylmaleimide, with various ketones (*299*) and with phenylmethanesulfonyl fluoride (*291*). Reversibility of the latter modification with dithiothreitol demonstrates that an *S*-sulfonate had been formed. It is surprising, therefore, that no inactivation is observed with di-isopropyl fluorophosphate, the only modification being phosphorylation of tyrosine-123 (*68, 69*).

From pH dependence in kinetic studies, either carboxylate or imidazole, or both have been invoked as coparticipants in the catalytic processes of acylation and deacylation (*298, 300*). The X-ray structure of papain shows clearly the proximity of Asp 64 and of His 158 to Cys 25 along the cleft of the molecule (*255, 301*). The simultaneous alkylation of His 158 and of Cys 25 by 1,3-dibromoacetone (*290*) confirms the 4–5-Å interfunctional distance calculated from the X-ray data.

Insufficient data are available to attempt to involve other residues in the cleft in substrate binding. Tryptophan-176 is highly exposed and may be implicated in function. Tetranitromethane has been reported to inactivate papain by nitration of a tryptophan residue (*126*) without any modification of tyrosine; inactivation is also achieved with a number of charge-transfer agents. A single tryptophan residue is also destroyed by *N*-bromosuccinimide or by photooxidation (*302*). It should be noted that modification of the active cysteine by these reagents has not been rigorously excluded. The possibility should be considered since 2-hydroxy-5-nitrobenzyl bromide apparently alkylates cysteine rather than tryptophan (*126*).

Exhaustive guanidination (*142*) or acetylation (*303*) of mercuripapain results in blocking of all but one lysine residue and a portion of N-terminal isoleucine without affecting activity. That the remaining lysine is not essential is suggested by the demonstration that, whereas mercuripapain is rapidly inactivated by nitrous acid, guanidinopapain is only slowly inactivated, the unreactive lysine residue surviving the treatment (*142*). The nonessentiality of the N-terminal residue, four to five lysines, and three tyrosines has also been shown by dinitrophenylation of mercuripapain crystals (*304*). Inspection of the model shows Lys 173 to be the

298. G. Lowe and A. Williams, *BJ* **96,** 194 and 199 (1965).
299. K. Morihara, *J. Biochem.* (*Tokyo*) **62,** 250 (1967).
300. J. R. Whitaker and M. L. Bender, *JACS* **87,** 2728 (1965).
301. The numbering of sequences is that proposed by Drenth *et al.* (*255*).
302. Y.-K. Sun and S.-L. Tsou, *Sci. Sinica* (*Peking*) **12,** 879 (1963).
303. W. A. Darlington and L. Keay, *Can. J. Biochem. Physiol.* **43,** 1171 (1965).
304. L. A. A. Sluyterman, J. Wijdenes, and B. G. Wolthers, *BBA* **178,** 392 (1969).

only nonsurface lysine residue present. Interestingly, Tyr 123 is highly reactive toward diisopropyl fluorophosphate in solution (*68*) but is not one of the three tyrosines blocked by dinitrophenylation of crystalline mercuripapain.

The conformation of papain is highly resistant to alteration. Full activity is observed in 8 *M* urea and in aqueous methanol, the more powerful reagent, guanidine, being required to effect denaturation (*305*). In aqueous solution, the three disulfide bridges are resistant to reduction by mercaptans; in 8 *M* urea, one bridge (56–95) is reduced selectively with loss of activity and full recovery upon reoxidation (*296*). Inspection of the X-ray model suggests this disulfide bridge to be the most exposed of the three. The isolation of a peptide with a 22–25 disulfide bridge from inactive papain suggests that disulfide interchange may be one cause of inactivation (*306*). Once again, inspection of the model suggests this interchange to be the most probable sterically.

C. Lysozyme

Lysozyme is a compact molecule with a substrate cleft approximately in the center (*307–309*). Early chemical studies, more recently confirmed, have eliminated any vital functions for amino groups. Treatment with *O*-methylisourea results in guanidination of all six ϵ-amino groups (no attack at the N-terminal), six equivalents of homoarginine being recovered following acid hydrolysis (*88*). Guanidinated lysozyme is fully active in lysis of cell walls of *Micrococcus lysodeikticus* (*310*), the pH activity profile duplicating that of the native enzyme. Methyl acetimidate was found to modify the α-amino group and three to five ϵ-amino groups; on the other hand, methyl benzimidate blocked half the ϵ-amino groups but not the α-amino group (*311*). In neither case was activity affected. Although fully ϵ-acetylated lysozyme shows only 15% activity at pH 7, the normal optimum, the pH-activity curve has, in fact, been shifted to provide a new optimum at pH 5, with about 50% activity (*88*). Because the bacterial cell wall has a net negative charge, excess positive charge

305. L. A. A. Sluyterman, *BBA* **139**, 418 (1967).

306. R. Arnon and E. Shapira, *JBC* **244**, 1033 (1969).

307. C. C. F. Blake, G. A. Mair, A. C. T. North, D. C. Phillips, and V. R. Sarma, *Proc. Roy. Soc.* **B167**, 365 (1967).

308. C. C. F. Blake, L. N. Johnson, G. A. Mair, A. C. T. North, D. C. Phillips, and V. R. Sarma, *Proc. Roy. Soc.* **B167**, 378 (1967).

309. D. C. Phillips, *Proc. Natl. Acad. Sci. U. S.* **57**, 484 (1967).

310. In this discussion, the activity of lysozyme (hen egg white) refers to lysis of bacterial cell walls, unless otherwise specified.

311. A. McCoubrey and M. H. Smith, *Biochem. Pharmacol.* **15**, 1623 (1966).

on the enzyme is needed for effective association. Thus, guanidination or amidination, by retaining positive charge, has little or no effect on binding. Acetylation, on the other hand, reduces binding ability by neutralization of charge and shifts the pH optimum. Toward the neutral substrate, glycol chitin, acetylated lysozyme is fully active, with only a small change in the pH optimum (*222*). Presumably, different forces are involved in binding the two macromolecular substrates. Despite the retention of activity (and optical properties) of amine-modified lysozyme, an increased susceptibility to proteolysis has been observed, possibly because of unfolding of terminal sequences. Total succinylation of amino groups also produces a conformational change, one disulfide bond becoming susceptible to reduction (*312*).

Of the three tyrosine residues present, only two are acetylated by acetylimidazole, with a loss of 70% activity. More thorough examination revealed, however, that three ε-amino groups were acetylated simultaneously, obscuring an estimate of the role of tyrosine (*88*). Similar treatment of the fully guanidinated enzyme resulted in acetylation of all three tyrosines, with full retention of activity. Since it has been suggested that Tyr 53 is hydrogen bonded to Asn 66 (*308*), the latter tyrosine may be the one resistant to acetylation. One of the tyrosine residues is also found to have an abnormally high p*K* (*313*) and to be resistant to cyanuric fluoride (*314*), iodine (*315*), and tetranitromethane (*189*). The unreactive residue has been identified as Tyr 53 in the latter two modification reactions (*189*). The presence of only 50% activity in dinitrolysozyme could result from conformational change or from a reduced ability to bind the bacterial polysaccharide because of the lowering of the phenolic p*K*. Dithionite reduction of the nitrophenols to aminophenols restores the phenolic p*K* to its original range (*189*). Since activity remains at the 50% level after such treatment, conformation change seems a more reasonable explanation. The conclusion is supported by the fact that both dinitro- and diaminolysozyme show enhanced ease of proteolysis and of disulfide bond reduction.

The single histidine residue of lysozyme is evidently nonessential, since it can be alkylated (at N-1) with iodoacetate, iodoacetamide, or 4-bromoacetamido-2-nitrophenol, all without effect on activity (*88*). In contrast is the report that inactivation occurs upon coupling of histidine with diazo-1H-tetrazole (*316*). Since the site and extent of diazo coupling

312. A. F. S. A. Habeeb, *ABB* **121,** 652 (1967).
313. Y. Inada, *J. Biochem.* (*Tokyo*) **49,** 217 (1961).
314. K. Kurihara, H. Horinishi, and K. Shibata, *BBA* **74,** 678 (1969).
315. I. Covelli and J. Wolff, *Biochemistry* **5,** 860 (1966).
316. H. Horinishi, Y. Hachimori, K. Kurihara, and K. Shibata, *BBA* **86,** 477 (1964).

was determined by spectral data alone, modification at spectrally insensitive sites cannot be excluded.

The action of oxidizing agents on lysozyme presents a confusing but fascinating array of specificities and some contradictions. In an early study of photooxidation (methylene blue, pH 7), the destruction of histidine and one tryptophan resulted in the retention of 30% activity (*317*). On the other hand, in a similar experiment using 84% acetic acid as solvent, the two methionine residues were the only groups modified, with a 95% loss of activity (*318*). Since X-ray models (*308*) show no reasonable role for methionine in binding or catalysis, a conformational alteration seems the most likely cause for the effect. It should be noted that the introduction of polar sulfoxide groups is more likely to effect a conformational change than the disruption of any preexisting interactions of methionine. The occurrence of a conformational change is shown by an increased ease of proteolysis and reduction of disulfide bonds as well as by a change in optical parameters. Activity was restored by reduction of the sulfoxides with mercaptoethanol. That different sensitizers can provide different specificities in photooxidation is revealed by the observation that the proflavin-sensitized reaction (in formic acid) leads to the oxidation not only of the methionine residues but also of all six tryptophans (*101*). In this case, reductive regeneration of methionine does not restore activity.

In a mixture of simple amino acids, tryptophan and methionine are oxidized by ozone with equal ease. With lysozyme in formic acid solution, however, the only oxidative event is the conversion of two tryptophans (Trp 108 and Trp 111) to formylkynurenine without loss of activity (*319*). With the temperature-dependent destruction of a third tryptophan residue, activity disappears. In the same solvent, methionine is reactive toward photosensitized oxygen (*101*) but not toward ozone. Clearly, the difference must reside in the mechanisms of reaction of the two reagents. Tryptophan-108 has been invoked as a key residue for substrate binding; the retention of activity after partial ozonolysis is surprising, since limited iodination at pH 5.5 oxidizes the same residue specifically to an oxindole with total loss of activity (*320*). One explanation proposed is that *N*-formylkynurenine may be able to substitute for the indole nucleus as a binding site, whereas oxindole cannot (*319*). A different tryptophan residue (62) is converted to an oxindole with a limited amount of *N*-

317. L. Weil, A. R. Buchert, and J. Maher, *ABB* **40,** 245 (1952).
318. G. Juri, G. Galiazzo, A. Marzotto, and E. Scoffone, *JBC* **243,** 4272 (1968).
319. A. Previero, M. A. Coletti-Previero, and P. Jollès, *JMB* **24,** 261 (1967).
320. F. J. Hartdegen and J. A. Rupley, *JACS* **89,** 1743 (1967).

bromosuccinimide at pH 4.5, again with loss of activity (*321*). The specific action at Trp 108 may be the result of an intramolecular oxidation, effected by the transient formation of an *N*-iodo derivative at the neighboring histidine residue (*2*).

Experience with *N*-bromosuccinimide oxidation of lysozyme offers an example of the dangers inherent in judging degree or site of modification on spectral changes alone. When amounts of reagent necessary to modify more than one tryptophan residue are employed, competitive reaction is observed at tyrosine and histidine as well (*171*).

Less than one residue of tryptophan is alkylated with 2-hydroxy-5-nitrobenzyl bromide, without effect on activity. In the presence of 10% 2-chloroethanol, conformational changes occur which permit two residues to be alkylated, still without loss of activity (*322*). As more perturbing solvent is introduced, additional residues become reactive, activity decreasing, with only 10% activity remaining following the alkylation of five residues in 20% 2-chloroethanol. It is interesting that two tryptophan residues are also modified by ozonolysis without loss of activity, although there has been no demonstration that the same two residues are modified in each case.

The involvement of two carboxyl groups in catalysis (Glu 35 and Asp 52), proposed on the basis of X-ray data (*308*), prompted efforts to effect selective blocking of carboxyls. Use of the classic method, ethanol-hydrogen chloride, offers little selectivity. Of the eleven carboxyl groups present, between five and six are rapidly esterified with almost total loss of activity. More specific esterification has been achieved by use of triethyloxonium fluoroborate in aqueous solution (*88*). At pH 4, and low concentration of reagent, one carboxyl group was esterified with retention of 57% activity. Upon storage at pH 7.2, the modified material reverted slowly to the fully active enzyme. At pH 4.5, a diester was obtained; following hydrolysis of the labile ester bond, the remaining monoester showed no activity.

By use of a water-soluble carbodiimide to activate carboxyl groups, and coupling with aminomethanesulfonic acid in the presence of virtual substrates, all carboxyls except Asp 52 and Glu 35 were blocked, 56% activity remaining (*71*). Following removal of the protecting substrate and reesterification, Asp 52 was blocked, with total loss of activity. The results of both these studies suggest that Glu 35 is resistant to esterification and that Asp 52 is involved either in binding or catalysis. Identification of the labile ester bond has not been accomplished, although Asp 18,

321. K. Kayashi, T. Imoto, G. Funatsu, and M. Funatsu, *J. Biochem.* (*Tokyo*) **58,** 227 (1965).

322. T. A. Bewley and C. H. Li, *Nature* **206,** 624 (1965).

which may be a determinant of conformation, is a possible candidate. It is tempting to invoke an intramolecularly catalyzed hydrolysis of the ester by the neighboring His 15 (*229*). Thus, an acylimidazole intermediate could be formed, which then hydrolyzes rapidly.

VIII. Conclusions

In this chapter, an attempt has been made to outline the ways in which chemical modification can be used to acquire information regarding structure and function of enzymes or proteins in general. Repeated admonitions have been included on the dangers of drawing unwarranted conclusions, conclusions based on superficial or incomplete data, or conclusions which fail to consider indirect influences. Acquaintance with the physical organic principles, by which interaction of reagent and enzyme are guided, may help in the design of more revealing experiments or of more selective reagents.

While the contributions of X-ray crystallography to protein chemistry are both impressive and invaluable, no single tool can provide a complete knowledge or understanding of enzymes. Information derived from chemical, physical, and biochemical techniques must be painstakingly and cautiously integrated with that from X-ray models. The increasing refinements in nonchemical methods should serve as an impetus for increased sophistication in chemical methods by providing independent audits on the results and conclusions. To date, the reagents used in chemical modification have been limited, largely, to very simple, readily available compounds. Increased specificity can be achieved by use of more complex, polyfunctional reagents designed according to some of the principles outlined herein. The concepts of active site and reporter labeling offer notable examples, which can be extended to binding, cofactor, and regulator sites. Control of modification by techniques such as rapid flow (*323*) or use of polymeric reagents (*324*) merit exploration. Practical uses of modification in the medical applications of immunochemistry and industrial applications of insoluble enzymes await development. Finally, chemical modification may contribute to an understanding of enzyme mechanisms by helping to bridge the wide gap between chemical models and active sites.

323. S. Isoe and L. A. Cohen, *ABB* **127,** 522 (1968).
324. L. A. Cohen, *J. Polymer, Sci.* **49,** 42 (1969).

4

Multienzyme Complexes

LESTER J. REED • DAVID J. COX

I. Introduction

Multienzyme complexes are aggregates of enzymes that catalyze two or more steps in a metabolic sequence. The multienzyme complexes that have been most thoroughly described are composed of two to six functionally related enzymes and have molecular weights ranging from a few hundred thousand to a few million. The particles contain definite amounts of each enzyme, and the component proteins are organized in a specific and regular way. They are not associated with lipids or nucleic acids, and they do not usually contain structural proteins which have no enzymic function. These multienzyme complexes thus represent a much simpler level of organization than the complex arrays of membrane-bound enzymes associated with subcellular organelles. Because they are relatively simple, they present some special opportunities for examining the principles governing cellular organization. The structures of a number of these aggregates have been described in considerable detail, and the components of several of them have been resolved. In a few cases, it has been possible to assemble particles resembling the native complexes from their separate parts.

Multienzyme complexes have been observed to participate in reactions involving a number of very different substrates, and they have been found in a variety of biological systems. It seems likely that this mode of subcellular organization is fairly common, and it is certainly not restricted to a narrow range of organisms. Statements regarding the role of multienzyme complexes in the cellular economy can be no more than speculative at present. It is clear, however, that the construction of a multienzyme complex provides opportunities, at least, for increased specificity and efficiency and for modes of control that would not be made available by simply confining structurally independent enzymes within the same compartment of the cell (*1*). A few comments on these points will be advanced later in the article.

What follows is a description of certain features of a few well-documented examples of multienzyme complexes. The discussion is intended to be illustrative rather than exhaustive, and we have omitted reference to a considerable number of cases in which preliminary data suggest the existence of multienzyme complexes, but the particles have thus far resisted isolation.

II. α-Ketoacid Dehydrogenase Complexes

Enzyme systems that catalyze a CoA- and DPN-linked oxidative decarboxylation of pyruvate and α-ketoglutarate have been isolated from *Escherichia coli* and from animal tissues as functional units with molecular weights in the millions. Two classes of complexes have been obtained, one specific for pyruvate, the other for α-ketoglutarate. Each of these multienzyme complexes has a distinctive structure and catalyzes a coordinated sequence of reactions (Fig. 1) (*1–3*). Several of them have been separated into their component enzymes and have been reassembled from the isolated enzymes.

1. L. J. Reed and D. J. Cox, *Ann. Rev. Biochem.* **35,** 57 (1966).
2. I. C. Gunsalus, *in* "The Mechanism of Enzyme Action" (W. D. McElroy and B. Glass, eds.), p. 545. Johns Hopkins Press, Baltimore, Maryland, 1954.
3. L. J. Reed, "The Enzymes," 2nd ed., Vol. 3, p. 195, 1960.

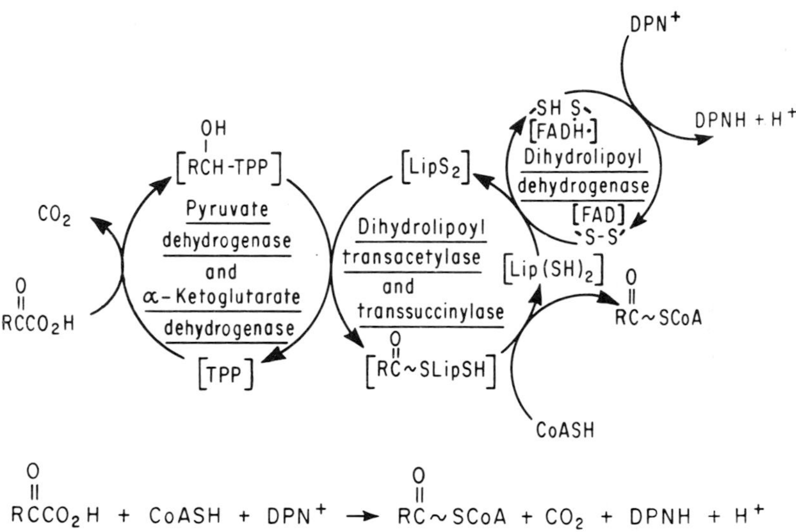

FIG. 1. Reaction sequence in pyruvate and α-ketoglutarate oxidation. The abbreviations used are; TPP, thiamine pyrophosphate; $LipS_2$ and $Lip(SH)_2$, lipoyl moiety and its reduced form; CoASH, coenzyme A; FAD, flavin adenine dinucleotide; DPN^+ and DPNH, diphosphopyridine nucleotide and its reduced form.

A. PYRUVATE DEHYDROGENASE COMPLEXES

1. *Composition and Organization*

The *E. coli* pyruvate dehydrogenase complex (molecular weight about four million) has been separated into three enzymes: pyruvate dehydrogenase, dihydrolipoyl transacetylase, and dihydrolipoyl dehydrogenase (a flavoprotein) (*4, 5*). When the three components are mixed at neutral pH, a large unit forms spontaneously; this unit resembles the native pyruvate dehydrogenase complex in composition, enzymic activities, and appearance in the electron microscope (*5–7*). The pyruvate dehydrogenase and the flavoprotein do not combine with each other, but each of

4. M. Koike, L. J. Reed, and W. R. Carroll, *JBC* **235**, 1924 (1960).
5. M. Koike, L. J. Reed, and W. R. Carroll, *JBC* **238**, 30 (1963).
6. H. Fernández-Morán, L. J. Reed, M. Koike, and C. R. Willms, *Science* **145**, 930 (1964).
7. C. R. Willms, R. M. Oliver, H. R. Henney, Jr., B. B. Mukherjee, and L. J. Reed, *JBC* **242**, 889 (1967).

these components does combine with the transacetylase. Apparently, the transacetylase plays both a catalytic and a structural role.

Biochemical and electron microscopic data (*8*) indicate that the *E. coli* pyruvate dehydrogenase complex is an organized mosaic of enzymes in which each of the component enzymes is so located as to permit efficient coupling of the individual reactions catalyzed by these enzymes. Electron micrographs of the pyruvate dehydrogenase complex negatively stained with phosphotungstate show a polyhedral structure with a diameter about 300 Å (Figs. 2A and 3G). Electron micrographs of the transacetylase, the pyruvate dehydrogenase, and the flavoprotein are shown in Fig. 2, parts B, C, and D, respectively.

The transacetylase consists of 24 identical polypeptide chains with a molecular weight of about 40,000, which are linked by noncovalent bonds (*7, 9, 10*). Each chain contains one molecule of covalently bound lipoic acid. The transacetylase has been reversibly dissociated into subunits; it is apparently a self-assembling system. The appearance of the transacetylase (Figs. 2B and 3C) indicates that it contains eight morphological subunits which are located at the vertices of a cube. Each morphological subunit apparently consists of three chains. In the model shown in Fig. 3, D through F, these subunits are represented as spheres, and each sphere is subdivided into three equal parts to represent the three chains.

Electron micrographs of the pyruvate dehydrogenase show a variety of triangular and rectangular images (Figs. 2C and 3A), which apparently represent different orientations of the molecule. The appearance of these images suggests that pyruvate dehydrogenase has a tetrahedron-like shape. An interpretative model is shown in Fig. 3B (*8*).

The basic geometry of the *E. coli* pyruvate dehydrogenase complex has been deduced from the biochemical and electron microscopic data (*8*). It appears that 24 pyruvate dehydrogenase units of molecular weight about 90,000 and 24 flavoprotein units of molecular weight about 55,000 are distributed in a regular manner along the edges of the transacetylase cube, one unit of each of the former components being attached to each transacetylase chain. An interpretative model of the complex is shown in Fig. 3, H through J.

The results of this structural and biochemical analysis suggest that the *E. coli* pyruvate dehydrogenase complex is built up by a stepwise self-assembly of subunits, involving specific noncovalent interactions between

8. L. J. Reed and R. M. Oliver, *Brookhaven Symp. Biol.* **21,** 397 (1968).
9. H. R. Henney, Jr., C. R. Willms, T. Muramatsu, B. B. Mukherjee, and L. J. Reed, *JBC* **242,** 898 (1967).
10. E. R. Schwartz and L. J. Reed, *JBC* **244,** 6074 (1969).

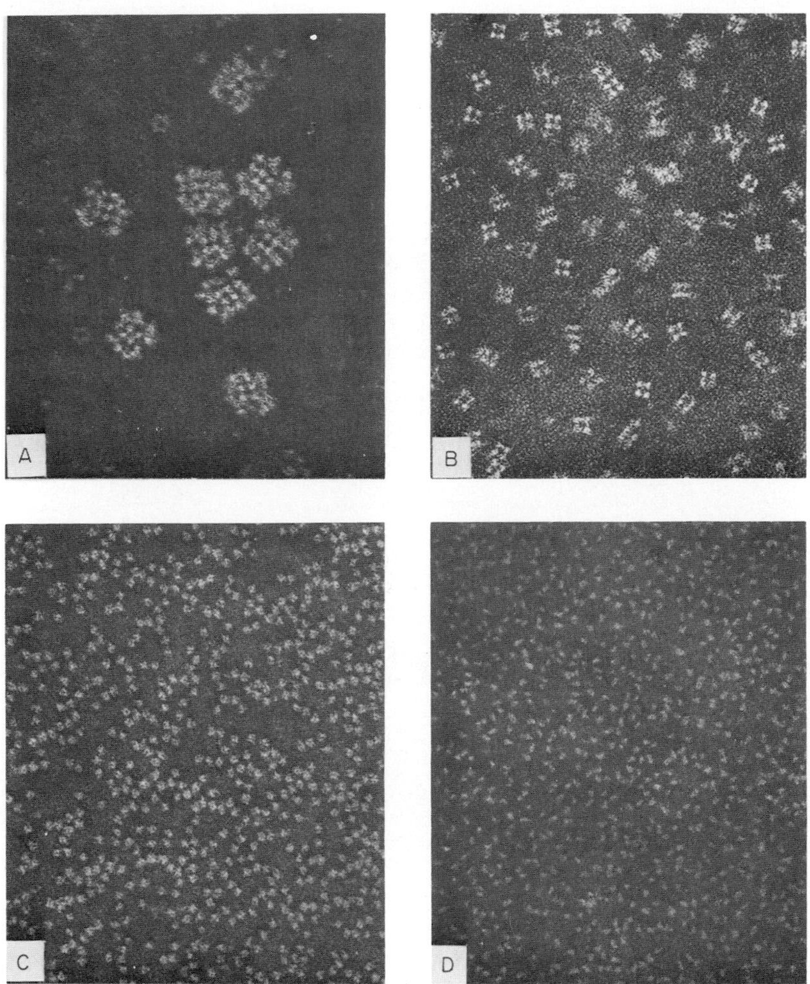

FIG. 2. Electron micrographs of the *E. coli* pyruvate dehydrogenase complex and its component enzymes, negatively stained with phosphotungstate. ×200,000. (A) Pyruvate dehydrogenase complex, (B) dihydrolipoyl transacetylase, (C) pyruvate dehydrogenase, and (D) dihydrolipoyl dehydrogenase.

these units, and that the size and shape of the complex are determined by the packing arrangement of its subunits. In this connection, it should be noted that Henning and co-workers (*11*) have found that the structural genes for the pyruvate dehydrogenase and for the transacetylase

11. U. Henning, G. Dennert, R. Hertel, and W. S. Shipp, *Cold Spring Harbor Symp. Quant. Biol.* **31**, 227 (1966).

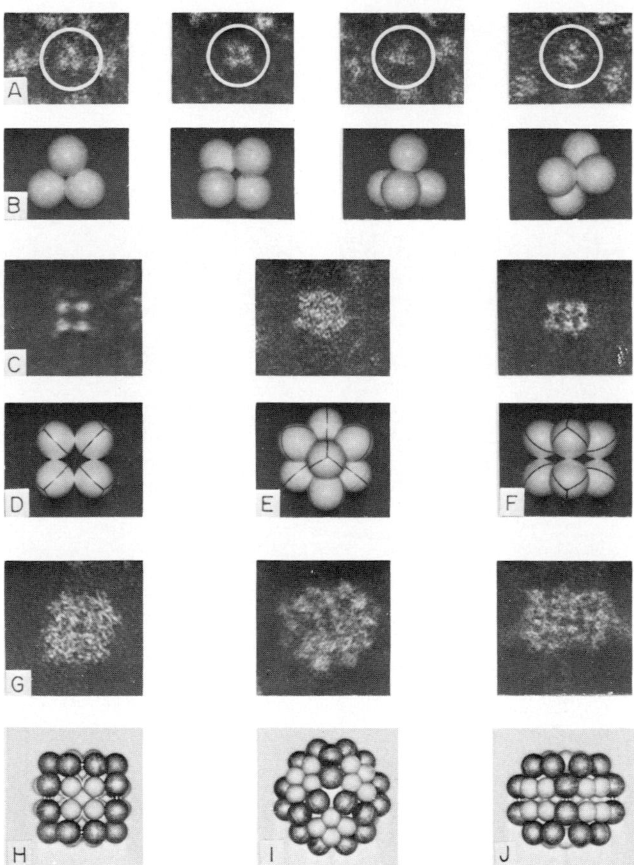

FIG. 3. Electron micrograph images and interpretative models of pyruvate dehydrogenase, dihydrolipoyl transacetylase, and the pyruvate dehydrogenase complex. (A, B) Individual images (×700,000) showing various orientations of pyruvate dehydrogenase, and corresponding views of the model. (C) Various orientations of the transacetylase. ×350,000. (D, E, F) Model of the transacetylase consisting of eight spheres at the vertices of a cube, viewed down a 4-fold, 3-fold, and 2-fold axis, respectively. (G) Various orientations of the pyruvate dehydrogenase complex. ×350,000. (H, I, J) Corresponding views of a model of the complex photographed down a 4-fold, 3-fold, and 2-fold axis, respectively, of the transacetylase cube. The 24 pyruvate dehydrogenase units (dark spheres) and 24 flavoprotein units (light spheres) are distributed in a regular manner along the edges of the transacetylase cube.

are closely linked on the *E. coli* chromosome and that biosynthesis of the complex begins with the biosynthesis of the pyruvate dehydrogenase component. The location of the structural gene for the flavoprotein is not yet known.

Interactions among the prosthetic groups of the three separate enzymes of the pyruvate dehydrogenase complex occur within a unit in which movement of the individual enzymes is restricted and from which intermediates do not dissociate. Highly favorable positioning of the three enzymic components and, by inference, of the prosthetic groups of these components must be assumed in order to account for the occurrence of the overall reaction. Although there is as yet no experimental evidence, the possibility has been considered that one or more of the component enzymes of the complex undergoes a conformational change during the coordinated reaction sequence bringing the bound prosthetic groups into juxtaposition. An alternative possibility is that the lipoyl moiety, which is bound covalently to the ϵ-amino group of a lysine residue in the transacetylase (Fig. 4) *(12)*, rotates between the prosthetic groups of the other two enzymes (Fig. 5).

Pyruvate dehydrogenase systems have been isolated from animal tissues as functional units with molecular weights in the millions *(13–17)*. The pyruvate dehydrogenase complex from pig heart muscle is reported to have a molecular weight of 7.4 million *(15)*. These complexes resemble the *E. coli* pyruvate dehydrogenase complex in that they contain a pyruvate dehydrogenase, a dihydrolipoyl transacetylase, and a dihydrolipoyl dehydrogenase (flavoprotein) *(15, 17, 18)*. However, in contrast to

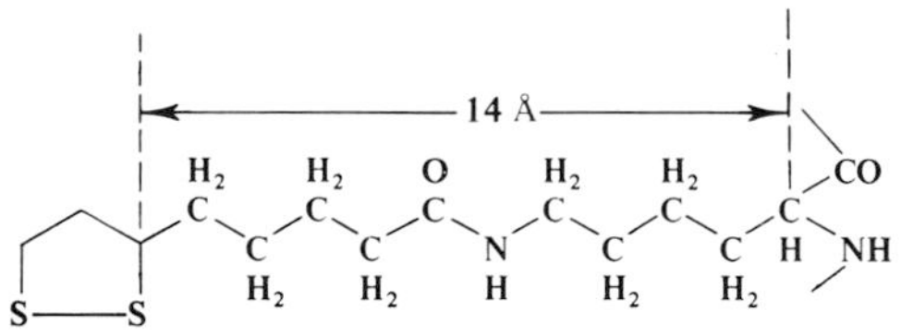

FIG. 4. Functional form of lipoic acid in the *E. coli* pyruvate and α-ketoglutarate dehydrogenase complexes. The carboxyl group of lipoic acid is bound in amide linkage to ϵ-amino group of a lysine residue. This attachment provides a flexible arm of approximately 14 Å for the reactive dithiolane ring, conceivably permitting rotation of the lipoyl moiety between the various active centers (see Fig. 5).

12. H. Nawa, W. T. Brady, M. Koike, and L. J. Reed, *JACS* **82,** 896 (1960).

13. R. S. Schweet, B. Katchman, R. M. Bock, and V. Jagannathan, *JBC* **196,** 563 (1952).

14. T. Hayakawa, M. Hirashima, S. Ide, M. Hamada, K. Okabe, and M. Koike, *JBC* **241,** 4694 (1966).

15. T. Hayakawa, T. Kanzaki, T. Kitamura, Y. Fukuyoshi, Y. Sakurai, K. Koike, T. Suematsu, and M. Koike, *JBC* **244,** 3660 (1969).

16. E. Ishikawa, R. M. Oliver, and L. J. Reed, *Proc. Natl. Acad. Sci. U. S.* **56,** 534 (1966).

17. T. C. Linn, F. H. Pettit, and L. J. Reed, *Proc. Natl. Acad. Sci. U. S.* **62,** 234 (1969).

18. A. A. Glemzha, L. S. Zil'ber, and S. E. Severin, *Biokhimiya* **31,** 1033 (1966).

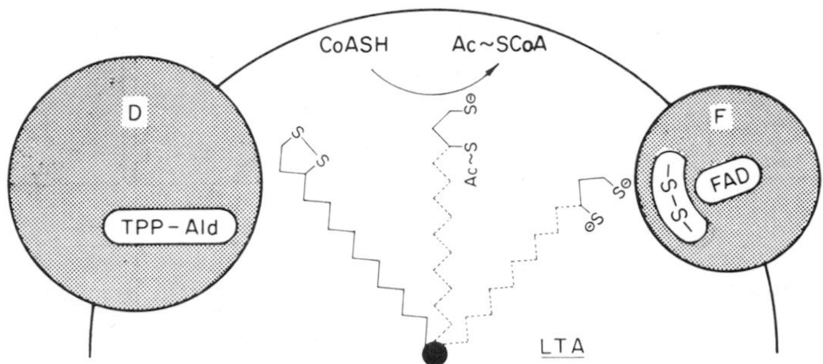

FIG. 5. A schematic representation of the possible rotation of a lipoyllysyl moiety between α-hydroxyethylthiamine pyrophosphate (TPP-Ald) bound to pyruvate dehydrogenase (D), the site for acetyl transfer to CoA, and the reactive disulfide of the flavoprotein (F). The lipoyllysyl moiety is an integral part of dihydrolipoyl transacetylase (LTA). The net charge on the lipoyl moiety during its cycle of transformations may be 0, minus 1, or minus 2. This change in net charge may provide the driving force for displacement of the lipoyl moiety from one site to the next within the complex.

the *E. coli* complex, the mammalian complex also contains two regulatory enzymes—a kinase and a phosphatase (*17, 19*).

As was found to be the case with the *E. coli* dihydrolipoyl transacetylase, the corresponding enzyme from animal sources serves both a catalytic and a structural function. The appearance of the mammalian transacetylase in negatively stained preparations (Fig. 6A) is quite different from that of the *E. coli* enzyme; its morphological subunits appear to be situated at the vertices of a pentagonal dodecahedron (Figs. 6B and C) (*8, 15, 16*). The molecules of pyruvate dehydrogenase, their associated regulatory enzymes, and the flavoprotein molecules appear to be attached to, and to surround, the transacetylase particle (Fig. 6D).

2. *Regulatory Features*

Pyruvate occupies a central position in metabolism (Fig. 7). Of particular importance are its oxidation to acetyl-CoA and its carboxylation to oxalacetate. Oxidation of acetyl-CoA via the tricarboxylic acid cycle leads to the generation of ATP. The conversion of pyruvate to oxalacetate is a major mechanism for the net synthesis of compounds of the tricarboxylic acid cycle which serve as primary biosynthetic intermediates and is, as well, the first step toward gluconeogenesis in liver and

19. T. C. Linn, F. H. Pettit, F. Hucho, and L. J. Reed, *Proc. Natl. Acad. Sci. U. S.* **64**, 227 (1969).

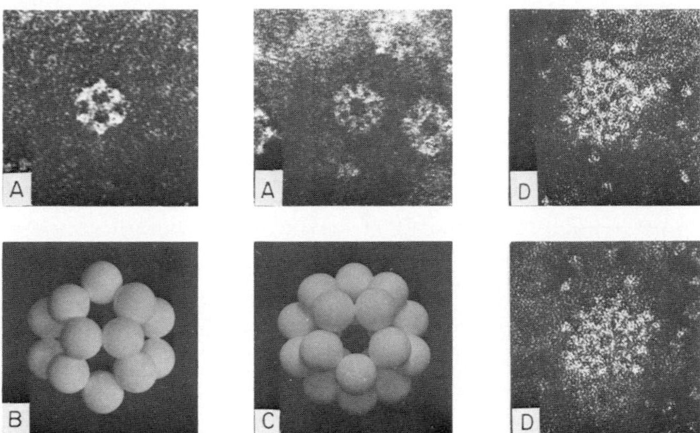

Fig. 6. Electron micrograph images of the mammalian pyruvate dehydrogenase complex and its transacetylase component, and interpretative model of the transacetylase: (A) Individual images (×300,000) showing two orientations of the transacetylase (diameter, about 210 Å). (B, C) Corresponding views of a model of the transacetylase photographed down a 2-fold and 5-fold axis, respectively. The model consists of 20 spheres at the vertices of a pentagonal dodecahedron. (D) Electron micrographs (×250,000) of the pyruvate dehydrogenase complex (diameter, 400–450 Å) from beef kidney mitochondria.

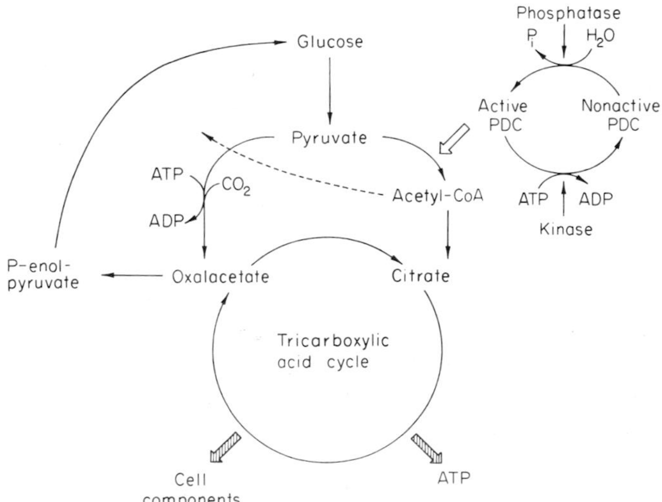

Fig. 7. Regulation of pyruvate metabolism in animal tissues. The open arrow indicates regulation of the activity of the pyruvate dehydrogenase complex (PDC) by phosphorylation and dephosphorylation (*17*, *19*). The site of this regulation is the pyruvate dehydrogenase component of the complex. The dashed line indicates activation of pyruvate carboxylase by acetyl-CoA (*20*).

kidney (*20–23*). These considerations suggest that the pyruvate dehydrogenase complex is a likely candidate for metabolic regulation, and mechanisms for control over the activity of the complex have indeed been found (*24*).

The overall activity of the pyruvate dehydrogenase complex from beef kidney mitochondria is subject to regulation by phosphorylation and dephosphorylation (*17*). Phosphorylation of the complex is accompanied by a proportional decrease in the overall activity, and dephosphorylation restores the activity. The site of this regulation is the pyruvate dehydrogenase component of the complex, which catalyzes the first and, apparently, rate-limiting step in pyruvate oxidation (Fig. 1). Phosphorylation and concomitant inactivation of pyruvate dehydrogenase are catalyzed by an ATP-specific kinase, and dephosphorylation and concomitant reactivation are catalyzed by a phosphatase. ADP is competitive with ATP (*19*). The phosphorylated pyruvate dehydrogenase contains two phosphoryl groups per molecule (molecular weight about 160,000) (*17*). The phosphoryl moieties appear to be attached to the protein in ester linkage.

The activity of the pyruvate dehydrogenase complex from pork liver mitochondria and from beef heart and liver mitochondria is also subject to regulation by phosphorylation and dephosphorylation (*19, 25*). Pyruvate dehydrogenase phosphatase has been separated from the other component enzymes of the kidney, heart, and liver pyruvate dehydrogenase complexes. The three phosphatases are functionally interchangeable (*19*). Pyruvate dehydrogenase kinase has been isolated from the kidney pyruvate dehydrogenase complex. It is functional with the heart and liver pyruvate dehydrogenase complexes. It appears that the concentration of Mg^{2+} and the ATP:ADP ratio play important roles in regulation of the phosphorylation–dephosphorylation reaction sequence. The kinase is active at concentrations of Mg^{2+} as low as 1 μM, whereas the phosphatase requires a Mg^{2+} concentration of about 10 mM for optimum activity (*17*).

There appears to be an inverse relationship between the regulation of the pyruvate dehydrogenase complex and that of pyruvate carboxylase. The former enzyme will tend to be active when the (intramitochondrial) ATP:ADP ratio is low, whereas the latter enzyme will tend to be active when the ATP:ADP ratio is high. This tendency may provide an expla-

20. D. B. Keech and M. F. Utter, *JBC* **238**, 2609 (1963).
21. H. A. Krebs, *Proc. Roy. Soc.* **B159**, 545 (1964).
22. P. Walter, V. Paetkau, and H. A. Lardy, *JBC* **241**, 2523 (1966).
23. M. C. Scrutton and M. F. Utter, *Ann. Rev. Biochem.* **37**, 249 (1968).
24. L. J. Reed, *in* "Current Topics in Cellular Regulation" (B. L. Horecker and E. R. Stadtman, eds.), Vol. 1, p. 233. Academic Press, New York, 1969.
25. O. Wieland and B. von Jagow-Westermann, *FEBS Letters* **3**, 271 (1969).

nation of reports in the literature (*22, 26–30*) that oxidation of fatty acids, acylcarnitines, and intermediates of the tricarboxylic acid cycle by liver and kidney mitochondria inhibits pyruvate oxidation and stimulates pyruvate carboxylation, and that this inhibition is prevented by uncoupling agents such as pentachlorophenol (*28*) and dinitrophenol (*29*). It would appear that oxidation of the competing substrate generates ATP which, in turn, inhibits the pyruvate dehydrogenase complex and stimulates conversion of pyruvate to oxalacetate. Uncoupling agents, by suppressing ATP formation, prevent this inhibition. It has been reported (*31–33*) that the activity of the pig heart pyruvate dehydrogenase complex is inhibited by the products of pyruvate oxidation, acetyl-CoA and DPNH, and that these inhibitions are reversed by CoA and DPN, respectively. The relative importance of this type of regulation as compared with the phosphorylation–dephosphorylation mechanism remains to be determined.

The *E. coli* pyruvate dehydrogenase complex, in contrast to the mammalian complex, does not undergo phosphorylation and dephosphorylation. However, the *E. coli* complex is subject to control by several metabolites (*34–36*). The complex is inhibited by the products of pyruvate oxidation, acetyl-CoA and DPNH, and these inhibitions are reversed by CoA and DPN, respectively (*34*). The site of inhibition by DPNH is the flavoprotein component of the complex. Acetyl-CoA appears to act, at least in part, as a feedback inhibitor of the pyruvate dehydrogenase component of the *E. coli* complex, and this inhibition is competitive with respect to pyruvate (*35*). The acetyl-CoA inhibition is reversed by nucleoside monophosphates and, to a lesser extent, by nucleoside diphosphates and inorganic orthophosphate. The activity of pyruvate dehydro-

26. R. J. Haslam, *in* "Regulation of Metabolic Processes in Mitochondria" (J. M. Tager *et al.*, eds.), p. 108. Elsevier, Amsterdam, 1966.

27. M. A. Mehlman, P. Walter, and H. A. Lardy, *JBC* **242,** 4594 (1967).

28. D. G. Nicholls, D. Shepherd, and P. B. Garland, *BJ* **103,** 677 (1967).

29. T. König and Gy. Szabados, *Acta Biochim. Biophys. Acad. Sci. Hung.* **2,** 253 (1967).

30. G. von Jagow, B. Westermann, and O. Wieland, *European J. Biochem.* **3,** 512 (1968).

31. P. B. Garland and P. J. Randle, *BJ* **91,** 6c (1964).

32. O. Wieland, B. von Jagow-Westermann, and B. Stukowski, *Z. Physiol. Chem.* **350,** 329 (1969).

33. J. Bremer, *European J. Biochem.* **8,** 535 (1969).

34. R. G. Hansen and U. Henning, *BBA* **122,** 355 (1966).

35. E. R. Schwartz, L. O. Old, and L. J. Reed, *BBRC* **31,** 495 (1968).

36. L. C. Shen, L. Fall, G. M. Walton, and D. E. Atkinson, *Biochemistry* **7,** 4041 (1968).

genase is also inhibited by GTP (*37*). This inhibition is noncompetitive with respect to pyruvate, and it is reversed by GDP. It appears that acetyl-CoA and GTP act at separate and independent sites on pyruvate dehydrogenase.

B. α-Ketoglutarate Dehydrogenase Complexes

The *E. coli* α-ketoglutarate dehydrogenase complex (molecular weight about 2.3 million) has been separated into three enzymes, analogous to those obtained from the *E. coli* pyruvate dehydrogenase complex, and it has been reassembled from the isolated enzymes (*38*). The three enzymes are α-ketoglutarate dehydrogenase, dihydrolipoyl transsuccinylase, and dihydrolipoyl dehydrogenase. As in the case of the transacetylase component of the pyruvate dehydrogenase complex, the transsuccinylase appears to serve a dual role, catalytic and structural. The transsuccinylase has a molecular weight of about one million, and its appearance in the electron microscope (Fig. 8A) suggests that it consists of eight morphological subunits arranged in a cubelike structure (*8*). The molecules of α-ketoglutarate dehydrogenase and flavoprotein appear to be distributed in a regular manner on the surface of the transsuccinylase (Fig. 8B).

The α-ketoglutarate dehydrogenase complexes from pig heart (molecular weight about 2.7 million) (*39–41*) and beef kidney (*16*) are composed of three enzymes analogous to those comprising the corresponding *E. coli* complex. The appearance in the electron microscope of the mammalian transsuccinylase (Fig. 8C) (*8*) is strikingly similar to that of the bacterial enzyme, and the general appearance of the mammalian α-ketoglutarate dehydrogenase complex (Fig. 8D) resembles that of the *E. coli* α-ketoacid dehydrogenase complexes. Neither the mammalian nor the bacterial α-ketoglutarate dehydrogenase complex is inhibited by ATP. No significant regulatory properties of these complexes have been uncovered thus far.

A comparison of the enzymes comprising the *E. coli* pyruvate and α-ketoglutarate dehydrogenase complexes has revealed that the two α-ketoacid dehydrogenases and the two transacylases are not functionally interchangeable, nor do these components form "hybrid" complexes (*38*). These biochemical data are in agreement with the genetic data of

37. E. R. Schwartz and L. J. Reed, *Biochemistry* **9,** 1434 (1970).
38. B. B. Mukherjee, J. Matthews, D. L. Horney, and L. J. Reed, *JBC* **240,** PC2268 (1965).
39. D. R. Sanadi, J. W. Littlefield, and R. M. Bock, *JBC* **197,** 851 (1952).
40. V. Massey, *BBA* **38,** 447 (1960).
41. M. Hirashima, T. Hayakawa, and M. Koike, *JBC* **242,** 902 (1967).

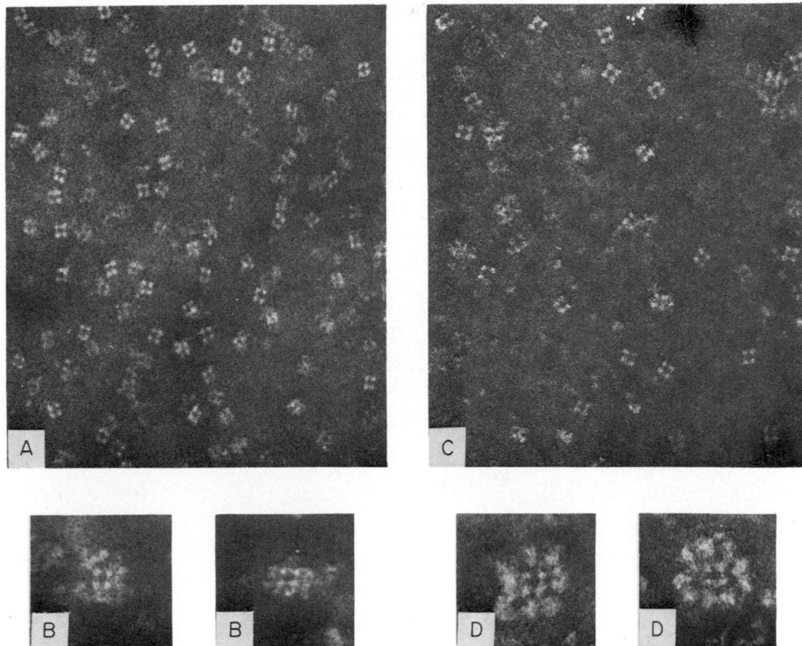

FIG. 8. (A, C) Electron micrographs of dihydrolipoyl transsuccinylase isolated from the *E. coli* and mammalian α-ketoglutarate dehydrogenase complexes, respectively. ×200,000. (B, D) Electron micrographs of the *E. coli* and mammalian α-ketoglutarate dehydrogenase complexes, respectively. ×300,000.

Henning and his co-workers (*11, 42*). They have found that the structural genes for the pyruvate dehydrogenase and for the transacetylase are closely linked (*ace* locus) and are located near the *leu* locus. The structural genes for the corresponding enzymes of the α-ketoglutarate dehydrogenase complex are located near the *gal* operon, i.e., about 20% of the total chromosomal length away from the *ace* locus. The two flavoproteins are interchangeable with respect to both function and complex formation, and they are indistinguishable with respect to several physical, chemical, and immunological parameters (*43*). The functional identity of the two flavoproteins may explain the failure to isolate mutants deficient in these components.

42. U. Henning, J. Dietrich, K. N. Murray, and G. Deppe, *in* "Molecular Genetics," p. 223. Springer, Berlin, 1968.
43. F. H. Pettit and L. J. Reed, *Proc. Natl. Acad. Sci. U. S.* **58**, 1126 (1967).

III. Fatty Acid Synthetases

The biosynthesis of long-chain saturated fatty acids from malonyl-CoA and acetyl-CoA is catalyzed by multienzyme systems referred to as "fatty acid synthetases" (*44–46*). The overall reaction may be represented by the following general equation:

$$\text{Acetyl-CoA} + n\ \text{malonyl-CoA} + 2n\ \text{TPNH} + 2n\ \text{H}^+ \rightarrow \text{CH}_3(\text{CH}_2\text{CH}_2)_n\text{CO-CoA} + n\ \text{CO}_2 + 2n\ \text{TPN}^+ + n\ \text{H}_2\text{O}$$

The yeast fatty acid synthetase produces a mixture of palmityl-CoA and stearyl-CoA, whereas the animal and bacterial synthetases form palmitate. The fatty acid synthetases of yeast and of avian and mammalian tissues have been isolated as relatively stable multienzyme complexes. The fatty acid synthetase systems of bacteria and plants, on the other hand, have been studied mostly in the form of the individual enzymes. However, a fatty acid synthetase of high molecular weight (about 1.7 million) has been isolated recently from *Mycobacterium phlei* (*46a*).

Lynen and his co-workers have described the fatty acid synthetase of baker's yeast in considerable detail (*47*). It has a molecular weight of about 2.3 million and exhibits a characteristic morphology. Electron micrographs of the yeast synthetase (Fig. 9) show oval particles with a diameter of 210–250 Å surrounded by an equatorial ring. Substructure is apparent in the images. Lynen and his colleagues (*47a*) have crystallized the yeast fatty acid synthetase, thereby opening the way for X-ray structural analysis. Although it has not been possible to separate the yeast synthetase into enzymically active components, chemical evidence suggests that this synthetase consists of three identical subcomplexes (molecular weight about 700,000), each a functional assemblage of about seven proteins. One of these proteins is apparently an acyl carrier protein, similar to the one found in *E. coli* (see below), and the component enzymes of the synthetase are presumed to be grouped around, and possibly bound to, this protein (*48*). It appears that the component enzymes of the yeast fatty acid synthetase are stabilized by their participation in the complex but become very labile when they are separated. At least, every effort to produce active subunits from the complex has failed.

44. F. Lynen, *Federation Proc.* **20,** 941 (1961).
45. S. J. Wakil, *Ann. Rev. Biochem.* **31,** 369 (1962).
46. P. R. Vagelos, *Ann. Rev. Biochem.* **33,** 139 (1964).
46a. D. N. Brindley, S. Matsumura, and K. Bloch, *Nature* **224,** 666 (1969).
47. F. Lynen, *BJ* **102,** 381 (1967).
47a. D. Oesterhelt, H. Bauer, and F. Lynen, *Proc. Natl. Acad. Sci. U. S.* **63,** 1377 (1969).
48. K. Willecke, E. Ritter, and F. Lynen, *European J. Biochem.* **8,** 503 (1969).

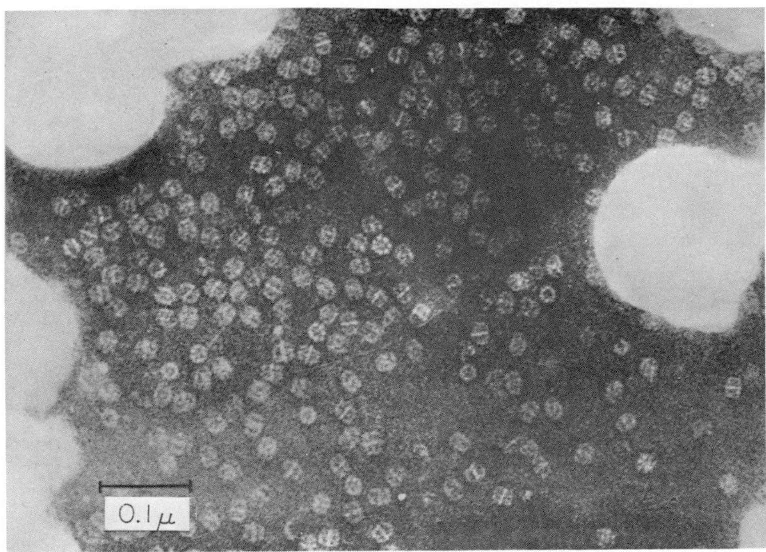

FIG. 9. Electron micrograph of the yeast fatty acid synthetase negatively stained with phosphotungstate (courtesy of Dr. F. Lynen).

The fatty acid synthetases of avian and rat liver have been isolated by Porter and his co-workers in a homogeneous state as organized units with molecular weights of about 450,000 and 540,000, respectively (*49, 50*). Although attempts to separate these synthetases into enzymically active components have been unsuccessful, considerable data on their composition and properties have been obtained. Both the avian and the mammalian fatty acid synthetases contain one mole of protein-bound 4′-phosphopantetheine per mole of synthetase. This finding suggests the presence of a protein similar to acyl carrier protein, and there appears to be one molecule of this protein per functional assemblage of molecular weight about 500,000. It seems possible that the isolated avian and mammalian synthetases correspond to the postulated subcomplex of the yeast synthetase mentioned above.

The fatty acid synthetase system isolated from *E. coli* in the laboratories of Vagelos and of Wakil consists of six structurally independent enzymes and an acyl carrier protein (*51*). All of the reactions of this system occur with the substrates bound as thiolesters to the acyl carrier protein (molecular weight, 9500). The prosthetic group and substrate

49. P. C. Yang, P. H. W. Butterworth, R. M. Bock, and J. W. Porter, *JBC* **242,** 3501 (1967).

50. D. N. Burton, A. G. Haavik, and J. W. Porter, *ABB* **126,** 141 (1968).

51. P. R. Vagelos, P. W. Majerus, A. W. Alberts, A. R. Larrabee, and G. P. Ailhaud, *Federation Proc.* **25,** 1485 (1966).

binding site of this protein is 4′-phosphopantetheine, which is bound as a phosphodiester to a serine residue in the molecule (*52, 53*). Thus the acyl binding site in acyl carrier protein is the same as that in coenzyme A. It appears that the polypeptide chain of acyl carrier protein possesses specific binding sites for the various enzymes of fatty acid biosynthesis (*54*).

The radically different organization of fatty acid synthetases in yeast and avian and mammalian liver on the one hand and in *E. coli* on the other is difficult to rationalize. One would expect that an organized particle with the growing fatty acid chain firmly bound to it could maintain a faster rate of fatty acid synthesis than could equivalent concentrations of independent enzymes with carrier-bound substrate shuttling between them. It may be that, in the intact cell, the bacterial system is organized or confined in some way that is no longer obvious when the cells are broken.

IV. Multienzyme Complexes in the Biosynthesis of Aromatic Amino Acids

Tyrosine, phenylalanine, and tryptophan are produced by a variety of bacteria and fungi according to the biosynthetic sequence shown schematically in Fig. 10. The synthesis of the aromatic acids is subject to sophisticated controls exerted through repression of enzyme synthesis and feedback inhibition by the ultimate products or by certain intermediates. Although the same sequence of reactions is found in all organisms that have been examined, the regulation of the pathway is not uniform at all. The species within a given genus are likely to control the pathway in the same way, but different genera have developed quite different and distinctive mechanisms for regulating the flow of metabolites along the common pathway and into the branches leading to the three amino acids (*55–57*).

Certain of the enzymes concerned with aromatic amino acid biosynthesis have been isolated as stable multienzyme complexes. The patterns of aggregation are even less uniform than the modes of metabolic control; the enzymes of the pathway may be quite differently organized in closely related species. It has been possible to develop attractive rationalizations

52. P. W. Majerus, A. W. Alberts, and P. R. Vagelos, *Proc. Natl. Acad. Sci. U. S.* **53**, 410 (1965).
53. E. L. Pugh and S. J. Wakil, *JBC* **240**, 4727 (1965).
54. P. W. Majerus, *Science* **159**, 428 (1968).
55. H. Tristram, *Sci. Progr. (London)* **56**, 449 (1968).
56. P. Truffa-Bachi and G. N. Cohen, *Ann. Rev. Biochem.* **37**, 79 (1968).
57. R. A. Jensen, D. S. Nasser, and E. W. Nester, *J. Bacteriol.* **94**, 1582 (1967).

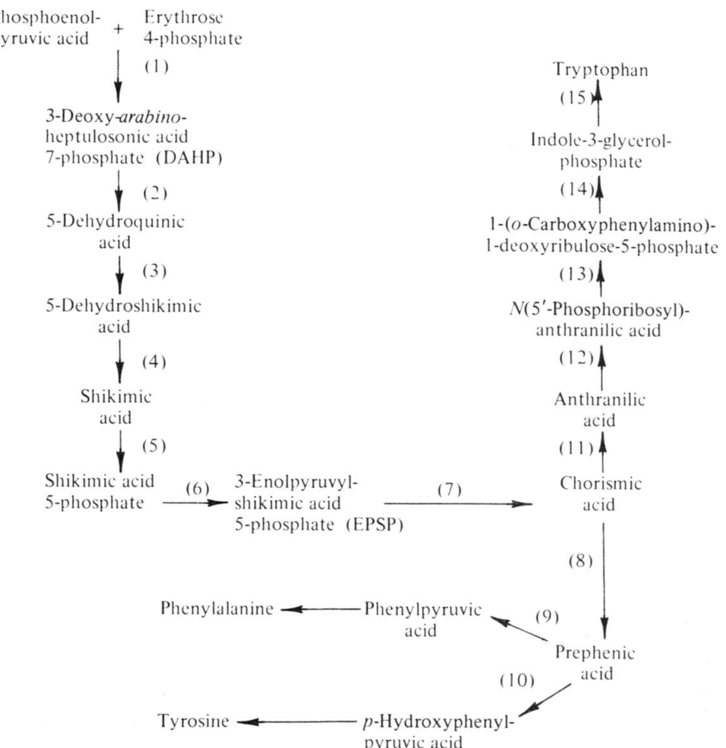

FIG. 10. Biosynthesis of aromatic amino acids. There is evidence (see text) for the physical association of the enzymes that catalyze the following reactions: (2) through (6); (1) + (8) + (5) (?); (8) + (9); (8) + (10); (11) + (12); (11) + (14); (11) + (13) + (14); (13) + (14); (15).

for the presence of aggregates in a few instances, but the functional significance of several of the complexes is still obscure.

Giles and his co-workers have isolated a multienzyme complex (*58*) from *Neurospora crassa* which contains the complete sequence of enzymes catalyzing the conversion of DAHP to EPSP [reactions (2) through (6), Fig. 10]. The structural genes which encode these five enzymes are closely linked, and the "gene cluster" has a number of features in common with bacterial operons (*59*). Sedimentation equilibrium measurements yield a molecular weight of 230,000 for the isolated complex, and this value is consistent with estimates derived from sucrose gradient centrifugation

58. N. H. Giles, M. E. Case, C. W. H. Partridge, and S. I. Ahmed, *Proc. Natl. Acad. Sci. U. S.* **58**, 1453 (1967).

59. M. E. Case and N. H. Giles, *Genetics* **60**, 49 (1968).

and Sepharose chromatography. Examination of the subunit structure of the complex is in progress (*60*).

Giles *et al.* have suggested that this multienzyme complex has developed in response to a discernible metabolic problem. They have found that *Neurospora* is capable of producing an inducible dehydroquinase in addition to the constitutive dehydroquinase [enzyme (3), Fig. 10] which is a component of the biosynthetic complex. The inducible enzyme presumably serves a catabolic function; it could, however, compete with the constitutive enzyme for dehydroquinate produced by the biosynthetic pathway. Giles and his colleagues suggest that because the constitutive dehydroquinase is a component of a multienzyme complex it can compete effectively for the dehydroquinate produced from DAHP and channel it preferentially down the biosynthetic pathway (*61, 62*). The possibility of segregating competing metabolic pathways by means of enzyme aggregates has been suggested speculatively before (*1*), but the aromatic enzyme complex of *Neurospora* is the first clearly documented instance of such a function for a multienzyme complex.

The five enzymes involved in the conversion of DAHP to EPSP apparently occur as a multienzyme complex not only in *Neurospora* but also in a number of other fungi (*63*). Ahmed and Giles have recently examined six species, and in each of them the five enzymes were inseparable by ammonium sulfate fractionation and by density gradient centrifugation. They also mention unpublished work which suggests the occurrence of a similar complex in *Saccharomyces cerevisiae.* The apparent sedimentation coefficients of the aggregates are close to 11 S, which indicates that they are similar in size to the *Neurospora* complex. Ahmed and Giles also report that a "catabolic" dehydroquinase, similar to that which serves so well to rationalize the function of the complex in *Neurospora,* was found in two of the six additional species of fungi. No dehydroquinase could be detected outside the complex in the other four species.

In marked contrast to the situation in the fungi, the enzymes catalyzing reactions (2) through (6) in the prechorismate segment of the aromatic amino acid pathway do not seem to be associated with each other in bacteria. Berlyn and Giles have examined six bacterial species drawn from different genera (*64*). The enzymes that catalyze reactions (2) through (6) in the biosynthetic sequence are clearly separated by centrifugation in sucrose density gradients. Moreover, the estimated molecu-

60. L. Burgoyne, M. E. Case, and N. H. Giles, *BBA* **191,** 452 (1969).

61. N. H. Giles, C. W. H. Partridge, S. I. Ahmed, and M. E. Case, *Proc. Natl. Acad. Sci. U. S.* **58,** 1930 (1967).

62. H. W. Rines, M. E. Case, and N. H. Giles, *Genetics* **61,** 789 (1969).

63. S. I. Ahmed and N. H. Giles, *J. Bacteriol.* **99,** 231 (1969).

64. M. B. Berlyn and N. H. Giles, *J. Bacteriol.* **99,** 222 (1969).

lar weights of the active species are nearly all below 80,000 and range downward to less than 20,000. Biochemical and genetic evidence suggests the presence of only one dehydroquinase in each of the six species of bacteria; this finding is of some importance in view of the channeling function proposed for the *Neurospora* complex.

Although the five enzymes of aromatic amino acid synthesis that form a complex in fungi do not do so in bacteria, Nester and his co-workers have found that two, and possibly three, other enzymes of the pathway do seem to be associated with each other in at least one bacterial species, *Bacillus subtilis* (*65*). Two of the enzymes involved are DAHP synthetase and chorismate mutase [enzymes (1) and (8)]; these two activities are not separated by sucrose gradient centrifugation or by chromatography on DEAE-cellulose, TEAE-cellulose, hydroxylapatite, or Sephadex G-100. The molecular weight of the complex was estimated by gel permeation chromatography to be 138,000. Two fractions with shikimate kinase activity [reaction (5)] were eluted during chromatography of *B. subtilis* extracts on DEAE-cellulose. One of these fractions was associated with the DAHP synthetase–chorismate mutase complex, and this portion of the shikimate kinase activity continued to travel with the complex during subsequent chromatography on Sephadex G-100. Berlyn and Giles did not observe a high molecular weight fraction with shikimate kinase activity in extracts from *B. subtilis* (*64*). They commented that the total shikimate kinase activity in their extracts was relatively low and suggested that the particular strain of *B. subtilis* they examined may lack a shikimate kinase isozyme present in the strain used by Nester *et al.*

The function of the *B. subtilis* complex is obscure. The reactions catalyzed by the three enzymes concerned do not follow each other directly along the metabolic pathway, and so rationalizations based on metabolic channeling or increased efficiency are not adequate (*66*). Nester *et al.*

65. E. W. Nester, J. H. Lorence, and D. S. Nasser, *Biochemistry* **6**, 1553 (1967).

66. Analogous difficulties are raised by the threonine-sensitive aspartokinase–homoserine dehydrogenase complex from *E. coli* that has been described by several laboratories [J-C. Patte, P. Truffa-Bachi, and G. N. Cohen, *BBA* **128**, 426 (1966); P. Truffa-Bachi, R. van Rapenbusch, J. Janin, C. Gros, and G. N. Cohen, *European J. Biochem.* **5**, 73 (1968); G. N. Cunningham, S. B. Maul, and W. Shive, *BBRC* **30**, 159 (1968); D. E. Wampler and E. W. Westhead, *Biochemistry* **7**, 1661 (1968); J. W. Ogilvie, J. H. Sightler, and R. B. Clark, *ibid.* **8**, 3557 (1969)]. This enzyme catalyzes the formation of β-aspartyl phosphate and the production of homoserine from β-aspartic semialdehyde but not the intermediate conversion of β-aspartyl phosphate to the semialdehyde. Some of the data suggest that this aggregate may be composed of identical polypeptide chains; it is, therefore, uncertain whether it should be regarded as a multienzyme complex [P. Truffa-Bachi, R. van Rapenbusch, J. Janin, C. Gros, and G. N. Cohen, *European J. Biochem.* **7**, 401 (1969)].

note that the three enzymes have in common an affinity for chorismate and prephenate; these two metabolites are the substrate and product of chorismate mutase, and both compounds exercise feedback control over DAHP synthetase and shikimate kinase. They suggest that the complex may provide more efficient control over entrance into the pathway leading to the aromatic amino acids than would structurally independent enzymes (*65*). In *Staphylococcus epidermidis,* the same pattern of feedback regulation is found, and Nester *et al.* have reported that chorismate mutase and DAHP synthetase from this organism are eluted together from DEAE-cellulose and appear in the same fractions after density gradient centrifugation (*65*). On the other hand, DAHP synthetase, shikimate kinase, and chorismate mutase are readily separable in extracts from *Bacillus licheniformis,* an organism in which the regulation of aromatic amino acid synthesis is generally similar to that in *B. subtilis* (*67*).

In many microorganisms, the activity of chorismate mutase [enzyme (8)] is regulated by both phenylalanine and tyrosine. In *E. coli* and in *Aerobacter aerogenes,* there are two chorismate mutase isozymes; one of the enzymes is inhibited only by phenylalanine and the other is inhibited only by tyrosine. Moreover, in both species, the phenylalanine-sensitive chorismate mutase is associated with prephenate dehydratase [enzyme (9)] while the tyrosine-sensitive isozyme is bound to prephenate dehydrogenase [enzyme (10)] (*68–70*). This arrangement is very well suited to a situation in which the organism has an adequate pool of either tyrosine or phenylalanine, but not both. In the first place, prephenate will be produced in smaller amounts than it would be if both amino acids were required. More importantly, the decreased quantities of prephenate that are produced will be steered preferentially in the appropriate direction. If the first enzymes beyond the branch point were not associated with the appropriate chorismate mutase isozymes, they would have to be under the direct feedback control of the respective end products. Otherwise, the reduced supply of prephenate would be converted arbitrarily to both the required amino acid and the unneeded one (*71*).

67. D. S. Nasser, G. Henderson, and E. W. Nester, *J. Bacteriol.* **98,** 44 (1969).
68. R. G. H. Cotton and F. Gibson, *BBA* **100,** 76 (1965).
69. J. Pittard and B. J. Wallace, *J. Bacteriol.* **91,** 1494 (1966).
70. R. G. H. Cotton and F. Gibson, *BBA* **160,** 188 (1968).
71. An unstable aggregate has recently been detected in yeast which may perform a function analogous to that postulated for the chorismate mutase complexes in *E. coli.* The enzymes involved in the yeast complex are aspartate transcarbamylase and a pyrimidine-inhibited carbamoylphosphate synthetase [P. F. Lue and J. G. Kaplan, *BBRC* **34,** 426 (1969)]. The purpose of this aggregate may be to channel carbamoylphosphate preferentially in the direction of pyrimidine biosynthesis when the intracellular concentration of pyrimidines is low.

The tryptophan branch of the aromatic amino acid pathway has been examined in great detail in a wide variety of microorganisms. Certain of the enzymes concerned may form simple aggregates, but they do so in a bewildering number of combinations, each one found in only a few organisms.

Anthranilate synthetase, which catalyzes reaction (11), is the most frequent participant in complex formation. It occurs in combination with another enzyme of the pathway in most fungi and in several bacteria (*72*). In *E. coli, Salmonella typhimurium,* and *A. aerogenes,* anthranilate synthetase is tightly bound to the phosphoribosyltransferase which catalyzes reaction (12). Ito and Yanofsky have described the *E. coli* complex in considerable detail (*73–75*). Their results are very similar to those obtained by Bauerle and Margolin with *S. typhimurium* (*76*) and by Egan and Gibson with *A. aerogenes* (*77*). The two enzymes cannot be separated easily, but they can be prepared separately from mutant strains which lack one component or the other (*74, 76, 78*). When the two components are mixed, they associate spontaneously to give a complex identical with that found in the wild type. This behavior has been of some use in the isolation of the complex (*78*). The anthranilate synthetase component isolated from mutants of *E. coli* (*74*) and *S. typhimurium* (*79*) has a molecular weight of 60,000. The partially purified *E. coli* complex is reported to have a sedimentation coefficient of 10.7 S (*80*), while the analogous complex from *S. typhimurium* has a sedimentation coefficient of about 13 S and a molecular weight estimated at 290,000 (*78*). The properties of anthranilate synthetase are changed in several ways by its association with the phosphoribosyltransferase. Either the free component or the complex can use ammonium ions in reaction (11), but only the complex can use glutamine (*75, 79*). The activating effect of complex formation on the glutamine reaction can be observed with extracts from mutants that do not produce an active phosphoribosyltransferase but which do produce a protein which is cross-reactive to antibody for the transferase. Even though both anthranilate synthetase and its complex with the transferase will convert chorismate to anthranilate in the presence of ammonium ions, the affinity of the com-

72. R. Hütter and J. A. DeMoss, *J. Bacteriol.* **94,** 1896 (1967).
73. J. Ito and C. Yanofsky, *JBC* **241,** 4112 (1966).
74. J. Ito, E. C. Cox, and C. Yanofsky, *J. Bacteriol.* **97,** 725 (1969).
75. J. Ito and C. Yanofsky, *J. Bacteriol.* **97,** 734 (1969).
76. R. H. Bauerle and P. Margolin, *Cold Spring Harbor Symp. Quant. Biol.* **31,** 203 (1966).
77. A. F. Egan and F. Gibson, *BBA* **130,** 276 (1966).
78. D. Smith and R. H. Bauerle, *Biochemistry* **8,** 1451 (1969).
79. H. Zalkin and D. Kling, *Biochemistry* **7,** 3566 (1968).
80. T. I. Baker and I. P. Crawford, *JBC* **241,** 5577 (1966).

plex for chorismate is considerably greater than that of the free enzyme. Anthranilate synthetase is also somewhat more sensitive to inhibition by tryptophan if the other member of the complex is present. Finally, the heat stability of anthranilate synthetase is considerably enhanced when it combines with the phosphoribosyltransferase (*75*). The transferase is also more heat stable in the complex than it is alone. Otherwise, the properties of the transferase are not radically altered by the formation of the complex; neither the substrate specificity nor the catalytic efficiency of the enzyme is changed. However, the activity of the complex in reaction (12) can be partially inhibited by tryptophan; the transferase is insensitive to tryptophan in the absence of anthranilate synthetase(*75*).

The complex found in *E. coli* and in *S. typhimurium* is not common to all bacteria. In fact, the limited data available suggest that the enzymes of the tryptophan branch in many bacteria form no complexes of any kind. The five enzymes in extracts of *Pseudomonas putida* (*81*), *Chromobacterium violaceum* (*82*), and *B. subtilis* (*83*) are clearly separable by gel filtration or by sucrose gradient centrifugation. Three of the five enzymes from *Serratia marcescens* are found together after density gradient centrifugation; however, the apparent sedimentation coefficient of the zone in which the activities are found is very low. It seems likely that no complex is present and that the appearance of the three enzymes in the same fraction is fortuitous (*84*).

In *E. coli,* the conversion of *N*-(5′-phosphoribosyl) anthranilate to indole-3-glycerolphosphate [reactions (13) and (14)] is carried out by a single enzyme. Since this enzyme apparently consists of a single polypeptide chain, it is not a multienzyme complex (*85*). A similar bifunctional enzyme is found in *N. crassa,* but here it is associated with anthranilate synthetase [enzyme (11)] (*86*). Gaertner and DeMoss have recently purified this enzyme complex (*87*). The most striking thing about it is that the three reactions it catalyzes [reactions (11), (13), and (14)] do not fall in succession along the metabolic pathway; the phosphoribosyltransferase [enzyme (12)] is missing. The complex has a molecular weight of 240,000 and consists of six polypeptide chains of at least two different kinds; it can be partially dissociated by careful treatment with *p*-mercuribenzoate. Dissociation of the complex by this method

81. T. Enatsu and I. P. Crawford, *J. Bacteriol.* **95,** 107 (1968).
82. J. Wegman and I. P. Crawford, *J. Bacteriol.* **95,** 2325 (1968).
83. S. O. Hoch, C. Anagnostopoulos, and I. P. Crawford, *BBRC* **35,** 838 (1969).
84. M. A. Hutchinson and W. L. Belser, *J. Bacteriol.* **98,** 109 (1969).
85. T. E. Creighton and C. Yanofsky, *JBC* **241,** 4616 (1966).
86. J. A. DeMoss and J. Wegman, *Proc. Natl. Acad. Sci. U. S.* **54,** 241 (1965).
87. F. H. Gaertner and J. A. DeMoss, *JBC* **244,** 2716 (1969).

leaves the phosphoribosylanthranilate isomerase [enzyme (13)] and indoleglycerolphosphate synthetase [enzyme (14)] activities intact but destroys the activity of the anthranilate synthetase.

Hütter and DeMoss have examined the aggregation patterns of the enzymes of tryptophan biosynthesis in a considerable number of fungi (*72*). The pattern found in *Neurospora* is the most common one, although, in a number of cases, anthranilate synthetase dissociates from the complex unless glutamine and EDTA are present.

In *S. cerevesiae*, the phosphoribosylanthranilate isomerase and indoleglycerolphosphate synthetase activities are carried by two different proteins which are readily separable (*72, 88*). Indoleglycerolphosphate synthetase [enzyme (14)], however, is associated with anthranilate synthetase [enzyme (11)]. The difference between the situation in yeast and that in *Neurospora* is that the yeast complex lacks the third enzyme of the tryptophan pathway as well as the second.

Tryptophan synthetase catalyzes the last reaction in the tryptophan pathway:

$$\text{Indole-3-glycerolphosphate} + \text{L-serine} \rightarrow \text{L-tryptophan} + \text{D-glyceraldehyde-3-phosphate} \quad (15)$$

The enzyme is also capable of catalyzing the following two half-reactions:

$$\text{Indole-3-glycerolphosphate} \rightarrow \text{indole} + \text{D-glyceraldehyde-3-phosphate} \quad (16)$$

$$\text{Indole} + \text{L-serine} \rightarrow \text{L-tryptophan} \quad (17)$$

Tryptophan synthetase from *E. coli* contains polypeptide chains of two different kinds, which are referred to as A protein and B protein (or α and β subunits) (*89, 90*). The composition of the complete enzyme is A_2B_2, and on dilution the enzyme dissociates partially: $A_2B_2 \rightleftharpoons 2A + B_2$ (*91*). The molecular weight of the A monomer is 29,000 (*92*) and that of the B dimer is about 100,000 (*93–95*). The two species are readily separated and each is catalytically active, the A protein in reaction (16) and the B protein in reaction (17). The catalytic efficiencies of the separate components are very much less than the corresponding activities

88. J. A. DeMoss, *BBRC* **18,** 850 (1965).
89. C. Yanofsky, *Bacteriol. Rev.* **24,** 221 (1960).
90. I. P. Crawford, J. Ito, and M. Hatanaka, *Ann. N. Y. Acad. Sci.* **151,** 171 (1968).
91. T. E. Creighton and C. Yanofsky, *JBC* **241,** 980 (1966).
92. U. Henning, F. C. Chao, D. R. Helinski, and C. Yanofsky, *JBC* **237,** 1523 (1962).
93. D. A. Wilson and I. P. Crawford, *JBC* **240,** 4801 (1965).
94. G. M. Hathaway, S. Kida, and I. P. Crawford, *Biochemistry* **8,** 989 (1969).
95. M. E. Goldberg, T. E. Creighton, R. L. Baldwin, and C. Yanofsky, *JMB* **21,** 71 (1966).

of the complete protein, however, and neither component has any activity in the overall reaction (15) (*96*). The complete reaction appears, at first glance, to be simply the sum of the half-reactions catalyzed inefficiently by the A and B proteins. However, free indole is not an intermediate in the complete reaction, and the rate of reaction (16) is considerably slower than that of the overall reaction. It is, therefore, generally believed that the conversion of indole-3-glycerolphosphate and serine to tryptophan is a concerted reaction in which both the A and B proteins participate and does not represent the successive operation of these two components on the substrate (*89*). The most obvious effects of the association between the two components are, first, the great enhancement of the activities found in the separate components and, second, the appearance of an activity that is absent from both components. The characteristic activity of the B protein [reaction (17)] can be markedly stimulated not only by the normal A protein but also by immunologically cross-reacting material isolated from mutants that do not produce a catalytically active A protein (*97*). The activity of the B protein in reaction (17) can also be stimulated to levels approaching that of the A_2B_2 aggregate by very high concentrations of ammonium ion (*98*). The B protein contains one pyridoxal phosphate prosthetic group per polypeptide chain (*93*); it is not too surprising, therefore, to find that the isolated B protein is able to catalyze the deamination of serine (*99*) and certain other side reactions that are commonly found with enzymes containing pyridoxal phosphate (*100*). What is surprising is that association of the A protein with the B protein inhibits the serine deaminase activity of the latter almost completely (*99*). The formation of the aggregate seems not only to stimulate those activities of the components that are involved in their normal function but also to suppress an irrelevant and possibly undesirable activity.

Proteins generally similar to the *E. coli* tryptophan synthetase and catalyzing the same reactions are found in a wide variety of microorganisms (*72*). The affinity of the proteins catalyzing reactions (16) and (17) for each other seems to vary from one organism to another. The A_2B_2 complex from *E. coli* is relatively stable in the presence of pyridoxal phosphate and serine. In dilute solutions of the enzyme lacking these metabolites, the subunits still have a considerable affinity for each

96. I. P. Crawford and C. Yanofsky, *Proc. Natl. Acad. Sci. U. S.* **44,** 1161 (1958).
97. C. Yanofsky and I. P. Crawford, *Proc. Natl. Acad. Sci. U. S.* **45,** 1016 (1959).
98. M. Hatanaka, E. A. White, K. Horibata, and I. P. Crawford, *ABB* **97,** 596 (1962).
99. I. P. Crawford and J. Ito, *Proc. Natl. Acad. Sci. U. S.* **51,** 390 (1964).
100. E. W. Miles, M. Hatanaka, and I. P. Crawford, *Biochemistry* **7,** 2742 (1968).

other, but the complex dissociates to an appreciable extent during sucrose gradient centrifugation or starch gel electrophoresis (*90, 91, 95*). The subunit structure of tryptophan synthetase from *N. crassa* is probably similar to that of the *E. coli* enzyme. The polypeptide chains of the mold enzyme, however, are much more firmly bound to each other than are those of the *E. coli* tryptophan synthetase. The *N. crassa* enzyme has been shown to dissociate only in concentrated guanidinium chloride (*101*). Independently active subunits have not been obtained, and it is not yet known whether the subunits of the *Neurospora* tryptophan synthetase are functionally analogous to those of the *E. coli* enzyme. The situation in *S. cerevisiae* is generally similar to that in *N. crassa,* although, in yeast, the possibility that the functions of the A and B proteins reside in the same polypeptides chain cannot be excluded (*102, 103*). In extracts from *P. putida,* on the other hand, enzymes closely analogous to the A and B proteins are clearly separated from each other by gel filtration, even in the presence of pyridoxal phosphate (*81*). If a complex tryptophan synthetase is present in this organism, it is much less stable than the one found in *E. coli.*

V. Biological Significance of Multienzyme Complexes

Most of the multienzyme complexes that have been described seem to have been encountered accidentally. In some cases, a reaction that appeared to occur in one step catalyzed by a single enzyme was found, once the enzyme was isolated, to be the net result of a sequence of reactions catalyzed by a complex. In other instances, procedures designed to isolate one enzyme have also been observed to co-purify another. On a few occasions, the effect of a point mutation on more than one enzymic activity has revealed the existence of a previously unsuspected complex. Once a multienzyme complex concerned with a particular metabolic pathway has been found in one organism, it has been a general practice to screen the analogous enzymes in other species in a more systematic search for aggregates. Despite the more or less haphazard way in which the known multienzyme complexes have been discovered, a considerable number of them have been found; it is reasonable to conclude that this relatively simple mode of subcellular organization is not unusual.

101. M. Carsiotis, E. Appella, P. Provost, J. Germershausen, and S. R. Suskind, *BBRC* **18,** 877 (1965).

102. W. Duntze and T. R. Manney, *J. Bacteriol.* **96,** 2085 (1968).

103. T. R. Manney, W. Duntze, N. Janosko, and J. Salazar, *J. Bacteriol.* **99,** 590 (1969).

A variety of speculative ideas have been advanced to account for the occurrence of functionally related enzymes in the form of complexes. A complex might, for example, be more efficient in catalyzing a multistep metabolic sequence than would a collection of independent enzymes. An intermediate produced by one enzyme would probably encounter the next enzyme in a sequence more rapidly if the two enzymes were physically associated with each other than if they were structurally independent. A complex would be particularly efficient if the metabolic intermediates were strongly bound to it and could not readily escape before the series of reactions was complete.

In at least a few cases, the activity of one component of a complex is modified by its association with the others. For example, the activities of the *E. coli* tryptophan synthetase components are altered by complex formation; the reactions involving substrates of the tryptophan pathway are catalyzed more efficiently, while the serine deaminase activity of the B protein is suppressed. The substrate specificity of anthranilate synthetase in *E. coli* is broadened by complex formation to include glutamine as an ammonia donor. This situation could arise if a catalytically active conformation were imposed on one enzyme by its association with another, or if residues of more than one polypeptide chain participated in the active site that catalyzes a particular reaction. There may be complexes in which the activity of one component depends absolutely on the presence of another; in such cases, it would be impossible to resolve the complex into independently active components. Resolution would also be difficult, of course, if the components of a complex were so strongly bound to each other that they could not be separated without doing irreversible damage to at least one of them. As a practical matter, these two situations would be difficult to distinguish from each other.

If the formation of a complex were necessary either to produce a species with any activity at all or to elevate the efficiency of a pathway to a physiologically useful level, then the ligand-controlled association of the members of a complex could provide a mechanism for the control of the pathway. The acetyl-CoA carboxylase of chicken liver may be controlled in this way; both the activity and the state of aggregation of this enzyme appear to be under the control of citrate (*104*). It is true that most of the known complexes are relatively stable and do not dissociate appreciably under physiologically plausible conditions. The possibility exists, however, that any complex whose dissociation is important as a mode of control would be of marginal stability and might not be readily recognized.

104. C. Gregolin, E. Ryder, R. C. Warner, A.. K. Kleinschmidt, and M. D. Lane, *Proc. Natl. Acad. Sci. U. S.* **56**, 1751 (1966).

Complex formation offers a way of segregating enzymes that would otherwise compete for the same metabolite, and such metabolic channeling has probably been invoked more frequently than any other function in speculations on the role of multienzyme complexes in the cellular economy. The argument is that an intermediate produced by an enzyme that is a member of a complex would be acted upon preferentially by another enzyme in the aggregate and would not be freely available to enzymes which are not present in the complex. This idea is particularly attractive in those cases in which potential competitors for a specific intermediate can be identified; the complexes involved in aromatic amino acid synthesis in fungi provide the best-documented example of this kind, but there are other complexes for which a channeling function seems plausible. If the intermediates formed along a metabolic sequence are strongly bound to the complex catalyzing the entire pathway, then the intermediates can be directed unambiguously in only one direction. This feature is exhibited by the α-ketoacid dehydrogenase complexes and by the yeast fatty acid synthetase. However, if intermediate products dissociate from an aggregate, the complex would be at least somewhat less effective as a channeling device. Before deciding that metabolic channeling is the primary function of a given multienzyme complex, it seems important to show that intermediates do not leave the complex and accumulate in quantities sufficient to allow them to be diverted by an enzyme outside the complex.

Implicit in the foregoing discussion has been the assumption that a rational function can be specified for each multienzyme complex and that these aggregates do not represent artifacts of isolation or evolutionary accidents of no particular use or disadvantage to the organism that produces them. If one adopts the position that multienzyme complexes should serve some useful purpose in the cells where they occur, two features of the available data are somewhat puzzling. First, it seems that the enzymes of a particular metabolic sequence may combine to form a stable aggregate in some species and not in others. The particle that catalyzes the biosynthesis of chorismic acid in fungi is presumably unnecessary in bacteria, which lack the catabolic dehydroquinase that could compete with the biosynthetic pathway. There is, as yet, no comparably persuasive explanation for the very different organization of fatty acid synthetases in yeast and avian and mammalian liver on the one hand and in higher plants and *E. coli* on the other. The irregular occurrence of simple aggregates involving enzymes of the tryptophan pathway is also difficult to account for.

The second difficulty is raised by aggregates that catalyze reactions that do not directly follow one another along metabolic pathways. Several complexes of this kind have been identified, and their occurrence cannot be explained by arguments based on efficiency or on metabolic channeling. In some cases, it has been observed that the substrate or product of one enzyme in a broken-sequence aggregate acts as an effector in another reaction catalyzed by the complex; behavior of this kind has led to the tentative suggestion that a member of a multienzyme complex could play a double role as a catalytic subunit for one reaction and as a regulatory subunit for another (*65*). Alternatively, a complex that catalyzes nonsuccessive reactions could be the remnant of a more elaborate aggregate from which subunits have dissociated in the course of isolation. It would be useful to know, both for the discontinuous aggregates and for the more obviously rational ones, whether the same complex can be isolated from cells that have been broken in different ways and extracted with different solvents. Detailed speculation on the physiological role of multienzyme complexes rests, for the moment, on the assumption that aggregates similar to or identical with those that have been isolated do occur in the cells. Ultimately, one would hope that ways will be found to demonstrate directly the presence of multienzyme complexes in intact cells; efforts in this direction will be of great interest.

5

Genetic Probes of Enzyme Structure

MILTON J. SCHLESINGER

I. Introduction

In 1958, F. H. C. Crick proposed that ". . . folding (of a protein) is simply a function of the order of the amino acids" (*1*). Results of intensive research in protein chemistry and microbial genetics in the ensuing ten years confirmed this hypothesis and further led to the more general concept that a protein's quaternary structure is derived directly from a linear sequence of nucleotide bases in a specific region of a cell's genome.

The experimental verification of a colinear relation between the primary structure of a polypeptide chain (the amino acid sequence) and the sequence of nucleotide bases in the gene was shown, almost simul-

1. F. H. C. Crick, *Symp. Soc. Exptl. Biol.* **12,** 138 (1958).

taneously, by careful peptide analyses and genetic mapping of *E. coli* mutants that had an altered tryptophan synthetase (*2*) and by the elegant experiments of Sarabhai *et al.* (*3*) who studied the head protein of the T4 coli phage. More recently, predictions of the consequences of certain specific mutations in the lysozyme gene of T4 which involved addition and deletion of bases have been beautifully confirmed by amino acid analyses of the peptides of this enzyme (*4*). The important work that led to the elucidation of the genetic code and the numerous experiments now in progress on *in vitro* synthesis of polypeptide chains leave little doubt about the relationship of gene to primary structure of a protein.

Anfinsen (*5*) was among the first protein chemists to offer experimental data that showed the final folded state of an enzyme depended on its primary sequence of amino acids. He and his co-workers observed the complete and accurate reformation of ribonuclease after the enzyme had been reduced to a random coil conformation. Since then many highly purified enzymes and proteins have been subjected to "total" denaturation conditions and found to be capable of reformation to their original native conformation. These include both enzymes composed of single polypeptide chains as well as multimeric proteins in which several polypeptide chains must interact in order to regain complete function.

The prediction of Crick's hypothesis is that some, but not necessarily all, alterations in the sequence of amino acids of the polypeptide chain should lead to changes in the final conformation of the protein and affect the function of that protein. On the basis of the relationship between gene and protein sequence, most of these alterations would be the result of chemical changes in the genome. In fact, before the concept elaborated above was formulated, there was direct evidence that a genetic trait was manifested by the substitution of a single amino acid. This was the discovery by Ingram (*6*) that sickle cell anemia was the result of a substitution of valine for glutamic acid at position 6 in the β chain of hemoglobin. This result confirmed Pauling's prediction made in 1954 that the genetically controlled disease of sickle cell anemia was associated with a protein abnormality (*7*).

2. C. Yanofsky, B. C. Carlton, J. R. Guest, D. R. Helinski, and U. Henning, *Proc. Natl. Acad. Sci. U. S.* **51,** 266 (1964).
3. A. S. Sarabhai, A. O. W. Stretton, S. Brenner, and A. Bolle, *Nature* **201,** 13 (1964).
4. G. Streisinger, Y. Okada, J. Emrich, J. Newton, A. Tsugita, E. Terzaghi, and M. Inouye, *Cold Spring Harbor Symp. Quant. Biol.* **31,** 77 (1966).
5. C. B. Anfinsen and E. Harber, *JBC* **236,** 1361 (1961).
6. V. M. Ingram, *Nature* **180,** 326 (1957).
7. L. Pauling, H. A. Itano, S. J. Singer, and I. C. Wells, *Science* **110,** 543 (1949).

It is precisely this close relation between gene defect, primary amino acid sequence, and quaternary structure that allows us to propose that mutants can be effectively used as an experimental method for the study of protein structure. It is the purpose of this chapter to set forth some of the principles involved; to present several specific examples in which mutations have provided important information about structure, function, and formation of proteins; and to point out the advantages and limitations of the approach.

II. Genes and Mutations

A. The Cistron and Types of Mutations

The original proposal by Beadle (*8*) that one gene equals one enzyme has had to be modified as we learned more about the chemical composition of enzymes and as the genome itself came under chemical analysis. In 1958, Benzer (*9*) introduced the *cistron* and defined it as a "functional region of the genome." He proposed that the cistron contains the information for the sequence of amino acids in a single polypeptide chain. More than one cistron may be required in order to completely elaborate a protein—obviously for those cases where the functional form of the protein contains several nonidentical chains. Also, at least one case is known where a single cistron appears to make a protein that carries out two different enzymic reactions (*10*). We might further qualify these remarks by noting that some cistrons, i.e., those for tRNA and rRNA, do not directly determine the sequence of a polypeptide chain and the sequences of certain small polypeptides (gramicidins and tyrocidines) are not determined directly by a cistron.

Chemical changes in the nucleotide bases of a cistron can lead to a variety of effects, most of which cause changes in the amino acid composition of a polypeptide chain and in turn alter the structure and function of the protein encoded by that cistron. The presence of a protein with an abnormal function in a cell often leads to a detectable change in the characteristics (phenotype) of the organism. The altered organism is referred to as a *mutant* and the lesion in the genome is a *mutation*.

As a consequence of a mutation in the cistron, the polypeptide may

8. G. W. Beadle, *Chem. Rev.* **37,** 15 (1945).
9. S. Benzer, *in* "The Chemical Basis of Heredity" (W. D. McElroy and B. Glass, eds.), p. 70. Johns Hopkins Press, Baltimore, Maryland, 1958.
10. J. C. Loper, *Proc. Natl. Acad. Sci. U. S.* **47,** 1440 (1961).

be altered in one of the following ways: (1) a single amino acid is replaced by another; (2) many amino acids are substituted; (3) the polypeptide chain prematurely terminates leading to fragments of partially completed chains; and (4) the sequence of amino acids for one polypeptide chain is covalently connected to the sequence of another polypeptide chain. Which event will occur depends on the specific chemical change in the genome. These can be simple substitutions of nucleotides which alter the triplet codons and lead to (1) and (3) above, the insertion or deletion of bases resulting in (2) or the deletion of large regions of the genome that leads to (4). The term *missense* mutation refers to those in which simple amino acid substitutions occur, and *nonsense* mutations describe those producing incompleted polypeptides.

It is possible to determine with high precision the relative location of a mutation in a cistron by the use of fine structure genetic mapping. Probably the best example of this type of analysis is the work of Yanofsky and co-workers (*2*, *11*) who mapped the alterations in adjacent nucleotides within the codon of a single amino acid residue (Fig. 1). Such analyses are complex and often fraught with experimental difficulties. We do not propose here to discuss this aspect of the problem but to point out that the location of mutations within cistrons has been successfully achieved for a number of enzymes. For those systems where deletion

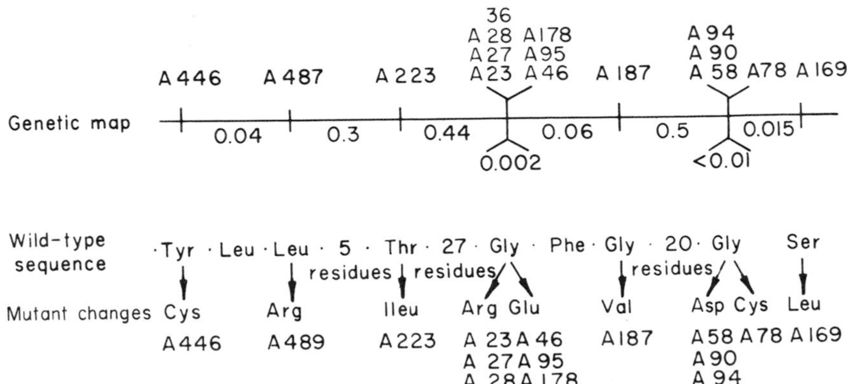

FIG. 1. Amino acid substitutions in mutants of the α subunit of *E. coli* tryptophan synthetase. The site of the mutation in the cistron is depicted together with the specific substitution and the relative position in the polypeptide chain (*11*).

11. D. R. Helinski and C. Yanofsky, *in* "The Proteins" (H. Neurath, eds.), 2nd ed., Vol. 4, p. 1. Academic Press, New York, 1966.

mutants can be readily isolated the experimental procedure has been greatly simplified (*12*).

B. Isolation of Mutants

Natural mutational events occur with very low frequencies (for bacteria, the number in most cistrons is of the order of one mutation in 10^6 to 10^8 individuals), but certain chemical and physical agents increase the frequency of mutations by several orders of magnitude. These include analogs of the nucleotide bases, alkylating agents, ultraviolet radiation, X-rays, heat, and ionizing radiation. Thus, in the laboratory organisms are first subjected to a mutagenic agent under conditions that increase mutational frequencies but prevent excessive killing of the organisms. Because one must work with small numbers of mutants among a large background of nonmutants the successful isolation of mutants depends on a fast, simple, sensitive procedure for looking at a few variants among large numbers; or better, a set of selective conditions that will permit the variant to grow but will eliminate the wild-type organism. For these reasons, microbes (viruses, bacteria, and yeasts) have been the organisms of choice for studying mutationally altered proteins.

A number of examples would be helpful to illustrate the point. Bacterial mutants defective in β-galactosidase formation can be easily detected by growing the bacterial cells on a medium containing a chromophoric compound which turns a red color around those colonies of cells that are able to ferment lactose (*13*). Bacterial mutants defective in β-galactosidase activity will appear noncolored on such a growth medium. A similar phenomenon can be applied to test for strains unable to utilize other sugars or carbon sources. Bacterial mutants defective in alkaline phosphatase formation can be detected by growing cells on a medium limiting in phosphate (conditions which lead to the induction of this enzyme) and then applying a histological stain that detects the enzyme's catalytic activity. Only those colonies that can make an active enzyme will react to the stain. In theory, any histological type of stain for enzymic activity can be used for ascertaining which bacterial colonies are able to form a specific enzyme, with the possible limitation that certain reagents are unable to penetrate into the cell and react with those enzymes localized in the cell's cytoplasm.

Other conditions are available which serve to select for a particular

12. A. J. Blume and E. Ballbinder, *Genetics* **53**, 577 (1966); W. Hayes, "The Genetics of Bacteria and their Viruses." Wiley, New York, 1965.
13. J. Lederberg, *J. Bacteriol.* **56**, 695 (1948).

type of mutant. Probably that most widely used is the penicillin selection procedure discovered by Lederberg (*14*) and modified by Gorini (*15*). This technique is based on the fact that penicillin will only kill growing bacteria. Thus any medium containing penicillin but lacking a factor that has become essential for bacterial growth as a result of the mutation will allow the mutant to survive while the wild-type unaffected cell will grow and commit suicide. Auxotrophs of bacterial cells with mutations in biosynthetic pathways for amino acids, purines, and pyrimidines can be easily isolated by this technique.

Although some of the first mutationally altered bacterial strains were selected on the basis of resistance to antibiotics, the widespread use of antibiotics as a selective reagent for obtaining specific mutants has only recently been exploited. The motivation for these studies was the discovery that resistance to many types of antibiotics often leads to mutational defects in those enzymes and cell components concerned with the synthesis of cell protein such as the ribosomes, DNA-dependent RNA polymerase and the factors controlling polypeptide chain initiation, elongation, and termination (*16*).

There are many enzymes like those just noted in which the product of the enzymic reaction is a vital macromolecule that cannot be supplied exogenously to the cells. In order to obtain mutants defective in these enzymes a condition must be found that will permit the mutant to grow; otherwise it will not survive. One highly successful technique has been the selection of temperature-sensitive mutants. Those organisms that can grow at normal temperatures (37°) but do not survive growth at higher temperatures (42°) are tested further for specific enzymic defects. Alternatively, a lower temperature (25°) can be used. Where this selection has been carried out using a complete growth medium for bacteria, mutants defective in enzymes required for DNA, RNA, and protein synthesis have been obtained. This technique has been successful for isolating mutants of bacterial viruses defective in enzymic activities crucial to viral development (*17*), and it also has proved valuable for detecting mutants of the oncogenic viruses (*18*).

There is another set of permissive and nonpermissive growth conditions that has been useful for the selection of a special class of mutants. One type of mutational event (nonsense mutation) converts a triplet of

14. J. Lederberg, *Methods Med. Res.* **3,** 5 (1950).
15. L. Gorini and H. Kaufman, *Science* **131,** 604 (1960).
16. P. Lengyel and D. Soll, *Bacteriol. Rev.* **33,** 264 (1969); D. Schlessinger and D. Apirion, *Ann. Rev. Microbiol.* **23,** 387 (1969).
17. R. S. Edgar and I. Lielausis, *JMB* **32,** 263 (1968).
18. M. Fried, *Proc. Natl. Acad. Sci. U. S.* **53,** 486 (1965).

bases from a codon specifying an amino acid to one for which no amino acyl tRNA is recognized. There are three such codons—UAG, UAA, and UGA. The first is termed an *amber* and the second an *ochre* mutation (*19*). The result of these specific mutations is to prematurely terminate the synthesis of the polypeptide chain. It has been found that certain strains of bacteria are "permissive" for these nonsense mutations; they contain cistrons that are able to partially correct or suppress (see below) the original mutation. Thus, many mutational defects that would normally be lethal in the nonpermissive (suppressor negative) strain can be rescued in the suppressor-positive strain.

Unfortunately, isolation of a mutant defective in a particular enzymic activity does not automatically provide a source of specifically modified protein. There are numerous examples of mutations that lead indirectly to loss of an enzymic activity without specifically modifying the protein itself. Such mutations are often pleiotropic in that more than one enzyme is affected. It is essential, therefore, to first establish that the mutation actually leads to an alteration of the particular enzyme that we plan to study. Then, to use genetics as a tool for probing structure and function of a protein, we must isolate the altered enzyme in much the same way as we purify an active enzyme.

Here again a good illustration is probably the best way to show how we can establish that a particular mutant carries a mutation in a specific protein. For this purpose, the study of a mutationally altered protein that affects the synthesis of *E. coli* alkaline phosphatase will be described. In these studies, Garen and Otsuji (*20*) isolated mutants that were constitutive for the production of alkaline phosphatase. In wild-type *E. coli*, this enzyme is produced only when the cells are deprived of inorganic phosphate in the growth medium (*21*), but, in the mutant, enzyme is produced under all conditions of growth. One of the first pieces of experimental data that suggested this mutation was in a cistron assigned to a phosphatase-regulator (R) protein was a genetic analysis of the mutant. Genetic mapping by recombination with several other mutants showing the same constitutive phenotype helped to establish that the lesion in the specific mutant occurred within the same cistron as several other constitutive mutants (*22*). Thus one of the criteria

19. S. Brenner, A. O. W. Stretton, and S. Kaplan, *Nature* **206**, 994 (1965); M. G. Weigert and A. Garen, *ibid.* p. 992; M. G. Weigert, E. Lanka, and A. Garen, *JMB* **23**, 391 (1967); S. Brenner, L. Barnett, E. R. Katz, and F. H. C. Crick, *Nature* **213**, 449 (1967) p. 449.
20. A. Garen and N. Otsuji, *JMB* **8**, 841 (1964).
21. A. Torriani, *BBA* **38**, 460 (1960).
22. H. Echols, A. Garen, S. Garen, and A. Torriani, *JMB* **3**, 425 (1961).

is a genetic one—the clustering of mutations within a small region of the genome. The next project was to prepare extracts from several of these mutants and to look for a difference in the pattern of protein bands on acrylamide gel between mutant and wild-type cells. In one mutant a difference appeared in one protein band. To verify that the mutant protein with an altered electrophoretic mobility was related to the wild type, two additional pieces of data were collected. First, a *revertant* of the original mutant was isolated. Cells of the constitutive strain were mutagenized and additional mutants selected in which a good regulatory protein was produced. These cells now became inducible under the same conditions as wild-type cells. Acrylamide gel analysis of extracts from the revertant showed that the putative regulator protein had again changed in its electrophoretic mobility. Second, the wild-type protein that was tentatively assigned as the product of the R cistron for this enzyme was isolated and antibodies directed against this protein were prepared. These antibodies were then tested for cross reactivity against the mutationally altered R protein and found to give a strong cross-reaction. These three sets of data—one genetic, one based on the electrophoretic properties of the protein, and one based on the protein's antigenic similarity—enabled Garen and Otsuji to assign a specific protein to a specific cistron and further provided methods for isolating additional mutationally altered proteins encoded by this cistron.

The above example describes the two most useful techniques now employed for identifying mutationally altered proteins. Antibodies have been successfully used in several systems for detecting mutants. Usually, antibodies are prepared against highly purified preparations of the wild-type enzyme and then used for screening extracts prepared from mutants. Mutationally altered proteins that cross-react with anti-wild-type antibodies are referred to as cross-reacting material or CRM. Figure 2 illustrates the interaction between antialkaline phosphatase antibodies and extracts from two phosphatase-negative mutants. For the alkaline phosphatase system, antibodies could be prepared against both the active dimeric enzyme and against a denatured subunit form of this protein (*23*). One phosphatase-negative mutant was found (see below) which reacted against the latter population of antibodies but not against the former. This fact points up one of the limitations of this technique, namely, that only a particular class of mutants can be detected using anti-wild-type enzyme antibodies. At least one attempt has been made to broaden the selection of mutants by preparing additional antibodies against mutationally altered proteins that were themselves only very

23. M. J. Schlesinger, *JBC* **242**, 1599 (1967).

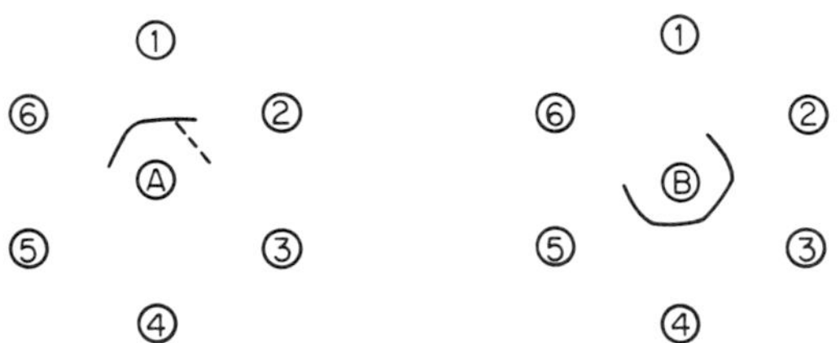

FIG. 2. Antibody–antigen precipitates in agar gel. Proteins from alkaline phosphatase-negative mutants of *E. coli* were tested with antibodies prepared against active enzyme (A) or denatured subunits (B). The peripheral wells contained (1) active enzymes, (2) extract from mutant S-6 (the cross-reaction of nonidentity is seen only when large amounts of extract are used), (3) subunits denatured by performic acid, (4) subunits denatured by reduction and alkylation, (5) subunits denatured by acidification and periodate, and (6) extract from mutant C10F35 (*42*).

weakly antigenic against wild-type antibodies (*24*). Probably further attempts toward this approach will extend the number and types of altered proteins that can be detected and purified.

Variations in the electrophoretic mobility of mutationally altered proteins have proved useful for the isolation and study of large numbers of variants of hemoglobin (see below). Such variants were detected by screening blood samples from individuals throughout the world. This technique has now been applied to the search for variants of certain enzymes whose activities can be readily detected in starch or acrylamide gel. These include phosphatases, dehydrogenases, and esterases (*25*).

To complete this discussion of the detection of mutationally altered proteins, one further type of analysis should be noted. In this procedure, proteins of wild-type and mutant organisms are labeled with the same amino acid but containing in one case a ^{14}C isotope and in the other a ^{3}H isotope. Extracts are prepared from both strains, mixed, and the proteins separated by standard chromatographic or electrophoretic techniques. The ratio of $^{14}C/^{3}H$ is determined for all proteins, and those that show a disparity in the average ratio are assumed to be altered by the mutation. This technique would be expected to be useful for the detection of peptide fragments that result from nonsense mutations.

C. Reversion, Suppression, and Complementation

Because the techniques employed for the isolation of mutants generally lead to a selection of mutations that make enzymes inactive and thus

24. S. Kaplan, S. Ensign, D. M. Bonner, and S. E. Mills, *Proc. Natl. Acad. Sci. U. S.* **51,** 372 (1964).
25. E. S. Vesell, *Ann. N. Y. Acad. Sci.* **151,** Art. I (1968).

difficult to detect and isolate, frequent use is made of revertants as a source for the study of modified proteins. By definition, a revertant is a mutant organism that has regained, at least qualitatively, the phenotype of the original wild-type strain. Often, the enzyme in the revertant will have regained sufficient activity to enable it to be purified and its properties studied. Reversion can occur by a number of different mechanisms. The most straightforward case, but probably the most infrequent, is a mutation in the same nucleotide base as the original mutation and leads simply to replacing the wild-type amino acid into the polypeptide. Alternatively, a second amino acid can be substituted in another region of the polypeptide chain and compensate for the amino acid replacement in the mutant—as if two wrongs can make a right. Such a revertant will produce polypeptides containing two modifications.

Mutations in other completely unrelated cistrons have been shown to restore function to a structurally modified enzyme. Cistrons in which this type of mutation occurs are referred to as *suppressors*. One of the first suppressor mutations described was found to affect the metal concentration in *Neurospora crassa* so that a particular mutationally altered tryptophan synthetase protein that had become ultrasensitive to metal was now able to function in the suppressor strain (*26*). Much more common are the suppressor cistrons for nonsense mutations (amber and ochre—see above) that lead to premature termination of polypeptide chains. Some of these cistrons are known to provide information for tRNA molecules in which single bases in the anticodon have been changed (*27*).

For those proteins and enzymes that require the association of several identical polypeptide chains in order to function, there is one additional genetic mechanism for correcting mutational defects. This phenomenon, referred to as *complementation,* is observed only in organisms that contain at least two copies of a cistron, i.e., in partial or complete diploid cells. Because the cistrons involved are allelic and encode information for identical polypeptides, the term intracistronic or interallelic complementation is used. The molecular basis for successful complementation has been shown to be the formation of hybrid molecules composed of differently mutated but otherwise identical polypeptide chains. Such heterologous multimeric proteins are not wild type and can usually be distinguished from the unaltered protein by differences in heat stabilities and kinetic properties.

A number of "hybrid" enzymes have been studied; among these are

26. S. R. Suskind and L. I. Kurek, *Proc. Natl. Acad. Sci. U. S.* **45,** 193 (1959).

27. J. D. Smith, J. N. Abelson, B. F. C. Clark, H. M. Goodman, and S. Brenner, *Cold Spring Harbor Symp. Quant. Biol.* **31,** 479 (1966).

the *E. coli* alkaline phosphatase (*28*) and β-galactosidase (*29*), and the glutamic dehydrogenase (*30*), tryptophan synthetase (*31*), α-isopropylmalate synthetase (*32*), and adenylosuccinase (*33*) from *N. crassa.* Hybrids of the latter enzyme showed a K_m distinct from wild type and were more heat labile and more sensitive to inhibitors and feedback effectors than wild-type enzyme.

Attempts have been made to correlate the genetic map of mutants in a single cistron with the complementation results. The rationale for this type of analysis presumes that the sites of mutations can provide information about the interactions between polypeptide chains in an oligomeric protein. Crick and Orgel (*34*) have presented an analysis of intracistronic complementation that argues against this type of correlation and the existing experimental data support their thesis.

Complementation is also known to occur in mutants that have mutations on different cistrons. This intercistronic complementation is observed for those enzymes composed of nonidentical subunits and the particular enzyme will generally be indistinguishable from wild-type protein. This results because there will be at least one unmodified copy of the different cistrons in the diploid cell, and good as well as bad copies of each of the polypeptide chains will be formed. The unmodified chains will associate to produce wild-type protein.

It is not clear to what extent complementation occurs as a meaningful device in nature for the repair of mutations. It has been suggested that one of the reasons for evolution to a diploid state was that in diploids complementation offered a very effective method for the escape from what otherwise would have been an excessive number of lethal mutations.

III. Specific Examples of Mutationally Altered Proteins

A. *E. coli* Alkaline Phosphatase

Garen and Levinthal (*35*) were the first to purify *E. coli* alkaline phosphatase and to isolate mutants defective in this enzymic activity.

28. M. Schlesinger and C. Levinthal, *JMB* **7,** 1 (1963).
29. D. Perrin, *Cold Spring Harbor Symp. Quant. Biol.* **28,** 529 (1963).
30. J. R. S. Fincham and A. Coddington, *Cold Spring Harbor Symp. Quant. Biol.* **28,** 517 (1963).
31. K. Suyama and D. M. Bonner, *BBA* **81,** 565 (1964).
32. R. E. Webster, C. A. Nelson, and S. R. Gross, *Biochemitry* **4,** 2319 (1965).
33. C. W. H. Partridge and N. H. Giles, *Nature* **199,** 304 (1963).
34. F. H. C. Crick and L. E. Orgel, *JMB* **8,** 161 (1964).
35. A. Garen and C. Levinthal, *BBA* **38,** 470 (1960); C. Levinthal, *Brookhaven Symp. Biol.* **12,** 76 (1959).

Fine structure genetic mapping led Garen to propose that there was only a single cistron which determines the structure of the alkaline phosphatase (*36*). A chemical analysis by Rothman and Bryne (*37*) of the pure enzyme showed that the enzyme contained two identical subunits; a finding consistent with the genetic data. Of the large number of structural gene mutants and revertants isolated, about one-fifth make a CRM while others fall into the category of suppressible mutants (*38*). Alkaline phosphatases from several of these latter suppressed mutants have been isolated and amino acid substitutions determined. Such analyses enabled Garen and co-workers to assign codons for the amber and ochre mutations (*39*).

One of the CRM mutants has been studied in detail and found to lack the zinc atoms that are essential for the activity of this enzyme (*40*). The interesting finding for this mutant was that the purified, inactive mutationally altered protein could be restored to an active enzyme by incubating the CRM with zinc. Restoration of activity was correlated with binding of zinc and this interaction led to a conformational change in the protein as detected by a change in the ultraviolet spectrum at 278 and 284 mμ (Fig. 3). This change was assigned to a removal of

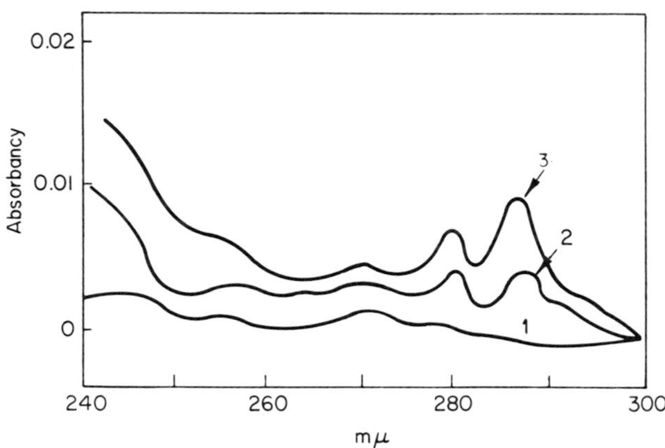

FIG. 3. Ultraviolet difference spectra generated as a result of treating CRM from alkaline phosphate-negative mutant U-47 with Zn^{2+} (*40*).

36. A. Garen and S. Garen, *JMB* **7,** 13 (1963); A. Garen, *Symp. Soc. Gen. Microbiol.* **10,** 239 (1960).

37. F. Rothman and R. Byrne, *JMB* **6,** 330 (1963).

38. A. Garen and O. Siddiqi, *Proc. Natl. Acad. Sci. U. S.* **48,** 1121 (1962).

39. M. G. Weigert, E Gallucci, E. Lanka, and A. Garen, *Cold Spring Harbor Symp. Quant. Biol.* **31,** 145 (1966).

40. M. J. Schlesinger, *JBC* **241,** 3181 (1966).

tyrosine residues from the solvent to a hydrophobic region in the protein. Another CRM mutant shows similar properties, but activation occurs only between pH 5 and 6 (*41*); this CRM is an inactive form of the dimeric protein. A third phosphatase-negative mutant was found to accumulate inactive subunits under normal growth conditions, but these subunits could be converted into a stable active enzyme by supplementing the growth medium with Zn^{2+} and inorganic phosphate. The enzyme was purified from this mutant and shown to have a turnover number very close to that of wild-type phosphatase (*42*).

Studies of these mutants have helped to establish the following properties of *E. coli* alkaline phosphatase: (1) some of those regions of the polypeptide chain that influence the interaction between subunits have no effect upon the catalytic site for the protein; (2) attachment of zinc to the dimer leads to a conformational change in the protein subunit and concomitant formation of a catalytic site; and (3) the formation of the active enzyme which is specifically localized to a region of the bacterial cell exterior to the cytoplasmic membrane (termed the *periplasmic space*) most likely proceeds according to the following scheme:

$$\underset{\text{(cytoplasm)}}{\text{Folded subunit}} \rightarrow \underset{\text{(periplasmic space)}}{\text{folded subunit}} \rightarrow \text{apo-dimer} \rightarrow \text{active } Zn^{2+}\text{-dimer}$$

Transport of the subunit from cytoplasm to the periplasmic space occurs prior to the association of subunits and activation by metal occurs after dimerization. Results of experiments with the wild-type protein substantiate these findings (*43*), although Torriani (*44*) has concluded that activation by zinc *in vivo* precedes dimerization.

In vitro studies with several purified CRM proteins from phosphatase-negative mutants were the first to clearly establish the molecular basis for genetic intracistronic complementation (*45*). By mixing subunits from one CRM or wild-type enzyme containing ^{35}S, ^{2}H, and ^{15}N isotopes with unlabeled subunits of another CRM, it was possible to separate and isolate the hybrid from the homologous dimers (Fig. 4). Hybrids composed of two CRMs were partially active whereas the homologous proteins had essentially no activity. Hybrids containing a CRM subunit and a wild-type subunit had a turnover number about half that of the

41. D. Lennette and M. J. Schlesinger, unpublished experiments (1970).

42. M. J. Schlesinger, *JBC* **242,** 1604 (1967).

43. J. A. Reynolds and M. J. Schlesinger, *Biochemistry* **6,** 3552 (1967); **8,** 588 (1969); M. J. Schlesinger, *J. Bacteriol.* **96,** 727 (1968); M. I. Harris and J. E. Coleman, *JBC* **243,** 5063 (1968).

44. A. Torriani, *J. Bacteriol.* **96,** 1200 (1968).

45. D. P. Fan, M. J. Schlesinger, A. Torriani, K. J. Barrett, and C. Levinthal, *JMB* **15,** 32 (1966).

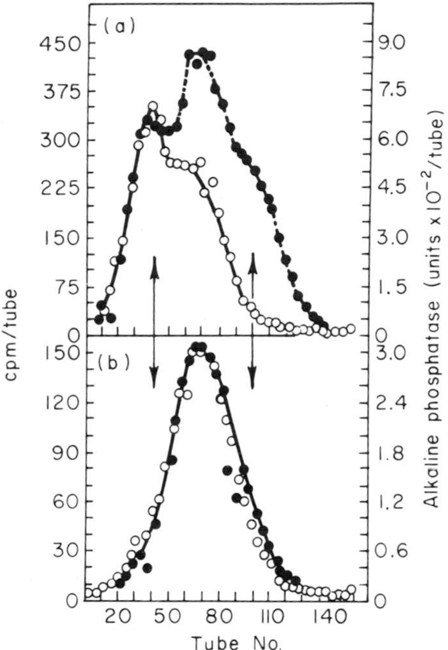

Fig. 4. Separation of hybrid alkaline phosphatases in rubidium chloride density gradients. Purified preparations of ^{15}N, ^{2}H, and ^{35}S wild-type enzyme were dissociated to subunits and mixed with (a) an equal amount of unlabeled wild-type subunits and (b) a 25-fold excess of subunits prepared from a phosphatase-negative CRM mutant. The arrows show the positions for pure heavy and pure light enzyme. (○) Radioactivity and (●) enzymic activity (*45*).

intact wild-type enzyme. Such hybrids may prove useful in the study of how this particular enzyme functions catalytically. Thus far, none of these CRMs has been analyzed to determine which amino acid has been substituted or where the replacement occurs in the polypeptide chain.

Mutations in cistrons other than that encoding the polypeptide chain have been shown to affect alkaline phosphatase activity. Three of these cistrons are assigned to "regulator proteins" which control the rate of formation of the alkaline phosphatase but do not affect the actual properties of the enzyme itself. As noted above, one of these proteins was detected by looking for a mutationally altered form from one of the mutants. A temperature-sensitive regulator protein for alkaline phosphatase has also been isolated (*46*).

46. J. Gallant and T. Spottswood, *Proc. Natl. Acad. Sci. U. S.* **52,** 1591 (1964).

B. *E. coli* β-GALACTOSIDASE

Genetic analyses have shown that several cistrons control the utilization of lactose by *E. coli* (*47*). The proteins encoded by these cistrons consist of β-galactosidase, a protein required for the transport of lactose into the cell, a transacetylase, and a repressor protein whose function is to regulate the synthesis of the previous three proteins. A large number of mutants in all four cistrons have been isolated, and studies with these mutants provided the data that led to the formulation of the Jacob–Monod model for negative regulation of protein synthesis and for the existence of repressor proteins (*48*). There are now very good biochemical data to support the Jacob–Monod model, and it is anticipated that further studies of mutants in this system will ultimately reveal the exact molecular mechanism for one type of control of protein synthesis.

Many mutants in the β-galactosidase cistron itself have been examined, particularly with regard to intracistronic complementation (*49*). One interesting aspect of these studies was the finding that only certain regions of the polypeptide chain were required in order for this large multimeric protein to function as an active enzyme. A catalytic site was restored by mixing extracts from a mutant that formed only about one-third of the chain (the carboxyl terminal section) with a second mutant that made an intact but modified polypeptide. This type of analysis may lead to the assignment of those regions of the polypeptide that are absolutely essential for the formation of a catalytic site and for the interaction of subunits.

Cell-free extracts from various missense, nonsense, and deletion mutants of β-galactosidase have been tested for cross-reactivity with antibodies directed against the intact tetrameric enzyme (*50*). There was a rough correlation between the size of the polypeptide chain and antigenic activity. Nonsense mutants in the N-terminal region of the gene (these produced short peptide fragments) showed poor reaction while those near the C-terminal gave good cross-reaction. A deletion mutant lacking the center portion of the polypeptide produced no cross-reacting protein. These studies aid in the identification of the antigenic determinants in the enzyme.

Mutationally altered repressor proteins, which in the wild type state

47. F. Jacob, *Science* **152,** 1470 (1966); J. Monod, *ibid.* **154,** 475 (1966); J. R. Beckwith, *ibid.* **156,** 597 (1967).
48. F. Jacob and J. Monod, *JMB* **3,** 318 (1961).
49. D. Zipser and D. Perrin, *Cold Spring Harbor Symp. Quant. Biol.* **28,** 533 (1963); A. Ullmann, F. Jacob, and J. Monod, *JMB* **32,** 1 (1968).
50. A. V. Fowler and I. Zabin, *JMB* **33,** 35 (1968).

has been shown to have high affinity for a specific region of DNA, offer valuable material for a study of protein–nucleic acid interactions. In fact, mutant repressor proteins were used as an important "control" in the experiments that established the nature of the "lac" repressor and its site of action (*51*).

The transport (M) protein for lactose was reported to be a component of the cytoplasmic membrane (*52*). Mutants in the cistron for this protein may provide important information on the interactions of proteins with cell membranes.

C. Tryptophan Synthetase

Mutants altered in the tryptophan synthetase of *E. coli* (*2*), *S. typhimurium* (*53*), and *N. crassa* (*24, 26*) have been studied in considerable detail. The *E. coli* enzyme readily dissociates into α and β_2 subunits that each have distinctive enzymic activities, but the physiologically important reaction occurs only with an $\alpha_2\beta_2$ tetramer (*54*). Yanofsky and co-workers isolated many mutants in the α subunit and identified both the amino acid substitutions and their relative positions in the polypeptide chain (cf. Fig. 1). The α subunit can convert indole 3-glycerolphosphate (InGP) to indole and glyceraldehyde-3-phosphate, and InGP protects α protein from denaturation by *N*-ethylmaleimide. From six CRM mutants defective in α activity, only two failed to show InGP protection (*55*). Both had substitutions in the same glycine residue of the wild-type chain—in one case to arginine and the other to glutamate. A substitution of isoleucine for glycine at this position gave an α subunit like wild type in the reaction with *N*-ethylmaleimide; thus, the presence of a charged amino acid in that position led to a profound effect on the binding of substrate.

The α subunit is normally a monomer but dimerizes after denaturation *in vitro* (*56*). Different mutant α subunits that were themselves inactive as a dimer formed heterologous, enzymically active dimers. In general, active protein resulted only when the two mutants had substitutions at opposite ends of the polypeptide chain, and nonsense mutants that mapped near the carboxyl terminal region of the cistron made fragments

51. W. Gilbert and B. Müller-Hill, *Proc. Natl. Acad. Sci. U. S.* **56,** 1891 (1966); *ibid.* **58,** 2415 (1967); A. D. Riggs and S. Bourgeois, *JMB* **34,** 361 (1968).
52. C. F. Fox and E. P. Kennedy, *Proc. Natl. Acad. Sci. U. S.* **54,** 891 (1965).
53. C. Yanofsky, *Bacteriol. Rev.* **24,** 221 (1960).
54. C. Yanofsky, D. R. Helinski, and B. D. Maling, *Cold Spring Harbor Symp. Quant. Biol.* **26,** 11 (1961).
55. J. K. Hardman and C. Yanofsky, *Science* **156,** 1369 (1967).
56. D. A. Jackson and C. Yanofsky, *JBC* **244,** 4526 and 4539 (1969).

that could complement α protein substituted in the amino terminal part of the polypeptide. Of particular interest was the fact that a cyanogen-bromide piece containing 82 residues from the amino terminal part of α could complement with an intact mutant α substituted in the amino terminal region to give a partially active dimer.

A number of revertants of the *E. coli* tryptophan synthetase have been analyzed (*57*) and two cases were found where the site of reversion was 36 residues from each of the original sites of amino acid substitutions (*58*). The amino acid replacements in the tryptophan-requiring mutants were separated by a single residue, thus implying a close functional relationship between two small regions in the α subunit.

Mutants producing CRM to the β subunit of the *E. coli* protein have been tested for their interaction with mutant and wild-type α subunit. One class of β-chain mutants has no activity alone but regains a catalytic site after reacting with α subunits (*59*).

D. Hemoglobin

Probably the most intensively studied among all proteins is hemoglobin, and it is not surprising that we know more about the structure and function of this protein than any other. The elegant X-ray studies of Perutz and co-workers have culminated in almost a complete understanding of the conformation of the subunits and their interactions in the tetrameric molecule (*60*). It was noted earlier in this chapter that the first genetic modification detected in a protein was in the hemoglobin isolated from persons having sickle cell anemia. Since that time over one-hundred altered forms of hemoglobin have been discovered and their properties examined. Table I is an abbreviated listing, taken from a review by Perutz and Lehmann (*61*), of defective hemoglobins containing substitutions in both the α and β subunits together with an analysis of how the replacement may lead to the structural alteration. As noted by Perutz and Lehmann, the properties of the different mutationally altered forms fits in extremely well with the assigned structure for normal hemoglobin. For example, approximately 60 interatomic contacts

57. M. K. Allen and C. Yanofsky, *Genetics* **48,** 1065 (1963).
58. C. Yanofsky, V. Thorn, and D. Thorpe, *Science* **146,** 1593 (1964).
59. I. P. Crawford, J. Ito, and M. Hatanaka, *Ann. N. Y. Acad. Sci.* **151,** 171 (1968).
60. M. F. Perutz, *European J. Biochem.* **8,** 455 (1969); M. F. Perutz, H. Muirhead, J. M. Cox, and L. G. Goaman, *Nature* **219,** 131 (1968); H. Muirhead and M. F. Perutz, *Cold Spring Harbor Symp. Quant. Biol.* **28,** 451 (1963).
61. M. F. Perutz and H. Lehmann, *Nature* **219,** 902 (1968).

TABLE I

AMINO ACID SUBSTITUTIONS IN HEMOGLOBIN

	Residue		Replacement			
Designation	No.	Position	From	To	Abnormal properties	Structural effects of replacement
					α Chain	
M Boston	58	E7	His	Tyr	Difficult to reduce; α chains do not combine with O_2	Phenol group of tyrosine probably forms ionic links with Fe^{2+}, thus stabilizing it in the ferric state
M Iwate	87	F8	His	Tyr		
J Capetown	92	FG4	Arg	Gln	High oxygen affinity; diminished heme–heme interaction	Arg FG4α_1 is in van der Waals contact with Arg C6β_2. Replacement may interfere with structural transition between oxy and deoxy forms
					β Chain	
Freiburg	23	B5	Val	—	High oxygen affinity	Deletion would draw the carboxyl group of Glu 22 (B4) into a crevice about 4 Å below the protein surface. It would also remove the last turn of the B helix and disturb the conformation of the AB corner
Gun Hill		Deletion of 5 residues between 91 and 97			No heme in β chain	Removes section of polypeptide chain which forms essential heme contacts and contacts with α subunits
Santa Ana	88	F4	Leu	Pro	No heme in β chain	Proline occurs at the amino end of helix F and may not cause any change in the conformation of the main chain, but the replacement removes the contact of the heme group with C_γ and C_δ of leu. This opens a crevice on the proximal side into which water could enter

between the heme groups and globin are nonpolar and invariant among all hemoglobins studied. That these contacts are essential for hemoglobin function is indicated by the observation that in one of the abnormal hemoglobins, a replacement of phenylalanine by valine produced an unstable α subunit, and substitution of a valine residue in this contact region by alanine leads to a loss of the heme. Methemoglobinanemia results when one of the histidines is replaced by tyrosine because the phenolic oxygen comes within close proximity of the iron atom and changes it to the ferric form. A number of other substitutions are known in which the amino acid replacements affect the stability of the tetramer and of the α dimer. For the latter, dissociation to free subunits is observed to occur much more easily in the mutant than in the wild-type hemoglobin. Many of these variants show that substitution at contacts between α and β subunits can also affect oxygen affinity and heme–heme interaction. The authors of this excellent review conclude ". . . the hemoglobin molecule is a model system for studying the effect of somatic mutation and species variations on the structure and function of a protein molecule."

Another class of mutations, detected because they cause the disease thalassemia, affects the rate of synthesis of the α- and β-polypeptide chain but does not lead to any alterations in the sequence of amino acids in the globin chains (*62*). In β-thalassaemia, there is a decrease in β-chain production and the α chains aggregate with stromal fractions of the cell producing inclusion bodies. Excess β chains accumulate in α-thalassemia to form hemoglobin H which precipitates in the red cell (*63*). Hemoglobin H is a tetramer of only β chains and shows no Bohr effect (*64*).

One additional type of variant of hemoglobin (Hb-Lepore) consists of a chain that is made up of sequences from both α and δ chains (*65*). It presumably arises as a result of a deletion or "unequal crossing over" in the genome such that the modified cistron is itself a hybrid of two partial cistrons.

E. Other Enzymes and Proteins

It is not intended to present in this chapter an exhaustive survey of all those proteins in which mutational defects have been discovered.

62. V. M. Ingram and A. O. W. Stretton, *Nature* **184,** 1903 (1959).
63. D. J. Weatherall, *Brit. Med. Bull.* **25,** 24 (1969).
64. R. E. Benesch, H. M. Ranney, R. Benesch, and G. M. Smith, *JBC* **236,** 2926 (1961).
65. C. Baglioni, *Proc. Natl. Acad. Sci. U. S.* **48,** 1880 (1962).

There are, however, a number of other mutationally altered enzymes and multienzyme (protein) complexes of considerable interest. For example, mutants of coli phage T4 have been isolated that make modified DNA polymerases and some of these mutations influence the base selection in DNA replication (*66*). Altered forms of the DNA-dependent RNA polymerase were found in mutants resistant to the antibiotic, rifampicin (*67*). Resistance to another antibiotic, streptomycin, has been shown to lead to a modification in one of the protein components of *E. coli* ribosomes (*68*), and this change affects ribosomal function (*69*).

An extensive genetic analysis of the coli phage T4 genome has provided considerable insight into how this virus attaches, infects, and develops inside its bacterial cell host (*70*). One class of T4 mutants is defective in the final assembly process and accumulates precursor particles such as heads, tails, and fibers (*71*). Such mutants are invaluable for studying the formation of this complex virus. Modifications in the amino acid sequence of the head-protein of T4 have been shown to produce osmotically fragile virus particles (*72*). A similar type of mutant has been studied in the tobacco mosaic virus, and the amino acid substitutions that occurred in the viral coat protein were identified (*73*).

An interaction between an enzyme and a cellular organelle has been revealed in studies with mutationally altered forms of *N. crassa* malate dehydrogenase and with respiratory-deficient mutants of this organism (*74*). Amino acid differences were detected in a protein believed to be

66. J. F. Speyer, J. D. Karam, and A. B. Lenny, *Cold Spring Harbor Symp. Quant. Biol.* **31,** 693 (1966); A. deWaard, A. V. Paul, and I. R. Lehman, *Proc. Natl. Acad. Sci. U. S.* **54,** 1241 (1965).

67. G. P. Tocchini-Valenti, P. Marino, and A. J. Colville, *Nature* **220,** 275 (1968); D. H. Ezekiel and J. E. Hutchins, *ibid.* **220,** 276 (1968); W. Wehrli, F. Knüsel, and M. Staehlin, *BBRC* **32,** 284 (1968).

68. M. Ozaki, S. Mizushima, and M. Nomura, *Nature* **222,** 333 (1969).

69. L. Gorini and J. R. Beckwith, *Ann. Rev. Microbiol.* **20,** 401 (1966); L. Luzzatto, D. Apirion, and D. Schlessinger, *JMB* **42,** 315 (1969); *J. Bacteriol.* **99,** 206 (1969).

70. R. H. Eptein, A. Bolle, C. M. Steinberg, E. Kellenberger, E. Boy de la Tour, R. Chevally, R. S. Edgar, M. Susman, G. H. Denhardt, and A. Lielausis, *Cold Spring Harbor Symp. Quant. Biol.* **28,** 375 (1963).

71. W. B. Woods, *Federation Proc.* **27,** 1160 (1968); R. S. Edgar and W. B. Woods, *Proc. Natl. Acad. Sci. U. S.* **55,** 498 (1966).

72. S. Brenner and L. Barnett, *Brookhaven Symp. Biol.* **12,** 86 (1959).

73. H. G. Wittman and B. Wittman-Liebold, *Cold Spring Harbor Symp. Quant. Biol.* **28,** 589 (1963); A. Tsugita, *JMB* **5,** 284 and 293 (1962).

74. D. O. Woodward and K. D. Munkres, *Proc. Natl. Acad. Sci. U. S.* **55,** 872 (1966); K. D. Munkres and D. O. Woodward, *ibid.* p. 1217; K. D. Munkres and F. M. Richards, *ABB* **109,** 457 (1965).

important to the structure of the mitochondrial membrane. In the mutant, this protein had binding affinities for NADH, ATP, and malate dehydrogenase significantly different from those of the wild-type protein. There was also a much greater effect on the *in vivo* function of malate dehydrogenase in the mutants than when the enzyme was studied in its purified state.

Several mutants of *N. crassa* that require α-amino compounds as growth factors make defective glutamic dehydrogenases (*75*). Several of these defective enzymes have been examined, and one of them could be partially activated in the presence of reagents such as succinate or EDTA or by mixing it with other defective proteins (*76*). The result of the latter was formation of a hybrid protein made up of differently mutated subunits.

Mutants have been particularly useful in elucidating the nature of several multienzyme complexes (*77*). Among these are the *E. coli* aspartate kinase-homoserine dehydrogenase complex (*78*), and the enzyme complexes required for the synthesis of tryptophan in *E. coli, S. typhimurium, N. crassa,* and yeasts (*79*).

There are several examples where mutations have altered one activity in a protein while leaving a second function of the enzyme intact. Enzymes with these dual functions are among the growing list of proteins whose activities are regulated by ligands unrelated to either substrate or products of the enzymic reaction. Noteworthy are the mutationally altered forms of the first enzyme in the biosynthetic pathway for histidine in *Salmonella typhimurium* that show decreased sensitivity for histidine as a feedback effector for this enzyme (*80*). One interesting mutant—a cold-sensitive organism—is unable to grow at temperatures less than 30° and has an affinity for histidine that is a thousand times greater than the wild-type enzyme (*81*). These are only a very small number of the more than one thousand mutants in *S. typhimurium* that are defective in histidine formation. The genetic locations of these

75. J. R. S. Fincham, *JMB* **4,** 257 (1962).
76. T. K. Sundaram and J. R. S. Fincham, *JMB* **10,** 423 (1964); **29,** 433 (1967).
77. H. E. Umbarger, *Ann. Rev. Biochem.* **38,** 323 (1969); L. J. Reed and D. J. Cox, *ibid.* **35,** 57 (1966).
78. P. Truffa-Bachi and G. N. Cohen, *Ann. Rev. Biochem.* **37,** 79 (1968).
79. J. Ito and C. Yanofsky, *JBC* **241,** 4112 (1966); R. H. Bauerle and P. Margolin, *Cold Spring Harbor Symp. Quant. Biol.* **31,** 203 (1966); J. A. DeMoss, *BBRC* **18,** 850 (1965).
80. D. E. Sheppard, *Genetics* **50,** 611 (1964).
81. G. A. O'Donovan and J. L. Ingraham, *Proc. Natl. Acad. Sci. U. S.* **54,** 451 (1965).

mutants have been assigned to nine cistrons, and in several cases, the altered enzymes have been studied (*82*).

Practically all of the mutants studied thus far have been confined to microorganisms primarily because it is easy to obtain selected types of mutants from the large populations of organisms that can be conveniently handled in the laboratory. It should be noted, however, that the screening of serum from large numbers of humans throughout the world has led to the discovery of many variants in serum proteins. These include the haptoglobins (*83*), glucose-6-phosphate dehydrogenase (*84*) from red cells, and phosphoglucomutase (*85*). Most of these variants were detected by their abnormal electrophoretic mobility in much the same manner as used for detecting hemoglobin variants. In a few cases, alterations in the enzymic properties of the abnormal enzymes have been reported (*85, 86*).

IV. Amino Acid Analogs

Mutations in general lead to single site-specific replacements of amino acids in a protein. It is possible to extend the use of mutants so that instead of a single replacement of an amino acid a particular amino acid will be substituted by an analog at every position where that amino acid appears in the polypeptide chain. For example, all of the tryptophan residues in a polypeptide chain could be replaced by 7-azatryptophan. The effect of this substitution on the structure and function of a particular enzyme would provide information on the importance of tryptophan per se in that protein. Theoretically, any amino acid analog that can be activated by the amino acyl tRNA synthetase and transferred to the appropriate tRNA can be utilized. The following list includes many of the analogs which have been shown to be incorporated into proteins: (natural amino acid in parentheses) 1,2,4-triazole-3-alanine, 2-methylhistidine (histidine); canavanine (arginine); azetidine carboxylic acid (proline), 3-fluorotyrosine (tyrosine), parafluorophenylanine (phenylalanine), norleucine, ethionine (methionine); trifluoroleucine (leucine).

82. J. C. Loper, M. Grabnar, R. C. Stahl, Z. Hartman, and P. E. Hartman, *Brookhaven Symp. Biol.* **17,** 15 (1964).

83. O. Smithies, G. E. Connell, and G. H. Dixon, *Nature* **196,** 232 (1962); J. A. Black and G. H. Dixon, *Nature* **218,** 736 (1968).

84. *World Health Organ. Tech. Rept. Ser.* **366** (1967); A. Yoshida, *Proc. Natl. Acad. Sci. U. S.* **57,** 835 (1967).

85. D. A. Hopkinson and H. Harris, *Ann. Human Genet.* **30,** 167 (1966). H. Harris, *Brit. Med. Bull.* **25,** 5 (1969).

86. L. Luzzatto and A. Afolayan, *J. Clin. Invest.* **47,** 1833 (1968).

Before describing specific applications of analogs, it is important to point out a number of serious limitations in their use. First, the analog must be incorporated into proteins. Second, in order to insure that all the particular natural amino acid will be substituted, some way must be found for eliminating the pool of the natural amino acid in the cell. For bacteria, this is easily achieved by isolating an auxotrophic mutant that requires the particular amino acid for growth. Such a mutant will have a block in the biosynthetic pathway for the amino acid. Because there is neglible turnover in bacterial cells and because the pool sizes of free amino acids are small, substitution of the analog for the amino acid in the growth medium insures almost total substitution. Third, the protein containing the analog must be purified and separated from the same protein that contains the natural amino acid. One way of achieving the latter condition is to work with an "inducible" enzyme so that the *de novo* synthesis of the protein can be initiated at the same time as the analog is added. For most analogs, the particular polypeptide chain will probably be sufficiently altered in its quaternary structure that an inactive enzyme results. For these cases, the detection of the protein would have to depend upon its antigenic activity. Fourth, almost all analogs eventually lead to death of the organism because of their cumulative effect on the many cell proteins; thus, only a limited amount of the protein containing the analog can be produced.

Despite these rather restrictive conditions, there is a great potential for the use of analogs. Under suitable conditions and with the proper analog it should be possible to isolate and "tailor make" enzymes. A very good example of this approach is the work of Anfinsen and Corley (*87*) who substituted norleucine for the methionine residues in a bacterial nuclease. Unlike methionine, the analog is resistant to cyanogen bromide (CNBr) and enzyme that still contained methionine was eliminated by treating the mixture with CNBr. The pure norleucine nuclease was enzymically active, thus suggesting that methionine residues per se are not essential for the catalytic site of this enzyme.

Analogs have been used extensively for probing the structure and function of *E. coli* alkaline phosphatase. Table II summarizes results of studies carried out thus far (*88–92*). It is interesting to note that the tryptophan residues per se are not crucial to structure or function of this

87. C. B. Anfinsen and L. G. Corley, *JBC* **244,** 5149 (1969).
88. M. H. Richmond, *JMB* **6,** 284 (1963).
89. S. Schlesinger and M. J. Schlesinger, *JBC* **242,** 3369 (1967); **244,** 3803 (1969).
90. J. Attias, M. J. Schlesinger, and S. Schlesinger, *JBC* **244,** 3810 (1969).
91. M. J. Schlesinger, unpublished experiments (1969).
92. S. Schlesinger, *JBC* **243,** 3877 (1968).

TABLE II
EFFECT OF AMINO ACID ANALOGS ON FORMATION AND FUNCTION OF *E. coli* ALKALINE PHOSPHATASE

Analog	Amino acid replaced	Type of enzyme produced	Reference
p-Fluorophenylalanine	Phenylalanine	Active enzyme	*88*
1,2,4-Triazole-3-alanine	Histidine	Inactive subunits	*89*
2-Methylhistidine	Histidine	Inactive partially folded subunits	*89*
Canavanine	Arginine	Inactive subunits and partially active enzyme	*90*
Azetidine-2-carboxylic acid	Proline	Inactive subunit	*91*
3-Fluorotyrosine	Tyrosine	Active enzyme	*91*
Azatryptophan and tryptazan	Tryptophan	Active enzyme	*92*

phosphatase. The histidine residues, however, which are essential for a proper catalytic site, also appear to be required for proper conformation of the subunit in order for them to associate. There is also the suggestion from the studies with canavanine that a few (possibly only one of the twenty-two) of the arginine residues are essential for both subunit association and for activity.

An ingenious application of analogs is found in a study of the formation of polio virus proteins in infected tissue culture cells (*93*). Isotope labeling patterns suggested that certain of the small viral proteins were formed from a very large polypeptide precursor. This precursor–product relationship was substantiated by experiments in which the feeding of ethionine, *p*-fluorophenylalanine, azetidine carboxylic acid, and canavanine to virus-infected cells inhibited the appearance of the viral proteins and led to an accumulation of a large polypeptide.

In another example, ethionine was shown to inhibit the formation of active β-galactosidase (*94*). When canavanine was fed to *E. coli*, profound changes were observed in cellular organization with disruption of the nuclear region and the accumulation of large electron-dense bodies attached to the cell membrane (*95*).

Amino acid analogs have generally been used as metabolic inhibitors (*96*), and they have proved highly successful for the isolation of mutants that contained alterations in those enzymes which react with the amino

93. M. F. Jacobson and D. Baltimore, *Proc. Natl. Acad. Sci. U. S.* **61,** 77 (1968).
94. J. Spizek and J. Janecik, *BBRC* **34,** 17 (1969).
95. C. F. Schachtele, D. L. Anderson, and P. Rogers, *JMB* **33,** 861 (1968).
96. L. Fowden, S. Neale, and H. Tristram, *Advan. Enzymol.* **29,** 89 (1967).

acids. These include the amino acyl tRNA synthetases and those amino acid biosynthetic enzymes that are feedback sensitive (*97*). Thus far the analogs tested have been those randomly selected among natural or synthetic organic compounds isolated for other purposes. It should be possible, however, to synthesize analogs with specific reactive groups that have properties which lend themselves to many of the techniques such as NMR, ESR, X-ray analysis, CD, and ORD which are applicable as probes for enzyme structure and function.

V. Advantages and Limitations of Mutants

One of the goals in seeking the molecular basis of protein function is to determine how the specific array of the amino acids in the polypeptide chain can provide for the unique structure and function of a protein. X-ray analysis of protein crystals provides the most powerful tool for ascertaining the role of amino acids in the final quaternary structure for only this technique can unambiguously assign the relationship of the different amino acids to each other. In conjunction with this technique are a variety of studies with proteins in solution which indicate how particular amino acid residues interact with solvent and with ligands. One of the most commonly used methods for analyzing the latter has been the selective chemical modification of specific amino acids. Despite its proven success this approach suffers certain limitations. Frequently, the reagents are not sufficiently restrictive or quantitative in their reaction with protein and the solution of reacted protein product contains a mixture of partially modified groups. For careful analysis it is essential to determine exactly which groups in the protein have been altered and what the alteration has been. Many of the reagents are only useful at pH or temperatures different from those essential for maintaining the native conformation of the protein with the result that modifications are only made on denatured forms of the protein.

The ideal approach for altering a specific amino acid is to obtain protein in which all molecules have only that one change. A single site mutation that leads to a single substitution of an amino acid will provide just this product. The problem is that at present it is not possible to direct the substitution of a particular amino acid at a specific residue in

97. I. N. Hirshfield, R. Dedeken, P. C. Horn, D. A. Hopwood, and W. K. Maas, *JMB* **35**, 83 (1968); F. C. Neidhardt, *Bacteriol. Rev.* **30**, 701 (1966); H. S. Moyed, *JBC* **236**, 2261 (1961); A. Szentirmai, M. Szentirmai, and H. E. Umbarger, *J. Bacteriol.* **95**, 1672 (1968).

the polypeptide. Conceivably, with the ability to chemically synthesize cistrons and to make polypeptide chains *in vitro,* such substitutions can be made. The straightforward chemical synthesis of polypeptide chains that has been successful for making ribonuclease (*98*) does provide a method for site-specific substitution.

Sites of mutations are, in general, random and the amino acid substitutions are not predictable. Certain regions in polypeptide chains have been found to be more susceptible than others, but these "hot spots" may reflect regions in the polypeptide chain more sensitive to the structure and function of the protein. Furthermore, the site of the defect and the particular substitution can only be ascertained after an analysis of the sequence of the protein—not a trivial experimental task. In addition, as noted above, particular mutants may be difficult to isolate and analyze and the modified proteins are often unstable and not easily obtainable in the high yields required for careful detailed studies.

Nevertheless, the use of mutants may be the most valuable approach for the study of multiple enzyme complexes and for determining the mechanisms by which protein subunits interact with themselves, with nucleic acids, and with cell organelles such as membranes. The examples cited here offer some indication of how genetics can be used as a probe to protein structure.

98. B. Gutte and R. B. Merrifield, *JACS* **91,** 501 (1969); R. G. Denke Walter, D. F. Veber, F. W. Holly, and R. Hirschmann, *JACS* **91,** 502 (1969).

6

Evolution of Enzymes

EMIL L. SMITH

I. General Considerations—Evolutionary and Structure–Function Relationships in Proteins

Comparison of the properties and structures of the proteins of various organisms has indicated that (a) those proteins which serve the same function in all organisms frequently manifest strong homology in structure but with some modifications in amino acid sequence, e.g., hemoglobins, insulins, cytochromes c, and various enzymes; (b) some enzymes present in lower organisms have disappeared from higher organisms, e.g., enzymes in the synthetic pathways for certain vitamins and the "essential" amino acids; and (c) many new proteins (enzymes) have appeared in higher organisms. These are the three generally recognized facets of protein evolution that we shall have to consider.

The evolution of enzymes can be assessed by comparisons of the properties and the sequences of the proteins in various organisms. Such comparisons have enriched our knowledge: first, of some of the genetic factors that have been involved in the changing structures of homologous enzymes; second, of some of the selective factors operative in the evolution of enzymes, leading to diverse structures manifesting new specificities and, ultimately, to the development of novel metabolic pathways (*1*); and, third, of the specific roles of amino acid residues in the mechanism of action and specificity of enzymes in relation to the more general problems of protein conformation. These topics are to a large degree interrelated but not always in a direct fashion.

Many changes in the genome are deleterious, producing lethal or semilethal effects; individuals possessing such mutations are rapidly eliminated from the population; however, there are occasional situations in which the heterozygous state has a selective advantage. Surviving species may be expected to have fully functional enzymes in which the changes in sequence have either produced an enhanced survival value or at least a "neutral" effect. Thus, survival value as applied to enzymes may be reflected in stability, mechanism of action, including kinetic efficiency and regulatory properties, as well as specificity and other factors.

Most investigators are in agreement on certain broad features of the biosynthesis of proteins and their structures. Although these premises may be self-evident it is desirable to state them explicitly. In almost all organisms (except certain viruses) the genotype is DNA and the sequence of the amino acids in the polypeptide chains of proteins is determined through a messenger RNA by a nucleotide triplet code, which appears to be universal in all organisms. Inasmuch as DNA usually consists of only four types of nucleotides and is similar in structure in all higher organisms, it is assumed that factors leading to changes in the DNA are probably the same. Thus the causes of alterations in the genome, which we shall not discuss here, are probably similar in all organisms. This does not rule out the possibility that some types of nucleotide modifications in the genome may be more frequent than others, e.g., in certain cases it appears that there is a preponderance of guanine to adenine mutations (at the messenger level) as compared to cytosine to uracil alterations (*2, 3*). Since the evidence for such non-

1. The problem of the evolution of metabolic pathways and the controls of such systems is outside the scope of this chapter and will not be considered.
2. W. M. Fitch, *JMB* **26**, 499 (1967).
3. E. Margoliash and W. M. Fitch, *Ann. N. Y. Acad. Sci.* **151**, 359 (1968).

random changes is based in part on data for various phyla, it may be assumed that such effects are universal for all types of DNA.

Natural selection involves the interaction of the organism (and the species) with the environment. For an enzyme or other protein, survival value may involve either the external or internal environment. Thus, it is not the genotype but the phenotype that is of primary concern to the evolutionist. Ultimately, the nature of the phenotype depends, directly or indirectly, on the genetically determined structures and hence the functions of the enzymes and other proteins, since practically all the phenotypic features of an organism depend on molecules which have been synthesized by enzymic catalysis.

In the following discussion it is also assumed, as is accepted at present, that the genome directly determines for proteins only the amino acid sequences of the peptide chains (*3a*). Thus, the folding of the individual chains into proper conformation occurs without further direct intervention of the genetic material (either DNA or RNA). Similarly, the interaction of individual subunits to produce multichain enzymes, and the interaction of enzymes with other proteins, as well as other cellular constituents to form complex organelles and other cellular structures, must also occur without further direct involvement of genetic determinants. The numerous observations that after denaturation or dissociation there is spontaneous reconstitution of simple and complex proteins, as well as organelles such as ribosomes (*4*) and the complex structures of viruses (*5*), attest to the probable general validity of these postulates. The fact that many organelles and membrane structures also contain nucleic acids, lipids, polysaccharides, etc., does not invalidate this view since these cellular constituents also have been synthesized through the action of genetically determined enzyme structures.

Translation of the genetic material into a polypeptide chain involves triplet codons for only twenty amino acids. Thus, all protein modification at a *direct* genetic level concerns only sequences containing these amino acids. (Other amino acids found in proteins have been demonstrated to be the result of secondary modification after formation of the peptide chains.) Further, the kinds of cofactors, prosthetic groups, and

3a. The genome also determines the sequences of RNA molecules that are involved in the mechanism of protein synthesis, i.e., the various types of transfer RNA and ribosomal RNA. Little is known at present regarding the evolution of such molecules, although it is already evident that species differences do exist and that selection must operate on these products of the genome also.

4. H. J. Vogel, J. O. Lampen, and V. Bryson, eds., "Organizational Biosynthesis." Academic Press, New York, 1967; see T. Staehelin, H. Raskas, and M. Meselson, *ibid.,* Part IX, p. 443; M. Nomura and P. Traub, *ibid.* p. 459.

5. D. J. Kushner, *Bacteriol. Rev.* **33,** 302 (1969).

coenzymes are also relatively limited in number. Thus with limited numbers of amino acids and cofactors, there are many different kinds of reactions and substrates. Yet these relatively few amino acids and cofactors have been utilized during the evolution of all of the complex catalytic functions in which enzymes participate, as well as all other roles that proteins can play in living organisms.

Of the twenty amino acids commonly occurring in proteins, clearly only a limited number of these can participate actively in the catalytic mechanisms of enzymes; namely, those with appropriate side chains that can act in proton transfer, as group acceptors, as nucleophiles, as electrophiles, etc. Some amino acids, particularly those with hydrophobic aliphatic side chains, may not be involved directly in enzymic catalysis per se, but these as well as all other amino acids do participate in determining the tertiary and quaternary structure of the protein, in binding substrates, regulatory effectors, coenzymes and other cofactors, and in forming larger complexes of structure such as organelles and membranes. For these reasons the possible roles of each amino acid should be considered briefly. For example, the imidazole group of histidine has been implicated as a participant in many different kinds of enzymic reactions since it can serve in proton transfer at neutral pH, as a ligand to metal ions and coenzymes, in the binding sites of substrates and regulatory effectors, and in other roles. Because histidine can play so many different roles, and so few other amino acids can do so, this cannot be taken as evidence that all the kinds of enzymes in which histidine participates are derived from a common ancestral protein structure. Although this has been shown to be the case in some instances, in others the development of a similar role for histidine has been the result of independent evolution. Evidence for evolutionary relationships can only be elucidated by examination of the structures of the enzymes themselves or some attribute of structure, such as immunological cross-reaction. True homology in this sense must be manifested at the level of primary sequence or in the overall conformation as judged by X-ray analysis. Some examples of these problems will be discussed below.

In order to clarify some of the subsequent discussion, it may be useful to indicate briefly a few additional problems in understanding amino acid function in proteins. Table I lists the twenty amino acid residues and indicates some of the roles which the side chains are known to have or are presumed to play in proteins in general, as well as in specific enzymic reactions. The list does not pretend to be exhaustive and, undoubtedly, our knowledge is still incomplete, but the listing serves to illustrate a few of the concepts and problems in understanding protein evolution.

TABLE I
AMINO ACID RESIDUES

Amino acid residue	Possible roles in protein and enzyme structure and function
Arginyl	Hydrophilic; electrostatic interactions
Lysyl	Hydrophilic; electrostatic interactions; attachment of prosthetic group or cofactor in amide bond; interacting to form Schiff base; ligand to metal ion
Histidyl	Hydrophilic or hydrophobic (depending on ionization); electrostatic interactions; proton transfer; ligand to metal ion; hydrogen bonding; acceptor in transfer reactions
Glutamyl Aspartyl	Hydrophilic; electrostatic interactions; proton transfer; ligand to metal ion; covalent linking in ester or amide through ω-carboxyl
Glutaminyl	Hydrophilic; hydrogen bonding
Asparaginyl	Hydrophilic; hydrogen bonding
Seryl	Hydrogen bonding; nucleophile; covalent linkage of OH in esters
Threonyl	Hydrogen bonding; nucleophile; covalent linkage of OH in esters; hydrophobic
Glycyl	Absence of side chain permits flexibility of folding and cross-hydrogen bonding
Alanyl Valyl Leucyl Isoleucyl Phenylalanyl	Hydrophobic interactions; determinants of steric and conformation specificity, e.g., numerous alanyl residues favor α-helix formation whereas numerous valyl or isoleucyl residues in sequence tend to inhibit formation of such structural features
Tyrosyl	Hydrophobic; hydrogen bonding; proton transfer; electrostatic interactions at high pH; ligand to metal ions
Tryptophanyl	Hydrophobic; hydrogen bonding
Cysteinyl	Nucleophile; acyl acceptor; hydrogen bonding; ligand to metal ions
Cystyl	Cross-linking through disulfide bonds
Methionyl	Hydrophobic; hydrogen bonding to S (?); ligand to metal ions
Prolyl	Interruption of α helix or β structures thus allowing irregular conformation; hydrophobic

Some amino acid residues can assume many functions whereas others are less versatile. Knowledge of amino acid substitutions in homologous enzymes can then permit some assessment of the role that a residue may play at a specific locus in a specific enzyme. Clearly, when histidine is replaced by any other amino acid without detectable change of function, we can assume that the histidyl residue at that locus in the enzyme does not participate in the catalytic site, as a ligand, etc.

Substantial variation in the amino acid sequences of homologous proteins can occur without major *apparent* alteration of function (see Section IV,A,2, however). Excellent examples are available for many proteins, particularly the hemoglobins (*6*) and cytochromes c (*7*, *8*). Some of the data for the latter and for other enzymes will be given below, and it will be indicated how such information can enrich our knowledge, not only of specific enzymes and their structures but also of the roles of the various amino acids.

In some instances replacement of certain amino acids at specific sites in the sequence is deleterious to function. Thus, it is assumed that an amino acid replacement of a residue which is critical for the catalytic mechanism, for the binding of a substrate, cofactor, or prosthetic group, or which is critically involved in determining conformation, will result in a loss or an alteration of enzymic activity. Such changes cannot be assessed in homologous proteins of different species. Much of our information is derived from genetic and sequence studies within a species, e.g., the tryptophan synthetase of *E. coli* (*9*). On the other hand, amino acid replacements which manifest no deleterious effect must be at sites in the protein which do not drastically modify any of these critical functions. From the few known examples where there has been a correlation of the three-dimensional structure, as determined by X-ray crystallography, and the sites of amino acid replacements, those "radical" replacements which have little effect are mainly on the outside of the molecule and remote from the active catalytic region. The many known substitutions of single amino acid residues in human hemoglobin provide examples of all types, including deleterious effects produced by changes in ligand binding, solubility, and stability (*10*). See also the discussion of the properties of the subtilisins and the pancreatic proteinases below as well as in Volume III of this treatise.

The terms *radical* and *conservative* in dealing with single amino acid substitutions (Section III,A) in homologous proteins have been useful in describing the properties of a given site in the sequence of a protein (*11*, *12*). Conservative substitution refers to pairs or groups of amino acid residues which possess similar side chains, e.g., glutamyl-aspartyl, lysyl-arginyl, seryl-threonyl, leucyl-isoleucyl-valyl, and phenylalanyl-tyrosyl. Less obvious at first sight are hydrophobic residues of diverse

6. G. Braunitzer, K. Hilse, V. Rudloff, and N. Hilschmann, *Advan. Protein Chem.* **19,** 1 (1964).
7. E. L. Smith and E. Margoliash, *Federation Proc.* **23,** 1276 (1964).
8. E. Margoliash and E. L. Smith, *in* "Evolving Genes and Proteins" (V. Bryson and H. J. Vogel, eds.), p. 221. Academic Press, New York, 1965.
9. S. Yanofsky and J. R. Guest, *Nature* **210,** 799 (1966).
10. M. F. Perutz and H. Lehmann, *Nature* **219,** 902 (1968).
11. E. L. Smith, *Proc. Natl. Acad. Sci. U. S.* **48,** 677 and 859 (1962).
12. E. L. Smith, *Harvey Lectures* **62,** 231 (1968).

structure including both aromatic and aliphatic side chains which can occupy the same site. Similarly, hydrophilic groups of different properties, usually on the outside of the molecule, may include anionic, cationic, and neutral side chains at the same site.

Threonine is unusual inasmuch as at some sites it is a replacement only for serine and at others it can replace the branched-chain aliphatic residues valine, isoleucine, and leucine (*12*). Thus, by virtue of its β-hydroxyl and its methyl group, threonine can assume at least two different roles as a conservative replacement in proteins. Another unusual example is methionine, frequently found at the same locus in homologous proteins as leucyl and other aliphatic residues. Thus, when the sulfur atom does not participate in a specific chemical interaction, methionine serves as an aliphatic residue similar to leucine. It is noteworthy that norleucine can be incorporated in proteins when the supply of methionine is limited (*13*).

Radical substitution indicates replacement by a residue of very different character. As noted above, such replacements may be presumed to occur mainly on the outside of the protein molecule when there is no significant change in properties. Indeed, this affords a method of deducing the location of such residues from comparative sequence studies (*7, 12*), but definitive proof must come from X-ray analysis or from chemical modification studies which do not influence activity or conformation. Clearly, all substitutions for cysteine, histidine, and tryptophan may be regarded as radical and imply, when there is no apparent functional change, that the unique side chains of these residues do not participate in any critical way in the activity or conformational properties.

Classification of residues as radical or conservative is not all-embracing and is useful only as a first approximation. Each of the twenty amino acids is unique. At the active site of an enzyme, glutamyl apparently cannot replace aspartyl or vice versa, nor has threonyl been found to replace seryl at the active sites of the "serine-histidine" group of proteinases and esterases. In other words, where the geometry is critical, at the active site, at a specificity site, or at loci important for conformation, probably all substitutions are deleterious and may be lethal. Presumably, we can detect conservative and radical substitutions only where the structure of the protein permits some tolerance. The extent of such tolerance will obviously depend on the structure-function relationships and the interactions of the particular protein. Some of the data from the

13. D. B. Cowie, G. N. Cohen, E. T. Bolton, and H. DeRobichon-Szulmajster, *BBA* **34**, 39 (1959).

various cytochromes c will be used to illustrate both the problems and the utility of these concepts (Section II).

Inasmuch as later volumes of this treatise will be concerned with all of the known enzymes for which detailed structural information is available, as well as with those enzymes for which comparative structures are known for several species, the discussion here will be devoted to those examples which serve to illustrate the general concepts under consideration and much of the detail concerning these and other enzymes will be left to later chapters. Regretfully, reference will be made only occasionally to other comparative studies; detailed discussion of many protein and peptide hormones as well as other proteins will be omitted. Several general reviews and symposia related to some phases of this chapter are available (*12, 14–19*).

II. Some Principles from the Study of Cytochrome c

In order to illustrate some of the principles concerning the evolution of homologous proteins, one may compare the sequences of the same protein from a large number of species of different taxa. The protein for which the most complete information is presently available, both as to numbers of sequences as well as types of organisms, is the cytochrome c universally occurring in all eukaryotic organisms. Problems in ascertaining homologous relationships of proteins have been fully discussed by others (*18–20*). Here we shall assume that the homology of these cytochromes in sequence and in function is obvious (*7, 8, 12, 21*).

The primary structures of cytochrome c are now known from a variety of vertebrates, some insects among the invertebrates, several Ascomycetes (yeasts), and some higher plants (*22*). The function of the protein has

14. V. Bryson and H. J. Vogel, eds., "Evolving Genes and Proteins." Academic Press, New York, 1965.

15. T. H. Jukes, "Molecules and Evolution." Columbia Univ. Press, New York, 1966.

16. N. van Thoai and J. Roche, eds., "Homologous Enzymes and Biochemical Evolution." Gordon & Breach, New York, 1968.

17. *Brookhaven Symp. Biol.* **21,** 1–428 (1969).

18. C. Nolan and E. Margoliash, *Ann. Rev. Biochem.* **37,** 727 (1968).

19. H. Neurath, K. A. Walsh, and W. P. Winter, *Science* **158,** 1638 (1967).

20. W. M. Fitch, *JMB* **16,** 9 (1966).

21. E. Margoliash and A. Schejter, *Advan. Protein Chem.* **21,** 113 (1966).

22. References to each of the cytochromes c not specifically cited can be found in the following reviews: Margoliash and Smith (*8*), Nolan and Margoliash (*18*), and Margoliash and Schejter (*21*).

remained the same in all of these organisms as indicated by certain of its properties. First, the cytochrome c from each of these species can react in equivalent fashion with a mammalian or yeast cytochrome oxidase. Although cytochrome c is not generally regarded as an enzyme, for present purposes we can consider it as such inasmuch as it exhibits one of the major properties of an enzyme, namely, specificity, it being the specific reductase for cytochrome oxidase. Second, the redox potential of approximately +250 mV is essentially the same for all these cytochromes. Third, the absorption spectra are essentially the same in both the oxidized and reduced forms from all species. From these similarities one can conclude that there have been no major changes in function or in conformation of cytochrome c during the long period of the evolution of presently known eukaryotes. These constant properties are in contrast to the wide diversity of absorption spectra and redox potentials found among other types of hemoproteins and, indeed, of other kinds of cytochromes (*21, 23*).

The homology of the various cytochromes c can be readily illustrated by comparing the amino acid sequences of two cytochromes c of different origin, namely, that from human heart tissue (*24*) and from wheat germ (*25*) (Fig. 1). Residues in identical positions are indicated by dashes and the differences in sequence are shown at the appropriate loci. The homology in sequence is indicated by the 69 sites that are occupied by the same residue in the two cytochromes. Inasmuch as all presently known cytochromes c show approximately equivalent or greater homology in sequence throughout the wide diversity of eukaryotic organisms, it is evident that all presently surviving eukaryotic life possesses a common ancestry (*7*). This information has also given the first evidence, from sequence studies of a protein, for the survival of a specific gene for the more than one billion years that are estimated to have elapsed since the divergence of plant and animal lines of evolution (*7, 25*). The close homology in sequence of a histone from plants and animals shows that the gene determining the structure of this histone (*26*) (see Section IV,A) has also been in existence for at least the same time span, and additional evidence suggests that this is true for other histones (*27*).

Certain features of the cytochrome c structure should be noted. First,

23. K. Okunuki, M. D. Kamen, and I. Sekuzu, eds., "Structure and Functions of Cytochromes." Univ. of Tokyo Press, Tokyo, 1968 (also University Park Press, Baltimore, Maryland, 1968).

24. H. Matsubara and E. L. Smith, *JBC* **237,** PC3575 (1962); **238,** 2732 (1963).

25. F. C. Stevens, A. N. Glazer, and E. L. Smith, *JBC* **242,** 2764 (1967).

26. R. J. DeLange, D. M. Fambrough, E. L. Smith, and J. Bonner, *JBC* **243,** 5906 (1968); **244,** 319 and 5669 (1969).

27. E. L. Smith, R. J. DeLange, and J. Bonner, *Physiol. Rev.* **50,** 159 (1970).

−8 −1 1
Acetyl-Ala-Ser-Phe-Ser-Glu-Ala-Pro-Pro-Gly-Asn-Pro-Asp-Ala-Gly-Ala-Lys-
Acetyl Asp-Val-Glu-Lys Lys

10 20
Ile-Phe-Lys-Thr-Lys-Cys-Ala-Gln-Cys-His-Thr-Val-Asp-Ala-Gly-Ala-Gly-His-
Ile-Met Ser Glu-Lys Gly-Lys

30 40
Lys-Gln-Gly-Pro-Asn-Leu-His-Gly-Leu-Phe-Gly-Arg-Gln-Ser-Gly-Thr-Thr-Ala-
Thr Lys-Thr Gln-Asn-Pro-

50 60
Gly-Tyr-Ser-Tyr-Ser-Ala-Ala-Asn-Lys-Asn-Lys-Ala-Val-Glu-Trp-Glu-Glu-Asn-
Thr Gly-Ile-Ile Gly Asp

70 80
Thr-Leu-Tyr-Asp-Tyr-Leu-Leu-Asn-Pro-Lys-Lys-Tyr-Ile-Pro-Gly-Thr-Lys-Met-
Met-Glu Glu

90
Val-Phe-Pro-Gly-Leu-Lys-Lys-Pro-Gln-Asp-Arg-Ala-Asp-Leu-Ile-Ala-Tyr-Leu-
Ile Val Ile Lys-Glu-Glu

100 104
Lys-Lys-Ala-Thr-Ser-SerCOOH
Asn-GluCOOH

FIG. 1. The amino acid sequence of wheat germ cytochrome c (*25*) and for comparison below this sequence, the residues which differ in human cytochrome c (*24*). Heme is attached by thioether bonds to the cysteinyl residues at positions 14 and 17. Residues indicated by negative numbers are absent in all vertebrate cytochromes c.

the covalent attachment of the heme to the cysteine residues at positions 14 and 17 is the same in these and all other species. Indeed, the heme-linked cysteine residues are separated by two other residues in other c-type cytochromes (*21*); this apparently results from spatial and steric properties of the porphyrin itself (*28–30*). Second, the numbering system is that derived from the vertebrates, all of which contain 103 or 104 residues and are acetylated at the amino terminus. The extra residues at the amino terminus occur in the cytochromes of all presently investigated invertebrates, Ascomycetes, and higher plants. The only known non-

28. A. Ehrenberg and H. Theorell, *Acta Chem. Scand.* **9,** 1193 (1955).
29. S. Sano, K. Ikeda, and S. Sakakibara, *BBRC* **15,** 284 (1964).
30. S. Inouye, S. Sakakibara, and S. Akabori, *Abstr., Proc. 6th Intern. Congr. Biochem., New York, 1964* Vol. 2, p. 83 (1964).

vertebrate cytochromes c that are *N*-acetylated are those from wheat germ (*25*) and other higher plants (*31*).

Inasmuch as the vertebrate cytochromes possess the shortest peptide chains, it is likely that chain shortening has occurred during the evolution of this protein. It is assumed, of course, that *N*-acetylation takes place subsequent to the formation of the completed peptide chain. Furthermore, the acetylation of vertebrate cytochromes, wheat germ, and other plant cytochromes c probably developed independently in plants and vertebrates.

Table II shows the amino terminal ends of some of the cytochromes c. Since there is no apparent change in properties *in vitro*, it appears that the extra residues are not strictly essential for the function of this protein as an electron carrier or as a reductase for cytochrome oxidase. Chain shortening during evolution by loss of residues at the amino or carboxyl ends is, therefore, certainly one process that must have occurred not only in the cytochromes but in other proteins as well (Section III,B).

The differences in sequence of the various cytochromes may be evaluated by comparing the total number of residues substituted or by comparing the presumed number of point mutations in the codons. For simplicity, we shall compare numbers of residues substituted, but it should be noted that extensive discussions have been presented of taxonomic and other relationships based on the minimal number of presumed codon differences. The discussion by Fitch and Margoliash (*32*) can be consulted for these points.

TABLE II
AMINO TERMINAL SEQUENCES OF CYTOCHROMES c

	−8 −7 −6 −5 −4 −3 −2 −1 1
Wheat[a]	Acetyl-Ala Ser-Phe-Ser-Glu-Ala-Pro-Pro-Gly
Candida	Pro-Ala-Pro-Phe-Glu-Gln-Gly
Yeast (iso-1)	Thr-Glu-Phe-Lys-Ala-Gly-
Neurospora	Gly-Phe-Ser-Ala-Gly-
Moths, flies[b]	Gly-Val-Pro-Ala-Gly-
Vertebrates	Acetyl-Gly-

[a] Other higher plant cytochromes also possess eight extra residues with the initial sequence acetyl-Ala-Ser-Phe (*31*).

[b] *Samia cynthia*, tobacco hornworm moth, screwworm fly (*Haematobia irritans*), and *Drosophila melanogaster*.

31. D. Boulter, M. V. Laycock, J. A. M. Ranshaw, and E. W. Thompson, *in* "Phytochemical Phylogeny" (J. B. Harborne, ed.). Academic Press, New York, 1970 (in press).

32. W. M. Fitch and E. Margoliash, *Science* **155,** 279 (1967).

If the sequences of various cytochromes c are compared to that of man, it is apparent that, in general, the more closely related the species is to man, the fewer the number of differences, in sequence. Examples of such comparisons are given in Table III. For these comparisons, we have included only the 104 residues usually present in vertebrate cytochromes. Thus, the cytochromes c of man and chimpanzee are identical, whereas, that of the rhesus monkey (*Macaca mulatta*) differs in a single residue. The cytochromes of other mammals show greater differences and the differences are still greater when human cytochrome is compared to those of reptiles, amphibians, fish, etc. Clearly, this illustrates the phenomenon of divergence (*7*, *12*), one of the best known principles of evolution (also see Section IV). Thus, as applied to homologous proteins the greater the length of time during the ancestral lines of the various species became separated, the greater the number of sequence (or mutational) differences. This principle applies to all proteins, but the degree of divergence varies for different proteins (Section IV,A,1). From more fragmentary information on other enzymes, it is evident that this is generally the case, and we may expect that in the next few years as the sequences of more proteins are determined, knowledge of this phenomenon will be greatly increased.

Divergence is, however, not the sole factor that can occur in evolution. Convergence is also a well-known phenomenon (Section IV). This is clearly based on the fact that under similar selection pressures certain features of organisms can tend to become more alike despite the fact that initially the ancestral types may have possessed different structures. A well-known morphological example is the fusiform shape of aquatic

TABLE III

NUMBERS OF VARIANT RESIDUES IN CYTOCHROMES COMPARED TO HUMAN[a]

Cytochrome compared to human	No. of residues	Cytochrome compared to human	No. of residues
Chimpanzee	0	Rattlesnake	14
Rhesus monkey	1	Turtle	15
Rabbit	9	Tuna	21
Kangaroo	10	Dogfish	23
Whale	10	Screwworm fly	25
Cow, pig, sheep[b]	10	Moth (*Samia cynthia*)	31
Dog	11	Wheat	35
Donkey	11	*Neurospora*	43
Horse	12	Yeast	44
Chicken, turkey[b]	13		

[a] Comparisons only for 104 residues in common.
[b] Cytochromes of identical sequence.

mammals being evolved independently to resemble the fusiform character of fishes; yet it is clear that several groups of aquatic mammals were derived from land-living relatives. It is now accepted that the whales (Cetacea) evolved from mammals closely related to the ungulates; this is evident from the closely related amino acid sequence of the cytochrome c of the California gray whale (*33*), which differs in only two residues from the cytochromes c of cow, pig, and sheep, which are identical.

A possible example of convergence in cytochrome c is illustrated in Table IV which shows a comparison of residue differences from wheat germ cytochrome c based once again on the 104 residues possessed in common (*12*). The data indicate that wheat germ cytochrome is more similar to those of primates than to those of species of such primitive plants as the Ascomycetes. The reasons for this are unknown, yet these limited data do suggest a possible convergence in properties, or simply a parallelism of mutations, as discussed below.

Parallelism (Section IV,C) has been manifested at all levels at which evolution can be viewed—at the behavioral level, at the organ level, at the tissue level, at the functional level of enzymes, etc. Parallelism can also be detected at the amino acid substitution level, thus rendering estimation of taxonomic and evolutionary relationships more difficult. For example, in the cytochromes c of the Ascomycetes (*34–36*) and wheat germ (*25*), residue 12 is threonine, suggesting that this has been a characteristic feature of the plant kingdom. Correspondingly, in the insects

TABLE IV
VARIANT RESIDUES IN CYTOCHROMES COMPARED TO WHEAT[a]

Cytochrome compared to wheat	No. of residues	Cytochrome compared to wheat	No. of residues
Neurospora	46	Turtle	38
Candida	42	Chicken, turkey	38
Tuna	41	Donkey	38
Dogfish	41	Cow, pig, sheep	37
Yeast (iso-1)	40	Rabbit	36
Kangaroo	39	Whale	36
Horse	39	Dog	36
Moth	38	Rhesus monkey	35
Rattlesnake	38	Man, chimpanzee	35

[a] Comparisons only for 104 residues in common.

33. A. D. Goldstone and E. L. Smith, *JBC* **241,** 4480 (1966).
34. J. Heller and E. L. Smith, *JBC* **241,** 3165 (1966).
35. Y. Yaoi, K. Titani, and K. Narita, *J. Biochem.* (*Tokyo*) **59,** 247 (1966).
36. K. Narita and K. Titani, *Proc. Japan Acad.* **41,** 831 (1965).

(*18*) and most vertebrates, residue 12 is glutamine, the exceptions being the rattlesnake (*37*) and the primates (*23, 38*) in which it is methionine. Thus, it would appear that glutamine is characteristic of animal cytochromes and that the presence of methionine in the snake and the primates represents a parallelism in structure. This development is interesting also in that substitution of methionine for glutamine represents a change of at least two nucleotide bases in the codon at this site.

Many other parallelisms can also be detected in the sequences of the cytochromes. An additional case of interest is that man and chimpanzee once again uniquely share a residue in common with the rattlesnake (namely, isoleucyl at position 58), whereas almost all other known animal cytochromes have a threonyl residue at this locus, except for the tuna (*39*) which has valyl. Here, however, all three residues represent point mutations in the codon.

As already noted (Section I), in a given protein we can expect that residues critical for function or conformation will remain constant in all of the homologous proteins of different species. Thus, an examination of the nature of such constant residues in relation to their possible functions (Table I), and the properties of the protein itself should contribute significantly to the understanding of structure–function relationships. In order to illustrate this approach, we shall examine a few of these features of cytochrome c.

For the 104 residues possessed by all types of cytochrome c, residues at 35 loci have been found to be constant (Fig. 2). Although this number may diminish slightly we cannot expect much further change from studies of more cytochromes of this type in view of the wide taxonomic range of organisms already investigated (*12*).

As shown in Fig. 2, there is only one sequence longer than two residues that has remained constant in all cytochromes c. The 11-residue sequence comprising residues 70 through 80 is obviously critical for this molecule. One feature of this sequence is the presence of two prolyl residues; these would prevent formation of an α helical or β structure.

Inasmuch as reduced cytochrome c is diamagnetic (*40*), it is evident that all ligand positions to the heme ion are filled. Early suggestions were that, aside from the bonds to the four pyrrole nitrogen atoms, these ligand positions were filled by the imidazole groups of two histidyl residues (*40*) or by one imidazole group and an ϵ-amino group of a lysyl

37. O. P. Bahl and E. L. Smith, *JBC* **240,** 3585 (1965).
38. J. A. Rothfus and E. L. Smith, *JBC* **240,** 4277 (1965).
39. G. Kreil, *Z. Physiol. Chem.* **334,** 154 (1963); **340,** 86 (1965).
40. H. Theorell, *JACS* **63,** 1820 (1941).

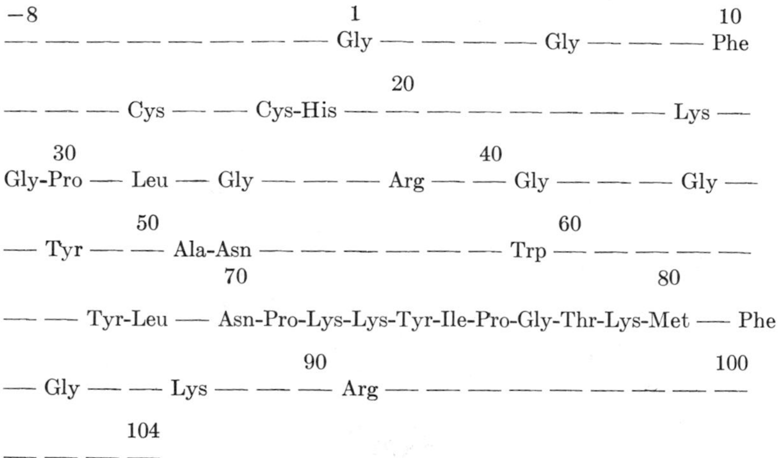

FIG. 2. Residues presently known to be constant in all cytochromes c investigated (*12*). The two Cys residues (14 and 17) indicate the positions of the heme attachment.

residue (*41, 42*). The latter view was excluded by the observation of Hettinger and Harbury (*43*) that all of the amino groups of cytochrome c could be guanidinated without loss of activity. When it was found that there is only one constant histidyl residue, Heller and Smith (*34, 44*) suggested that the sixth ligand may be to a nonnitrogeneous residue. Harbury *et al.* (*45*) indicated, from studies of mixed hemochromogens, that the sixth ligand may involve the sulfur atom of the only constant methionyl residue at position 80. This was consistent with earlier reports (*46, 47*) that carboxymethylation of Met 80 renders the protein autoxidizable and disrupts the structure generally. X-ray diffraction evidence has now clearly confirmed the conclusion that the methionine sulfur atom is the ligand to the heme iron (*48*).

41. E. Margoliash, N. Frohwirt, and E. Wiener, *BJ* **71**, 559 (1959).
42. E. Margoliash, *Brookhaven Symp. Biol.* **15**, 266 (1962).
43. T. P. Hettinger and H. A. Harbury, *Biochemistry* **4**, 2585 (1965).
44. J. Heller and E. L. Smith, *Proc. Natl. Acad. Sci. U. S.* **54**, 1621 (1965).
45. H. A. Harbury, J. R. Cronin, M. W. Fanger, T. P. Hettinger, A. J. Murphy, Y. P. Myer, and S. N. Vinogradov, *Proc. Natl. Acad. Sci. U. S.* **54**, 1658 (1965).
46. K. Ando, H. Matsubara, and K. Okunuki, *Proc. Japan Acad.* **41**, 79 (1965).
47. H. Matsubara, K. Ando, and K. Okunuki, *Proc. Japan Acad.* **41**, 408 (1965).
48. R. E. Dickerson, M. L. Kopka, J. E. Weinzierl, J. C. Varnum, D. Eisenberg, and E. Margoliash, *JBC* **242**, 3015 (1967); unpublished observations (1970).

Okunuki *et al.* (*49*) reported that bovine cytochrome c which was treated to produce a mono-trinitrophenyl derivative contains this substituent on the ε-amino group of Lys 72 (or 73). The activity of this cytochrome derivative with cytochrome oxidase was greatly diminished. It appears, therefore, that not only is residue 72 (or 73) the most reactive of the lysine residues but also it is probably involved in the reaction with the oxidase.

DeLange *et al.* (*50*) identified the novel residue, ε-*N*-trimethyllysine (lysine betaine) in the cytochromes c of wheat germ and of *Neurospora crassa,* the former containing two residues of lysine betaine at positions 72 and 86 and the latter, one residue at position 72. *Saccharomyces* and *Candida* cytochromes c also possess a single residue of this amino acid located at position 72 (*51*). This unusual amino acid residue has now been found also in additional plant cytochromes (*31*); mung bean, castor bean, sunflower and sesame cytochromes c also contain two residues of this amino acid. Thus, an unusual phylogenetic distribution of lysine betaine occurs in the cytochromes. The Ascomycetes possess a single residue of the amino acid at position 72, higher plants have two residues, presumably at positions 72 and 86 in all cases, although determined thus far only for the cytochromes of wheat and mung bean, and none of the animal cytochromes has been found to contain this amino acid.

Higher plant cytochromes were presumably derived from a common ancestor as indicated by the finding that all have eight extra residues at the amino terminus with the initial sequence acetyl-Ala-Ser-Phe, as in wheat germ cytochrome (Table II). Wheat and mung bean cytochromes have 19 residues in common which do not occur at these loci in other cytochromes (*31*).

The trimethylation of residues 72 and 86 occurs after the polypeptide chain is formed, as is the case for all modified amino acid residues in proteins. Indeed, Scott and Mitchell (*52, 53*) have identified two forms of *Neurospora* cytochrome c, the one in early cultures possesses a normal lysyl residue at position 72, and the modified cytochrome c, isolated from later cultures, contains the novel residue at position 72. By pulse labeling experiments, they were able to show that the early cytochrome is transformed to the modified form in the intact protein (*53*).

49. K. Okunuki, K. Wada, H. Matsubara, and S. Takemori, *in* "Oxidases and Related Redox Systems" (T. E. King, H. S. Mason, and M. Morrison, eds.), Vol. 2, p. 549. Wiley, New York, 1965.

50. R. J. DeLange, A. N. Glazer, and E. L. Smith, *JBC* **244,** 1385 (1969).

51. R. J. DeLange, A. N. Glazer, and E. L. Smith, *JBC* **245,** 3325 (1970).

52. W. A. Scott and H. K. Mitchell, *Neurospora Newsletter* **8,** 9 (1965).

53. W. A. Scott and H. K. Mitchell, *Biochemistry* **8,** 4282 (1969).

Although basic residues in cytochrome c are important for the interaction with the oxidase (*54*) and residue 72 presumably plays an important role in this interaction (*49*), it is not evident why trimethylation is essential. Although the trimethylammonium group is a stronger base (higher pK value) than the ϵ-ammonium group, both would be cationic at neutral pH, but the trimethyl compound would retain the charge at higher pH values and, of course, also offers a greater hydrophobic surface for interaction. In any case, in view of the high energetic cost of such specific trimethylations, it would appear that this modification of a normal lysyl residue has conferred some additional functional or survival value in fungal and plant cytochromes.

Thus, at least two residues in the constant 11-residue sequence, Lys 72 and Met 80, possess important functional properties. Among other constant residues in cytochrome c, the following may be mentioned: The two cysteine residues are involved in covalent linkage to the heme; His 18 is a ligand to the iron (*21, 40*); eight glycine and three prolyl residues are implicated in determining conformational properties (*7, 12*); one tyrosyl and the only constant tryptophanyl are hydrogen bonded to the propionic acid side chains of the porphyrin (*48*), etc. The roles of the two constant arginyl residues are unknown; other constant residues may be important solely for conformational interactions. It may be noted that on the basis of the distribution of constant glycyl and prolyl residues, and of β-branched and bulky residues it was suggested that cytochrome c would be found to contain little if any α helical or β structure (*7, 12*); this has been confirmed by the X-ray work on horse cytochrome c (*48*). The relatively constant pattern of distribution of hydrophobic residues in clusters has been conserved in all cytochromes (*8*); most of these residues are concerned only with conformation since they are mainly in the interior of the molecule (*48*), as is the situation in all globular proteins.

Figure 3 shows the amino acid sequence of wheat germ cytochrome c (*25*) and at each site all other residues that have been identified in the cytochromes that have been reported to date; this is the longest polypeptide chain yet identified among the homologous proteins of this type. An inspection of the residues at each locus indicates that some substitutions are conservative and some radical; some examples of the former are listed in Table V and of the latter in Table VI. Utilization of this information in attempts to ascertain the role of some of the side chains at some loci, in relation to functional characteristics, has been presented elsewhere (*12, 55*).

54. These observations by many investigators have been reviewed by Margoliash and Schejter (*21*).

55. E. L. Smith, *in* "Homologous Enzymes and Biochemical Evolution" (N. van Thoai and J. Roche, eds.), p. 43. Gordon & Breach, New York, 1968.

Ala
Lys Ile Ala Val
Thr Pro Phe Glu Gln Ser Val Glu Lys Thr Thr
Pro Ala Gly Val Ser Ala Asp Ser Lys Asn Lys Asn Leu
Acetyl-Ala-Ser-Phe-Ser -Glu-Ala -Pro-Pro-Gly-Asn-Pro-Asp-Ala-Gly-Ala-Lys-Ile-Phe-
−8 1 10

Thr Ile Gly Thr Thr
Ile Met Ser Leu Glu Gly Asn Leu Pro Ile
Val Gln Arg Glu Glu Gly Cys Glu Lys Asn Gly Lys Gln Val
Lys-Thr -Lys-Cys-Ala-Gln-Cys-His-Thr-Val -Asp-Ala-Gly-Ala -Gly-His-Lys-Gln-Gly-Pro-
20 30
—HEME—

Glu
Asn Val
Tyr Gln Glu
Trp Ile Ile His Ser Val Asp Thr Asp
Ala Ser Phe Tyr Ser Lys Thr Gln Ala Pro Phe Ala Thr Asn
Asn-Leu-His-Gly-Leu-Phe-Gly-Arg-Gln-Ser-Gly-Thr-Thr-Ala-Gly-Tyr-Ser-Tyr-Ser -Ala-
40 50

Asn
Lys Lys
Arg Val Gln Arg
Gln Ile Ala Gln Glu Met
Ser Asn Leu Asp Asn Pro Ser Ile Thr
Ile Ala Ala Gly Ile Thr Gly Asp Asp Asn Met Phe Glu Glu
Ala-Asn-Lys-Asn-Lys-Ala-Val-Glu-Trp-Glu-Glu-Asn-Thr-Leu -Tyr-Asp-Tyr-Leu-Leu-Asn-
60 70

Ser
Ala
Asp
Thr Gly
Thr Lys Thr
Val Glu Glu
Ile Gly Asp Lys
Ala Ala Ile Ser Ala Asn Glu
Pro-Lys-Lys-Tyr-Ile-Pro-Gly-Thr-Lys-Met-Val-Phe-Pro-Gly-Leu-Lys-Lys-Pro-Gln-Asp-
80 90

Gln
Lys
Thr
Val D
Glu Asp Lys D
Gly Glu Ser D Asn Glu
Asp Gln Thr Ala Cys Ala
Asn Asn Ile Val Thr Phe Met Leu Ser Lys Cys Ala Lys
Arg-Ala -Asp-Leu-Ile -Ala -Tyr-Leu -Lys-Lys-Ala -Thr-Ser -SerCOOH
100

TABLE V
EXAMPLES OF CONSERVATIVE SUBSTITUTION IN CYTOCHROMES[a]

Type	Residue No.	Amino acid residues found
Cationic	13	Arg, Lys
Anionic	90	Asp, Glu
β-Hydroxy	40, 49	Ser, Thr
Aromatic[b]	46, 97	Phe, Tyr
Aliphatic	57, 95	Val, Ile
Aliphatic	64, 98	Leu, Met
Aliphatic	85, 94	Ile, Leu
Nonpolar	9	Ile, Val, Leu, Thr
Nonpolar	35	Leu, Ile, Phe
Nonpolar	36	Tyr, Ile, Phe
Polar (+ or 0)[c,d]	54	Asn, Gln, Lys, Arg, Ser, Ala
Polar (+, 0, −)	60	Glu, Asp, Asn, Gln, Gly, Ala, Lys
Polar (−, 0)	61	Glu, Asp, Gln, Asn
Polar (+, 0, −)	89	Lys, Gly, Ser, Thr, Ala, Gln, Asn, Asp, Glu

[a] From Smith (*12*).
[b] It is assumed that the sites with hydrophobic residues classified as aromatic, aliphatic, and nonpolar are probably internal.
[c] It is assumed that the sites with various polar residues are probably external.
[d] Parenthetical notations refer to charge on the side chains.

TABLE VI
EXAMPLES OF RADICAL SUBSTITUTION IN CYTOCHROMES[a]

Residue No.	Amino acid residues found[b]
3	Ile, Val, Pro, Ala, Ser
33	Gln, Asn, Ser, His, Tyr, Trp
39	Gln, Lys, His
44	Val, Glu, Gln, Ala, Asp, Pro
65	Arg, Ser, Tyr, Phe, Met
83	Val, Thr, Pro, Ala, Gly
92	Lys, Asp, Glu, Gln, Asn, Thr, Ala, Gly, Val

[a] Adapted from Smith (*12*).
[b] It is assumed that residues of such mixed types could not be accommodated in the interior of the protein without drastic change in conformation. Thus these side chains are presumably on the outside of the molecule as is the situation at sites where only polar residues are found.

FIG. 3. All amino acid substitutions known in eukaryotic cytochromes c investigated to date. D marks positions for which gaps may occur. References to all species included are given by Nolan and Margoliash (*18*).

III. Expressions of Genetic Phenomena in Protein Structure

A. Point Mutations

Undoubtedly, the most common as well as the simplest type of genetic modification is the point mutation, i.e., the replacement of a single nucleotide in a triplet codon, resulting in the substitution of a single amino acid in a peptide chain. This is borne out by the study of the "spontaneously" occurring abnormal human hemoglobins in which most of the changes that have been detected are of this kind (*10*). This is true also for the artificially induced mutations resulting in amino acid substitutions in the coat protein of tobacco mosaic virus (*56*) and in the A protein of the tryptophan synthetase of *E. coli* (*9, 57, 58*).

Comparison of the homologous proteins of different species also shows that most of the changes can be explained as resulting from accumulation of point mutations (*59*). This includes the eukaryotic cytochromes c, hemoglobins, fibrinopeptides, insulins, and subtilisins. A good example among enzymes is represented by the two subtilisins for which the complete sequences are known (*60, 61*). These two varieties of the same enzyme differ only slightly in specificity and in kinetic properties (*62, 63*) but differ very strikingly in sequence (Fig. 4).

Subtilisin BPN′ contains 275 residues in its single peptide chain whereas subtilisin Carlsberg has 274 residues (*61*). Aside from the deletion (or addition) of a residue, assumed to be at residue 56 in order to give greatest homology in sequence, the two enzymes differ in 84 residues or 30.6% of the sequence. From the presently assigned nucleotide codons, 61 of the changes can be ascribed to single base changes and 23 to double

56. H. G. Wittman and B. Wittmann-Liebold, *Cold Spring Harbor Symp. Quant. Biol.* **31**, 163 (1966).

57. C. Yanofsky, J. Ito, and V. Horn, *Cold Spring Harbor Symp. Quant. Biol.* **31**, 151 (1966).

58. C. Yanofsky, G. R. Drapeau, J. R. Guest, and B. C. Carlton, *Proc. Natl. Acad. Sci. U. S.* **57**, 276 (1967).

59. Much of the earlier information has been reviewed by Jukes (*15*).

60. F. S. Markland and E. L. Smith, *JBC* **242**, 5198 (1967).

61. E. L. Smith, R. L. DeLange, W. H. Evans, M. Landon, and F. S. Markland, *JBC* **243**, 2184 (1968).

62. E. L. Smith, F. S. Markland, and A. N. Glazer, *in* "Structure-Function Relationships of Proteolytic Enzymes" (P. Desnuelle, H. Neurath, and M. Ottesen, ed.), p. 160. Munksgaard, Copenhagen, 1970.

63. F. S. Markland and E. L. Smith, "The Enzymes," 3rd ed., Vol. 3, 1970 (in press).

FIG. 4. The amino acid sequence of subtilisin BPN′. The residues that differ in subtilisin Carlsberg are given above the corresponding residue. The dash at residue 56 indicates a deletion in the Carlsberg enzyme. The reactive serine is residue 221; the active histidine is residue 64. From Smith *et al.* (*61*).

base changes on the basis of the assumption that the minimal number of base changes has occurred and ignoring codons that would require assuming a greater number of mutational events. If an equal probability of point mutation in the 84 altered codons is assumed, we should expect 30.6% alterations in these or 25.7 double base changes, equivalent to 25.7 residues. (Mutational events that do not alter the sequence are not counted.) The calculated 25.7 residues are in reasonable accord with the 23 amino acid replacements that have been assigned to double base changes. Although the agreement does not prove that the two homologous genes for these subtilisins evolved from the common ancestral gene solely by the accumulation of point mutations, it does indicate that no other assumptions are required to explain the data.

Although the amino acid differences in the two subtilisins are found throughout the sequence, it is noteworthy that near the reactive seryl, histidyl, and aspartyl residues the sequences are similar (*62*). Further, the model of subtilisin BPN′ constructed from X-ray diffraction data suggests that most of the substitutions are found on the outside of the molecule (*64*). Thus, the similarity of enzymic and many other properties is not influenced in a major way by most of the known alterations in sequence.

A similar comparison can be made for two pancreatic ribonucleases of known sequence. Figure 5 shows the sequences for the bovine (*65*) and rat (*66*, *66a*) enzymes. The rat enzyme has three additional residues at the NH_2 terminus. It is striking that the sequences around three of the

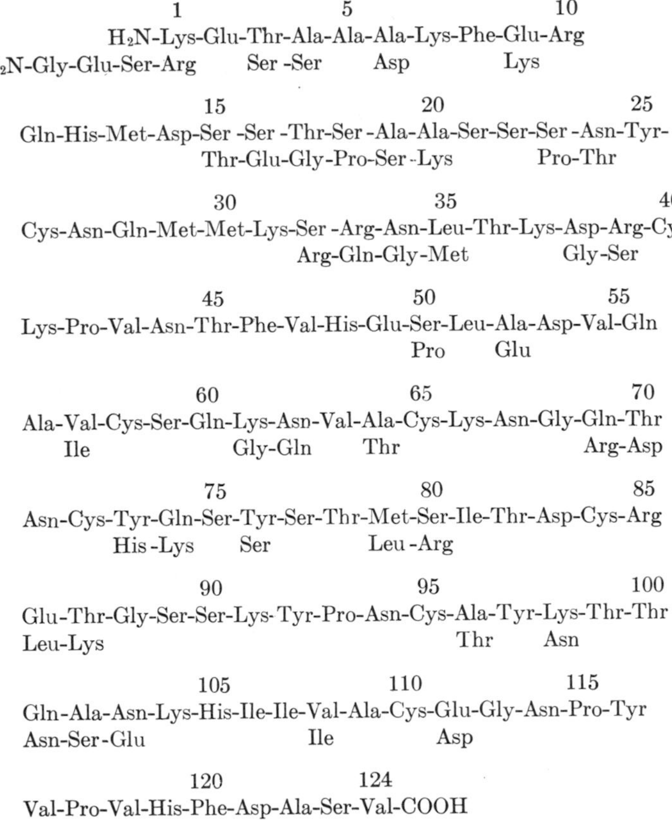

FIG. 5. Amino acid sequences of pancreatic ribonucleases: upper line, bovine (*65*); lower line, rat (*66*).

64. C. S. Wright, R. A. Alden, and J. Kraut, *Nature* **221,** 235 (1969).
65. D. G. Smyth, W. H. Stein, and S. Moore, *JBC* **238,** 227 (1963).
66. J. J. Beintema and M. Gruber, *Abstr., Proc. 7th Intern. Congr. Biochem., Tokyo, 1967* Vol. IV A, p. 117. Sci. Council Japan, Tokyo, 1968.

66a. J. J. Beintema and M. Gruber, *BBA* **147,** 612 (1967).

residues implicated in the reaction mechanism (*67–71*), His 12, His 119, and Lys 41, are identical. The homology of the two enzymes is evident since 83 of the 124 residues are in the same positions; these identities also include the eight half-cystyl residues, implying that the four disulfide bridges are also the same although these have not yet been determined for the rat enzyme. The two ribonucleases differ in 41 of the 124 residues (33.6%) present in both enzymes. Of these 31 can represent single base changes and 10 are double base changes, calculated on the same basis as for the two subtilisins discussed above. The expected number of double base changes is 33.6% of the 41 total changes or 13.8 residues, in fair agreement with the 10 actually found. This implies, possibly, that a higher percentage of the expected double base changes might be deleterious, in contrast to the situation in the subtilisins, or alternatively, that changes in the cistron other than point mutations might have been involved. It has been noted that none of the residues which differ in the two enzymes is in contact with substrate or inhibitors as judged from the model of the bovine enzyme constructed from X-ray analysis (*71*).

As expected, ovine ribonuclease (*72*) is very similar to the bovine enzyme, the former containing threonine instead of serine at residue 3, lysine in place of glutamic acid at residue 37, and a third, undetermined difference in the region between residues 99 and 104. A partial sequence of equine ribonuclease has also been reported (*66a*).

The complete sequences of the glyceraldehyde-3-phosphate dehydrogenases from lobster (*73*) and porcine muscle (*74*) show differences in sequence of the same order as already presented for the subtilisins and pancreatic ribonucleases. The dehydrogenases from the two species (Fig. 6) are alike in 241 residues (72%), excluding one deletion and two extra residues at the carboxyl terminus in the lobster enzyme. Of the 90 residue differences (28%), 54 can be ascribed to single-base changes, 35 to double-base changes, and one (Met-Cys at residue 130) to a triple-base change (*74*). As in the enzymes discussed above, the sequences around the reactive residues Cys 149 and Lys 183 are the same in both dehydrogenases. Further, the nine tyrosyl, three tryptophanyl, and eleven of the

67. E. A. Barnard and W. D. Stein, *JMB* **1,** 339, 350 (1959).
68. H. G. Gundlach, W. H. Stein, and S. Moore, *JBC* **234,** 1754 (1959).
69. A. M. Crestfield, W. H. Stein, and S. Moore, *JBC* **238,** 2958 and 2965 (1963).
70. C. H. W. Hirs, *Brookhaven Symp. Biol.* **15,** 154 (1962).
71. H. W. Wyckoff, *Brookhaven Symp. Biol.* **21,** 253 (1969).
72. C. B. Anfinsen, S. E. G. Aqvist, J. P. Cooke, and B. Jönsson, *JBC* **234,** 1118 (1959).
73. B. E. Davidson, M. Sajgò, H. F. Noller, and J. I. Harris, *Nature* **216,** 1181 (1967).
74. J. I. Harris and R. N. Perham, *Nature* **219,** 1025 (1968).

```
                                  10                                  20
   Val-Lys-Val-Gly-Val-Asp-Gly-Phe-Gly-Arg-Ile-Gly-Arg-Leu-Val-Thr-Arg-Ala-Ala-Phe-Asn-Ser-Gly-
Ac-Ser     Ile    Ile                                         Leu               Leu-Ser -Cys

                   30                                  40
Lys-Val-Asp-Ile-Val-Ala-Ile-Asn-Asp-Pro-Phe-Ile-Asp-Leu-His-Tyr-Met-Val-Tyr-Met-Phe-Glu-Tyr-Asp-
  —Ala-Gln-Val        Val                  Ala      Glu                        Lys

       50                                  60                                  70
Ser-Thr-His-Gly-Lys-Phe-His-Gly-Thr-Val-Lys-Ala-Glu-Asp-Gly-Lys-Leu-Val-Ile-Asp-Gly-Lys-Ala-Ile
              Val      Lys     Glu         Met            Ala        Val            Lys

                             80                                  90
Thr-Ile-Phe-Gln-Glu-Arg -Asp-Pro-Ala-Asn-Ile-Lys-Trp-Gly-Asp-Ala-Gly-Thr-Ala-Tyr-Val-Val-Glu-Ser-
    Val     Asn     Met-Lys     Glu        Pro      Ser -Lys       Ala -Glu   Ile

                100                                  110
Thr-Gly-Val-Phe-Thr-Thr-Met-Glu-Lys-Ala-Gly-Ala-His-Leu-Lys-Gly-Gly-Ala-Lys-Arg-Val-Ile-Ile-Ser-
                        Ile                Ser        Phe                    Lys      Val

120                                  130                                  140
Ala-Pro-Ser-Ala-Asp-Ala-Pro-Met-Phe-Val-Met-Gly-Val-Asn-His-Glu-Lys-Tyr-Asp-Asn-Ser -Leu -Lys-Ile-
                                        Cys               Leu            Ser -Lys-Asp-Met-Thr-Val

                  *  150                                  160
Val-Ser-Asn-Ala-Ser-CYS-Thr-Thr-Asn-Cys-Leu-Ala-Pro-Leu-Ala-Lys-Val-Ile-His-Asp-His-Phe-Gly-Ile-
                                                   Val                Leu    Glu-Asn    Glu
```

```
     170                                  180         *                         190
Val-Glu-Gly-Leu-Met-Thr-Thr-Val-His-Ala-Ile-Thr-Ala-Thr-Gln-LYS-Thr-Val-Asp-Gly-Pro-Ser-Gly-Lys-
                                        Val                                             Ala

                               200                                   210
Leu-Trp-Arg-Asp-Gly-Arg-Gly-Ala-Ala-Gln-Asn-Ile-Ile-Pro-Ala-Ser-Thr-Gly-Ala-Ala-Lys-Ala-Val-Gly-
Asp         Gly                                         Ser

            220                                   230
Lys-Val-Ile-Pro-Glu-Leu-Asp-Gly-Lys-Leu-Thr-Gly-Met-Ala-Phe-Arg-Val-Pro-Thr-Pro-Asn-Val-Ser-Val-
                                                                            Asp

240                                    250                                 260
Val-Asp-Leu-Thr-Cys-Arg-Leu-Glu-Lys-Pro-Ala -Lys-Tyr-Asp-Asp-Ile-Lys-Lys-Val-Val-Lys-Gln-Ala-Ser-
                Val         Gly     Glu-Cys-Ser                     Ala-Ala-Met     Thr

                       270                             280
Glu-Gly-Pro-Leu-Lys-Gly-Ile-Leu-Gly-Tyr-Thr-Glu-Asp-Gln-Val-Val-Ser-Cys-Asp-Phe-Asn-Asp-Ser -Thr-His-Ser-
                Gln     Phe                         Asp             Ser             Ile Gly-Asp-Asn-Arg

290                                  300                                 310
Ser-Thr-Phe-Asp-Ala-Gly-Ala-Gly-Ile-Ala-Leu-Asn-Asp-His-Phe-Val-Lys-Leu-Ile-Ser-Trp-Tyr-
    Ile             Lys             Gln     Ser -Lys-Thr            Val -Val

                               320                                330
Asp-Asn-Glu-Phe-Gly-Tyr-Ser-Asn-Arg-Val-Val-Asp-Leu-Met-Val-His-Met-Ala -Ser -Lys-Glu-
                            Gln         Ile         Leu -Lys        Gln-Lys-Val -Asp-Ser-Ala
```

FIG. 6. Amino acid sequence of glyceraldehyde-3-phosphate dehydrogenase of pig muscle; residues below the continuous sequence indicate those which differ in the lobster muscle enzyme. Residue 24 (——) is a deletion in the lobster enzyme. From Harris and Perham (*74*).

twelve prolyl residues are in corresponding positions. Once again, these data indicate that the conformation and mechanism of these homologous enzymes is similar. Harris and Perham (*74*) mention unpublished and incomplete data on the rabbit, bovine, and yeast enzymes, the mammalian ones being very similar and all homologous with those of lobster and porcine muscle.

The amino acid sequences of pencillinases from *Staphylococcus aureus* (*74a*) and *Bacillus licheniformis* (*74b*) show identity in 43% of the sequence, 110 of the 255 residues present in both, allowing for deletions at residues 211 and 225 (Fig. 7). It is noteworthy that the enzyme from *B. licheniformis* contains five additional residues at each end of the molecule as compared to that from *S. aureus*.

B. Chain Shortening

It has already been noted that chain shortening at the amino terminus must have occurred during the evolution of cytochrome c (Section II). Loss of a residue by chain termination at the carboxyl end may have occurred several times (*12*). Chain termination by means of a point mutation to one of the three known termination codons (*75*) can occur, presumably, anywhere in the sequence of the gene. If such a mutation occurs so as to result in a peptide chain, shorter at either end, without loss of function of the resulting enzyme (or other protein), it may be assumed that the mutant type would have increased survival value insofar as the mutant strain would utilize less energy than the original wild type, in making nucleotides, nucleotide bonds, amino acids, and peptide bonds (see Section IV,D).

Inasmuch as there is good evidence for the elongation of peptide chains during evolution, as discussed below (Section III,C), one must suppose that during such elongation unnecessary or dispensable segments were also added to the resulting polypeptides. These would then be "trimmed off" eventually by chain-terminating mutations until the protein was reduced to essentially the minimal size that retained full effectiveness. In a general way, this notion is supported by the numerous observations that relatively few residues can be removed from the ends of the peptide chains of enzymes without impairment of catalytic activity. Thus, it would also be expected that homologous proteins from different species

74a. R. P. Ambler and R. J. Meadway, *Nature* **222,** 24 (1969).
74b. R. J. Meadway, *BJ* **115,** 12P (1969).
75. S. Brenner, L. Barnett, E. R. Katz, and F. H. C. Crick, *Nature* **213,** 449 (1967).

would only rarely differ greatly in size. Note the similarity of the two ribonucleases (Fig. 5) and the two glyceraldehyde-3-phosphate dehydrogenases (Fig. 6). However, the two penicillinases differ somewhat (Fig. 7).

C. Deletions, Additions, Chain Extensions

In comparing the sequences of structurally related proteins, it has been obvious that to obtain maximal homology of the sequences, it is frequently necessary to assume gaps of one or more residues in certain chains. Abnormal human hemoglobins Freiburg (*76*) and Gun Hill (*77*) show β-chain deletions of one and five residues, respectively, as compared to the normal β chain. The hemoglobin chains of different types in the same species, as well as in different species, also indicate that gaps occur (*6, 78*). Other homologous proteins similarly manifest such gaps varying from a single residue in the two subtilisins (Fig. 4) and in glyceraldehyde-3-phosphate dehydrogenases (Fig. 6), to two single residue gaps in the penicillinases (Fig. 7), to longer gaps among the immunoglobulins (*79*) and the pancreatic proteinases (*80–82*), and others.

The deletion of a single nucleotide would produce a frame shift, resulting in a completely altered polypeptide sequence carboxyl terminal to the deletion (*83, 84*). A compensatory addition of a nucleotide would be required to restore the homology. Such a double change would seem an unlikely or rare event, but it has been observed in artificially induced situations (*83, 84*). It is more likely that deletions in the genome would have occurred in groups of three nucleotides, whatever the responsible

76. R. T. Jones, B. Brimball, T. H. J. Huisman, E. Kleihauer, and K. Betke, *Science* **154,** 1024 (1966).

77. T. B. Bradley, R. C. Wohl, and R. F. Rieder, *Science* **157,** 1581 (1967).

78. G. Braunitzer, V. Braun, K. Hilse, G. Hobom, V. Rudloff, and G. von Wettstein, *in* "Evolving Genes and Proteins" (V. Bryson and H. J. Vogel, eds.), p. 183. Academic Press, New York, 1965.

79. Reviewed by Nolan and Margoliash (*18*).

80. K. A. Walsh and H. Neurath, *Proc. Natl. Acad. Sci. U. S.* **52,** 884 (1964).

81. B. S. Hartley, *Nature* **201,** 1284 (1964).

82. H. Neurath, K. A. Walsh, and W. P. Winter, *Science* **158,** 1638 (1967).

83. G. Streisinger, Y. Okada, J. Emrich, J. Newton, and A. Tsugita, *Cold Spring Harbor Symp. Quant. Biol.* **31,** 77 (1966).

84. W. J. Brammer, H. Berger, and C. Yanofsky, *Proc. Natl. Acad. Sci. U. S.* **58.** 1499 (1967).

1 10 20 25
H_2N-Lys-Glu-Leu-Asn-Asp-Leu-Glu-Lys-Lys-Tyr-Asn-Ala-His-Ile-Gly-Val-Tyr-Ala-Leu-Asp-
H_2N-Lys-Thr-Glu-Met-Lys -Asp-Asp-Phe-Ala -Lys Glu-Gln-Phe-Asp Lys-Leu Ile -Phe

26 30 40 50
Thr-Lys-Ser-Gly-Lys-Glu-Val-Lys-Phe-Asn-Ser-Asp-Lys-Arg-Phe-Ala-Tyr-Ala-Ser-Thr-Ser-Lys-Ala-Ile -Asn
Gly-Thr-Asn-Arg-Thr Ala -Tyr-Arg-Pro Glu Phe Ile Leu-Thr-

51 60 70 75
Ser -Ala -Ile-Leu-Leu-Glu-Gln-Val -Pro-Tyr-Asn-Lys-Leu-Asn-Lys-Lys-Val-His -Ile -Asn-Lys-Asp-Asp-Ile-Val-
Val-Gly-Val Gln Lys-Ser -Ile -Glu-Asp Gln-Arg-Ile -Thr-Tyr-Thr-Arg Leu

76 80 90 100
Ala-Tyr-Ser-Pro-Ile-Leu-Glu-Lys-Tyr-Val-Gly-Lys-Asp-Ile-Thr-Leu-Lys-Ala-Leu-Ile -Glu-Ala-Ser-Met-Thr-
Asn Asn Thr His Asp-Thr-Gly-Met Glu Ala-Asp Leu -Arg-

101 110 120 125
Tyr-Ser-Asp-Asn-Thr-Ala-Asn-Asn-Lys-Ile-Ile-Lys-Glu-Ile-Gly-Gly-Ile -Lys-Val-Lys-Lys-Gln-Arg-Leu-Lys
Ala Gln Leu Leu Gln Pro-Glu-Ser -Leu Lys-Glu Arg-

```
126                 130                                      140                                  150
Glu-Leu-Gly-Asp-Lys-Val-Thr-Asn-Pro-Val-Arg-Tyr-Glu-Ile-Glu-Leu-Asn-Tyr-Tyr-Ser-Pro-Lys-Ser -Lys-Lys-
Lys-Ile             Glu                     Glu      Phe       Pro              Glu-Val -Asn    Gly-Glu-Thr-Glu

151                                    160                                      170              175
Asp-Thr-Ser-Thr-Pro-Ala-Ala-Phe-Gly-Lys-Thr-Leu-Asn-Lys-Leu-Ile -Ala -Asn-Gly-Lys-Leu-Ser -Lys-Glu-Asn
                    Ala-Arg    Leu-Val -Thr-Ser       Arg-Ala -Phe-Ala-Leu-Glu-Asp          Pro-Ser      Lys-

176                 180                                         190                                200
Lys-Lys-Phe-Leu-Leu-Asp-Leu-Met-Leu-Asn-Asn-Lys-Ser-Gly-Asp-Thr-Leu-Ile-Lys-Asp-Gly-Val-Pro-Lys-Asp-
Arg-Glu-Leu         Ile      Trp      Lys-Arg      Thr-Thr          Ala         Arg-Ala             Asp-Gly-

201                                    210                                     220           225
Tyr-Lys-Val-Ala-Asp-Lys-Ser-Gly-Gln-Ala-Ile-Thr-Tyr-Ala -Ser-Arg-Asn-Val-Ala-Phe-Val-Tyr-Pro-Lys-
Trp-Glu                      Thr     Ala       — Ser      Gly-Thr          Ile      Ile -Ile -Trp      —

226                 230                                      240                                   250
Gly-Gln-Ser -Glu-Pro-Ile-Val-Leu-Val-Ile -Phe-Thr-Asn-Lys-Asp-Asn-Lys-Ser -Asp-Lys-Pro -Asn-Asp-Lys-Leu-
Pro -Lys-Gly-Asp      Val          Ala-Val-Leu-Ser -Ser -Arg      Lys      Asp-Ala      Tyr-Asp

251                                   260                              267
Ile-Ser-Glu-Thr-Ala-Lys-Ser-Val-Met-Lys-Glu-Phe-COOH
   Ala       Ala -Thr      Val               Ala -Leu-Asn-Met-Asn-Gly-Lys-COOH
```

FIG. 7. Amino acid sequence of the penicillinase from *Staphylococcus aureus* (*74a*) shown in the continuous sequence. Residues below this line are those that differ in the enzyme of *Bacillus licheniformis* (*74b*); the bars indicate deletions at residues **211** and **225** in the enzyme of the latter.

mechanism, in order to retain the functional properties of the protein and the homology in the remainder of the protein. At least two mechanisms have been proposed, one involving simply the excision of a loop in the gene with a joining at the two broken ends. Such deletions are known for the excision of segments of chromosomes involving whole genes, as demonstrated in classic genetic and cytological studies.

Another mechanism involves unequal crossing-over between allelic genes, one allele becoming longer by the introduction of a repeating segment of nucleotides and hence of amino acid residues in the protein, and the other allele becoming shorter (*85*). It is difficult to prove which mechanism has been responsible for deletions; however, many proteins have now been found to contain repeating segments of residues, thus, unequal crossing-over can be assumed to have been responsible, at least in some cases, for both deletions and additions of residues in the internal portions of polypeptide chains.

The first instance of internal chain repetitions in sequence was reported for human haptoglobin, in which a 75-residue chain becomes a 140-residue polypeptide with losses of segments from the amino and carboxyl ends of the original chains (*86*). An instance has also been reported for a partial triplication of sequences in a type of haptoglobin with a chain almost three times as long as the original parent type of chain (*85*).

In several human hemoglobins of the Lepore type, the unusual sequences have been ascribed to unequal crossing-over between the closely linked, but nonallelic genes for the δ and β chains (*87–90*). In two such instances, the amino ends were derived from the δ chains and the carboxyl ends from the β chains. Alignment of the δ and β genes to permit such crossing-over is possible only because of the extensive similarity in the sequences of the two genes as indicated by the similarity of sequences in the two polypeptides (*6*).

Repetitions of sequences within other proteins have been repeatedly noted in both the light and heavy chains of the immunoglobulins (*91–93*), in bacterial ferredoxins (*94–96*), and, possibly, in the clupeines (*97*). In all of these proteins, there have also been other changes in sequences

85. G. H. Dixon, *Essays Biochem.* **2,** 147 (1966).
86. O. Smithies, G. E. Connell, and G. H. Dixon, *Nature* **196,** 232 (1962).
87. D. Labie, W. A. Schroeder, and T. H. J. Huisman, *BBA* **127,** 428 (1966).
88. C. Baglioni, *Proc. Natl. Acad. Sci. U. S.* **48,** 1880 (1962); *BBA* **97,** 37 (1966).
89. C. C. Curtain, *Australian J. Exptl. Biol. Med. Sci.* **42,** 89 (1964).
90. J. Barnabas and C. J. Muller, *Nature* **194,** 931 (1962).
91. S. J. Singer and R. F. Doolittle, *Science* **153,** 13 (1966).
92. R. L. Hill, R. Delaney, R. E. Fellows, and H. E. Lebovitz, *Proc. Natl. Acad. Sci. U. S.* **56,** 1762 (1966).

—substitution of single residues, deletions (or additions), etc. Thus, all of the various types of changes in the genome can be observed as part of the evolutionary history of presently existing proteins.

Few examples of chain elongations by repetitions of sequence have been noted as yet in enzymes. An interesting and unusual instance is apparent in the subtilisins (*12, 61*). Large segments in the amino terminal portion of the two subtilisins at residues **39–42**, **67–75**, and **82–94** are very similar to those at residues **238–241**, **226–233**, and **126–136**, respectively in the carboxyl portion of the proteins (Fig. 8). Thus, the corresponding similar parts of the sequence occur in inverse order in the two portions of the molecule. In addition, since these repeated sequences are separated by different numbers of residues, the process of evolution of the subtilisins from smaller precursors must have been complex, indeed. Further, one part of the sequence appears to be repeated twice and the same pentapeptide sequence occurs four times. We may assume, therefore, that the relatively long, single chain of the subtilisins (275 residues) somehow evolved from much shorter chains by a series of complex processes.

It has been suggested (*98*) that *Neurospora* cytochrome (*34*) shows a repetition of sequence although other eukaryote cytochromes do not. The diheme *Chromatium* cytochrome cc′ also shows evidence of a repetition in sequence (*99*).

In the above-noted examples of deletions and additions of residues within polypeptide chains and hence in the DNA of the cistrons, it is evident that mechanisms exist for rejoining the broken linkages of the polynucleotide chains in the DNA. Presumably, the responsible enzymes are ligases, identical or similar to those known to correct ordinary breaks in DNA, or those known to effect the joining of ends to make the circular, continuous DNA of microorganisms and viruses (*100–103a*).

---

93. F. W. Putnam, *Brookhaven Symp. Biol.* **21,** 306 (1969).
94. M. Tanaka, T. Nakashima, A. M. Benson, H. Mower, and K. T. Yasunobu, *Biochemistry* **5,** 1666 (1966).
95. A. M. Benson, H. F. Mower, and K. T. Yasunobu, *ABB* **121,** 563 (1967).
96. H. Matsubara, T. H. Jukes, and C. R. Cantor, *Brookhaven Symp. Biol.* **21,** 201 (1969).
97. J. A. Black and G. H. Dixon, *Nature* **216,** 152 (1967).
98. C. R. Cantor and T. H. Jukes, *Proc. Natl. Acad. Sci. U. S.* **56,** 177 (1966).
99. K. Dus, R. G. Bartsch, and M. D. Kamen, *JBC* **237,** 3083 (1962).
100. M. Gellert, *Proc. Natl. Acad. Sci. U. S.* **57,** 148 (1967).
101. B. Weiss and C. C. Richardson, *Proc. Natl. Acad. Sci. U. S.* **57,** 1021 (1967).
102. B. M. Olivera and I. R. Lehman, *Proc. Natl. Acad. Sci. U. S.* **57,** 1426 (1967).
103. M. L. Gefter, A. Backer, and J. Hurwitz, *Proc. Natl. Acad. Sci. U. S.* **58,** 240 (1967).
103a. P. Howard-Flanders, *Ann. Rev. Biochem.* **37,** 175 (1968).

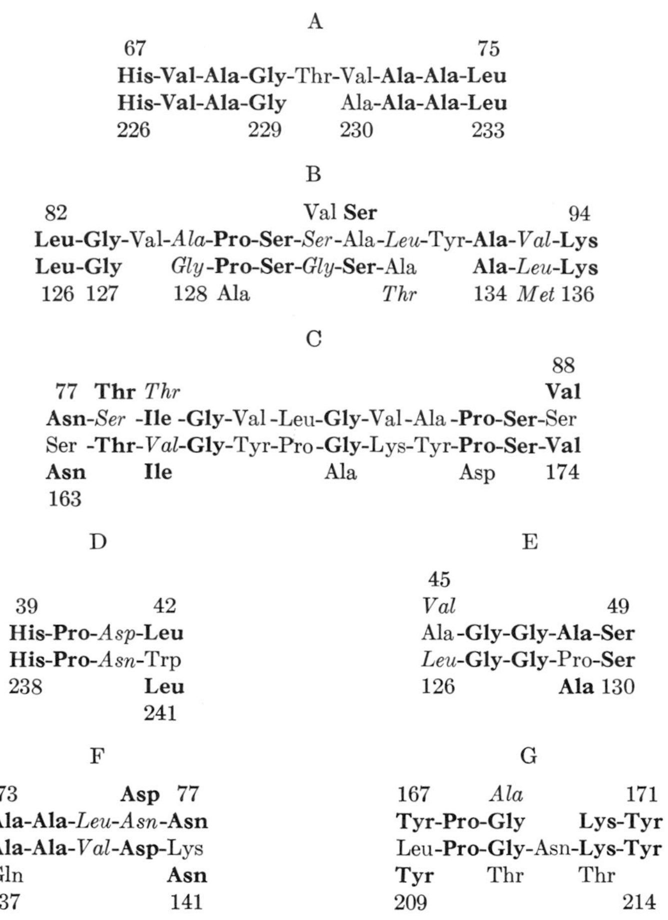

FIG. 8. Apparent repetitions in seven segments of the sequences of the two subtilisins (Fig. 4). Residues in boldface type are identical; those in italics appear to represent conservative replacements. Continuous sequences are from BPN′ or both types of enzymes; residues above or below these sequences represent differences found in the Carlsberg enzyme (*61*).

Some evidence suggests that both nonallelic as well as allelic genes can be joined to form longer peptide chains. In the immunoglobulins (see below) there may be a joining process between the constant portion and the variable portions of the same polypeptide chain. It has been postulated that there is one gene (and a number of allelomorphs) for the constant part and many genes for the variable. If this is so, joining could be accomplished at the level of the messenger RNA or the polypeptide chains (*104, 105, 105a*).

An interesting situation is represented by the enzyme tryptophan synthetase. In *E. coli*, there are two types of peptide chains in the complete enzyme: A with a molecular weight of about 29,000 (*106*) and B of about 120,000 (*107*); the two chains are readily separated and reassociated (*108*). Furthermore, the responsible cistrons are independent but adjacent. In contrast, the tryptophan synthetases of *Neurospora crassa* and yeast consist of a single long polypeptide chain of molecular weight 110,000 (*109*), which possesses both of the kinds of activities in the separable A and B chains of *E. coli*. In *Neurospora*, the two adjacent distinct genes presumably were joined to produce a single long cistron with the loss of a portion of the DNA. Inasmuch as the two chains of the *E. coli* enzyme possess distinctive properties and catalytic activities, but yield maximal activity only when they interact, it would appear that the *Neurospora* enzyme is a later evolutionary development with full activity in a single polypeptide component. Table VII summarizes the information concerning the properties of the various tryptophan synthetases and a hypothetical evolutionary scheme from completely independent components to a single component.

The above situation suggests that many enzymes with long and complex peptide chains may have been produced by the joining of adjacent genes to form a single longer gene yielding one polypeptide chain with a more integrated activity and more easily subject to regulation both as to enzymic activity and to biosynthesis. Biosynthesis of two polypeptide chains in equivalent amounts would appear to involve a more complex, regulatory process than the production of a single polypeptide chain.

It may be noted in passing that not all peptide chain modifications may have a direct genetic basis. Vertebrate insulins consist of two polypeptide chains derived from a single chain precursor, proinsulin, by specific proteolysis and loss of an internal polypeptide segment. Inasmuch as proinsulin prossesses insulin activity (*109a*), this may well be an evolutionary precursor of insulin, the proteolysis being a secondary

---

104. Theories concerned with this process have been reviewed by Putnam (*93*) and by Edelman and Gally (*105*).

105. G. M. Edelman and J. A. Gally, *Brookhaven Symp. Biol.* **21,** 328 (1969).

105a. G. M. Edelman and W. E. Gall, *Ann. Rev. Biochem.* **38,** 415 (1969).

106. U. Henning, D. R. Helinski, F. C. Chao, and C. Yanofsky, *JBC* **237,** 1523 (1962).

107. D. A. Wilson and I. P. Crawford, *Bacteriol. Proc.* p. 92 (1964).

108. I. P. Crawford and C. Yanofsky, *Proc. Natl. Acad. Sci. U. S.* **44,** 1161 (1958).

109. D. M. Bonner, J. A. DeMoss, and S. E. Mills, *in* "Evolving Genes and Proteins" (V. Bryson and H. J. Vogel, eds.), p. 305. Academic Press, New York, 1965.

109a. O. K. Behrens and E. L. Grinnan, *Ann. Rev. Biochem.* **38,** 83 (1969).

TABLE VII

Types of Tryptophan Synthetases in Microorganisms[a]

Components	Reaction	Activity	Organism
COMPLETELY INDEPENDENT COMPONENTS			No known organism
(A)	Ind GP → Ind + GAP		
(B)	Ind + Ser → Trp + H_2O		
(A) + (B)	Ind GP + Ser → Ind → Trp + GAP		
	↓		
PARTIALLY DEPENDENT COMPONENTS			
(B)	Ind + Ser → Trp + H_2O	Independent activity	*Bacillus subtilis*
(A) + (B)	Ind GP → Ind + GAP	Interaction produces maximal activity	*Anabaena*
	Ind GP + Ser → Trp + GAP		*Chlorella*
	↓		
COMPLETELY DEPENDENT COMPONENTS			
(A) + (B)	Ind GP + Ser → Trp + GAP	Interaction produces maximal activity	*Escherichia coli*
	Ind GP → Ind + GAP		*Salmonella typhimurium*
	Ind + Ser → Trp		
	↓		
SINGLE COMPONENT			
(AB)	Ind + Ser → Trp + H_2O		*Neurospora crassa*
	Ind GP + Ser → Trp + GAP		*Saccharomyces cerevisiae*
	Ind GP → Ind + GAP		

[a] A hypothetical evolutionary scheme for tryptophan synthetases. Ind GP, indolylglycerol 3-phosphate; Ind, indole; GAP, glyceraldehyde 3-phosphate. Adapted from Bonner *et al.* (*109*).

phenomenon. The situation here is different from that in the pancreatic proteolytic enzymes inasmuch as their zymogen precursors are enzymically inactive.

D. Gene Duplication

Sufficient data have been obtained to indicate that, in general, more complex organisms possess a greater total genome than do simpler organisms, although within a class there are large variations. A summary of much of this information is shown in Fig. 9 (*110*). The amount of DNA concerned with the information for synthesis of RNA (ribosomal and transfer) and for coding for specific proteins (structural genes) is not generally known; however, it is apparent that during evolution,

110. R. J. Britten and E. H. Davidson, *Science* **165,** 349 (1969).

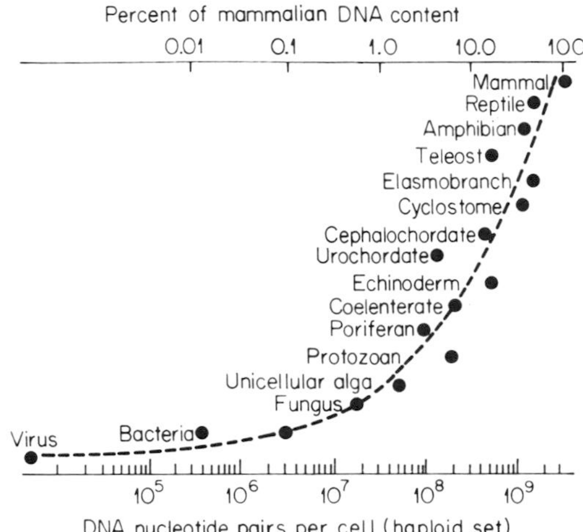

FIG. 9. The minimal amount of DNA that has been observed for various species in the types of organisms listed. Each point represents the measured DNA content per cell for a haploid set of chromosomes. The ordinate scale and the shape of the curve is arbitrary. From Britten and Davidson (*110*).

more and more proteins of different types have been evolved. Furthermore, it is obvious that higher organisms possess many enzymes for metabolic pathways concerned with the elaboration of specialized compounds which do not exist in lower organisms, as well as many more genes concerned with structure and regulation, particularly where there is differentiation into specialized tissues and organs.

If we exclude the possibility in later evolution of the completely novel assembly of nucleotides to form new genes there would seem to be only a few ways in which genes for new enzymes (or other proteins) could have arisen, considering the strict type of copying during replication of DNA. One method could be that a specific gene underwent alteration to such a degree that the old function of the resultant protein was changed, provided that this function was no longer needed. Such a set of changes might lead to a new function possessing greater survival value for the organism but would not yield any increase in the genome.

Considerable evidence is now on hand that probably the most important and perhaps the only method for increasing the genome and hence the number of enzymes (proteins) with novel functions has been through the process of gene duplication, followed by subsequent independent evolution of one or both pairs of genes. If one pair of genes maintains the synthesis of the original protein, the other pair can undergo changes leading to the synthesis of a protein with a modified specificity or a totally new function. If the new function enhances the survival value, the organism can evolve still further and occupy a different ecological niche.

Gene duplication has long been known in classic genetics as a result of the processes of unequal double crossing-over, translocation, and doubling of the chromosomes (polyploidy). The process could also occur by repeated copying of the same gene during replication (single gene heterosis) through slippage of the replicating enzyme. Hybridization of different species followed by chromosome doubling appears to have been a significant factor in plants but not in animals (*111*). Inasmuch as the presence of extra chromosomes and sizable translocations generally lowers viability, it has been assumed that single locus duplication is presumably the main route for this phenomenon (*112, 113*).

Studies of the amino acid sequences of proteins have provided strong evidence that gene duplication has been a frequent event in the development of new enzymes and other proteins. Among the examples first recognized through sequence studies were the hemoglobins. Myoglobin and the α and β chains of adult vertebrate hemoglobins manifest great homology in sequences as do also the γ chains of fetal hemoglobins and the modified chains of various minor hemoglobin components. Ingram (*114, 114a*) proposed that a single pair of genes, by successive duplications, eventually gave rise to all of these types of polypeptide chains (*115*). The synthesis of each of these chains is under separate gene control, and it has been demonstrated that the genes for the α and β chains are not linked in the same chromosome; however, evidence from the Lepore types of abnormal hemoglobins indicate that the β and δ chains are adjacent (Section III,C). The individual genes for the various chains could have arisen only by gene duplication followed by subsequent

111. E. Mayr, "Animal Species and Evolution," Chapter 6. Harvard Univ. Press (Belknap), Cambridge, Massachusetts, 1963.

112. J. B. Spofford, *Am. Naturalist* **103,** 407 (1969).

113. Spofford (*112*) assumes that gene duplication ("single locus heterosis") would be advantageous in the synthesis of multiple chain enzymes with identical subunits; one of the duplicated genes would thus be free to mutate. The hypothesis is attractive, but it should be noted that there is considerable evidence for gene duplications followed by divergence to explain the origin of many single chain enzymes and hormones. As discussed further, these include lysozymes, pancreatic proteinases and others; see also Section IV,E,2.

114. V. M. Ingram, *Nature* **189,** 704 (1961).

114a. V. M. Ingram, "The Hemoglobins in Genetics and Evolution." Columbia Univ. Press, New York, 1963.

115. These studies have been reviewed by Zuckerkandl and Pauling (*14* p. 97) and by Nolan and Margoliash (*18*).

sequence modifications occurring as a result of point mutations, deletions (or additions), etc.

The same phenomena are also shown by the many immunoglobulins, each of these proteins possessing a segment with essentially similar constant sequence and a segment with variable portions of the sequence (*93*). It has been assumed that the variable portions represent products of numerous gene duplications which subsequently become joined to the constant portion (*93, 105, 105a*).

In passing, we may note a few of the instances where related, homologous structures have also been recognized among the polypeptide and protein hormones. All of these exhibit different functional properties and must have arisen through gene duplications, e.g., the vasopressins, oxytocins and their relatives (*116*); melanophore-stimulating and corticotrophic hormones (*117*); gastrin (*117a*), cholecystokinin (*118*), and caerulin (*118a*); secretin (*119*) and glucagon (*119a*); human growth hormone (*120*) and placental somato-mammotropin (*120a*); one chain of thyrotropin and of luteinizing hormone (*121, 122*).

1. *Pancreatic Proteinases*

Gene duplications have undoubtedly been responsible for the production of the many distinct proteolytic enzymes of the pancreas. The striking homology of the partial sequences of trypsin and chymotrypsin was first noted by Šorm *et al.* (*123*). From the complete sequences of

116. R. Acher, *Angew. Chem.* **5,** 798 (1966).

117. C. H. Li, L. Barnefi, M. Chrétien, and D. Chung, *Nature* **208,** 1093 (1965).

117a. R. A. Gregory, P. M. Hardy, D. S. Jones, G. W. Kenner, and R. C. Sheppard, *Nature* **204,** 931 (1964).

118. V. Mutt and J. E. Jorpes, *European J. Biochem.* **6,** 156 (1968).

118a. A. Anastasi, V. Erspamer, and R. Endean, *ABB* **125,** 57 (1968).

119. M. Bodanszky, M. A. Ondetti, S. D. Levine, V. L. Narayanan, M. V. Saltza, J. T. Sheehan, N. J. Williams, and E. F. Sabo, *Chem. & Ind.* (*London*) p. 1757 (1966).

119a. W. W. Bromer, L. G. Sinn, and O. K. Behrens, *JACS* **79,** 2807 (1957).

120. C. H. Li, J. S. Dixon, and W-K. Liu, *ABB* **133,** 70 (1969).

120a. K. J. Catt, B. Moffat, and H. D. Niall, *Science* **157,** 321 (1967).

121. D. N. Ward, C. M. Sweeney, G. N. Holcomb, W. M. Lamkin, and M. Fujino, *Proc. 3rd Intern. Congr. Endocrinol., Mexico City, 1968* Intern. Congr. Ser. No. 157, p. 385 Excerpta Med. Found., Amsterdam, 1968.

122. T.-H. Liao, G. Hennen, S. M. Howard, B. Shome, and J. G. Pierce, *JBC* **244,** 6458 (1969); S. M. Howard and J. G. Pierce, *JBC* **244,** 6468 (1969); J. G. Pierce and T.-H. Liao, *Fed. Proc.* **29,** 600 (1970); T.-H. Liao and J. G. Pierce *JBC* **245,** 3275 (1970).

123. F. Šorm, B. Keil, V. Holeyšovsky, B. Meloun, O. Mikeš, and J. Vaněček, *Collection Czech, Chem. Commun.* **23,** 985 (1958).

several of these enzymes and the partial structures of others, it is now evident that there is great homology in the sequences of trypsin (*124, 125*); chymotrypsins A (*126–129*) and B (*130, 130a*); elastase (*131, 132, 132a*); in the enzyme of blood clotting, thrombin (*133, 134*); and in the enzyme responsible for clot lysis, plasmin (*135*). Presumably, additional enzymes of this type will also prove to be related to these. Further, the homologies extend to many species as well (*136*). In each of these enzymes, there is not only homology in primary sequence but in the location of the disulfide bridges, in the conversion by proteolysis from an inactive zymogen precursor to an active proteolytic enzyme; however, the specificity with respect to residues at the sensitive peptide bonds in substrates is different in each case.

For those zymogens made in the pancreas which are transformed to active enzymes participating in the hydrolysis of dietary protein, it is apparent that there are clear advantages in survival value in possessing a multiplicity of enzymes of different specificities. These, in conjunction with other hydrolases of peptide bonds, will produce much more rapid and complete breakdown of proteins to amino acids than would one or a few enzymes with restricted specificity. Thus, the organism obtains a more efficient utilization of its ingested proteins, a critical need for species that cannot make many of the amino acids essential for protein synthesis and are dependent on dietary amino acids as the major source of organic nitrogen.

124. K. A. Walsh and H. Neurath, *Proc. Natl. Acad. Sci. U. S.* **52,** 889 (1964).

125. O. Mikeš, V. Tomášek, V. Holeyšovsky, and F. Šorm, *BBA* **117,** 281 (1966).

126. B. S. Hartley, *Nature* **201,** 1284 (1964).

127. B. S. Hartley and D. L. Kauffman, *BJ* **101,** 229 (1966).

128. B. Meloun, I. Kluh, V. Kostka, L. Morávek, Z. Prusík, J. Vaněček, B. Keil, and F. Šorm, *BBA* **130,** 543 (1966).

129. B. S. Hartley, J. R. Brown, D. L. Kauffman, and L. B. Smillie, *Nature* **207,** 1157 (1965).

130. L. B. Smillie and B. S. Hartley, *JMB* **12,** 933 (1965).

130a. L. B. Smillie, A. Furka, N. Nagabhushan, K. J. Stevenson, and C. O. Parkes, *Nature* **218,** 343 (1968).

131. J. R. Brown, D. L. Kauffman, and B. S. Hartley, *BJ* **103,** 497 (1967).

132. D. M. Shotton and B. S. Hartley, *Nature* **225,** 802 (1970).

132a. B. S. Hartley, *Phil. Trans. Roy. Soc. London* **B257,** 77 (1969).

133. J. A. Gladner and K. Laki, *JACS* **80,** 1263 (1958).

134. S. M. Magnusson, *in* "Structure-Function Relationships of Proteolytic Enzymes" (P. Desnuella, H. Neurath, and M. Ottesen, eds.), p. 138, Munksgaard, Copenhagen, 1970.

135. W. R. Groskopf, L. Summaria, and K. C. Robbins, *JBC* **244,** 3590 (1969).

136. The current status of work on proteolytic enzymes has been summarized by various authors at a recent symposium (*137*).

137. P. Desnuelle, H. Neurath, and M. Ottesen, eds., "Structure-Function Relationships of Proteolytic Enzymes." Munksgaard, Copenhagen, 1970.

TABLE VIII
HOMOLOGY OF SEQUENCES AROUND THE ACTIVE SERYL RESIDUE IN PANCREATIC PROTEINASES AND RELATED ENZYMES[a]

Bovine trypsin	Cys-Gln-Gly-Asp-*Ser*-Gly-Gly-Pro-Val
Bovine chymotrypsin A	Cys-Met-Gly-Asp-*Ser*-Gly-Gly-Pro-Leu
Bovine chymotrypsin B	Cys-Met-Gly-Asp-*Ser*-Gly-Gly-Pro-Leu
Porcine elastase	Cys-Gln-Gly-Asp-*Ser*-Gly-Gly-Pro-Leu
Bovine thrombin	Cys-Glu-Gly-Asp-*Ser*-Gly-Gly-Pro-Phe
Human plasmin	Cys-Gln-Gly-Asp-*Ser*-Gly-Gly-Pro-Leu
Sorangium proteinase	Gly-Arg-Gly-Asp-*Ser*-Gly-Gly-Ser-Trp
Streptomyces griseus proteinase	Cys-Gln-Gly-Asp-*Ser*-Gly-Gly-Pro-Val

[a] Italic *Ser* is reactive with diisopropylfluorophosphate and similar reagents.

For chymotrypsin and elastase the homology in sequence is reflected also in overall conformation as indicated by the three-dimensional structures obtained by X-ray diffraction methods (*138, 139*). Furthermore, the catalytic mechanisms and the residues in the active sites are also clearly homologous involving the cooperative effect of interacting residues of aspartic acid, histidine, and serine (*140*). The homology of the sequences around these three active residues is shown in Tables VIII, IX, and X.

Since the specificities of these enzymes are different, each showing preferences for particular side chains in the residues bearing the carboxyl group of the sensitive peptide (or ester) bond, this must be reflected in the portion of the enzyme structure responsible for such specific interaction. For trypsin, the cationic group in the substrate (arginyl or lysyl)

TABLE IX
SEQUENCES AROUND THE REACTIVE HISTIDYL RESIDUES OF PANCREATIC PROTEINASES AND RELATED ENZYMES[a]

Bovine trypsin	Ser-Ala-Ala-*His*-Cys-Tyr
Bovine chymotrypsin A	Thr-Ala-Ala-*His*-Cys-Gly
Bovine chymotrypsin B	Thr-Ala-Ala-*His*-Cys-Gly
Porcine elastase	Thr-Ala-Ala-*His*-Cys-Val
Bovine thrombin	Thr-Ala-Ala-*His*-Cys-Leu
Sorangium proteinase	Thr-Ala-Gly-*His*-Cys-Gly
Streptomyces griseus proteinase	Thr-Ala-Ala-*His*-Cys-Val

[a] Italic *His* is in the active site of these enzymes.

138. D. M. Shotton and H. C. Watson, *Nature* **225**, 811 (1970).

139. The extensive homology in sequences of the other proteinases of this type suggests that the conformations of all these enzymes must be similar (*19, 136, 140*).

140. D. H. Blow, J. J. Birktoft, and B. S. Hartley, *Nature* **221**, 337 (1969).

TABLE X
HOMOLOGY OF SEQUENCES AROUND THE PRESUMED ACTIVE ASPARTYL RESIDUE IN PROTEINASES[a]

Bovine trypsin	Thr-Leu-Asn-Asn-*Asp*-Ile-Met-Leu-Ile-Lys
Bovine chymotrypsin A	Thr-Ile-Asn-Asn-*Asp*-Ile-Thr-Leu-Leu-Lys
Bovine chymotrypsin B	Thr-Val-Arg-Asn-*Asp*-Ile-Thr-Leu-Leu-Lys
Porcine elastase	Ala-Ala-Gly-Tyr-*Asp*-Ile-Ala-Leu-Leu-Arg
Bovine thrombin	Asn-Leu-Asp-Arg-*Asp*-Ile-Ala-Leu-Leu-Lys
Sorangium proteinase	Phe-Pro-Gly-Asn-*Asp*-Arg-Ala-Trp-Val-Ser

[a] *Asp* is assumed to function in a charge-relay system (*140*). In chymotrypsin A and elastase this residue is in contact with the active histidyl residue (Table IX). *Sorangium* proteinase contains only two aspartyl residues, one adjacent to the reactive seryl (Table VIII) and the one in the sequence shown above (*147*), which shows little homology with the mammalian enzymes.

interacts with the anionic group of an aspartyl residue 192 (*141*), whereas in chymotrypsin the residue in the corresponding locus is serine (*140*). Many of the differences in sequences among these enzymes may not be concerned with the changes in specificity and reflect only the usual divergence since the time that the gene duplications occurred.

It is noteworthy that the resemblance of the pancreatic proteinases to many of the enzymes involved in the complex chain of events leading to blood clotting involves for the latter group first, a series of conversions from inactive precursors (zymogens) by proteolytic cleavage to yield active proteolytic enzymes, and, finally, the initiation of the clot by hydrolysis of fibrinogen to produce fibrin through the specific action of thrombin. Interestingly, thrombin is more specific in its action than trypsin since the former acts only on arginyl bonds (*142*). Plasmin, produced from its inactive precursor, is also homologous with these enzymes (*135*). Thus, while some of these enzymes now perform distinct biological functions in digestion, clotting, or clot dissolution, they still retain structural and functional homology (*19*).

The differentiation of the above series of enzymes must have occurred long ago since homologous pancreatic proteinases of the several mammalian types are found in the dogfish, lungfish, starfish, and other species (*143*). Even more striking is the presence of a chymotrypsin-like

141. E. Shaw, *in* "Structure-Function Relationships of Proteolytic Enzymes" (P. Desnuelle, H. Neurath, and M. Ottesen, eds.), p. 70. Munksgaard, Copenhagen, 1970.

142. S. Sherry and W. Troll, *JBC* **208,** 95 (1954).

143. H. Neurath, R. A. Bradshaw, and R. Arnon, *in* "Structure-Function Relationships of Proteolytic Enzymes" (P. Desnuelle, H. Neurath, and M. Ottesen, eds.), p. 113. Munksgaard, Copenhagen, 1970.

enzyme in the sea anemone, *Metridium* (*144*). In *Streptomyces griseus* there is an enzyme similar in specificity and homologous in the presently reported partial sequence with mammalian trypsin; however, the latter contains six disulfide bridges whereas the former contains only three (*145*). The complete sequence and the three disulfide bridges of the α-lytic proteinase of a *Myxobacterium, Sorangium,* also show homology with these enzymes (*146*); its enzymic properties are similar to those of porcine elastase (*147*). The enzyme, cocoonase, from silkworms, manifests the same specificity and many other similarities to mammalian trypsins, including amino acid composition, and many also prove to be homologous in sequence (*148*). Presumably, many other related enzymes may be widely dispersed in various phyla (*19, 138, 148a*).

One characteristic feature of all of these homologous proteinases is the unique serine residue which is reactive with diisopropylfluorophosphate and similar reagents. It will be interesting to determine how many of the esterases also known to be sensitive to this reagent, e.g., choline esterase, prove to be related in structure; however, not all enzymes sensitive to this reagent are homologous with the pancreatic proteinases. As will be discussed further below (Section IV,C), the subtilisins resemble the latter in the reactive residues in the active site and in mechanism of action but appear to be of completely independent evolutionary origin (*61, 62*).

2. *Sulfhydryl Proteinases*

Many plants contain enzymes of this type. For papain, essentially the complete sequence is known (*149, 150*) and it is now evident that the sequence near the active thiol is similar in ficin (*151*) and bromelain

144. D. Gibson and G. H. Dixon, *Nature* **222,** 753 (1969).

145. L. Jurasek, D. Fackre, and L. B. Smillie, *BBRC* **37,** 99 (1969).

146. M. O. J. Olson, N. Nagabhushan, M. Dzwinill, L. B. Smillie, and D. R. Whitaker, in press.

147. D. R. Whitaker, "Methods in Enzymology," 1970 (in press).

148. F. C. Kafatos, A. M. Tartakoff, and J. H. Law, *JBC* **242,** 1477 and 1488 (1967).

148a. W. P. Winter and H. Neurath, *Federation Proc.* **27,** 492 (1969).

149. A. Light, R. Frater, J. R. Kimmel, and E. L. Smith, *Proc. Natl. Acad. Sci. U. S.* **52,** 1276 (1964); R. E. J. Mitchel, I. M. Charken, and E. L. Smith, *JBC* **245,** 3485 (1970).

150. J. Drenth, J. N. Jansonius, R. Koekoek, H. M. Swen, and B. G. Wolthers, *Nature* **218,** 929 (1968).

151. R. C. Wong and I. E. Liener, *BBRC* **17,** 470 (1964).

TABLE XI
SEQUENCES AROUND THE ACTIVE CYSTEINYL RESIDUE OF PLANT PROTEINASES[a]

Papain	Pro-Val-Lys-Asn-Gln-Gly-Ser-Cys-Gly-Ser-*Cys*-Trp
Ficin	Pro-Ile-Arg-Gln-Gln-Gly-Gln-Cys-Gly-Ser-*Cys*
Bromelain	Asn-Gln-Asp-Pro-Cys-Gly-Ala-*Cys*-Trp
Chymopapain	Lys-Arg-Val-Pro-Asp-Ser-Gly-Glu-*Cys*-Tyr

[a] The italicized cyteinyl residue is in the active site of these enzymes. In papain the nearby Cys is linked in a disulfide bond (*149*), and this may also be the case for ficin and bromelain.

(*152*, *153*) (Table XI). These, and presumably other similar plant enzymes, will probably prove to be a homologous group. Papaya latex contains other sulfhydryl proteinases, and the sequence of the peptide containing the active thiol of chymopapain (*154*) shows some similarities to that of papain; further work on the sequence is essential, however, before it can be certain that these two enzymes found in the same plant represent a gene duplication.

It has also been shown that the sequence first reported near the active histidine residue of papain (*149*, *150*) is similar in ficin and bromelain (*154a*).

3. *Pepsin, Rennin (Chymosin), and Gastricsin*

Presently available sequence information, particularly around the disulfide bridges, indicates that these enzymes are structurally related (*155*, *155a*) and thus are presumably products of gene duplication. Human gastricsin also appears to be related in structure to pepsin and rennin (*156*).

4. *Carboxypeptidases*

The complete sequence of bovine carboxypeptidase A (*157*) has revealed that this enzyme exists in two forms which differ at three posi-

152. L. Chao and I. E. Liener, *BBRC* **27,** 100 (1967).

153. T. Murachi and N. Takahashi, *in* "Structure-Function Relationships of Proteolytic Enzymes" (P. Desnuelle, H. Neurath, and M. Ottesen, eds.), p. 298. Munksgaard, Copenhagen, 1970.

154. J. N. Tsunoda and K. T. Yasunobu, *JBC* **241,** 4610 (1966).

154a. S. S. Husain and G. Lowe, *BJ* **117,** 333, 341 (1970).

155. B. Foltmann and B. S. Hartley, *BJ* **104,** 1064 (1967).

155a. B. Foltmann, *Phil. Trans. Roy. Soc. London,* **B257,** 147 (1970).

156. W.-Y. Huang and P. S. Tang, *JBC* **245,** 2189 (1970).

157. R. A. Bradshaw, L. H. Ericsson, K. A. Walsh, and H. Neurath, *Proc. Natl. Acad. Sci. U. S.* **63,** 1389 (1969).

tions in the sequence (*158*). These forms are presumably products of allelomorphic genes since examination of individual pancreas glands has shown that half contained both variants and half showed one or the other type, in accord with simple Mendelian theory (*159*). Carboxypeptidase B, which differs in specificity, shows homology with the A enzyme, where the sequence is known (*143, 159*). Thus, the A and B enzymes are presumably products of genes which were derived from a common ancestral form by duplication. Here, as in the case of the ethanol dehydrogenases (see below), there is some overlapping specificity, A and B liberating some COOH-terminal residues of the same type, but only B being able to attack bonds containing terminal arginine and lysine residues (*160*).

5. *Ethanol Dehydrogenases*

Mammalian liver contains two alcohol dehydrogenases both catalyzing the reversible conversion of ethanol to acetaldehyde. One of these is specific for ethanol, but the other also catalyzes the dehydrogenation of 3β-hydroxy-5β-steroids and retinol (*161*). Each of these two enzymes is made up of two identical peptide chains, but another form containing one chain of each type possesses properties intermediate between the other two. The study of the two enzymes from equine liver by Jörnvall (*162, 163*) has shown that their sequences are almost identical. The ethanol-specific enzyme (E) possesses at least five residues which differ in the steroid dehydrogenase (S) as shown in the accompanying tabulation. Each of these five differences presumably results from a point mutation since the respective codons differ only in a single nucleotide, yet this has been sufficient to alter the specificity of the two "isoenzymes." Indeed,

Residue No.	E Chain	S Chain
17	Glu	Gln
94	Thr	Ile
101	Arg	Ser
110	Phe	Leu
366 ± 5	Glu	Lys

158. P. H. Pétra, R. A. Bradshaw, K. A. Walsh, and H. Neurath, *Biochemistry* **8,** 2762 (1969).
159. R. A. Bradshaw, H. Neurath, and K. A. Walsh, *Proc. Natl. Acad. Sci. U. S.* **63,** 406 (1969).
160. D. J. Cox, E. Wintersberger, and H. Neurath, *Biochemistry* **1,** 1069 (1962).
161. H. Theorell, S. Taniguchi, Å. Åkeson, and L. Skursky, *BBRC* **24,** 603 (1966).
162. H. Jörnvall, *BBRC* **35,** 542 (1969).
163. H. Jörnvall, *Nature* **225,** 1133 (1970).

	1 5 10 15
LADH(EE):	Acetyl-*Ser*-Thr-Ala-*Gly*-Lys-Val-Ile-Lys-Cys-Lys-Ala-Ala-Val-*Leu*-Trp-
GPDH(Lobster):	Acetyl-*Ser*-Lys-Ile-*Gly*-Ile-Asp-Gly-Phe-Gly-Arg-Ile-Gly-Arg-*Leu*-Val-
GPDH(Pig):	H_2N-Val-Lys-Val-*Gly*-Val-Asp-Gly-Phe-Gly-Arg-Ile-Gly-Arg-*Leu*-Val-
YADH:	

20 25 30 35

Glu-Glu-Lys-Lys-Pro-Phe-*Ser*-Ile-Glu-Glu-Val-Glu-*Val-Ala*-Pro-Pro-Lys-Ala-His-Glu

Leu-Arg-Ala-Ala-Leu-Ser-Cys-Gly Ala-Gln-Val-*Val-Ala*-Val-Asn-Asp-Pro-Phe-Ile

Thr-Arg-Ala-Ala-Phe-Asn-*Ser*-Gly-Lys-Val-Asp-Ile-*Val-Ala*-Ile-Asn-Asp-Pro-Phe-Ile

40 45 * 50 57

Val-Arg-Ile-Lys-*Met-Val*-Ala-Thr-Gly-Ile-Cys-Arg-*Ser*-Asp-Asp-His-*Val*-Val-Ser-*Gly-Thr*-Leu

Ala-Leu-Glu-Tyr-*Met-Val-Tyr*-Met-Phe-Lys-Tyr-Asp-*Ser*-Thr-His-Gly-*Val*-Phe-Lys-*Gly*-Glu-Val

Asp-Leu-His-Tyr-*Met-Val-Tyr*-Met-Phe-Glu-Tyr-Asp-*Ser*-Thr-His-Gly-Lys-Phe-*His-Gly-Thr*-Val

-*Tyr*-Ser-Gly-Val-Cys-His-Thr-Asp-Leu-His-Ala-Trp-*His-Gly*-Asp-

*

FIG. 10. Comparison of the amino terminal sequences of glyceraldehyde-3-phosphate dehydrogenases (GPDH) from lobster and pig muscle (*74*) and of alcohol dehydrogenases of equine liver [LADH(EE)] and yeast (YADH) (*164*). In the initial 57 residues, 12 are identical (21%). The active cysteine (residue 46*) in ADH is not the same as in GPDH.

not all of these different residues in the steroid dehydrogenase may be concerned with the affinity for steroids or retinol. An interesting question is whether the two enzymes should be named as distinct enzymes or classified as isoenzymes. In any case, the two dehydrogenases must have arisen because of a gene duplication, the subsequent mutation(s) altering the specificity. It is likely that the gene for the ethanol-specific enzyme must be the older of the two, since enzymes similar in specificity are present in organisms that are lacking in steroids.

It is also noteworthy (*164*) that a substantial portion of the amino terminal sequence of these alcohol dehydrogenases is similar to that of glyceraldehyde-3-phosphate dehydrogenase (Fig. 10). It would appear, therefore, that the two distinct types of dehydrogenases also arose by gene duplication very early in evolution, as judged by the large differences in sequences, in contrast to the recent divergence of the two ethanol dehydrogenases, as indicated by their remarkable similarity. In view of these relationships it is to be expected that other DPN-linked dehydrogenases may prove to contain homologous sequences.

6. *Isoenzymes*

Many different enzymes are now known which exhibit similar specificity but manifest different electrophoretic or chromatographic properties

164. H. Jörnvall, private communication (1970).

(*164a*). For many of these isoenzymes, it is now evident that these are related varieties of the same enzyme, as demonstrated by partial sequences, by immunological cross-reaction, by similar peptide patterns, or by amino acid composition, etc. Presumably, some of these different forms of the same enzyme may simply be allotypes but others have undoubtedly resulted from gene duplication and may be regarded, where there has been no demonstrable change in specificity, as representing a potential for further divergence in amino acid sequence and hence in activity and specificity. The situation with the two ethanol dehydrogenases described above illustrates the evolutionary possibilities for the various isoenzymes. Clearly, further work on the sequences and enzymic properties of the many kinds of isoenzymes should be very revealing in this regard. The organ specificity of some isoenzymes indicates the importance of this phenomenon in differentiation (*165, 166*).

The partial sequences of the carbonic anhydrases indicate strong homology but with many differences in sequence that influence the activity of these isoenzymes (*167*).

The work on the lactate dehydrogenases, demonstrating differences in tissue distribution and in inhibitory properties, has led to the view that these represent adaptations to the different metabolic demands of cardiac and skeletal muscle tissue (*168*). The differences in distribution of the two types in skeletal muscles of ground-living and flight birds also appear to represent important functional adaptation of the two types (*169*). Although sequences have not been reported as yet, there is no question as to the homology of the two kinds of peptide chains in view of their ability to form enzymes of mixed types (*170*).

7. *Lysozyme and α-Lactalbumin*

Homologous lysozymes similar to that of hen egg white are widely distributed in other kinds of bird eggs and in many tissues and secretions

164a. "Multiple Molecular Forms of Enzymes" *Ann. N. Y. Acad. Sci.* **151,** 1–689 (1968).

165. C. L. Markert and F. Miller, *Proc. Natl. Acad. Sci. U. S.* **45,** 753 (1959).

166. C. L. Markert and E. Appella, *Ann. N. Y. Acad. Sci.* **94,** 678 (1961).

167. P. O. Nyman, L. Strid, and G. Westermark, *European J. Biochem.* **6,** 172 (1968); B. Andersson, P. O. Göthe, T. Nilsson, P. O. Nyman, and L. Strid, *ibid.* p. 190.

168. N. O. Kaplan and R. D. Cahn, *Proc. Natl. Acad. Sci. U. S.* **48,** 2123 (1962).

169. A. C. Wilson, R. D. Cahn, and N. O. Kaplan, *Nature* **197,** 331 (1963).

170. N. O. Kaplan, *in* "Evolving Genes and Proteins" (V. Bryson and H. J. Vogel, eds.), Academic Press, New York, 1965. p. 243.

of vertebrates, e.g., milk, tears, saliva, spleen, placenta, serum, and leukocytes (*171*).

Of great interest was the report by Brew *et al.* (*172*) that α-lactalbumin, long known as a constituent of cow's milk, shows extensive homology with egg white lysozyme in primary sequence and in the location of the four disulfide bridges (*173*) (Fig. 11). Of the 129 residues in lysozyme and the 123 residues in α-lactalbumin, 42 have been located in the corresponding positions and up to eight more may also be identical. In addition, 23 substitutions are conservative replacements (*173*). The two proteins are probably similar in conformation also (*174*), despite the deletions and substitutions of residues. α-Lactalbumin is, however, completely devoid of lysozyme activity, possibly because the catalytically active glutamyl at position 35 has been replaced by a threonyl or histidyl residue.

Independently, Brodbeck, Ebner, and co-workers (*175, 176*) discovered that α-lactalbumin is the B protein of bovine lactose synthetase, this enzyme being made up of two distinct peptide chains. Subsequently (*177*), the A chain of the synthetase was discovered to be similar, if not identical, to the UDPgalactose:*N*-acetylglucosamine galactosyltransferase concerned with elongation of chains in glycoproteins (*178*). The two reactions are:

$$\text{UDP-D-galactose} + \text{D-glucose} \xrightarrow[\text{(A + B chains)}]{\text{lactose synthetase}} \text{UDP} + \text{lactose} \qquad (1)$$

$$\text{UDP-D-galactose} + N\text{-acetyl-D-glucosamine} \xrightarrow[\text{(A chain)}]{\text{transferase}} \text{UDP} + N\text{-acetyllactosamine} \qquad (2)$$

In reaction (2) the rate is slow with free *N*-acetylglucosamine but rapid when this residue is attached to a polypeptide chain. When α-lactalbumin (chain B) is added, reaction (2) is inhibited, and synthesis of lactose proceeds rapidly in the presence of glucose as shown in reaction (1).

From the above information the following evolutionary events have

171. P. Jollès, *Angew. Chem.* **8,** 227 (1969).

172. K. Brew, T. C. Vanaman, and R. L. Hill, *JBC* **242,** 3747 (1967).

173. R. L. Hill, K. Brew, T. C. Vanaman, I. P. Trayer, and P. Mattock, *Brookhaven Symp. Biol.* **21,** 139 (1969).

174. W. J. Browne, A. C. T. North, D. C. Phillips, K. Brew, T. C. Vanaman, and R. L. Hill, *JMB* **42,** 65 (1969).

175. K. E. Ebner, W. L. Denton, and V. Brodbeck, *BBRC* **24,** 232 (1966).

176. V. Brodbeck, W. L. Denton, N. Tanahashi, and K. E. Ebner, *JBC* **242,** 1391 (1967).

177. K. Brew, T. C. Vanaman, and R. L. Hill, *Proc. Natl. Acad. Sci. U. S.* **59,** 491 (1968).

178. E. J. McGuire, G. W. Jourdian, D. M. Carlson, and S. Roseman, *JBC* **240,**

1 10
H_2N-Glu-Gln-Leu-Thr-Lys-*Cys*-*Glu*-Val-Phe-Arg-Glu-Leu-*Lys*- -Asp-*Leu*-Lys-Gly-*Tyr*-Gly-

1 10 20
H_2N-Lys-Val-Phe-Gly-Arg-*Cys*-*Glu*-Leu-Ala-Ala-Ala-Met-*Lys*-Arg-His-Gly-*Leu*-Asp-Asn-*Tyr*-Arg-

20 30 40
Gly-Val-*Ser*-*Leu*-Pro-Glu-*Trp*-*Val*-*Cys*-Thr-Thr- -*Phe*-His-Thr-*Ser*-Gly-Tyr-*Asx*-*Thr*-*Glx*-*Ala*-

30 40
Gly-Tyr-*Ser*-*Leu*-Gly-Asn-*Trp*-*Val*-*Cys*-Ala-Ala-Lys-*Phe*-Glu- -*Ser*-Asn-Phe-*Asn*-*Thr*-*Gln*-*Ala*-

50 60
Ile-Val-Glx-*Asx*- -*Asx*(Glx, *Ser*, *Thr*)*Asx*-*Tyr*-*Gly*-Leu-Phe(*Glx*, *Ile*, *Asx*, Asx)Lys-Ile -*Trp*-*Cys*-

50 60
Thr-Asn-Arg-*Asn*-Thr-*Asp*-Gly-*Ser*-*Thr*-*Asp*-*Tyr*-*Gly*-Ile -Leu-*Gln*-*Ile*-*Asn*-Ser- Arg-Trp-*Trp*-*Cys*-

70 80
Lys-*Asx*-Asx-Glx-Asx-*Pro*-His-*Ser*-Ser-Asx-Ile -*Cys*-*Asn*-*Ile*-Ser-*Cys*-Asp-Lys-Phe-*Leu*-Asx-Asx-

70 80
Asn-*Asp*-Gly-Arg-Thr-*Pro*-Gly-*Ser*-Arg-Asn-Leu-*Cys*-*Asn*-*Ile*-Pro-*Cys*-Ser-Ala-Leu-*Leu*-Ser-Ser-

90 100
Asx-Leu-*Thr*-Asx-Asx-Ile-Met-*Cys*-Val-*Lys*-*Lys*-*Ile*-Leu- -*Asp*-Lys-Val-*Gly*-Ile -*Asn*-Tyr-*Trp*-

90 100
Asp-Ile -*Thr*-Ala-Ser-Val-Asn-*Cys*-Ala-*Lys*-*Lys*-*Ile*-Val-Ser-*Asp*-Gly-Asp-*Gly*-Met-*Asn*-Ala-*Trp*-

110 120 123
Leu-*Ala*-His-Lys-Ala-Leu-*Cys*-Ser-Glu-Lys-Leu-Asp-*Gln*- -*Trp*-Leu- -*Cys*-Glu-Cys-*Leu*-COOH

110 120 129
Val-*Ala*-Trp-Arg-Asn-Arg-*Cys*-Lys-Gly-Thr-Asp-Val-*Gln*-Ala-*Trp*-Ile-Arg-Gly-*Cys*- -Arg-*Leu*-COOH

FIG. 11. Homology in sequences of bovine α-lactalbumin (upper sequence) and hen egg white lysozyme (lower sequence). Residues present in identical positions are italicized. Gaps permit proper alignment of the homologous regions. From Hill *et al.* (*173*).

been postulated (*173*). Initially, there must have been a duplication of the gene for lysozyme, each of these genes then being free to evolve independently. [In duck eggs, there are three distinct lysozymes (*171*), the sequences of the two that have been studied showing two amino acid substitutions.] At some later stage in evolution one of these proteins lost its catalytic activity as a lysozyme but must have retained its binding affinity for glycosidic compounds, the common feature being retention of an affinity for the $\beta 1 \rightarrow 4$ glucopyranosyl structure. In addition, it developed an affinity for the A protein (transferase), present in various animal tissues. Furthermore, the B protein (now α-lactalbumin) came under a new set of controls—one, in being produced only in mammary tissue and, two, in being under hormonal control so that its production is very low until the onset of lactation. At least three hormones (insulin, hydrocortisone, and prolactin) are required for synthesis of the A protein and α-lactalbumin in mouse mammary gland explants (*179*).

This situation provides an interesting example of evolution at the mammalian level. Since α-lactalbumin is synthesized only in the mammary gland, lactose is not made in other tissues. Thus, the combination of a modified protein (α-lactalbumin) and a functional enzyme (the transferase) has resulted in the formation of a new enzyme, lactose synthetase. It is noteworthy that mammary tissue also retains its ability to make lysozyme, a constituent of milk (*180*).

Other examples among enzymes of such a functional change, as exhibited in the lysozyme to α-lactalbumin transformation, are still unknown. Brew *et al.* (*173*) have noted, however, that this situation has some features similar to the tryptophan synthetase of *E. coli* (Section III,C) in which each chain has distinctive catalytic properties but a new enzymic property emerges on the combination of the two (also see Section IV,E).

IV. Expression of Evolutionary Factors in Protein Structure

A. Divergence

As already noted for the cytochromes c, the degree of divergence in the sequence of a homologous protein is, in general, related to the phylo-

PC4113 (1965).

179. R. W. Turkington, K. Brew, T. C. Vanaman, and R. L. Hill, *JBC* **243**, 3382 (1968).

180. H. A. McKenzie, *Advan. Protein Chem.* **22**, 56 (1967).

genetic relations of the species possessing the protein. These relationships are clearly evident whether the numbers of variant residues (*8*) or the number of mutational differences (*15*) are considered. By using the minimal observable mutational differences, Fitch and Margoliash (*181*) have constructed a phylogenetic tree for the cytochrome c gene (Fig. 12). The calculated mutational distances are on the whole in accord with classic concepts of phylogeny. For cytochrome c, a relatively "old" gene (see below), the data are mainly useful in comparing different groups of vertebrates or different phyla. Comparisons within closely related groups of animals do not show sufficient differences in sequence to be useful for such comparisons. For example, since cow, sheep, and pig cytochromes c are identical, study of this protein in the artiodactyl ungulates is useless for taxonomic purposes; in contrast, the great variability of the fibrinopeptides has permitted detailed studies to be made among many species in this group of mammals (*182*). Extensive investigations of homologous hemoglobins in a wide variety of species have also been reported (*183*). Sequences of several mammalian myoglobins have been studied (*184, 185*).

As yet, there are only a few proteins for which the complete sequences have been established in a sufficient variety of organisms to be of value for assessing taxonomic and evolutionary relationships. With the availability of improved methods of sequence determination the field may be expected to grow rapidly. At this juncture, however, it may be useful to mention a few caveats regarding such studies. First, in large enzymes obtainable only in small amounts, studies of partial sequence are frequently made by the isolation of peptides containing labeled "active" residues. Although such studies are important for assessing homology and for understanding enzyme mechanisms, the inherent conservatism around active sites may preclude any deductions regarding evolutionary divergence. It is precisely in these portions of the enzyme sequence that "constant residues" will be most prominent, as in the 11-residue constant sequence of cytochrome c (Section II), in the sequences around the active histidyl or seryl residues of the pancreatic proteinases and their relatives (Section III,D,1), and in the reactive residues of the subtilisins (Section III,A), ribonucleases (Section III,A), and glyceraldehyde-3-

181. W. M. Fitch and E. Margoliash, *Science* **155,** 279 (1967).
182. G. A. Mross and R. F. Doolittle, *ABB* **122,** 674 (1967).
183. See various articles in Bryson and Vogel (*14*); also Braunitzer *et al.* (*6*).
184. R. L. Hill, C. M. Harris, J. F. Naylor, and W. M. Sams, *JBC* **244,** 2182 (1969).
185. R. A. Bradshaw and F. R. N. Gurd, *JBC* **244,** 2167 (1969).

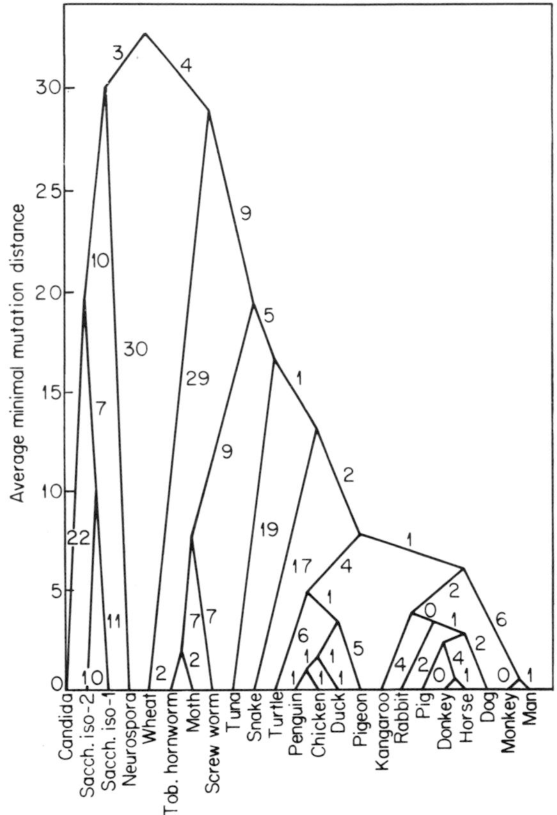

FIG. 12. Phylogenetic tree based on the minimal mutation distances among the eukaryotic cytochromes c. Each number on the figure is the mutation distance along the line of descent as calculated from the best fit of the data. Each apex is placed at an ordinate value which is the sum of all mutations in the lines of descent from that apex. From Fitch and Margoliash (*32*).

phosphate dehydrogenases (Section III,A) of different species. Second, for proteins, which are already suspected to be homologous in sequence, it is most desirable to initiate comparative studies by selecting species, as distantly related as possible, in order to assess the degree of change. As discussed below for the histone IV (Section IV,A,1) of a higher plant and a higher animal, the differences in sequence are so few, that little or nothing concerning phylogenetic relationships could possibly be gleaned by investigation of more closely related species.

1. *Factors Influencing Rate of Divergence*

It has been sometimes assumed that the rate of amino acid substitution is approximately constant. On this basis attempts have been made to relate the numbers of amino acid substitutions (or mutational changes in the codons) to the approximate time of divergence of the various taxa. At best such calculations can be only grossly approximate, there being many difficulties with the underlying tacit assumptions, except for comparisons within a small taxonomic group (*12*). First of all, the generation time in many organisms is usually annual or nearly so. Higher animals deviate from this only by approximately one order of magnitude. In contrast, under optimal conditions microorganisms may reproduce as often as 10^4 or 10^5 times per year. Second, the rate of penetrance of a mutational change into a species will depend not only on the magnitude of the difference in survival value of the wild-type and mutant strain but also on whether the organisms are haploid or diploid or possess alternating vegetative and sexual reproduction cycles. From the comparisons of the various cytochromes presently known the maximal number of amino acid differences between any two mammals is 12 residues (man and horse) whereas the two Ascomycetes, *Saccharomyces* and *Neurospora,* differ in 39 residues. Presumably, these differences in divergence may be related not only to time of divergence of the respective common ancestral forms but also to several or all of the aforementioned factors. A few additional cases among the cytochromes c may be cited to indicate that "time of divergence" is not the only factor involved in rate of divergence. The cytochromes of the two reptiles, turtle and rattlesnake, differ in 21 residues whereas those of tuna and dogfish differ in only 14 residues, despite the presumed paleontological evidence of the earlier divergence of the lines leading to modern elasmobranchs and teleosts than of the different reptiles.

It is well known that not all features of organisms have evolved at the same rate. Some "old" features of organisms have changed relatively little in the evolution of recent species whereas "new" features of organisms may be undergoing relatively rapid changes. Examples of the former among mammals are the relatively constant anatomical and physiological features of such organs as the eye and the heart in contrast to the marked changes in the size of different parts of the brain, differences in integument, in methods of locomotion—aquatic, aerial, bipedal, quadripedal, and in other features. Indeed, some groups of organisms appear to have changed little during long periods of time whereas other groups have undergone a series of rapid transformations. Yet the DNA of all higher organisms possesses the same structure and we may assume that mutation rates do not differ greatly in different organisms or in different genes. What does differ is the degree of adaptation and the selection processes operative on the population of a given species. Only recently has it become evident that similar phenomena are evident in the evolution of homologous proteins. The structures of the hemoglobins, cytochromes c, and a histone may be cited as examples of proteins manifesting diverse amounts and rates of change.

Myoglobins and hemoglobins are present in many different phyla but it is not clear whether all of these oxygen carriers containing ferroheme evolved independently or whether all of them are directly related in a homologous sense, since there is no clear continuity of the presence of these proteins among the small, scattered groups of arthropods, nematodes, platyhelminthes, annelids, mollusks, protozoa, root nodules of plants, etc., that possess hemoglobins. In contrast, hemoglobins and myoglobins are present in all classes of vertebrates and are lacking in only a few fish (Section IV,D).

Among the vertebrate hemoproteins there are very large differences in sequences with relatively few constant residues, but the myoglobins and hemoglobins investigated to date all show homology in sequence and conformation. The large differences in sequences may be a reflection of rapid evolutionary changes in adapting to the greatly different physiological requirements for oxygen transport or storage in the many kinds of environments in which the vertebrates are dispersed, e.g., fetal life, aquatic life of different kinds, various temperatures, terrestrial life at different altitudes, etc. Further, the differences in internal environment may also be important in view of the changes in body temperature, ionic environment, and the metabolic differences in the nucleated erythrocytes of lower vertebrates and enucleated cells of mammals (see also Section IV,A,2). While the "substrate" for hemoglobin is a relatively simple and small molecule and the structural requirements of the protein may not be as stringent as for enzymes with complex reaction mechanisms and specificity requirements for organic substrates and cofactors, the differences among myoglobins and hemoglobins in the magnitude of the Bohr effect, the shape of the O_2 binding curve, and the affinity for O_2 are indeed substantial (Section IV,A,2). Thus, the hemoglobins of vertebrates may be regarded as relatively "young proteins," still undergoing substantial alterations in adaptation to the changing metabolic requirements of aerobic life.

In the case of the cytochromes c, for the 104 residues common to the sequence of all species, 35 residues or approximately one-third have been found to be constant in the species presently investigated (see below).

There are 44 sites which are occupied by different residues in the cytochromes of man and yeast. It is evident that there has been much less variation tolerated in the structure of cytochrome c than in the vertebrate hemoglobins. Since the only known role of cytochrome c is in the respiratory chain where it functions in conjunction with O_2 and cytochrome oxidase, it would seem that the cytochrome system antedates the myoglobins and the hemoglobins, the main functions of these proteins being to store or to transport oxygen for terminal oxidation by the cytochrome system. Thus the cytochromes c must be "older proteins" than the hemoglobins, their major structural features having evolved before the divergence of all presently existing forms of eukaryotic life. Since major parts of the cytochrome c sequence had become adjusted to functional and conformational requirements, as well as to interactions in the mitochondrion with both the oxidase and a reductase system, no further substitutions were possible at these constant loci (*12*). In other words, at all or most of the 35 constant loci of the cytochromes c, all mutational changes have been lost, presumably because survival value was lowered, that is, the changes were lethal or semilethal. At other loci, many of the substitutions are of the conservative type, as already discussed (Section II).

An interesting case of a "very old protein" is histone IV (glycine-arginine-rich or GAR histone). The amino acid sequences of this protein from bovine thymus and from pea seedlings (*26*, *27*) show only two differences in the 102 residues in the chain (Fig. 13). Both of these changes are explicable as point mutations and are conservative, involving an arginine-lysine and an isoleucine-valine pair of changes. (In addition, there are secondary differences in degree of acetylation and methylation, but these are not dependent on differences in the sequences of the DNA in the genes for this histone.) For the histone then, little change has occurred since the long period of divergence from early eukaryotes to present animal and plant life as represented by the histone IV from the aforementioned two species. Evidently, the structural character of this histone became so precisely adjusted to its functions in the early evolution of eukaryotes, before the separation of the plant and animal lines of evolution, that little change in the sequence has been tolerated. As in the case of cytochrome c, it is evident that the gene for histone IV has been in existence during the entire period of eukaryotic life. Furthermore, the great similarities in properties of other types of histones obtained from both pea seedlings and bovine thymus suggest that these proteins will manifest equally extensive homology (*27*).

These three examples serve to illustrate an important principle that has emerged from the study of the evolution of homologous proteins.

```
1                                       10                             16           20
Ac-Ser-Gly-Arg-Gly-Lys-Gly-Gly-Lys-Gly-Leu-Gly-Lys-Gly-Gly-Ala-Lys(Ac)-Arg-His-Arg-Lys(Me)1,2

21                                      30                                          40
Val-Leu-Arg-Asp-Asn-Ile-Gln-Gly-Ile-Thr-Lys-Pro-Ala-Ile-Arg-Arg-Leu-Ala-Arg-Arg-

41                                      50                                          60
Gly-Gly-Val-Lys-Arg-Ile-Ser-Gly-Leu-Ile-Tyr-Glu-Glu-Thr-Arg-Gly-Val-Leu-Lys-Val-
                                                                                Ile

61                                      70                              77          80
Phe-Leu-Glu-Asn-Val-Ile-Arg-Asp-Ala-Val-Thr-Tyr-Thr-Glu-His-Ala-Lys-Arg-Lys-Thr-
                                                                    Arg

81                                      90                                          100
Val-Thr-Ala-Met-Asp-Val-Val-Tyr-Ala-Leu-Lys-Arg-Gln-Gly-Arg-Thr-Leu-Tyr-Gly-Phe-

    102
Gly-Gly-COOH
```

Fig. 13. Amino acid sequence of histone IV (glycine-arginine-rich histone) of bovine thymus (continuous sequence); residues that differ in the homologous pea seedling histone are given below (*26, 27*).

In each instance the functions appear to be the same but the structures have altered in different amounts. Nevertheless, the functions may have altered in subtle ways. Only in the case of the histone can one conclude that the precise functional role of this protein may not have altered in any major way since it became fully adapted to its interactions with the DNA of the nucleus. The fact that the amounts of change in sequence in different homologous proteins differ so profoundly, as in the three cases described above, appears to be consonant with similar conclusions derived from the study of whole organisms or their morphological features (*111*).

2. *Speciation of Homologous Proteins: Genetic Drift*

In the divergence of various evolutionary lines, it is now clear that in a given species practically all individuals possess a fully functional ("normal" or "wild-type") protein with the identical sequence in the polypeptide chain(s). Thus, despite the discovery of many "abnormal" hemoglobins the structure of hemoglobin A is accepted as the normal type. Similarly, in other proteins studied to date, a single sequence is accepted as the type structure for a species. Yet this is in contrast to the well-recognized wide variations of genotypes in a given species and the existence of considerable heterozygosity. How much variability exists for structural genes is not evident from the few proteins for which the sequences are presently known. Indeed, how much of the apparent variation results from genes that control the quantitative production of proteins by influencing regulatory processes in development and in the adult animal is unknown. Thus, while many genetically determined metabolic disorders result from lack of an enzyme or production of an inactive or faulty enzyme (Garrod's "inborn errors"), some metabolic errors appear to result from quantitative and regulatory factors. For example, though there is a clearly recognized genetic influence in the production of diabetes mellitus (*186*), there is no evidence that the insulin produced by such individuals is different in structure (*187*).

Thus we have to accept the view that although there is always heterozygosity within a species, for some proteins at least, the variants are relatively few in number and the structure of the normal protein can readily be established for a given species. Even in comparing species, some proteins exhibit large differences in sequence and others none, al-

186. A. E. Renold and G. F. Cahill, Jr., *in* "The Metabolic Basis of Inherited Disease" (J. B. Stanbury, J. B. Wyngaarden, and D. S. Frederickson, eds.), 2nd ed., p. 69. McGraw-Hill, New York, 1966.

187. L. F. Smith, *Am. J. Med.* **40,** 662 (1966).

though the species are remarkably different. A striking case is represented by the identical insulins of pig, sperm whale, and dog (*187*). Yet pig, cow, and sheep insulins differ from each other (*187*) whereas the cytochromes c of these three species are identical (*7*). Speciation is, of course, dependent on the complete genotype and not on single proteins.

What then are the factors that have led to the apparent stabilization of a specific amino acid sequence for a given species when the homologous proteins of other species possess apparently identical functional properties yet may differ in amino acid sequence (*188*)? There appear to be at least three different ways in which this could be explained. First, from a rather rigid Darwinian viewpoint of evolution, it could be assumed that the amino acid substitution became a characteristic of the species because in the total genotype there was enhanced survival value, whatever that may have been—increased stability of the protein, more efficient interaction with other components of the cell or the organism as a whole, greater efficiency as an enzyme *in vivo*, better regulation, etc. This requires no further comment since it is simply what is generally viewed as the most general factor of evolution itself. Obviously, any changes resulting in a diminished survival value would be eliminated.

A second possible reason for the spread of new residues in a given protein of a species is that there was a greater survival value for other alterations in the genome and that "neutral" mutations were carried along passively. This implies that the development of the new species occurred in a small, relatively isolated clone or group of organisms in which heterozygosity was present. This has been termed the *founder principle* (*111*). Since speciation represents the accumulation of novel genetic (and phenotypic) features, this is a conceivable route for the incorporation of neutral substitutions in a new species; however, the amount of such substitution is difficult to evaluate (*189*).

A third view is that neutral substitutions became incorporated simply through genetic drift. This may be defined as representing a fluctuation in the frequency of allelic genes of apparently equal survival value de-

188. It should be emphasized that all reports of identical function of homologous proteins of different species are based on tests performed *in vitro;* for example, although all eukaryote cytochromes c appear to be equivalent in activity and can function in the oxidase systems of other species (*21*), there is no evidence of identical activity in the whole organism.

189. "How important such populations [few original founders] are for speciation and ultimate evolution is still rather obscure. . . . it may clarify evolutionary discussions if authors would refrain from invoking 'genetic drift' as a cause of evolution" (*111*, p. 214).

pending on factors of isolation, rates of reproduction of the various types, etc.

If two structural genes producing proteins differing in one or a few residues really were *exactly* equivalent in survival value, no such neutral mutations would ever disappear from the population. *All such neutral types should always be present.* Thus, we should expect considerable microheterogeneity of proteins; i.e., at all loci in homologous proteins where species differences have been observed in diploid species, two or more residues should be encountered of equivalent survival value. There are no data to support this concept at the present time; indeed, all of the evidence points to the opposite conclusion. The protein that has been examined in more individuals of a single species than any other is human hemoglobin. Most of the abnormal hemoglobins that have been described were detected solely because of the "abnormal" properties and many of these showed altered functional properties that were detected in the hematology clinic. Lehmann and Huntsman (*190*), in their review of human hemoglobins, point out that the only variant hemoglobins found in large numbers of individuals are β-chain variants, all of which in the heterozygous state confer enhanced survival value in malignant malarial infections; these are hemoglobins S, C, D_{Punjab}, and E. Thus, these β-chain variants (and various thalassemias) persist because of the selective advantage in the heterozygous condition. Other abnormal hemoglobins are exceedingly rare and many have been detected only in a single individual or family. Thus, in this most thoroughly studied of all proteins, there is no evidence for genetic drift as judged by microheterogeneity involving neutral substitutions.

The same conclusion can be reached on the basis of the structural studies on many proteins for which large pools of individual animals were used. Heterogeneity has been only rarely detected (*191*).

In the study of many cytochromes in the author's laboratory, special care was taken to determine whether such microheterogeneity existed. Only in the case of human cytochrome c was a single peptide found which indicated heterogeneity and the low yield suggested that a single heterozygous individual, from a sample comprising approximately 70 individuals, had a different cytochrome (*24*). The change involved a substitution of leucine for methionine at residue 65. It is striking that this

190. H. Lehmann and R. G. Huntsman, "Man's Haemoglobins." North-Holland Publ., Amsterdam, 1966.

191. Where such heterogeneity has been documented, further investigation has revealed the presence of allelomorphic genes, e.g., for the α chain of rabbit hemoglobin (*192*).

substitution has not been found in any other cytochrome c to date. To assess possible heterogeneity, some 16 samples of cytochrome c were isolated from individual human hearts by Matsubara and Smith (*193*). All of these preparations of cytochrome c showed the same chymotryptic peptide pattern and gave the identical amino acid composition. Admittedly, this is a small sample of the human population but it did include individuals of European, Asiatic, and African origin, and some individuals who died of unexplained cardiac failure (*193*). Similar observations have been noted for equine and human cytochromes c by Margoliash (*194*).

Interestingly enough, samples of the above size have been sufficient to detect heterogeneity in certain proteins. Such allotypes have been reported for bovine carboxypeptidase A (Section III,D,4), in bovine β-lactoglobulin (*180*), in some ungulate and other hemoglobins (*18, 192*), in immunoglobulins (*93*), etc. (*195*). It is noteworthy that bovine insulin, produced commercially on a large scale in many parts of the world and investigated in many laboratories, has, to the best of our knowledge, never been reported to show heterogeneity, although mammalian insulins do show species differences and apparently equivalent hormonal activity in the same test animal. Rat insulin (*187*) does contain two distinct allotypes, but this is the only instance in which this is presently known to occur.

Thus, as in the case of the abnormal human hemoglobins of wide distribution, it would appear that Darwinian survival value is the predominant factor in the spread of a point mutation or other genetic change. Proteins with only equivalent survival value could spread slowly only into small clones and could not overtake or replace the predominant "wild-type" protein. Thus, within a species, class, or order, stability of sequence will most commonly indicate satisfactory adjustment of the structural features of the protein to its function and to its environment.

In any case, the introduction of "neutral" substitutions into a protein represents only a minor phase of the evolution of proteins and cannot offer the changes in properties that reflect novel adaptations. This point will be discussed further below (Section IV,E).

192. T. Hunter and A. Munro, *Nature* **223**, 1270 (1969).

193. H. Matsubara and E. L. Smith, unpublished studies (1962).

194. E. Margoliash, *Brookhaven Symp. Biol.* **21**, 258 (1969).

195. It is noteworthy that the domestic cow is one of the few instances in which there appears to have been repeated domestication [see Mayr (*111*)]. Could the allotypes of carboxypeptidase A, of β-lactoglobin, and of certain hemoglobins result from di- or polyphyletic origin followed by extensive cross-breeding?

Recently, a number of authors have been concerned with the problem of neutral mutations and genetic drift in protein structure, e.g., Epstein (*196*), Kimura (*197*), King and Jukes (*198*), Arnheim and Taylor (*199*), and Margoliash *et al.* (*200*).

Inasmuch as King and Jukes (*198*) have been strong proponents for the importance of neutral mutations in homologous proteins, it is desirable to consider some of their arguments. In brief, they conclude that 74 to 81 sites in cytochrome c are variable (*200a*) "and substitutions in the variable sites . . . seem to follow the Poisson distribution, which would indicate that there is very little restriction on the type of residue that can be accommodated at most of the variable sites." They suggest only seven specific sites, however, that are "neutral rather than adaptive, and that many other neutral substitutions exist in the cytochromes c, particularly at sites where there are many interchanges." They conclude that "the fact that some variable amino acid sites are more subject to change than others in a set of homologous proteins is an expression primarily of the variable nature of point mutations and only secondarily of protein function." These authors also note that "the correspondence between the frequencies of the 20 amino acids in proteins and their frequencies in the genetic code table are indirect evidence for the preponderance of selectively neutral changes in evolution at the molecular level."

It is obvious that these conclusions, only briefly stated here, are not in agreement with many of the viewpoints expressed in this chapter as well as in previous publications by the present author (*12*) and many others [see, for example, Mayr (*111*)]. The views already presented will not be recapitulated, but a few points will be made more explicit. What King and Jukes are discussing seems to be not evolution at all but only the species character of a homologous protein. They concede that natural selection operates on deleterious mutations, which cannot be observed since such changes have been eliminated, but they do not consider that observable changes must also be subject to natural selection. Indeed, their view leads to the anti-evolutionary concept that all pro-

196. C. J. Epstein, *in* "Homologous Enzymes and Biochemical Evolution" (N. van Thoai and J. Roche, eds.), p. XXI. Gordon & Breach, New York, 1968.

197. M. Kimura, *Nature* **217,** 624 (1968).

198. J. L. King and T. H. Jukes, *Science* **164,** 788 (1969).

199. N. Arnheim and C. E. Taylor, *Nature* **223,** 900 (1969).

200. E. Margoliash, W. M. Fitch, and R. E. Dickerson, *Brookhaven Symp. Biol.* **21,** 259 (1969).

200a. Note that 35 residues are constant in the 104 of vertebrates, thus only 69 variable sites are known (Fig. 3).

teins (enzymes) were created almost initially with full functional capacity and would continue to exist without reflecting further refinement or improvement in adaptation to the internal and external environments of the species. This bias is reflected in the amazing statement that "the majority of the interspecies differences between the hemoglobins are functionally neutral." This is contrary to the well-established data that among vertebrate hemoglobins: (a) the apparent dissociation constants for O_2 binding (P_{50}) differ in different animals (*201, 202*), and that P_{50} varies with the size of the mammal (*203*); (b) the magnitude of the Bohr effect differs in different species (*204*); (c) in many species, fetal and adult forms differ substantially in properties, in some cases in the free proteins, in others in the specific erythrocyte environments (*205*); and (d) in different groups, regulatory effectors in the erythrocytes differ—2,3-diphosphoglycerate in mammals, phytate in birds, and many other phosphate compounds—presumably in all species (*206*) these effectors aid in the rapid unloading of oxygen from oxyhemoglobin, etc.

Part of their argument is that changes in DNA are the prime reasons for variation in protein structure. This is of course true, but what remains in the DNA and in the protein is subject to natural selection. The "correspondence between the frequencies of the 20 amino acids in proteins and their frequencies in the genetic code table" seems to this author to be based on a false assessment (*207*). This is tantamount to saying that because the total amino acid compositions of some species are similar to those of many other species that these organisms must be alike. What appears more meaningful from an adaptive view is:

(1) Proteins of different functions not only have different compositions but also totally different sequences. Compare collagens, insulins, equine cytochrome c, wool keratin, histones, bovine ribonuclease, etc.; in each of these proteins one or more amino acids is totally absent, and the compositions are far from being in statistical agreement with codon frequencies.

(2) Despite the universality of the genetic code, the total amino acid composition of plants differs from that of animals, e.g., the paucity of lysine and methionine, and the high content of arginine in domesticated plants present well-known problems in nutrition.

201. A. C. Redfield, *Biol. Rev.* **9,** 175 (1934).

202. C. L. Prosser and F. A. Brown, Jr., "Comparative Animal Physiology," 2nd ed., pp. 210–211. Saunders, Philadelphia, Pennsylvania, 1961.

203. K. Schmidt-Nielsen and J. Larimer, *Am. J. Physiol.* **195,** 424 (1958).

204. See Epstein (*196,* p. 215).

205. C. Manwell, *Ann. Rev. Physiol.* **22,** 191 (1960).

206. R. Benesch and R. E. Benesch, *BBRC* **26,** 162 (1967); R. Benesch, R. E. Benesch, and C. I. Yu, *Proc. Natl. Acad. Sci. U. S.* **59,** 526 (1968).

(3) The DNA of bacteria differs markedly in relative amounts of G + C and A + T. Sueoka (*208*) has shown that while the contents of some amino acids in these organisms show no significant trend, the relative amounts of alanine, lysine, glycine, isoleucine, and others show positive or negative linear correlation with the G + C content. Thus, despite the universality of the code, *the amino acid composition mirrors the nucleotide composition of the DNA* and not the relative frequency of the codons for these amino acids.

(4) Within a series of homologous proteins, e.g., cytochrome c, there appear to be trends of change in amino acid composition. The content of isoleucine in higher vertebrate cytochromes is 6–8 residues whereas in Ascomycetes and wheat germ, it is half as much, 3–4 residues; the lysine content of all mammalian and bird cytochromes is 18–19 residues; in the insects, Ascomycetes, and plants it is 12–16 residues. We need not know the reasons for these trends to recognize their existence.

(5) Animals require many preformed amino acids in the diet whereas plants and many microorganisms can synthesize all the amino acids. The advantages of loss of function are discussed below (Section IV,D). When "neutral" substitutions are possible in specific loci of proteins, it would be advantageous for the organism to utilize preformed amino acids available from the diet rather than those that require energy for synthesis. Although the presumed gain would be small for the synthesis of some proteins, it could be substantial for those proteins synthesized in large quantities or for all the proteins of the organisms. This may be one factor which could aid in explaining the substantial differences in the gross amino acid compositions of animal as compared to plant and bacterial proteins.

(6) One reservation regarding "neutral" mutations, as reflected in protein structure and function, must be considered, namely, that certain nucleotide bases are preferable in the sequence for the structure of the DNA or the messenger RNA. Adams *et al.* (*209*) have reported that, for a portion of the nucleotide sequence of the RNA phage R17, 19 of the possible 24 base pairs can form hydrogen bonds and this is postulated to provide a mechanism for close packing of the polynucleotide in the complete viral assembly. If such factors are of general importance in the chromosome structure of eukaryotes, as well as the structure of viruses,

---

207. The selection of 53 vertebrate proteins certainly reflects a bias against the remainder of the living world.

208. N. Sueoka, *in* "Evolving Genes and Proteins" (V. Bryson and H. J. Vogel, eds.), p. 479. Academic Press, New York, 1965; *Proc. Natl. Acad. Sci. U. S.* **47**, 1141 (1961); *Cold Spring Harbor Symp. Quant. Biol.* **26**, 35 (1961).

209. J. M. Adams, P. G. N. Jeppesen, F. Sanger, and B. G. Barell, *Nature* **223**, 1009 (1969).

mutations could be neutral for the protein but not necessarily for the DNA or RNA. Indeed, it was noted (*209*) that the phenomenon of base pairing may explain the use of alternate codons for the same residue and an advantage of degeneracy in the code. Thus the Darwinian survival value would result from structural factors in the DNA or the messenger unit rather than the protein; however, this would still not be a reflection of statistically random fluctuations mirroring the numbers of codons.

Perhaps the best way of concluding this discussion is to quote the two statements to which King and Jukes take exception.

> The consensus is that completely neutral genes or alleles must be very rare if they exist at all. To an evolutionary biologist, it therefore seems highly improbable that proteins, supposedly fully determined by genes, should have nonfunctional parts, that dormant genes should exist over a period of generations, or that molecules should change in a regular but nonadaptive way. . . . [natural selection] is the composer of the genetic message, and DNA, RNA, enzymes, and other molecules in the system are successively its messengers (*210*).

and

> One of the objectives of protein chemistry is to have a full and comprehensive understanding of all the possible roles that the 20 amino acids can play in function and conformation. Each of these amino acids must have a unique survival value in the phenotype of the organism—the phenotype being manifested in the structures of the proteins. This is as true for a single protein as for the whole organism (*12*).

B. Convergence

It has already been noted (Section II) that convergence is a familiar phenomenon in considering physiological and anatomical features of evolution. It may be expected, however, that unless a homologous protein is studied from a rather large number of species, it will be difficult or impossible to assess whether similar sequences are due to retention of homology (lack of divergence) or to convergence. There is some evidence, however, which suggests convergent features in the evolution of the sequences of the cytochromes c with regard to the presence of certain residues in distantly related species. Also, the slightly greater resemblance of the sequence of the wheat germ protein to those of the primates than of the former to those of invertebrates and Ascomycetes may have been due to convergence (Section II).

Convergence, as exemplified by the evolution of a similar enzymic

210. G. G. Simpson, *Science* **146,** 1535 (1964).

mechanism with nonhomologous sequences and conformations, is shown by the subtilisins and the pancreatic proteinases; this is discussed below in Section IV,C.

C. Parallelism in Evolution

The phenomenon of parallelism in evolution is well known and is usually defined as representing the evolution of similar functional adaptation based on different physiological and morphological devices, for example, the development of wings in insects and birds and the development of fins and flippers in aquatic vertebrates.

From the viewpoint of enzyme chemistry it is clear that different structures and mechanisms have developed to perform the same biochemical function but not necessarily the same physiological role [see, for example, Neurath *et al.* (*19*)]. One of the best known situations of this sort is represented by the proteolytic enzymes. It has long been recognized that there are at least four different classes of proteolytic enzymes in terms of the structures in the active sites (*211*). One group includes proteinases of the serine-histidine type, including chymotrypsin, trypsin, subtilisins, and others; these are usually recognized by reaction with such inhibitors as diisopropylfluorophosphate and similar reagents. A second type comprises those enzymes in which inhibition is produced by a variety of agents which react with thiol groups, including mercurials, alkyl halides, and oxidizing agents. Among the best known enzymes of this sort are papain, ficin, and various cathepsins of animal tissues. A third type includes the enzymes containing an essential metal ion. Such enzymes are, of course, strongly inhibited by reagents which bind to or remove the metal ion from the protein; these include various carboxypeptidases, aminopeptidases, and others. A fourth group is typified by pepsin, rennin, and gastricsin in which it is now clear that at least one carboxyl group is a participant in the active site. There may, of course, be additional as-yet unrecognized classes of proteolytic enzymes.

All of these enzymes catalyze peptide bond hydrolysis, but some may, in addition, catalyze transfer reactions as their major physiological function (*212*). Here is a case in which parallelism is clearly manifested at the functional level. Inasmuch as essentially the complete sequences of

211. E. L. Smith, "The Enzymes," 2nd ed., Vol. 4, p. 1, 1960.

212. A. Meister, "Biochemistry of the Amino Acids," 2nd ed., Vol. 1, p. 473. Academic Press, New York, 1965.

trypsin (*124*), chymotrypsin (*126*), elastase (*132*), subtilisins (*61*), papain (*149, 150*), carboxypeptidase A (*159*), and partial sequences of pepsin (*213–215*) and rennin (*155a*) are known, it is evident that there is no significant homology in either the primary structures or the conformations of these four different types of proteolytic enzymes.

It is evident that the four classes of proteinases discussed above are unrelated. Within a class, however, parallelism has also occurred. At present, the most convincing example of this is represented by the bacterial subtilisins and the pancreatic proteinases of the trypsin-chymotrypsin type (*61*). These two groups of enzymes manifest no evidence of similarity in primary structure (*61*). Furthermore, the subtilisins are completely lacking in disulfide bridges, whereas the pancreatic proteinases and their homologs in lower organisms all contain disulfide bridges. In addition, the three-dimensional structures as deduced from X-ray crystallography, clearly indicate that chymotrypsin, for example, possesses an entirely different structure from that of subtilisin BPN′ (*64*). From this evidence one must conclude that the two structures have evolved from different precursor polypeptide chains. Nevertheless, the two classes of enzymes show an interesting convergence since the same mechanism must have developed independently. The similarities of mechanism are evident in the conformation of the active sites insofar as the active histidyl residue is situated between aspartyl and seryl residues. The similarity, of course, is evident from observations that diisopropylfluorophosphate and similar reagents inhibit and react with seryl residues in both types and from findings that the pH activity curves reflect the participation of a histidyl residue (*62*). The latter has also been shown by the use of covalently bound inhibitors (*62, 216*). Here is an example of *parallelism* in developing the same catalytic activity from different sequences and conformations, on the one hand, and *convergence* in evolving the same mechanism with the participation of the identical three residues, on the other.

In addition to the proteinases there is also evidence of evolutionary parallelism among other groups of enzymes which possess the same cata-

213. W. H. Stein, *in* "Structure-Function Relationships of Proteolytic Enzymes" (P. Desnuelle, H. Neurath, and M. Ottensen, eds.), p. 253. Munksgaard, Copenhagen, 1970.

214. B. Keil, *in* "Structure-Function Relationships of Proteolytic Enzymes" (P. Desnuelle, H. Neurath, and M. Ottesen, eds.), p. 102. Munksgaard, Copenhagen, 1970.

215. R. S. Bayliss, J. R. Knowles, and G. B. Wybrandt, *BJ* **113,** 377 (1969).

216. F. S. Markland, E. Shaw, and E. L. Smith, *Proc. Natl. Acad. Sci. U. S.* **61,** 1440 (1968).

lytic function. Three classes of lysozymes are now recognized. The first is exemplified by the lysozyme of hen egg white for which the sequence and conformation are fully known. This type of enzyme is also recognized as occurring in homologous form in a wide variety of vertebrates as based on immunological cross-reactions, amino acid compositions, and in a few instances by complete sequences (*171*). As an example, the partial sequence of human lysozyme (*217*) shows strong homology to hen egg white lysozyme. A second class of lysozymes includes those of the various bacteriophages (*218, 219*). The complete sequences of the lysozymes of phages T4 and T2 show no homology to and lack the disulfide bridges of the vertebrate lysozymes. A third class is represented by the lysozymes of certain plants; these enzymes differ in many properties and although only partial sequences are known, there is no evident homology in size, sequence, or functional attributes to the enzymes of the other two types. Best recognized of these enzymes at the present time are the lysozymes of papaya (*220, 221*) and Ficus (*222*).

In the same manner, the sequences of pancreatic ribonuclease (*65*) and the nucleases of *Staphylococcus aureus* (*223*) and *Aspergillus oryzae* (*224*) show no homology in structure.

There are, of course, many other groups of enzymes that on the basis of very incomplete evidence may manifest the same kind of parallelism, although it is clear, for example, that among the dehydrogenases, glyceraldehyde-3-phosphate dehydrogenase bears some resemblance in amino acid sequence to the alcohol dehydrogenases (see Section III,D). It is premature, however, to infer that all other dehydrogenases are homologous in structure and have evolved from a common precursor.

There are two classes of fructose-1,6-diphosphate aldolases (*225*): One type—present in bacteria, yeast, other fungi, and blue-green algae—requires a divalent metal ion for activity; the other type—present in animals, plants, and green algae—exhibits no metal ion requirement but

217. S. Kammerman and R. E. Canfield, *Federation Proc.* **28**, 343 (1969).

218. A. Tsugita and M. Inouye, *JMB* **37**, 201 (1968).

219. M. Inouye and A. Tsugita, *JMB* **37**, 213 (1968).

220. E. L. Smith, J. R. Kimmel, D. M. Brown, and E. O. P. Thompson, *JBC* **215**, 76 (1955).

221. J. B. Howard and A. N. Glazer, *JBC* **242**, 5715 (1967); **244**, 1399 (1969).

222. A. N. Glazer, A. O. Barel, J. B. Howard, and D. M. Brown, *JBC* **244**, 3583 (1969).

223. H. Taniuchi, C. B. Anfinsen, and A. Sodja, *JBC* **242**, 4752 (1967).

224. K. Takahashi, *JBC* **240**, PC4119 (1965).

225. W. J. Rutter *in* "Evolving Genes and Proteins" (V. Bryson and H. J. Vogel, eds.), p. 279. Academic Press, New York, 1965.

involves Schiff base formation of the substrate with a lysyl residue (*226*). Interestingly, *Euglena* and *Chlamydomonas* possess both types of enzymes. The two types of enzymes exhibit marked differences in properties, and the evolution of these appears to represent an interesting type of parallelism in function (*225*). The muscle and liver enzymes of the rabbit show relative differences in specificity and in kinetic constants; these two types of the animal enzyme show partially similar peptide patterns, suggesting that both originated from a common precursor but have diverged in properties (*225*).

At present, we lack information regarding many of the more complex types of enzymes. Complete sequences of enzymes utilizing a similar coenzyme or prosthetic group are available in only a few instances. For the hemoproteins, the class represented by myoglobins and hemoglobins shows no relationship whatsoever in either sequence or conformation to the eukaryotic cytochromes c (*7, 8*) nor do either of these two types of hemoproteins show any resemblance to the primary amino acid sequence of catalase (*227*). The cytochrome c_2 of *Rhodospirillum rubrum,* a photosynthetic bacterium, does show considerable homology with the eukaryotic cytochromes c (*227a*). It is noteworthy, however, that the most striking feature of cytochrome c, the constant sequence of residues 70 through 80 (Section II) is only partially present in c_2. The sequence of *R. rubrum* c_2 also shows some resemblances to that of the cytochrome c of *Pseudomonas fluorescens* (*227a*).

Undoubtedly, many more examples of parallelism will be discovered in terms of utilization of similar residues for the same mechanism. As already discussed (Section I), the fact that so few amino acid residues can participate actively in enzymic catalysis ensures that parallelisms in mechanisms will have developed and that convergence will also have occurred. Utilization of a similar prosthetic group, such as heme, for a variety of distinct functions is not readily classifiable in classic biological terms.

D. Loss of Enzymic Function and Survival Value

Many auxotrophic microorganisms and all higher animals are dependent on the environment for certain organic nutrients which are provided

226. B. L. Horecker, P. T. Rowley, E. Grazi, T. Cheng, and O. Tchola, *Biochem. Z.* **238,** 36 (1963).

227. W. A. Schroeder, J. R. Shelton, J. B. Shelton, B. Robberson, and G. Apell, *ABB* **131,** 653 (1969).

227a. K. Dus, K. Sletten, and M. D. Kamen, *JBC* **243,** 5507 (1968).

by the synthetic activities of plants and other organisms. These substances include many amino acids, vitamins, and certain fatty acids, all of which are readily available from dietary sources. As Lwoff (*228*) first noted in his studies of parasitism, the loss of synthetic ability must be advantageous to such organisms since this will permit utilization of chemical energy or carbon compounds for other needs. This view has been convincingly demonstrated by the work of Zamenhof and Eichhorn (*229*). They grew a mixture of two strains of *E. coli,* one positive and one negative for the enzyme tryptophan synthetase, in the presence of tryptophan. The two strains differed only in a point mutation for the active and inactive enzymes. The results clearly showed much more rapid growth for the tryptophan-negative strain under these conditions. Even greater advantage was found for a strain negative for anthranilate (a precursor earlier on the pathway for tryptophan formation) grown with wild type in the presence of tryptophan. Thus, the more biosynthetic steps eliminated, the more advantage for the defective strain in competition with the wild type. The effects are undoubtedly complex in that the presence of tryptophan represses synthesis of unrequired intermediates by feedback inhibition and by gene repression depresses synthesis of the enzymes as well.

In cases like the above, there is a definite survival value for the loss of an unnecessary function since the diet provides the required compounds. Interestingly enough, however, there is no information available as to the fate in evolution of such "functionless" genes. One possibility is that these were lost, possibly by chain-terminating processes (Section III,B), since not only would there be no advantage to retention by the organism of genes producing inactive proteins but also the organism would indeed gain by no longer utilizing energy or valuable intermediates to make the DNA of such genes, the messenger RNA's, and the proteins. This has been found to be the case for some tryptophan-requiring wild-type strains of *Streptococcus faecalis,* where one or more genes for tryptophan biosynthesis are missing (*229*). Another possibility for the fate of such genes is that they could mutate to provide proteins with new functions which did possess survival value, but there is presently no evidence for such an event.

An interesting situation of loss of a specific protein is represented by some fish of the order Chaenichthyidae which live in Antarctic waters (*230*). These poikilotherms, which do not have any hemoglobin whatso-

228. A. Lwoff, "L'evolution physiologique, étude des pertes de fonctions chez les microorganismes." Hermann, Paris, 1944.

229. S. Zamenhof and H. H. Eichhorn, *Nature* **216,** 456 (1967).

230. J. T. Rund, *Nature* **173,** 848 (1954).

ever, live at temperatures near 0° and the low rate of metabolism at these temperatures is satisfied by the oxygen carried in the plasma. These fish are, of course, confined to such waters, since at higher temperatures their metabolic requirements and the lower solubility of oxygen would render their existence impossible. Other members of this order, living in warmer waters, do possess hemoglobin. Clearly, ahemoglobinemia could be viable and possess survival value only in this unusual ecological situation.

The loss in higher animals of synthetic ability for vitamins and certain amino acids is reflected in the loss of the entire synthetic pathways for these compounds. Presumably, the losses of all these enzymes are old events since the dietary requirements are shared by many classes and phyla. Only in the case of ascorbic acid does the loss of an enzyme essential in the synthetic pathway appear to be a relatively recent occurence; the requirement for this compound in the diet is known only for man and other primates and the guinea pig. Only one enzyme is known to be missing, L-gulonolactone oxidase, which converts L-gulonolactone to 3-keto-L-gulonolactone (L-xylohexulonolactone) (*231*).

Although insects and mammals share many dietary requirements for amino acids and vitamins, the need for preformed carnitine (*232*) and cholesterol (*233*) and the absence of the entire pathway for synthesis of the latter (*234*), in at least some insects, suggest the general importance of loss of function in complex organisms.

E. Development of Novel Enzymic Properties

The most important and interesting problem concerning the evolution of enzymes is how new types of enzymes originated. This is, of course, a corollary of the more general question of the emergence of all evolutionary novelties. Several cases have been presented of changes in enzyme function, particularly in Section III. Most of the known situations are explicable on the basis of gene duplication, followed by alterations in sequence that have led to a new specificity, e.g., the ethanol dehydrogenases, pancreatic proteinases, and carboxypeptidases A and B.

At this point, it is useful to recall that a point mutation within a codon can yield nine other codons, some of which may be for the same amino acid, some for conservative replacements, and some for radical replace-

231. I. B. Chatterjee, G. C. Chatterjee, N. C. Ghosh, J. J. Ghosh, and B. C. Guha, *Biochem. J.* **74,** 193 (1960); **76,** 279 (1960).

232. G. Fraenkel, *Ann. N. Y. Acad. Sci.* **77,** 267 (1959).

233. H. Lipke and G. Fraenkel, *Ann. Rev. Entomol.* **1,** 17 (1956).

234. J. Clark and K. Bloch, *JBC* **234,** 2578 (1959).

ments. Clearly, the first type would yield no change in protein sequence and, in general, the second would be expected only rarely to produce any major change in enzymic properties. In contrast, radical substitution could produce not only deleterious changes but also could result in modification in specificity and other properties. Thus mutations leading to substitution of radically different amino acid side chains would also appear to offer the greatest opportunity for those adaptive changes which could result in enhanced survival value, and it is such rare changes that are of great importance in evolution (*11*). Two instances may be recalled as illustrating this concept. At the substrate binding sites, the critical difference between trypsin and chymotrypsin appears to be the presence of aspartyl vs. seryl, respectively (Section III,D,1), a radical difference. Similarly, in the two ethanol dehydrogenases some of the known differences in sequence are of the radical type (Section III,D,5). These examples are satisfying insofar as they reveal how specificity can change, but in each of these situations there has been no modification of the type of reaction catalyzed or in the mechanism of the reaction (*234a*). The evolution of other kinds of alterations of enzyme behavior—in regulation, in mechanism, in development of completely novel reactions, etc.—remains to be elucidated. Although evidence is still fragmentary, a few cases, including some already mentioned, will be reviewed briefly in order to indicate possible future lines of progress.

1. *Intrachain Repetitions*

As already noted, a feature of the sequences of some proteins is the presence of repetitive sequences yielding evidence that longer polypeptide chains evolved from smaller ones (Section III,C). What is unknown, however, is whether or not the smaller ancestral chains possessed the same function. Hopefully, some of the "fossilized" smaller chains may still remain in some organisms. Discovery of these would be of great value in understanding the development of such proteins as the immunoglobulins with their repetitions of sequences in both light and heavy chains.

2. *Multichain Enzymes*

Another aspect of protein evolution concerns the origin of those enzymes with multiple peptide chains. The simplest situation is represented by those enzymes which possess identical subunits. Clearly the same

234a. Many of the hormones that are related in structure (Section III,D) do show marked differences in physiological function, but the nature of the chemical action is for the most part unknown.

gene or several identical genes can determine repetitively the formation of the same chain and these subunits, without further genetic intervention, aggregate to form the holoenzyme. This is clearly the case for glyceraldehyde-3-phosphate dehydrogenase (*73*), alcohol dehydrogenases (*162*), glutamate dehydrogenase (*235, 235a*), lactate dehydrogenases (*170*), catalase (*227*), and others. Since "simple" enzymes consist of only one chain, the novel development was the substitution of residues responsible for the binding forces leading to the defined aggregation of the subunits.

Inasmuch as single chain enzymes do not appear to be subject to allosteric regulation, the development of multichain enzymes represented a major step in the evolution of regulatory processes. This may be illustrated by vertebrate myoglobin which consists of a single chain and does not exhibit an S-shaped curve for oxygen binding or a Bohr effect. Lamprey hemoglobin in the deoxygenated state is a dimer with identical chains, but its oxygen-binding curve is S shaped and the protein dissociates into its two subunits on oxygenation. In contrast, other vertebrate hemoglobins are tetramers consisting of two types of chains which originated after duplication of a common ancestral gene (*114a*). All of these exhibit both an S-shaped curve and a Bohr effect.

Some enzymes contain two types of chains which have totally different structures and possess different functional properties. Aspartate transcarbamylase (*236*) is a good example inasmuch as one chain possesses the specific catalytic activity and the other functions as a regulatory subunit with a high binding affinity for cytidine triphosphate, a negative feedback inhibitor. Here functional regulation is achieved by two distinct types of chains rather than one, as in glutamate dehydrogenase, for example.

A somewhat different situation is exhibited by lactose synthetase, since the B chain (α-lactalbumin) is catalytically inert but was derived presumably by duplication of a gene related to lysozyme. In this case, the inherent transferase activity is in the A subunit, but its specificity is modified by the presence of the B subunit (Section III,D). Here the novel specificity of the mammary gland enzyme is derived by the combination of two apparently unrelated kinds of subunits. This is also the situation with the tryptophan synthetase of *E. coli*, but the further evolution of the enzyme in eukaryotes led to their fusion into a single peptide chain (Section III,D).

235. C. Frieden, *JBC* **237,** 2396 (1962).

235a. E. Appella and G. M. Tomkins, *JMB* **18,** 77 (1966); H. Eisenberg and G. M. Tomkins, *ibid.* **31,** 37 (1968).

236. J. C. Gerhart and H. K. Schachman, *Biochemistry* **4,** 1054 (1965).

The examples mentioned all indicate that many major alterations in properties of enzymes developed not only by modifications within a single polypeptide chain but also by novel combinations of apparently unrelated types of proteins. It may be remarked that among presently known enzymes, the majority of single chain, small enzymes are mainly extracellular in function, whereas among known intracellular enzymes, most seem to be of the multichain variety with identical subunits in some cases and different subunits in others (*237*).

F. Epilogue

From what is presently known concerning protein structure it is now clear that almost all of the variety of genetic events originally observed by classic studies of discrete genes can now be deduced as having occurred also in the cistrons for single polypeptide chains. In other words, the presently observed sequences in polypeptide chains can be explained on the basis that individual genes have undergone point mutations, extension by repetition of sequence, chain shortening, deletions, additions, translocations, duplications, etc. (Section III). Again, as in classic genetics, most of such alterations can be presumed to have been lethal, only occasionally "neutral," and even more rarely, possess enhanced survival value. It is the last kind of change, of course, that we define as "progressive evolution" in the Darwinian sense.

There have been discussions in the past as to whether "mutations" have occurred at a sufficient rate to have accounted for the evolution of all forms of life since its origin, and controversy as to whether "mutations" could lead in reasonable time to truly novel functions and devices. Many of these discussions and calculations were presented before the chemical nature of the gene and of point mutations was understood. Inasmuch as all presently existing organisms appear to possess fundamentally similar types of structures and mechanisms for protein synthesis, there would appear to have been only relatively minor evolutionary changes in this regard since these mechanisms were elaborated. This implies that major aspects of evolutionary developments should be ascribed to the nature and numbers of structural genes—those coding for proteins and hence for phenotypes.

237. These generalizations are highly tentative inasmuch as the kinds of enzymes which function intra- and extracellularly are so different in types of reactions catalyzed. Of course, the zymogens of pancreatic or gastric origin are single chain but may be converted to two or more chains by intrachain proteolytic scission; these do not represent multichain enzymes in the same sense as the various dehydrogenases, etc.

If these views represent a reasonable approximation, calculations of mutation rates, as point mutations, as the sole or major method of variation in the genome, would no longer possess the significance formerly ascribed to them. The origin of the myriads of different proteins (and enzymes) of all living organisms need not be explained only on the basis of point mutations important as these are. Other processes of change in protein structure, such as those discussed in the preceding pages, offer at least equal and perhaps greater opportunities for the development of novel metabolic and morphological characteristics. Presently, there appears to be no useful way to calculate rates by which chain extensions, deletions, repetitions, duplications, etc., can occur, or when two distinct types of protein subunits will associate to produce a novel protein.

Although the importance of gene duplication for the evolution of new function has long been recognized, its prevalence is only now becoming apparent. The increasing volume of reports of the discovery of new isoenzymes is a partial recognition of this phenomenon, but duplicate genes which still produce the identical polypeptide chain also represent a potential for further evolution. Indeed, drastic changes in structure of a single gene would probably be almost invariably lethal, but when duplicates exist the chances of survival for such a change could be significant in a "redundant" gene until a novel function appeared. Direct evidence for the existence of duplicated identical genes is sparse (*238*), but the discovery of large amounts of "redundancy" of structure in DNA of eukaryotic organisms (*241*) suggests that even if only a part of this "redundancy" is concerned with structural genes (*242*) we are at the threshold of new approaches to a better understanding of evolutionary

238. The observation (*239*) of new strains of *E. coli* which possessed duplicated genes for β-galactosidase when these organisms were grown in a medium low in lactose suggests a possible selective advantage for organisms with gene duplication. These strains developed as much as four times the amount of enzyme as the original strain. Inasmuch as this β-galactosidase consists of four identical chains (*240*) such single gene heterosis is favorable for enhancing the rate of synthesis of these long chains and would also support the view of Spofford (*112*) that this type of heterosis would increase the rate of synthesis of identical chains of multichain enzymes.

239. A. Novick and T. Horiuchi, *Cold Spring Harbor Symp. Quant. Biol.* **26,** 239 (1961).

240. J. L. Brown, S. Koorajian, J. Katze, and I. Zabin, *JBC* **241,** 2826 (1966).

241. R. J. Britten and D. E. Kohne, *Science* **161,** 529 (1968).

242. It is recognized that much of this redundancy in DNA is "ribosomal DNA" (*243*); however, what portion of the redundant DNA codes for structural genes remains to be determined.

243. M. Birnstiel, M. Grunstein, J. Speirs, and W. Hennig, *Nature* **223,** 1267 (1969).

potential and rates. Nevertheless, not all genes can exist in multiplicated form as witnessed by the laws of Mendelian segregation and the existence of homozygous lethals.

The ability of proteins and other cellular components to associate spontaneously into structures with novel functions indicates a potential for evolution which classic theory could not readily assess, although it was recognized that changes in single genes had multiple effects. Thus, our present and still very fragmentary view indicates that many kinds of changes can occur in the genome, but all of these still represent only one aspect of the evolutionary potential. The ability of all varieties of macromolecules, once made, to form complex associations with each other and with small molecules as well, represents another level from which to view the evolution of enzymes and their regulation.

7

The Molecular Basis for Enzyme Regulation

D. E. KOSHLAND, JR.

I. Introduction

One of the axioms of basic biological research is that the correlation of structure and function will lead to insight into the fundamental processes of nature. The early pioneers in biochemistry were as aware of this principle as the current disciples of "molecular biology." The structures of the simple molecules—the carbohydrates, the amino acids and the nucleotides—had to be established, however, before the structures of the complex macromolecules. If there is any fundamental difference in the era of molecular biology and the era of "classic biochemistry," it is the result of the availability of macromolecular structures today and the fact that the emphasis in correlation of structure and function has moved to these larger structures. Just as the structures of the low molecular weight metabolites were essential for the elucidation of the metabolic pathways, so the structures of these macromolecules are necessary to understand the genetic code, the behavior of membranes, and the regulation of enzymes.

It is in the latter area, the control of the dynamic processes through the structure of the protein molecule, to which this chapter will be devoted. The value of developing models for enzyme regulation is largely accepted because of the success of previous structure–function correlations. Aside from this historical precedent, however, there are two major reasons why the delineation of such models is of key importance. In the first place the kinetics of regulatory proteins are unusually complex. The simple Michaelis–Menten kinetics cannot be assumed as a necessary concomitant of any new enzyme and in general the kinetics of many proteins important in regulation deviate from the Michaelis–Menten equation. This means that the addition of an activator or inhibitor can change not only the activity of the enzyme but also the entire shape of the velocity-substrate curve. Moreover, each enzyme may be affected by more than one inhibitor and activator making literally impossible the description of all possible combinations on an empirical basis. It will be necessary to understand the basic kinetic features of all effectors on each enzyme and to integrate the kinetics of each control enzyme into an overall picture which will describe regulation in the cell.

In the second place the search for appropriate models to explain allosteric control inevitably leads to a closer examination of the structure of a protein. The subunit composition and the overall design of the protein are obviously intimately related to this behavior. Thus, the need for models has led to new insight into protein design and knowledge of protein structure has aided in the elucidation of kinetic models.

It is apparent that control through conformational changes is not the only regulatory device in the cell. The specificity patterns of enzymes provide control properties. Product inhibition, substrate levels, enzyme induction and repression, transport across barriers, etc., are all devices used by the living system to control dynamic processes. Conformational changes are not the only mechanism, but it is amazing how universal they are. Already they have been implicated in active sites, repressors, membranes, and nerve receptors. Thus, enzyme models which explain such behavior cannot be the whole answer to regulation but they provide the framework for explaining these related processes.

A number of recent reviews have been presented describing the role of conformational effects in allosteric proteins and the basic features of molecular models designed to explain individual enzymes (*1–5*). Other chapters in this volume will deal with the properties of individual regulatory enzymes (Stadtman, Chapter 8) and the role of enzymes in metabolic regulation (Atkinson, Chapter 9). This chapter will therefore emphasize our current knowledge of the relation between molecular models and the dynamics of allosteric control. We shall describe (a) the qualitative features of the models which allow one to interpret allosteric phenomena in terms of protein structure, (b) the quantitative features by which detailed kinetic patterns can be explained in terms of "molecular parameters" directly related to protein structure, (c) the various diagnostic tests which can be applied to develop a molecular model for an individual enzyme, and (d) some of the deduced molecular properties of allosteric proteins.

To simplify the presentation of this chapter a glossary is provided in which the definitions and terms used in this article are defined. A brief description of a term will in some cases be presented when the term is introduced, but precise definitions of complex terms or simple terms frequently used will be described in the glossary to provide precision without delaying the reader in the main text.

1. D. E. Atkinson, *Ann. Rev. Biochem.* **35,** 85 (1966).
2. E. R. Stadtman, *Advan. Enzymol.* **28,** 41 (1966).
3. H. E. Umbarger, *Science* **146,** 674 (1964).
4. D. E. Koshland, Jr. and K. E. Neet, *Ann. Rev. Biochem.* **37,** 359 (1968).
5. D. E. Koshland, Jr., *in* "Current Topics in Cellular Regulation" (B. L. Horecker and E. R. Stadtman, eds.), Vol. 1, p. 1. Academic Press, New York, 1969. 1969.

II. Qualitative Features of Molecular Models

It is probably fair to state that the qualitative features of allosteric control are well understood today. The fact that conformational changes occur in a protein provide a mechanism by which metabolites which are not themselves directly involved in the reactions catalyzed by a particular enzyme can control the activity of that enzyme (*6–8*). Such ligand-induced conformational changes appear to be a universal feature of proteins documented in many cases (*4*) including the elegant evidence produced by X-ray crystallography. Although it is apparent that many of the properties of enzymes can still be explained by a template-type enzyme, even the smallest hydrolytic enzymes exhibit ligand-induced conformational changes. For example, Trentham and Gutfreund have recently demonstrated that alkaline phosphatase undergoes ligand-induced conformational changes (*9*). Perutz and co-workers have shown that oxygen induces conformational changes in hemoglobin (*10*), Phillips *et al.* have provided evidence for small but significant changes in lysozyme (*11*), and Lipscomb *et al.* have shown striking changes induced by substrates in carboxypeptidase (*12*). Each of these proteins with the exception of hemoglobin is small, degradative, and not yet identified with regulatory processes. When this evidence is added to the massive documentation from solution studies in the case of the multisubunit protein, it appears that conformational changes are the rule rather than the exception.

In Fig. 1, therefore, we have shown schematically the effects of a substrate, S, an inhibitor, I, and an activator, J, on a protein in which conformational changes are induced. In order for a substrate to be effective it must not only bind to the protein but must also induce a conformational change which provides a proper alignment for the catalytic groups A and B. This of necessity means that some of the binding energy

6. D. E. Koshland, Jr., *Proc. Natl. Acad. Sci. U. S.* **44,** 98 (1958); "The Enzymes," 2nd ed., Vol. 1, p. 305, 1958; *J. Cellular Comp. Physiol.* **54,** 245 (1959); *Cold Spring Harbor Symp. Quant. Biol.* **28,** 473 (1963).

7. J. Monod, J.-P. Changeux, and F. Jacob, *JMB* **6,** 306 (1963).

8. J. Gerhart and A. B. Pardee, *JBC* **237,** 891 (1962).

9. S. Halford, N. G. Bennett, D. R. Trentham, and H. Gutfreund, *BJ* **114,** 243 (1969).

10. M. F. Perutz, M. G. Rossman, A. F. Cullis, H. Muirhead, G. Will, and A. C. T. North, *Nature* **185,** 416 (1960).

11. C. C. F. Blake, D. F. Koenig, G. A. Mair, A. C. T. North, D. C. Phillips, and V. R. Sarma, *Nature* **206,** 757 (1965).

12. M. L. Ludwig, J. A. Hartsuck, T. A. Steitz, H. Muirhead, J. C. Coppola, G. N. Reeke, and W. N. Lipscomb, *Proc. Natl. Acad. Sci. U. S.* **57,** 511 (1967).

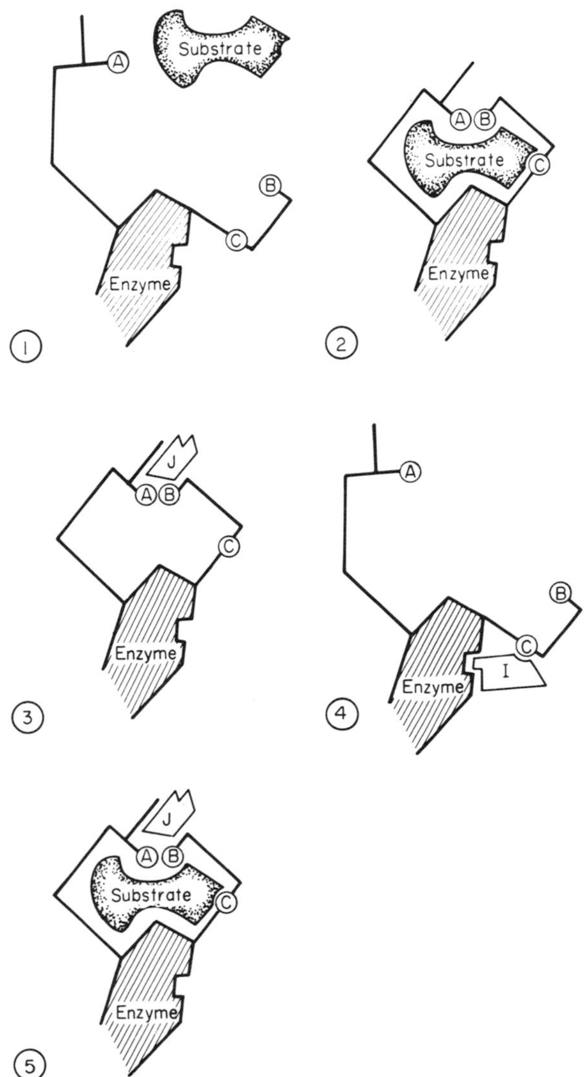

FIG. 1. Schematic illustration of ligand-induced conformational changes in a monomer. (1) Protein conformation prior to ligand binding. (2) Protein conformation after substrate has induced conformational change leading to proper alignment of catalytic groups A and B. (3) Activator J induces conformational change aligning catalytic groups properly in the absence of substrate. (4) Inhibitor I induces conformational change maintaining peptide chain with B group in position such that substrate has too low affinity to bind effectively. (5) Substrate and activator bind to protein in active conformation.

is expended in inducing the conformational change. An activator, J, could also bind to this protein in such a way as to lead to the proper alignment of these groups in the absence of substrate or could lead to the stabilization of this structure in the presence of substrate. Its binding to the protein could therefore result in an apparent increase in affinity of the substrate for the protein, reflected by a decreased apparent K_m. This in turn would lead to an increase in activity when the substrate is present in less than saturating concentrations. An inhibitor, I, might induce a conformation which would prevent the proper alignment of binding groups and decrease the affinity for the substrate (cf. Fig. 1). The inhibitor would in effect decrease the concentration of protein which can bind substrate and act as a competitive inhibitor. Induced conformational changes which affect V_M as well as K_m are, of course, also possible.

It will be convenient to signify conformational changes in a schematic way without drawing detailed diagrams of catalytic groups; therefore, the schematic device shown in Fig. 2 was devised to signify by circles, squares, and triangles the different conformational states of the protein. In Fig. 2 the illustrative example of Fig. 1 is repeated in this simplified form to indicate that the same conformation of the molecule is induced by the substrate and activator whereas a different conformation is induced by the inhibitor. It will be convenient in derivations in the text to refer to these as conformations A, B, and C as defined in the glossary.

It is important to recognize that the conformations designated here refer to functionally distinct states. There may be many conformations of a protein which bear little relation to its function. A single lysine residue may have no relation to catalytic or regulatory properties but might occupy a number of positions on the surface of the protein. In fact in some crystallographic studies some lysine side chains are a blur because

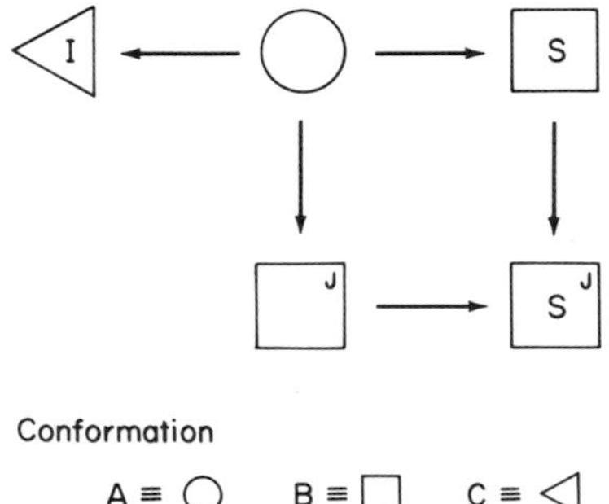

FIG. 2. Schematic illustration of conformational changes of Fig. 1. Here circle, square, and triangle are used to indicate different conformational states without specifying specific peptide alignments.

of this tendency to occupy many positions. It would clearly be unwise to designate each separate position of each lysine as a new conformational state when one is considering regulatory behavior. On the other hand, an extremely small change in a crucial residue can change the catalytic activity (*13*) of a protein. Such a conformation change though small in magnitude produces a functionally distinct conformation. The conformations designated by A, B, and C refer to these functionally distinct identities but do not imply either that a large change in structure has occurred or that the nonfunctional residues occupy a single position.

If one plots a velocity vs. (S) plot for the relationship involved in Fig. 2, classic Michaelis–Menten kinetics would be followed. If the polypeptide chain designated in Figs. 1 and 2 is part of a multisubunit structure, the conformational changes induced by ligands could affect the neighboring active sites and alter these kinetic patterns. Such a multisubunit structure is shown in Fig. 3 in order to illustrate the qualitative effects of these relationships. Figure 3a is a simple illustration in which the ligand induces a conformational change in the subunit to which it is

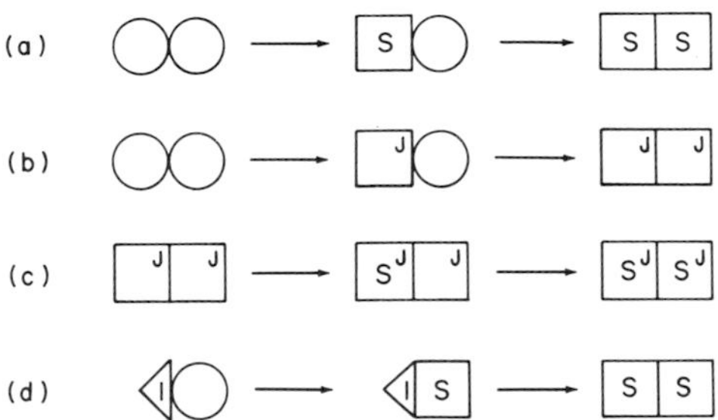

FIG. 3. Schematic illustration of conformational changes in a dimer in which ligand-induced conformational changes of S, J, and I are the same as in Fig. 2. (a) The sequential binding of ligand S where only the subunit to which S is bound changes conformation (simplest sequential model). (b) The same conformational changes induced by ligand J in the absence of S. If adjacent squares (B conformations) are more attractive than adjacent A conformations (K_{BB} greater than K_{AA}), a positive homotropic effect will be observed on binding of S and J. (c) The binding of S to a molecule already saturated with J. Since no further conformational change is induced, the binding of S would follow Michaelis–Menten kinetics and $S_{0.5}$ would be observed at lower concentrations than in the absence of J. (d) A protein containing sufficient I in solution to saturate one of the subunits in the absence of ligand S. If adjacent B and C conformations are more attractive than AB or AA, the inhibitor will aid the binding of the first molecule of S through the attractive subunit interactions. As more S is added, however, it will displace I in a competitive reaction so that high concentrations will find I behaving as an inhibitor.

13. K. E. Neet and D. E. Koshland, Jr., *JBC* **243,** 6392 (1968).

bound. If the energy of the interaction between subunits is changed by this conformational change, e.g., if there is a net stabilizing energy when there are two adjacent B conformations, then it can be seen that a second molecule of substrate will bind more readily than the first as more and more substrate is bound. This is the basis for the cooperative effect observed so frequently in allosteric proteins. If J induces the same conformational change as the substrate, it should have the same cooperative effect on itself, i.e., the first molecule of J will make it easier for the subsequent molecules of J to bind. Moreover, the presence of J on a protein will make it easier to bind S since the energy-consuming conformational change has already been induced during the binding of J (cf. Fig. 3c). On the other hand, the inhibitor may induce another conformation (C) which binds neither S nor J, as shown in Fig. 3d.

Only a few of the possible molecular species are shown in Fig. 3, but it is immediately apparent that the relationships become very complex. The addition of J not only affects the binding of S to the subunit to which J is bound, as in the simple case of Fig. 2, but also affects a manner in which the subsequent molecules of J can bind to the same protein and the manner in which the molecules of S may be bound to adjacent subunits.

The same interrelated effects can be observed for the inhibitor in which conformation C may introduce new subunit interactions. Even without mathematical derivations this diagram makes it apparent that extremely complex kinetic patterns should result and moreover that the pattern of behavior between S and J will be affected by the amount of inhibitor bound, i.e., by the concentration levels of I in solution. Surprisingly, these complex interactions can be expressed relatively simply using the molecular parameters discussed in the next section.

III. Quantitative Molecular Parameters for Regulatory Proteins

A. Derivation of General Equations

To describe regulatory proteins in quantitative terms it is necessary to use parameters which can be evaluated experimentally. In addition, it is desirable to choose these parameters on the basis of molecular features of the protein so that they eventually can be related to the protein structure. A procedure has been devised that provides parameters which are descriptive in molecular terms and can apply to every situation involving substrates, inhibitors, and activators in regulation (*14, 15*). If any simple model (see Section III,B) had been universally correct, it would have been possible to express the mathematics of allosteric proteins in a few simple constants. The evidence already in the literature, however, demonstrates that no simple model will explain the behavior of all allosteric proteins. It is necessary therefore to develop the mathematical tools for the general case and then explain how the simpler models fit into this framework. To make the conceptual ideas easier to grasp the essential features of a mathematical term will be described when it is first used, but the detailed definition and appropriate references will be presented in the glossary.

In essence the molecular parameters are three: (a) a constant to express the free energy of the conformational transition K_t; (b) a constant to express the intrinsic affinity of a ligand for a given conformational state of the subunit K_X; and (c) a constant to reflect the change in subunit interactions K_{AB}, K_{AA}, etc. (Although X will be used to indicate a ligand which can be S, J, or I, we shall frequently use S to illustrate general principles because of its better recognition value.)

The derivation of a mathematical equation for some complex phenomenon such as a v vs. S plot involves three steps: (a) the designation of the molecular species assumed to be present in solution, (b) the derivation of the individual molecular parameters for each molecular species, and (c) the summation of these individual cases into an overall equation.

The designation of the molecular species which might be present is the vital step in the design of any particular model since the fundamental assumptions are made in these selections of molecular species. For example, in Fig. 4 some of the molecular species which might be present in a dimer are shown. In column I are shown the species (A_2, ABS, and B_2S_2) which might be postulated to be present in the absence of any added activator or inhibitor. The listing of these forms is a schematic way of making the following assumptions: (a) there are two binding sites for S, (b) when S binds a new conformational state is induced, (c) there is no significant concentration of molecular species in which S binds to proteins in other conformations, and (d) there is no significant concentration of B conformation in the absence of bound S. These, of course, are not necessary assumptions for all proteins but they illustrate that the designation of molecular species is the key step in the derivation of any particular equation for an individual protein.

The next step, the expression of the individual species in terms of the molecular parameters, is then routine as shown in the figure. Each of

14. D. E. Koshland, Jr., G. Nemethy, and D. Filmer, *Biochemistry* **5**, 365 (1966).
15. M. E. Kirtley and D. E. Koshland, Jr., *JBC* **242**, 4192 (1967).

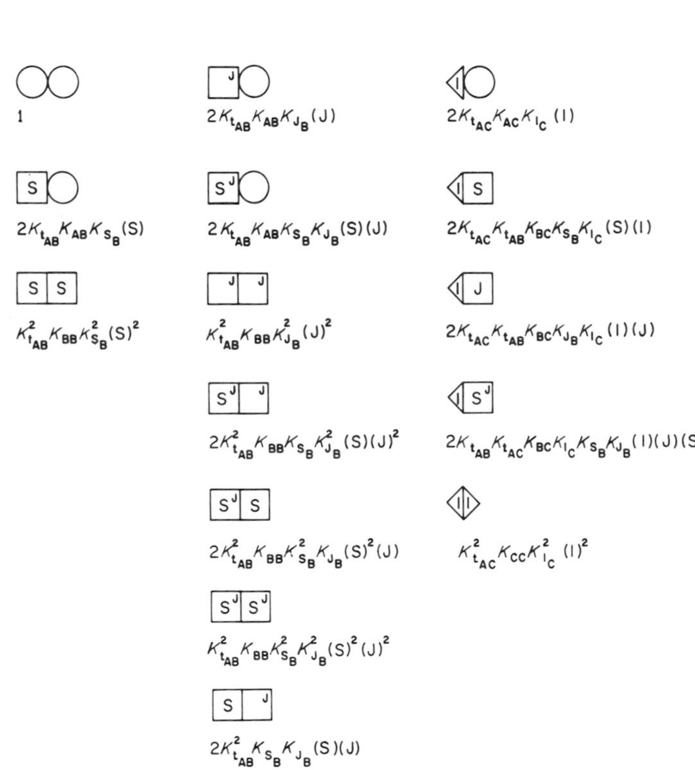

FIG. 4. Molecular species postulated to be present in a dimer for the calculation of particular binding and velocity curves. Column I shows the species which might be present when only S is added to the system, assuming S induces a conformation change from the A to the B conformation in the subunit to which it is bound. Column II shows the additional species which would be present if J were added to the system assuming that J is an activator and induces the same conformation as S but is bound to a site which allows independent binding of S and J. Column III shows the additional species which might be present on addition of an inhibitor I to a solution containing S and J. It is assumed that I induces a conformation change different from that of S or J. Beneath each molecular species are the molecular parameters needed to calculate the concentration of that species relative to the concentration of A_2.

the terms is the factor by which the concentration of the unliganded (A_2) species must be multiplied in order to obtain the new molecular species. For example, the terms for ABS can be imagined to be split up into the processes of Eq. (1). One step involves the energy of the conformation change $K_{t_{AB}}$. The next step involves the binding of the ligand to the B conformation K_{S_B}. The third involves the association of two subunits,

$$\begin{array}{l} \bigcirc \xrightarrow{K_{tAB}} \square \xrightarrow{K_{SB}} \boxed{S} \xrightarrow{K_{AB}} \boxed{S}\bigcirc \\ \quad \llcorner \xrightarrow{\qquad\qquad K_{AA} \qquad\qquad} \bigcirc\bigcirc \end{array} \tag{1}$$

one in the A conformation and one in the B conformation K_{AB}. Although it would be quite possible to leave this expression in terms of the association of K_{AB} representing the dimerization of two subunits from monomer reactants, in most actual cases association–dissociation phenomena are not encountered. It has become convenient therefore to use the standard state as the associated polymer and define the reference point as the unliganded state (A_2 in this case); thus, the equilibrium expression for the change in subunit interactions is designated as K_{AB}/K_{AA}. Hence, the complete equation for the process of Eq. (1) is shown below. A factor of 2 is included to account for the fact that there are two

$$\frac{(\mathrm{ABS})}{(\mathrm{A_2})} = 2K_{tAB}K_{SB}\frac{K_{AB}}{K_{AA}}(\mathrm{S}) \tag{2}$$

statistical ways ($\boxed{S}\bigcirc + \bigcirc\boxed{S}$) of arranging ABS, whereas there is only one way of arranging A_2. It will be convenient to designate K_{AA} as 1 in all cases because this simplifies the mathematical expressions and refers all subunit interactions to the standard state, i.e., the unliganded polymer. (It is quite easy as shown below to extend this method to enzyme association–dissociation phenomena, but for the moment we shall deal only with those cases in which the state of association does not change.) Using the same principle the molecular parameters for each of the species can be derived by mere inspection. Thus, the species $B_2S_2J_2$ in Fig. 4 involves K_{tAB}^2 because two subunits are changed from the A to the B conformation; a statistical factor of 1 because there is only one way of arranging the species; exponents of 2 for the affinity constants (K_{SB} and K_{JB}) and the concentration terms (S and J), because two molecules of each ligand are involved.

The next step is to combine these terms into a proper overall equation. This will, of course, depend on the experiment being devised. For example, if one is measuring the binding of S in the absence of I and J, the expression N_s in schematic terms is given by Eq. (3) and in terms of molecular parameters by Eq. (4). If one is measuring N_s in the presence

$$N_s = \frac{\Sigma(\text{terms involving S}) \times (\text{No. of molecules of S in each term})}{\Sigma(\text{all terms})}$$

$$= \frac{\boxed{S}\bigcirc + 2\boxed{S}\boxed{S}}{\bigcirc\bigcirc + \boxed{S}\bigcirc + \boxed{S}\boxed{S}} \tag{3}$$

$$N_s = \frac{2K_{t_{AB}}K_{S_B}K_{AB}(S) + 2K_{t_{AB}}{}^2K_{S_B}{}^2K_{BB}(S)^2}{1 + 2K_{t_{AB}}K_{S_B}K_{AB}(S) + K_{t_{AB}}{}^2K_{S_B}{}^2K_{BB}(S)^2} \tag{4}$$

of S and J, then all the molecular species in columns I and II of Fig. 4 are required and the resulting expression is shown in Eq. (5). If J is 0, this equation simplifies to Eq. (4). Analogous expressions can

$$N_s = \frac{(ABS + 2(B_2S_2) + (ABSJ) + (B_2SJ) + 2(B_2S_2J) + (B_2SJ_2) + 2(B_2S_2J_2)}{1 + (ABS) + (ABJ) + (ABSJ) + (B_2S_2) + (B_2J_2) + (B_2SJ_2) + (B_2S_2J) + (B_2S_2J_2) + (B_2SJ)} \tag{5}$$

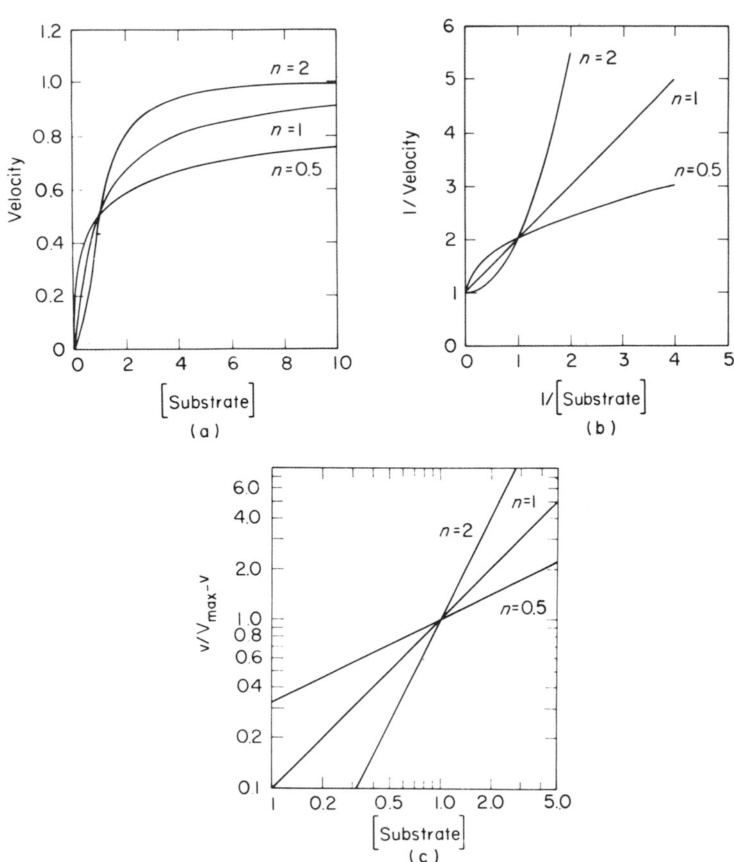

FIG. 5. Michaelis–Menten, double reciprocal, and Hill plots for independent binding, positive cooperativity, and negative cooperativity. The data calculated from equation $v = V_m(S)^{n_H}/(K + S)^{n_H}$ in which $V_m = 1$, $K = 1$, $n_H = 0.5$, 1 and 2. The same data are then plotted in (a) a classic Michaelis–Menten plot, (b) a double reciprocal plot, and (c) a Hill plot.

be derived using the same principles for N_s in the presence of S and I, for N_I in the presence of S and J, and so forth. There are many experimentally realizable situations, and the procedure for the derivation of the appropriate molecular formulas is extremely rapid and easy for each experimental test once the molecular species have been designated.

If one now wishes to test a new model slightly different from the first, it is quite easy to test the quantitative effect of the new assumptions. For example, if binding to the A conformation as well as B is to be considered, one adds terms for such molecular species as A_2S and A_2S_2 (○Ⓢ and ⓈⓈ). The fit of the new experimental equation can be compared with the previous theoretical expression to see if it deviates significantly and to the experimental facts to determine which fits the data better. If the equation with the new experimental terms is in closer agreement with the experimental facts, these added terms clearly need to be included as potential molecular species. If they do not contribute under any experimental situation, one can conclude that they do not make a significant quantitative contribution to the molecular species present. One cannot conclude, of course, that they do not play a kinetic role as transient intermediates.

Kinetic expressions are readily derivable from equilibrium expressions by assigning catalytic constant terms for the decomposition into products of any molecular species. Thus, the velocity expression for the case of Eq. (4) might be

$$v = \frac{k_1(\mathrm{ABS}) + 2k_2(\mathrm{B_2S_2})}{1 + (\mathrm{ABS}) + (\mathrm{B_2S_2})} \tag{6}$$

where k_2 may be, but is not necessarily, equal to k_1. If J were a compulsory activator such that only species in which S and J were on the same subunit were active and all such species had the same k_{cat}, then the velocity expression corresponding to Eq. (5) would be

$$v = \frac{k_{cat}[(\mathrm{ABSJ}) + (\mathrm{B_2SJ_2}) + (\mathrm{B_2S_2J}) + 2(\mathrm{B_2S_2J_2})]}{1 + (\mathrm{ABS}) + (\mathrm{ABJ}) + (\mathrm{ABSJ}) + (\mathrm{B_2S_2}) + (\mathrm{B_2J_2}) + (\mathrm{B_2SJ_2}) + (\mathrm{B_2S_2J}) + (\mathrm{B_2S_2J_2}) + (\mathrm{B_2SJ})} \tag{7}$$

It is also possible to include rate constants between molecular species and this will be essential in some cases. The equations become more complex in that case and no new principles are involved so these complexities can be ignored temporarily in illustrating the approach.

B. Simple Models

The above is a general approach based on assumptions of ligand-induced changes which can be applied to any conceivable situation. It is clear, however, that the number of permutations in a multisubunit protein with several effectors is large. The desire for some arbitrary restrictions which would simplify the mathematics is very great. There are two basic ways of approaching such a simplification and both have been tried (*14, 16*). Since these models have been discussed at length elsewhere (*5*), this account will be brief.

The first approach is to look for some underlying feature of protein design which will limit the molecular species to be expected. Monod *et al.* (*16*) considered the expected properties of proteins and deduced that symmetry and the conservation of symmetry were two such principles. Briefly stated this meant that subunits in a protein would arrange so that symmetry features were apparent and that these features would be conserved in any changes of the subunits. If these principles were coupled with a third one, that there are equilibria between preexisting states of the protein, a model for cooperativity is obtained. A simple illustration of this model is given in Eq. (8) where the nomenclature above the diagrams is that of Monod *et al.* and the nomenclature below the diagrams

$$\begin{array}{cccccccccccc} T_0 & & R_0 \text{ state} & & R_1 & & R_2 & & R_3 & & R_4 \\ \text{(4 circles)} & \underset{}{\overset{L}{\rightleftarrows}} & \begin{array}{|c|c|}\hline \ & \ \\ \hline \ & \ \\ \hline\end{array} & \rightleftarrows & \begin{array}{|c|c|}\hline S & \ \\ \hline \ & \ \\ \hline\end{array} & \rightleftarrows & \begin{array}{|c|c|}\hline S & S \\ \hline \ & \ \\ \hline\end{array} & \rightleftarrows & \begin{array}{|c|c|}\hline S & S \\ \hline \ & S \\ \hline\end{array} & \rightleftarrows & \begin{array}{|c|c|}\hline S & S \\ \hline S & S \\ \hline\end{array} \\ A_4 & & B_4 & & B_4S & & B_4S_2 & & B_4S_3 & & B_4S_4 \end{array} \qquad (8)$$

that of this chapter. If the equilibrium expressed by L is far to the left in the absence of ligand, and if S binds preferentially to the R state, increasing concentrations of S will shift the equilibrium to the right and sigmoid $\bar{Y}$ vs. (S) plots will be observed. Because symmetry principles exclude conformational hybrids such as A_3B and A_2B_2, the number of conformational states is greatly reduced. The model is far more than a device for mathematical simplicity, however, since it is based on fundamental principles of protein design.

The second approach is to accept the probability that many molecular species will be present but to make arbitrary assumptions for simple models to serve as first approximations. This approach was used by Koshland *et al.* (*14*). Their examination of proteins proceeded from the induced-fit approach, and they assumed ligand-induced distortions would produce a wide variety of species. The mathematical tools (which are those described above) were designated to deal with the complex general case, but a simple model could be devised based on arbitrary assumptions. Thus if one assumes that (a) only two conformational states are available to any one subunit and (b) only the subunit to which the

16. J. Monod, J. Wyman, and J.-P. Changeux, *JMB* **12**, 88 (1965).

ligand is bound changes shape, the simplest sequential model is obtained. An illustrative example for a tetrameric protein is shown in Eq. (9). In this case the simplest model is not required by any feature

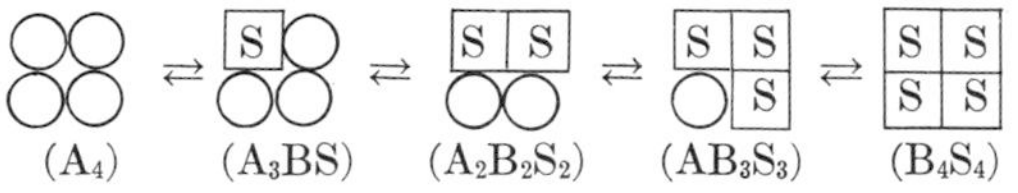

$$(A_4) \rightleftarrows (A_3BS) \rightleftarrows (A_2B_2S_2) \rightleftarrows (AB_3S_3) \rightleftarrows (B_4S_4) \quad (9)$$

of the theory (which in fact suggests diversity) but rather from arbitrary assumptions in a desire to produce simplicity. Nevertheless a calculation (*17*) showed that this simplest sequential model would be a good first approximation in many cases for reasonable values of the subunit interaction terms.

Each of these simple models can be modified somewhat without doing violence to their fundamental postulates. Thus, the Monod *et al.* model does not require only two final states or exclusive binding to the R state only. Too many final states, however, seems antithetical to the fundamental principles of the model—preexisting states and a predesigned symmetry relationship. Similarly, preexisting equilibria and nonexclusive binding can be considered in the simplest sequential model but addition of other conformational states for an individual subunit makes the model appear like the general sequential model described above. Certain features of these simpler models are basic however to their underlying assumptions of protein design and lead to predictions which can be tested by the diagnostic tools described below (cf. also reference *5* for experimental ways of comparing these simple models).

IV. Diagnostic Tests

It is believed that the molecular parameters and procedures outlined in the previous section are sufficient to describe any theoretical situation. However, it is clearly ridiculous to write all possible molecular species *a priori*. In the dimer example chosen for illustrative purposes only three conformational states of individual subunits were assumed and still the number of molecular species involved was large. Thus, the possibility of deriving a completely general detailed equation to be used in all cases is impossible. Some derivations for more complex cases have been published (*15*, *18*, *19*) but in any given experimental situation it will be

17. J. E. Haber and D. E. Koshland, Jr., *Proc. Natl. Acad. Sci. U. S.* **58**, 2087 (1967).

18. C. Frieden, *JBC* **242**, 4045 (1967).

19. J. Wyman, *Advan. Protein Chem.* **4**, 407 (1948).

important to limit the species being considered to manageable proportions. There are two ways of doing this. One is to assume that one of the simpler models described above is operative and test predictions based on this hypothesis. A second way is to assume that the general approach described in Section III will eventually be needed and to design experiments which indicate the species present. Thus diagnostic tests are needed to indicate whether the simple models are pertinent and, if not, what molecular species of the model contribute significantly for an individual protein. In this section therefore we shall consider some individual diagnostic tests, the manner in which they are applied, and the conclusions which can be derived from them.

A. $\bar{Y}$ versus (X) and Double Reciprocal Plots

A saturation curve for a protein can be defined as a plot in which some function of the fraction of sites occupied ($\bar{Y}_X$) is plotted against some function of the concentration of the ligand (X). In a true binding expression either the fraction saturation, which varies from 0 to 1 or N_X (the number of molecules bound per molecule of protein which varies from 0 to n_M) is plotted against the ligand concentration. If it is assumed that the velocity of the reaction is proportional to the fraction of sites occupied, a v/V_M vs. (X) plot has the nature of a $\bar{Y}_X$ vs. (X) plot; and, indeed, it is this assumption which leads to the Michaelis–Menten equation. In a complex system v/V_M may not be proportional to $\bar{Y}$ but such plots have been useful, particularly when only small amounts of enzyme are available and can be used to develop hypothesis provided the investigator is well aware that an assumption in regard to mechanism is being made. A saturation plot will therefore be defined as any plot in which the investigator assumes that a quantity proportional to site saturation is plotted versus some function of ligand concentration.

Such saturation plots can be made in different ways, and each has certain diagnostic values. The $\bar{Y}$ vs. (X) or N_X vs. (X) in the classic Michaelis–Menten case is a hyperbolic curve as shown in Fig. 5a (see p. 352). In such plots it is common practice to estimate V_M or n_M by the asymptote at high substrate concentrations, to estimate K_m from the midpoint of the curve, and to use the hyperbolic shape of the curve as an indication of independence of sites.

It was soon recognized that the quantitative evaluation of V_M and K_m from such a plot is poor, and the Lineweaver–Burk or double reciprocal method (*20*) was introduced as shown in Fig. 5b. If subunit interactions occur, there will be deviations from each of these types of plots as shown

20. H. Lineweaver and D. Burk, *JACS* **56**, 658 (1934).

in Figs. 5a and b. In cases of positive cooperativity or positive homotropic effects a sigmoid appearance of the $\bar{Y}$ vs. (X) plot is observed (*8, 21*) and the double reciprocal plot is concave upward. Both of these deviations have been used as diagnostic tests. In "negative cooperativity," i.e., when the first molecule of ligand bound makes it more difficult for the next to bind, these curves will deviate in a different way (*22*). The $\bar{Y}$ vs. (X) plots will look hyperbolic but will approach a limiting value at a much slower rate than the normal curve (cf. Fig. 5a) and the Lineweaver–Burk plots will show a concave downward deviation (Fig. 5b). For negative cooperativity the classic $\bar{Y}_X$ vs. (X) plot is of little diagnostic value since it looks qualitatively like a normal curve; in fact, negative cooperativity was not recognized for years undoubtedly for this reason. On the other hand, the double reciprocal plot makes the phenomenon quite apparent by its concave downward deviation. Since such curves can be obtained from several polymorphic forms, from electrostatic repulsions of a template-type enzyme, and from several other causes (*22*), a concave downward deviation is an indication rather than proof of negative cooperativity. However, in a purified enzyme which is homogeneous, one is unlikely to find a number of polymorphic forms and electrostatic repulsions cannot in most cases lead to large deviations. The deviation on the double reciprocal plot therefore provides a reasonable working hypothesis for negative cooperativity resulting from subunit interactions.

B. $\bar{Y}_X$ or N_X versus Log (X) Plots

A plot of the fraction saturation versus the log (X) has certain advantages which are not present in the $\bar{Y}_X$ vs. (X) plot. In the first place the shape of the curve depends strongly on the choice of abscissa units when the linear scale is used (*1*). A curve with a K_m of 10^{-7} will look far steeper than a curve with a K_m of 10^{-3} even though each are of equal cooperativity. A plot of log (X), however, gives curves which are of equal steepness for equal cooperativity (*14*). All Michaelis–Menten curves are sigmoid and are similar in shape. Varying K_m's are reflected in a difference in midpoint, but the curves are superimposable if translated along the x axis. Using a log plot, a far wider range of ligand concentrations can be covered and the steepness of the curve is an immediate visual index of cooperativity.

21. C. Bohr, K. Hasselbach, and A. Krogh, *Skand. Arch. Physiol.* **16,** 402 (1904).
22. A. Levitzki and D. E. Koshland, Jr., *Proc. Natl. Acad. Sci. U. S.* **62,** 1121 (1969).

C. Cooperativity Index and $X_{0.5}$

The ratio of two ligand concentrations needed to produce different levels of fractional saturation has been suggested as a measure of cooperativity (*23*). The ratio between 90 and 10% saturation symbolized by R_s ($S_{0.9}/S_{0.1}$) was shown to have diagnostic value (*14*). This R_s will be 81 for all curves following Michaelis–Menten kinetics, less than 81 for positively cooperative curves, and greater than 81 for negatively cooperative curves. Thus, the diagnostic tool can readily be read off any plotted curve and involves a minimum of extra calculation. In a velocity plot, this cooperativity index does not indicate that a 90% saturation point has indeed been reached, but in a true binding curve where the 10 and 90% points can be established unequivocally it can be a simple and quick diagnostic test.

It is convenient in allosteric proteins to use the term ($X_{0.5}$) to indicate the ligand concentration at half-saturation since a simple K_m cannot be calculated for curves which deviate from Michaelis–Menten kinetics (*14*); $S_{0.5}$ becomes equal to K_m when Michaelis–Menten kinetics is observed.

D. The Hill Plot

One of the particularly useful techniques was initially applied by Hill (*24*) to the binding curve of oxygen to hemoglobin. Hill showed that a plot of the type $\bar{Y}_S = S^n/(K + S^n)$ could give a sigmoid $\bar{Y}$ vs. (X) plot, and this equation could be written in logarithmic form as Eq. (10). A plot of log $\bar{Y}$ vs. log S gives a straight line of slope n_H where n_H is

$$\log \frac{\bar{Y}}{1 - \bar{Y}_S} = n_H \log S - \log K \tag{10}$$

defined as the Hill coefficient. This equation has the form of a general binding equation except that all intermediate terms such as ES_1, ES_2, and ES_{n-1} are assumed to be negligible. This in effect means that an assumption of infinite cooperativity has been made in the derivation of this equation, i.e., the first molecule of ligand bound is immediately followed by the binding of all subsequent molecules of ligand. In such a situation n_H will equal n_M but in fact this is a theoretical limit which is never completely satisfied. In hemoglobin which contains four binding sites the Hill coefficient is 2.8 (*19, 24*).

In most other cases the Hill coefficient is well below the maximum number of sites, and this empirical Hill coefficient therefore represents

23. K. Taketa and B. J. Pogell, *JBC* **240**, 651 (1965).
24. A. V. Hill, *J. Physiol.* (*London*) **40**, IV–VIII (1910).

a combination of the number of sites and the strength of the interaction between sites. A comparison of n_H with n_M then is an indication of the extent of cooperativity of the protein. Thus, in the case of CTP synthetase Hill coefficients of 3.8 and 3.5 were found for ATP and UTP (*25*) and the four binding sites for each of these nucleotides have been established by binding studies (*26*). This indicates a very high degree of cooperativity.

In a binding study the Hill coefficient will theoretically establish the minimum number of sites, i.e., a coefficient of 2.8 will indicate three or more sites. It cannot establish a maximum number of sites (*1*).

If velocity is plotted instead of fraction of sites occupied, the n_H could be greater than n_M, but it is intriguing that in most of the enzymes that have been studied so far n_H is less than n_M. As a first approximation, therefore, Hill coefficients derived from velocity studies can be used as an indication of the minimum number of sites. Another form of this equation plots the log $v_0/(v_I - 1)$ vs. log I. This should give the number of interacting sites for the inhibitor and can be compared with similar values for substrates and activators.

The Hill plots of the same data as Figs. 5a and b are shown in Fig. 5c. A Hill coefficient greater than one indicates positive cooperativity, less than one negative cooperativity, and equal to one noninteracting sites. It has been observed that Hill plots may give two slopes around the midpoint, and this may also be of diagnostic value (*27, 28*).

A different way of plotting the Hill equation has been proposed by Wyman (*29*) and is shown in Fig. 6. In this case the data give a displacement such that the perpendicular between the two unit slopes extrapolations can give an indication of the interaction energy. Thus, in the case of hemoglobin shown in the figure the perpendicular between the two curves multiplied by $RT\sqrt{2}$ will give the interaction energy. This perpendicular gives the interaction energy only if the two limiting slopes are equal to one, which will occur in cases of simple positive cooperativity when the saturation curve is symmetrical about the $S_{0.5}$ slope.

E. The Scatchard or Klotz Plots

Plots of binding data in the forms of $(X_B/E_T)/(X)$ vs. (X_B/E_T) were suggested by Scatchard (*30*) and Klotz (*31*) as methods of determining

25. C. W. Long and A. B. Pardee, *JBC* **242,** 4715 (1967).
26. C. W. Long, A. Levitzki, and D. E. Koshland, Jr., *JBC* **245,** 80 (1970); A. Levitzki and D. E. Koshland, Jr., in preparation.
27. R. A. Cook and D. E. Koshland, Jr., *Biochemistry* **9,** 3337 (1970).
28. D. Dalziel and P. C. Engel, *FEBS Letters* **1,** 349 (1968).

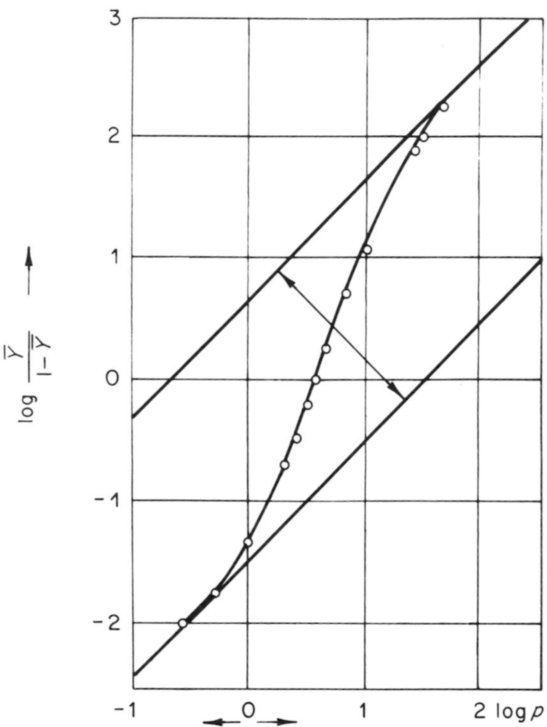

FIG. 6. A Hill plot of the oxygen equilibrium of sheep hemoglobin according to the method of Wyman.

the total number of binding sites in a protein [(X_B/E_T) represents moles of ligand bound per mole of enzyme]. Such plots are particularly useful in determining the number of sites n_M when there is no cooperativity. In that case a slope proportional to the binding constant and a limiting intercept equal to the number of sites is obtained. When negative or positive cooperativity is present, the slopes are no longer linear as shown in Fig. 7. In a positively cooperative case the curve is concave downward and in a negatively cooperative case concave upward. Cooperativity makes extrapolation to n_M hazardous, but the curvature can be used as an indication of the type of cooperativity, and frequently rough values of n_M can be obtained. Deviations which appear to be the same as negative cooperativity could occur if there is heterogeneity of

29. J. Wyman, *JACS* **89,** 2202 (1967).
30. G. Scatchard, *Ann. N. Y. Acad. Sci.* **51,** 660 (1949).
31. I. M. Klotz, F. M. Walker, and R. B. Pivan, *JACS* **68,** 1486 (1946).

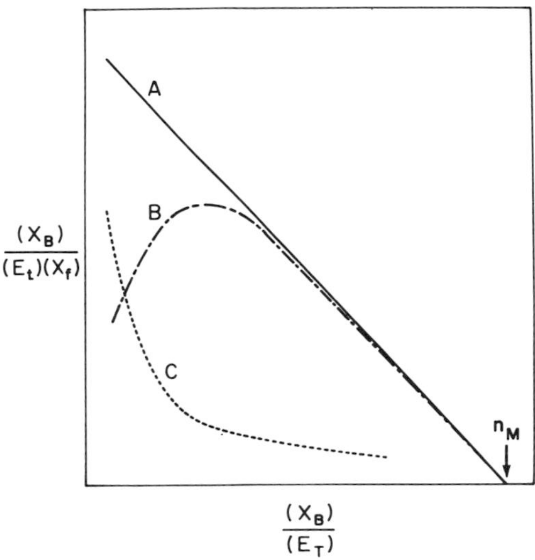

FIG. 7. A Scatchard plot of binding data for A, independent binding sites; B, positive cooperativity; and C, negative cooperativity.

binding sites. If negative cooperativity exists, a replotting of the data with appropriate assumptions can give a straight line relationship (*32*) and this may be useful on occasion.

F. Fitting Saturation Curves

Because the two early molecular models applied to hemoglobin both fitted the saturation curve (*14*) for that protein, many investigators have concluded that saturation curves per se are of little diagnostic use in determining molecular models. The first evidence against this came in the case of rabbit muscle glyceraldehyde-3-phosphate dehydrogenase in which detailed determination of individual binding constants allowed a delineation of mechanism (*32*). In this case the negative cooperativity provided an easy access to the binding constant, but such simplicity is not available for positive cooperative cases in which no portion of the saturation curve can be identified with a simple constant. Now, however, a procedure has been developed which allows curve fitting to any saturation curve and which provides diagnostic value in relation to molecular models (*33*).

32. A. Conway and D. E. Koshland, Jr., *Biochemistry* **7**, 4011 (1968).
33. A. Cornish-Bowden and D. E. Koshland, Jr., *Biochemistry* **9**, 3325 (1970).

The binding of a single ligand to a multisite enzyme may be expressed by the equilibrium equations (11)–(14). These are the true constants

$$\mathrm{E} + \mathrm{X} \rightleftarrows \mathrm{EX} \qquad K_1 = \frac{(\mathrm{EX})}{(\mathrm{E})(\mathrm{X})} \tag{11}$$

$$\mathrm{EX} + \mathrm{X} \rightleftarrows \mathrm{EX}_2 \qquad K_2 = \frac{(\mathrm{EX}_2)}{(\mathrm{EX})(\mathrm{X})} \tag{12}$$

$$\mathrm{EX}_{i-1} + \mathrm{X} \rightleftarrows \mathrm{EX}_i \qquad K_i = \frac{(\mathrm{EX}_i)}{(\mathrm{EX}_{i=1})(\mathrm{X})} \tag{13}$$

$$\mathrm{EX}_{n-1} + \mathrm{X} \rightleftarrows \mathrm{EX}_n \qquad K_m = \frac{(\mathrm{EX}_\mathrm{M})}{(\mathrm{EX}_{n-1})(\mathrm{X})} \tag{14}$$

for the binding of the first, second, third, fourth, etc., molecules of ligand. In a positive cooperativity or a mixed system of positive and negative cooperativity, it is not easy to disentangle the various constants since it is rare that EX_1 is formed completely before some EX_2 is present. In principle four points and four unknowns should suffice to determine the individual constants. In practice many more points are usually necessary because there is some experimental error. To make the experimentally observed constants K_i more meaningful in theoretical terms they can be converted to the "intrinsic" binding constants K_1', K_2', etc., by multiplying by statistical factors as shown in Eq. (15). The fraction satura-

$$K_i = \frac{(n + 1 - i)}{i} K_i' \tag{15}$$

tion $\bar{Y}_\mathrm{X}$ or N_s for a protein is then given by Eq. (16). Equation (17) is

$$N_\mathrm{X} = n_\mathrm{M}\bar{Y}_\mathrm{X} = \frac{\Sigma_1^n i(\mathrm{EX}_i)}{\Sigma_1^n(\mathrm{EX}_i)} \tag{16}$$

$$N_\mathrm{X} = \frac{(\mathrm{EX}) + 2(\mathrm{EX}_2)}{(\mathrm{E}) + (\mathrm{EX}) + (\mathrm{EX}_2)} \tag{17}$$

the application of this equation to the dimer case treated previously. If the concentrations of each species are expressed in terms of the concentration of the free enzyme (E), then E will cancel out of the numerator and denominator and Eqs. (18) and (19) will express N_X in terms of the

$$N_\mathrm{X} = \frac{K_1\mathrm{X} + 2K_1K_2(\mathrm{X})^2}{1 + K_1(\mathrm{X}) + K_1K_2(\mathrm{X})^2} \tag{18}$$

$$= \frac{2K_1'(\mathrm{X}) + 2K_1'K_2'(\mathrm{X})^2}{1 + 2K_1'(\mathrm{X}) + K_1'K_2'(\mathrm{X})^2} \tag{19}$$

measured constants K_1, K_2, K_3, etc. From these the intrinsic constants K_1', K_2', K_3', etc., can be determined.

For curve fitting procedures the equation for N_X is rewritten in the more general form of Eq. (20) and the ψ's which are conglomerates of the

$$N_X = \frac{\psi_1(X) + 2\psi_2(X)^2}{1 + \psi_1(X) + \psi_2(X)^2} \qquad (20)$$

constants K_1, K_2, K_3, etc. It is readily seen that a unique determination of these ψ values will give the intrinsic constants which are illustrated for the dimer case in Eq. (21).

$$K_1' = \psi_1/2 \qquad K_2' = 2\psi_2/\psi_1 \qquad (21)$$

In Table I some intrinsic constants are calculated for various models that have been applied to cooperative proteins. Michaelis–Menten binding is indicated if there are equal independent sites, i.e., $K_1' = K_2' = K_3' = K_4'$. In the "tetrahedral" and "square" versions of the simplest sequential model the constants have a fixed relationship to each other, i.e., $K_1'K_4' = K_2'K_3'$. In the symmetry model of Monod *et al.* the constants are again limited in relation to each other, i.e., $K_1' < K_2' \leq K_3' \leq K_4'$. In other cases a mixture of positive and negative cooperativity leads to a different relationship of intrinsic constants. Thus in the case of yeast glyceraldehyde-3-phosphate dehydrogenase it was found that $K_1' < K_2' > K_3' > K_4'$ (*33*). This showed that this enzyme follows neither the symmetry model nor the simplest sequential model but requires a more complex sequential model (cf. Fig. 8). Determination of the ψ values by curve fitting and calculation of intrinsic constants can therefore delineate between models (*33*).

Two limitations of this diagnostic approach are apparent. First, if two different theories give the same relation of ψ values, they cannot be distinguished from each other. Thus, the square and tetrahedral versions of the simplest sequential model both give $K_1'K_4' = K_2'K_3'$ and hence such a finding is compatible with either alternative. Second, two different theoretical curves may provide equally good fits to actual data in a particular case and therefore a choice cannot be made. Such was the case of the fit of saturation curves to the hemoglobin data (*14*). In principle more accurate data could distinguish between models but in practice this may be difficult.

Two features of Table I are worth emphasis. The first is that determination of the intrinsic constants immediately puts limits on the molecular parameters K_t, K_{AB}, etc. In many cases these appear as products which must be further analyzed by other experiments, but in any case saturation curves provide important steps in determining the molecular parameters. In the second place the relation of the intrinsic constants emphasizes that a binding equation with $\psi_1\psi_2\psi_3\psi_4$ may not necessarily have four independent parameters. In the simplest Monod–Wyman–Changeux case with exclusive binding there are two independent parameters. In the simplest sequential square case there are also two. As further complexity is introduced into the models, e.g., nonexclusive binding and more than two

TABLE I

INTRINSIC CONSTANTS CALCULATED FOR FOUR POSSIBLE MODELS FOR A TETRAMERIC PROTEIN[a]

Subunit arrangement	Intrinsic constants [cf. Eqs. (1) and (2) in text]			
	K_1'	K_2'	K_3'	K_4'
Michaelis–Menten	$K_{S_B}K_{t_{AB}}$	$K_{S_B}K_{t_{AB}}$	$K_{S_B}K_{t_{AB}}$	$K_{S_B}K_{t_{AB}}$
Tetrahedral	$K_{AB}^3K_{S_B}K_{t_{AB}}$	$K_{AB}K_{BB}K_{S_B}K_{t_{AB}}$	$\frac{K_{BB}^2}{K_{AB}}K_{S_B}K_{t_{AB}}$	$\frac{K_{BB}^3}{K_{AB}^3}K_{S_B}K_{t_{AB}}$
Square	$K_{AB}^2K_{S_B}K_{t_{AB}}$	$\frac{(K_{AB}^2 + 2K_{BB})}{3}K_{S_B}K_{t_{AB}}$	$\frac{3K_{BB}^2K_{S_B}K_{t_{AB}}}{(K_{AB}^2 + 2K_{BB})}$	$\frac{K_{BB}^2}{K_{AB}^2}K_{S_B}K_{t_{AB}}$
Concerted[b]	$\frac{K_{t_{AB}}^4K_{BB}^4K_{S_B}}{1 + K_{t_{AB}}^4K_{BB}^4}$	K_{S_B}	K_{S_B} .	K_{S_B}

[a] All equations derived for case of (a) exclusive binding of S to B conformation and (b) all subunits identical. The saturation equation will be

$$N_S = \frac{4K_1'(S) + 12K_1'K_2'(S)^2 + 12K_1'K_2'K_3'(S)^3 + 4K_1'K_2'K_3'K_4'(S)^4}{1 + 4K_1'(S) + 6K_1'K_2'(S)^2 + 4K_1'K_2'K_3'(S)^3 + K_1'K_2'K_3'K_4'(S)^4}$$

[b] Constants K_1', K_2', K_3', K_4' in terms of Monod *et al.*, nomenclature would be $K_R/(1+L)$, K_R, K_R, and K_R, respectively.

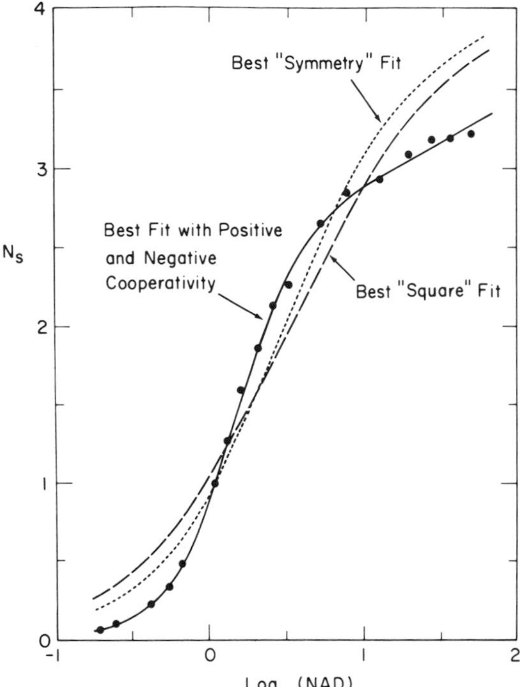

FIG. 8. Illustration of curve fitting procedures for data from yeast glyceraldehyde-3-phosphate dehydrogenase. Heavy line indicates computer selected best fit assuming both positive and negative cooperativity during binding. Dotted line indicates best fit assuming concerted model of Monod *et al.* (*16*). Dashed line indicates best fit with simplest sequential model of Koshland *et al.* (*14*) using square geometry. Data are from yeast glyceraldehyde-3-phosphate dehydrogenase (*27*).

conformational states, added parameters are introduced, but since one is dealing with molecular models these parameters are frequently interrelated. The ψ values are true experimental quantities whose interaction may reveal a relationship between fundamental molecular parameters. This curve fitting procedure is particularly important since the data can be obtained with small amounts of protein and even on occasion by kinetic analysis.

G. Equations of State

In many cases it will not be possible to distinguish models by analysis of the saturation curves alone, and in many cases where such a distinction can be made further data are needed to factor intrinsic constants into appropriate values of K_t, K_S, etc. In such situations a second type of comparison can be made by calculating an "equation of state," i.e., the fraction of the subunits in a new conformation compared to the total molecular species. Thus, the fraction of the total number of subunits present in the B conformation can be designated as $\bar{B}$ and this can be plotted as a function of the moles of ligand added or of $\bar{Y}$ in precisely the same way as a $\bar{Y}$ vs. (X) plot is obtained. Such a $\bar{B}$ equation for the species shown in columns I and II of Fig. 4 is given in Eq. (22). If no inhibitor or activator is added, the $\bar{B}$ equation simplifies to that of Eq.

$$\bar{B} = \frac{\begin{array}{c}(\boxed{S}\bigcirc + \boxed{S^J}\bigcirc + \boxed{J}\bigcirc) \\ + 2(\boxed{S}\boxed{S} + \boxed{J}\boxed{J} + \boxed{S^J}\boxed{J} + \boxed{S^J}\boxed{S} + \boxed{S}\boxed{J} + \boxed{S^J}\boxed{S^J})\end{array}}{\text{(All species in columns I and II of Fig. 4)}} \tag{22}$$

$$\bar{B} = \frac{\boxed{S}\bigcirc + 2\boxed{S}\boxed{S}}{\bigcirc\bigcirc + \boxed{S}\bigcirc + \boxed{S}\boxed{S}} \tag{23}$$

(23) whose right-hand side is precisely the equation observed for $\bar{Y}_S$ derived earlier [Eq. (3)]. In other words, for this model the fraction of sites in the B conformation equals the fraction of sites saturated, i.e., $\bar{B} = \bar{Y}$.

In the symmetry model for a dimer case [Eq. (8)] the $\bar{B}$ vs. (S) and $\bar{Y}$ vs. (S) curves would be given by Eqs. (24) and (25), and it is seen

$$\bar{B} = \frac{2\boxed{}\boxed{} + 2\boxed{S}\boxed{} + 2\boxed{S}\boxed{S}}{\bigcirc\bigcirc + \boxed{}\boxed{} + \boxed{S}\boxed{} + \boxed{S}\boxed{S}} \tag{24}$$

$$\bar{Y} = \frac{\boxed{S}\boxed{} + 2\boxed{S}\boxed{S}}{\bigcirc\bigcirc + \boxed{}\boxed{} + \boxed{S}\boxed{} + \boxed{S}\boxed{S}} \tag{25}$$

that $\bar{B}$ does not equal $\bar{Y}$ at equal (S) values. Thus, in Fig. 9 the straight line would be typical of the equation of state predicted by the simplest sequential model and the curved line that predicted by the symmetry model. It was precisely such an approach that Ogawa and McConnell used in their spin label studies on hemoglobin (*34*).

It is extremely important in such analyses to calculate the molecular species expected to be present in order to ascertain what kind of a curve is to be expected. If, for example, the $\bar{Y}$ vs. (S) curve indicated that only species $\bigcirc\bigcirc$ and $\boxed{S}\boxed{S}$ were important in the symmetry model, it too would obviously give a straight line on the $\bar{B}$ vs. $\bar{Y}$ plot. This is an improbable situation but the constants determined from the $\bar{Y}$ vs. (S) plot will indicate whether the $\bar{B}$ vs. $\bar{Y}$ plot should be linear or curved.

This point, however, has occasionally been forgotten by those who con-

34. S. Ogawa and H. McConnell, *Proc. Natl. Acad. Sci. U. S.* **58**, 19 (1967).

clude that a curved $\bar{B}$ vs. $\bar{Y}$ plot excludes a sequential mechanism. In Eq. (26) a sequential mechanism of the type used to explain rabbit mus-

$$\bigcirc\!\bigcirc \rightarrow \boxed{\mathrm{S}}\!\triangleright \rightarrow \boxed{\mathrm{S}\,|\,\mathrm{S}} \tag{26}$$

cle glyceraldehyde-3-phosphate dehydrogenase is shown. Whether a curved or linear equation of state is obtained in this case will depend on several factors. If the conformation C indicated in the BCS species has properties midway between A and B conformations as shown by some conformational probes such as spin label or ORD, then the $\overline{BC}$ vs. (S) [Eq. (27)] would be

$$\overline{BC} = \frac{1.5(\boxed{\mathrm{S}}\!\triangleright) + 2\boxed{\mathrm{S}\,|\,\mathrm{S}}}{\bigcirc\!\bigcirc + \boxed{\mathrm{S}}\!\triangleright + \boxed{\mathrm{S}\,|\,\mathrm{S}}} \tag{27}$$

This would clearly not give a linear $\overline{BC}$ vs. $\bar{Y}$ plot. Thus, a nonlinear plot will eliminate a simplest sequential model in which only A and B conformations are present but clearly does not exclude a general sequential model involving more than two conformational states. Proteolytic digestion (*35*), optical rotation (*36*), the ultracentrifuge (*37*), and amino acid

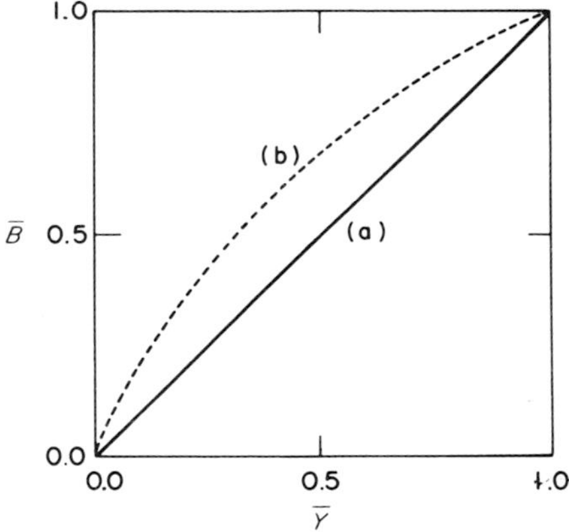

FIG. 9. Illustrative $\bar{B}$ vs. $\bar{Y}$ plot. Curved line would be obtained for concerted model if B_4S, B_4S_2, and B_4S_3 terms contribute and for general sequential models. Straight line for simplest sequential model (cf. text).

35. D. McClintock and G. Markus, *JBC* **243,** 2855 (1968); **244,** 36 (1969).
36. S. R. Simon and C. R. Cantor, *Proc. Natl. Acad. Sci. U. S.* **63,** 205 (1969).
37. J. C. Gerhart, H. K. Schachman, and J.-P. Changeux, *Biochemistry* **7,** 531 (1968).

reactivity (*38*) have all been used for such analyses of $\bar{B}$ vs. $\bar{Y}$ and offer one of the more significant tools for evaluation of molecular parameters.

H. Velocity Curves

The binding curve represents a true thermodynamic situation. When velocities are used in equivalent plots, the analysis is theoretically less certain. When the expression v/V_M is proportional to $\bar{Y}$ all of the statements made in regard to the binding curves will apply to the velocity curves. Because a velocity expression introduces added complexity one cannot *a priori* assume a proportionality to $\bar{Y}$. In the first place conformational transitions prior to the rate determining step may not all be rapid and reversible. Second, it cannot be assumed that the rate determining step in the decomposition of ES complexes are the same for all molecular species. Thus, if one converts Eq. (4), a binding equation, to a kinetic expression [Eq. (6)], the constants k_1 and k_2 are the turnover numbers for the protein containing one molecule and two molecules of ligand, respectively. It is not a necessary feature of all proteins that these two rate constants equal each other. Only when they do will v/V_M be proportional to the fraction saturation. In many cases velocity plots have been very useful despite their potential difficulties and have revealed properties of the enzyme which have been verified by binding studies. For example, the fact that the Hill slopes from kinetic data have in almost every case given a minimum number of subunits indicates that the approximations may be true in many cases. It has, for example, been shown in the case of glyceraldehyde-3-phosphate dehydrogenase that the v vs. (S) and $\bar{Y}$ vs. (S) plots are essentially identical (*39*). Thus, as a working hypothesis, velocity data can be used as long as the investigator is well aware of the limitations of the procedure.

Figure 10 shows a series of plots which parallel the plots of Fig. 6 but in which the rate constants k_1 and k_2 of Eq. (6) are not equal to each other. It is seen that these changes affect the slopes and shapes of the curve.

If velocity plots are compared with binding plots and agree well, one has prima facie support for assumptions (a) that the conformational changes are rapid and reversible prior to the rate determining step and (b) that the constants for various molecular species, ES_1, ES_2, etc., are the same. If the binding and kinetic data disagree then a rate determining

38. D. H. Meadows and O. Jardetsky, *Proc. Natl. Acad. Sci. U. S.* **61**, 406 (1968).
39. J. Teipel and D. E. Koshland, Jr., *BBA* **198**, 183 (1970).

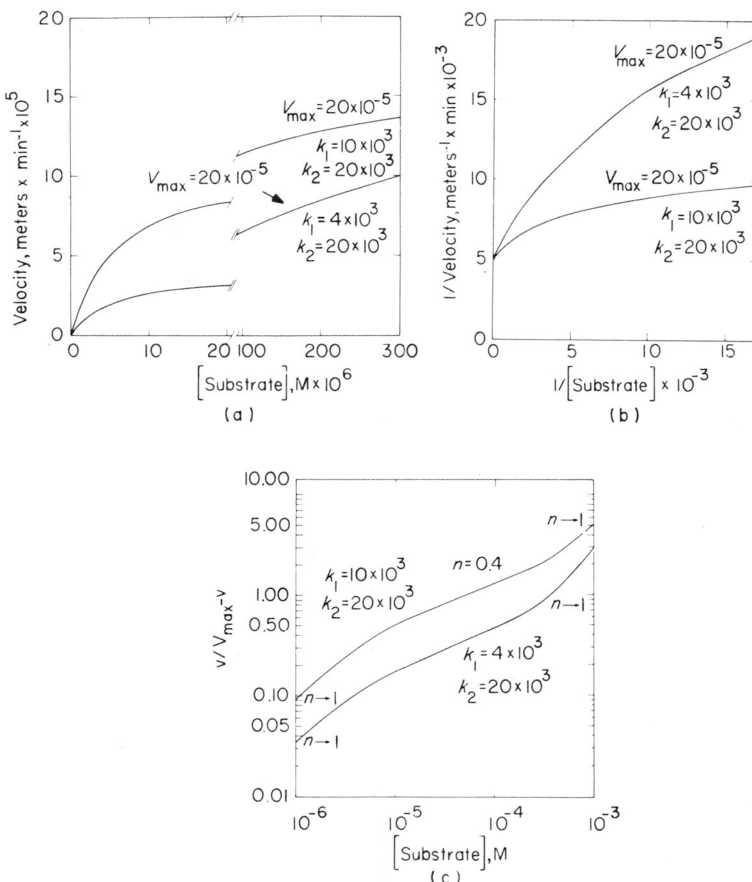

FIG. 10. Michaelis–Menten, double reciprocal, and Hill plots for negative cooperativity in which k_{cat} is increasing. Data plotted for equation $v = k_1(ES) + k_2(ES_2)$ in which (a) $k_1 = 0.4 \times 10^4$ min^{-1}, $k_2 = 2.0 \times 10^4$ min^{-1}, and (b) $k_1 = 1.0 \times 10^4$ min^{-1} and $k_2 = 2.0 \times 10^4$ min^{-1}. Concentrations of ES and ES_2 calculated based on dissociations constants $K_1 = 5 \times 10^{-6}\,M$ and $K_2 = 5 \times 10^{-4}\,M$.

conformational change may be indicated or the turnover number of different molecular species may differ. Frequently, one may wish to plot other functions such as one in which only those species containing S and J in the same subunit register activity. This would make J a compulsory activator, and a curve derived in this manner which fit the experimental data would support the hypothesis. Some applications of this approach to activator and inhibitor data have already been made (*15*).

I. Frozen Conformational States

Even when a theoretical saturation curve is fitted to experimental data, a number of ambiguities remain. Several theories may give a good approximation to the same data. In addition, the molecular parameters may always appear as product terms and thus be incapable of being distinguished individually. To factor out the contribution of individual terms and also resolve between models it seemed reasonable to attempt to "freeze" some of the conformational states thereby isolating some of the molecular factors and limiting the number of molecular species which need to be analyzed.

Hemoglobin offered a particularly advantageous situation in this regard. The nitroso derivatives and the oxidized cyanomet forms of the subunits appeared to be extremely similar to the conformations of the oxygen liganded subunits (*40–42*). In the former case the nitroso derivatives were stable for relatively short periods of time but were sufficiently stable for Gibson and co-workers to study fast reaction kinetics of conformational changes induced by oxygen (*43*). The oxidized ferric or cyanomet derivatives are stable for long intervals and allowed two groups (*44, 45*) to analyze the interactions between like chains by this procedure.

As seen in Fig. 11, either the α chains or the β chains can be converted to the cyanomet forms. This form of a subunit does not bind oxygen and has been shown by X-ray crystallography (*46*) to be in a conformation which mimics the oxygen liganded form of the subunit, here shown as the B conformation. The conformation changes during binding of oxygen to $\alpha_2^{+CN}\beta_2$ therefore changes the $\alpha_1\beta_1$, $\alpha_1\beta_2$, and β–β interactions whereas binding to the $\alpha_2\beta_2^{+CN}$ changes the $\alpha_1\beta_1$, $\alpha_1\beta_2$, and α–α interactions. Since the $\alpha_1\beta_1$ and $\alpha_1\beta_2$ interactions are the same for each step in the oxygen binding, cooperativity would be observed only if there were appreciable changes in β–β interactions for the $\alpha_2^{+CN}\beta_2$ case and α–α interactions for the $\alpha_2\beta_2^{+CN}$ case. When Hill plots of oxygen binding to these species were followed, a Hill coefficient of 1.3 was observed for the $\alpha_2^{+CN}\beta_2$ case showing appreciable changes in β–β interactions. This calculated to a value of 3.45 of $K_{BB\beta}$ or 700 cal for the β–β interaction. A Hill coefficient of 1.0 for $\alpha_2\beta_2^{+CN}$ showing negligible changes in α–α interactions was obtained.

40. J. J. M. DeVijlder and E. C. Slater, *BBA* **167,** 23 (1968).
41. E. Antonini, M. Brunori, J. Wyman, and R. W. Noble, *JBC* **241,** 3236 (1966).
42. I. Tyuma, R. E. Benesch, and R. Benesch, *Biochemistry* **5,** 2957 (1966).
43. Q. H. Gibson and L. J. Parkhurst, *JBC* **243,** 5521 (1968).
44. R. Banerjee and R. Cassoly, *JMB* **42,** 351 (1969).
45. J. E. Haber and D. E. Koshland, Jr., *BBA* **194,** 339 (1969).
46. M. F. Perutz, *Proc. Roy. Soc.* **B173,** 113 (1969); **B141,** 69 (1953).

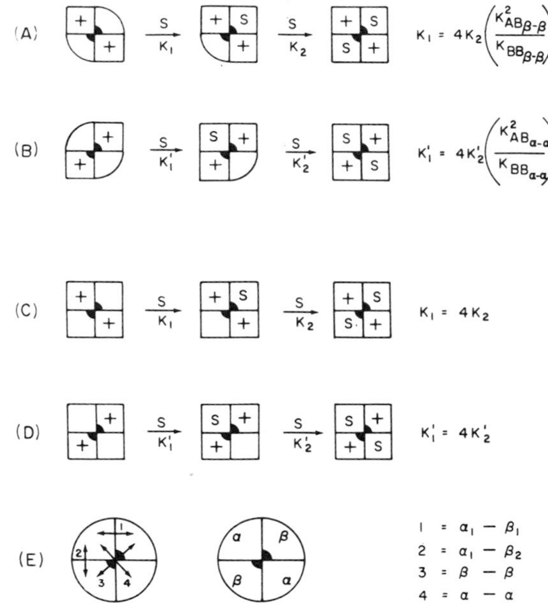

FIG. 11. Schematic illustration of use of frozen conformational states to determine subunit interactions. (A) Tetrameric protein with α subunits oxidized (illustrated by "+") so that they are frozen in a conformation which mimics the oxygen liganded state. Binding of oxygen to this protein will involve the same α–β-subunit changes in each step but will involve different β–β interactions in step 1 than in step 2. (B) The same protein in which the β subunits are frozen in the form which mimics the oxygen liganded state in which case the α–α interactions vary from step 1 to step 2. (C) The situation if the concerted model is followed, in which case the constants for the binding of ligand will be related simply by statistical factors. (D) The same situation is shown when the other subunits are frozen. The K's in (C) and (D) do not necessarily have to be equal since the net affinity of α subunits may not be identical to that of the β subunits but the slopes in the Hill plots of these cases must be 1.0. (E) Illustration of subunit interactions.

Both coefficients would have been 1.0 if a concerted model applied. This procedure therefore is able to (a) provide evidence in regard to sequential versus concerted models, (b) determine whether a tetrahedral model or a square model is needed to explain binding, and (c) place a quantitative value on some of the molecular parameters.

This result is of interest not only for hemoglobin but also may offer a general approach since affinity labels which glue a substrate analog into some of the subunits of an allosteric enzyme could provide the same role of freezing the conformational state as does the oxidation of the chains in hemoglobin.

J. Minimal Substrate Technique

A suggestion for the determination of more accurate saturation curves has been made (*39*). Using a minimal substrate (i.e., one which reacts but has so little structure that it induces a small, if any, conformational change) may make it possible to establish the limits of a saturation curve more precisely than when equilibrium dialysis techniques are used. At very high and very low substrate concentrations the discrimination between models for enzymic action is greatest, but it is precisely in this region that the equilibrium dialysis technique involves the subtraction of two large numbers and is hence inaccurate. In the case of glyceraldehyde-3-phosphate dehydrogenase it was found that the velocity of acetaldehyde reaction as a function of DPN concentration paralleled the binding of DPN very closely (*39*). This would suggest that acetaldehyde which is a very poor substrate for the enzyme (glyceraldehyde-3-phosphate is the true substrate) reacts at the subunit to which DPN is bound but otherwise causes no perturbation of the conformational states of the protein. Thus, it is in essence an indicator of the binding of DPN. However, since it is a velocity measurement, it can readily be carried to both low and high substrate concentrations. The procedure, for example, would involve reaction of a minimal substrate (acetaldehyde) kept at constant concentration with a normal substrate (NAD) which is varied in concentration. A measurement of the binding of the normal substrate in the middle range where equilibrium dialysis is accurate should then be correlated with the turnover of the enzyme. If these two results are parallel, extension to the extremes of substrate concentration can be made using reaction velocity as an indicator of binding. The procedure has some elements of peril. The middle range may be largely dominated by ES_2 and ES_3, and it just might happen that ES_4 has a much higher turnover number than ES_3. One can usually carry out equilibrium dialysis studies so there is some overlap with ES_4; thus a good working hypothesis, if not a definitive proof of mechanism, may be obtained by such an approach.

K. Possible Procedure for Evaluation of an Allosteric Protein

Although each protein is an entity unto itself, the diagnostic tests listed above have now been utilized sufficiently that an operating procedure for the investigation of a new protein might be outlined. The procedures which would give valuable information might be as follows:

(1) Purify the protein and determine the numbers and kinds of subunits.

(2) Measure the binding saturation curve with each single ligand if possible.

(3) Establish from curve fitting procedures the intrinsic constants K_1', K_2', etc., and check the relationship between these constants to establish whether any simple mechanism is operating or if a simple mechanism is a close first approximation to the data.

(4) Plot the same data in a number of different ways as double reciprocal plots, Hill plots, log(X) vs. N_X plots, etc. Each of these plots can be revealing in different ways, and checking their properties against those discussed in the sections above and the original literature will indicate phenomenological features of the enzyme. Table II summarizes briefly the working hypothesis that can be drawn from these diagnostic tests.

(5) From the working hypothesis deduced from the model studies postulate the molecular species which are likely to be present and check against other diagnostic procedures; for example, predict the $\bar{B}$ vs. $\bar{Y}$ plot and using some appropriate physical tool, check the theory against experimental fact.

(6) Combine more than one ligand and determine whether the mixed situation is adequately described by the molecular species deduced from the studies of individual ligands. If deviations occur, introduce the appropriate modifications. In some cases the deviations will provide valuable new information. For example, if ligand X_1 shows no tendency to bind to the protein in the absence of ligand X_2 but does provide a good binding curve in the presence of ligand X_2, a compulsory binding order is indicated. Thus, from this a prediction of the binding pattern of X_1 and X_2 in various combinations can be made and checked experimentally.

(7) When the effect of one ligand on another is being explored, the extremes will usually be most easy to examine first. Thus, by varying the "variable" ligand keeping the "fixed" ligand at extreme values (i.e., complete saturation or very low concentrations), initial working hypothesis in the presence of both ligands can be obtained.

(8) Check association–dissociation phenomena by varying the concentration of enzyme and repeating some of the plots outlined above. A change dependent on protein concentration would suggest association–dissociation phenomena.

(9) Finally, check velocity curves against the saturation curves to determine whether the velocity studies parallel the binding studies. Check various combinations of activators, inhibitors, and substrates to determine whether the molecular species postulated are adequate to describe these varying situations.

No detailed procedure is possible for all cases. In some cases the structural information may be present in advance of the detailed binding

TABLE II
SUMMARY OF DIAGNOSTIC TESTS FOR PROTEIN WITH IDENTICAL SUBUNITS[a]

Diagnostic test	Observation	Conclusion
$\bar{Y}$ vs. (X) plot	Hyperbolic curve	Noninteracting sites[b]
	Sigmoid curve	Positive cooperativity
	"Bumps"	Possibility of positive and negative cooperativity (if metal ion, etc., not limiting)
Cooperativity index, R_X	$R_X = 81$	Noninteracting sites[b]
	$R_X < 81$	Positive cooperativity
	$R_X > 81$	Negative cooperativity
Hill plot	$n_H = 1$	Noninteracting sites[b]
	$n_H > 1$	Positive cooperativity
	$n_H < 1$	Negative cooperativity
	$n_H > 1$	$n_M > n_H$
Double reciprocal plot	Straight line	Noninteracting sites, V_M and K_m determined
	Concave upward	Positive cooperativity
	Concave downward	Negative cooperativity
Scatchard plot	Straight line	Noninteracting sites[b]
	Concave downward	Positive cooperativity
	Concave upward	Negative cooperativity
	Extrapolation at high (X)	Intercept gives n_M
Curve fitting procedure to obtain intrinsic constants	Intrinsic constants all equal	Noninteracting sites[b]
	For a tetramer:	
	$K_1'K_4' = K_2'K_3'$	Simplest sequential square or tetrahedral model
	$K_1' \leq K_2' = K_3' = K_4'$	Symmetry model
	$K_1' \leq K_2' \leq K_3' \leq K_4'$	Simple positive cooperativity
	$K_1' \geq K_2' \geq K_3' \geq K_4'$	Simple negative cooperativity
	$K_1' \leq K_2' \geq K_3' \geq K_4'$, etc.	Mixed positive and negative cooperativity
Equation of state	$\bar{B} = \bar{Y}$	Simplest sequential model or symmetry in which A_{n_M} and $A_{n_M}B_{n_M}S_{n_M}$ are only species
	$\bar{B} > \bar{Y}$	General sequential or symmetry model
Order of binding	X_1 must be bound to protein in order for X_2 to bind	X_1 induces conformational change

TABLE II (*Continued*)

Diagnostic test	Observation	Conclusion
Activator	J which is not a substrate increases velocity or affinity of X_2	J probably induces conformational change—in special cases, e.g., metal ion, may take direct part in catalytic action
Inhibitor	Competitive	May compete directly at active site without conformational change. If acts at distant site induces conformational change
	Noncompetitive	Induces conformational change

[a] The limitations of each diagnostic test are discussed in the text and in the original literature.

[b] Noninteracting sites can arise either (a) by the absence of a conformational change and no direct site–site interactions or (b) an induced conformational change in which the distortion is not transmitted to neighboring subunits, e.g., $K_{AB} = K_{BB} = 1$. Thus evidence for noninteracting sites coupled with some evidence for a conformational change requires the ligand-induced sequential model.

studies. It is even possible that X-ray studies and the geometrical relation of subunits could provide insight into potential models. In other cases the enzyme may be in a low state of purification, and velocity studies may give tentative working hypotheses even before the protein is purified.

V. Molecular Properties of Allosteric Proteins

The diagnostic tests described above and the physical tools which have been developed to evaluate these models have begun to give us insight into the behavior of regulatory proteins. A few illustrative examples in which the mathematical and conceptual approaches allow us to explain observed phenomena in molecular terms will be examined.

A. Molecular Basis of Cooperativity

The mathematical analyses described above allow one to put qualitative understanding of regulatory control on a quantitative basis. In addition, they give a more precise understanding of the molecular basis of cooperativity. It is seen that Michaelis–Menten kinetics will occur if the intrinsic binding constants of a saturation equation are all equal to each other. This will occur if there are no conformational changes as in a template-type enzyme. It will also be found in an enzyme in which conformational changes are occurring in individual subunits but where there is no transmission of the ligand-induced distortion to neighboring subunits. Such an observation would be made if the subunit interaction parameters K_{AB}, K_{BB}, etc., are equal to one. One enzyme in which conformational changes within the subunit have been demonstrated conclusively but in which the two active sites apparently act independently is alcohol dehydrogenases studied by Theorell (*47*). Undoubtedly many other proteins operate in a similar manner.

On the other hand, cooperativity is obtained if K_{AB} or K_{BB} are not equal to one, i.e., there are ligand-induced changes which affect subunit interactions. Since the theory of ligand-induced changes provided no reason to assume that stability relationships with neighboring subunits should always increase or stay the same, all such values were considered in the computer analysis of saturation curves. From these relations the ligand-induced model predicted negative cooperativity as well as positive cooperativity and this was subsequently found experimentally (*22, 27, 28, 40, 48*). In Fig. 12 the schematic illustration of subunit interactions which lead to no change, greater or less stability is shown together with the subunit interaction terms which will be greater, equal to, or less than one. The fact that these qualitative considerations at the protein level and the quantitative evaluations at the mathematical level fit so well and cover such a wide range of enzyme phenomena leads to considerable confidence that ligand-induced changes are a general and important mechanism and probably the fundamental explanation for allosteric effects in regulatory proteins.

Extending this example to other cooperative phenomena is easy, and some simple examples are illustrated in Fig. 13 for a tetramer. If adjacent B subunits are more attractive than adjacent AA, AB, or BC subunits, the binding of S will be cooperative and will show a sigmoid v vs. (S) curve (line 1). If J induces the same conformational change, it will show a simple positive homotropic cooperativity (line 2). If prior to the addition of substrate, sufficient activator is added to saturate all sites which will bind J and thereby convert the protein completely to the B_4 conformational state, the binding of S can then proceed without further conformational change (line 3). Thus, S will not only bind more readily (a decreased apparent K_m), but the binding of S will now follow a simple Michaelis–Menten pattern since no changes in subunit interactions occur

47. H. Theorell, *Harvey Lectures* **5**, 17 (1966).
48. L. M. Corwin and G. Fanning, *JBC* **243**, 3517 (1968).

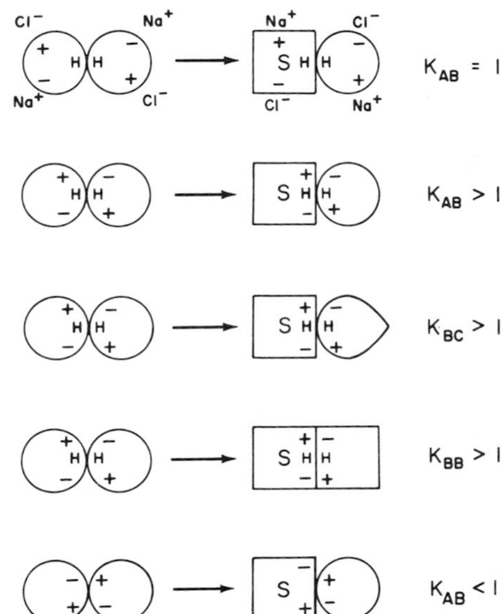

FIG. 12. Alternative forms of the change in subunit interactions as a result of ligand-induced conformational changes. H stands for hydrophobic bonds which attract the two subunits to each other and are, in this case, presumed not to change during the conformation changes. The "+" and "−" signs represent electrostatic charges but could equally well represent hydrogen bond attractions or hydrophobic attractions. In the top line the situation in which the changes in conformation do not cause any net change in subunit interactions are illustrated. Sodium ions and chloride ions are depicted to indicate that distant positive and negative charges can be neutralized by counterions in the solution. In this case the electrostatic charges are too far apart to provide stabilization of the subunits. Thus, $K_{AB} = 1$ because there is no change in the subunit interactions. In line 2 the positive and negative charges are brought near enough to each other to give added stabilization by the change in one subunit, but there is no change in the shape of the unliganded subunit. It is presumed that counterions such as sodium and chloride ions are present in this system also, of course, but the intersubunit charges in this case are close enough to have some positive interaction. In the third line there is again a net stabilization because of the added attraction of the adjacent electrostatic charges, but the adjacent unliganded subunit is changed in shape as well as stability. In the fourth line the neighboring subunit is changed in shape so that it is essentially identical to that of the liganded subunit because of the strong interactions of the adjacent charges. In the final line the induced conformation separates attractive opposite charges so that subunit interaction is decreased.

when (S) is bound. Thus, the activator J has changed the shape as well as the midpoint of the activity versus (S) curve.

There is nothing in line 1 of Fig. 13 to indicate that the ligand-induced changes must lead to positive cooperativity. If there is no change in subunit interactions ($K_{AB} = K_{BB} = 1$) (cf. line 1 of Fig. 12) there will be

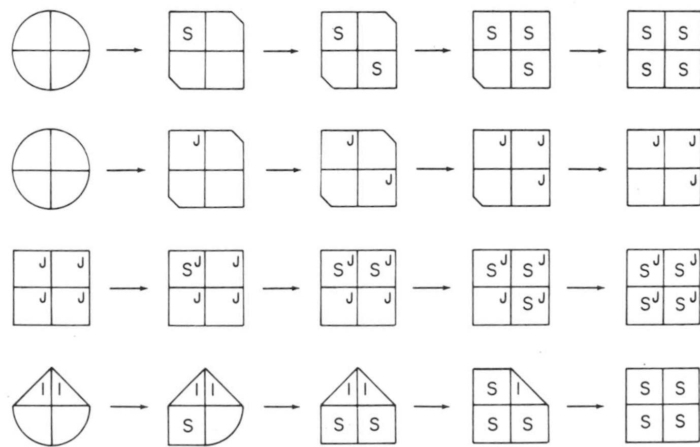

FIG. 13. Tetrameric interactions analogous to those of the dimer in Fig. 3 except ligand-induced distortions of neighboring subunits are included.

no cooperativity, and Michaelis–Menten kinetics will be observed. If there is a net decrease in stability between adjacent B conformations ($K_{BB} < 1$, K_{AB}, etc., $= 1$) then negative cooperativity will be observed. The mathematical effects of varying K_{AB} and K_{BB} while keeping K_S and K_t constant have been shown (*14, 32*).

It is also worth noting that these figures explain how ligand-induced changes may lead to association or dissociation in some proteins and not others. As seen in Fig. 12 the ligand-induced distortions can lead either to strengthening or weakening of subunit interactions relative to adjacent A subunits. If the K_{AA} value is very strong, then a change in subunit interactions even if it tends to destabilize the subunit attractions will not be enough to lead to dissociation. If K_{AA} is only moderate, the binding of ligand may provide the marginal decrement leading to dissociation of the protein. Conversely, the added ligands may lead to a strengthening of subunit interactions and to an association of subunits. Reichard's work (*49*) on thioredoxin reductase provides an elegant example of the influence of allosteric effectors on the association–dissociation properties of a regulatory protein.

On the fourth line of Fig. 13 is shown the binding of S to a protein which contains two molecules of I initially. It is conceivable that an inhibitor which prevents the binding of S to that subunit affects subunit interactions in the same way as S and J. In that case the binding of I to two of the subunits will make the first molecule of S easier to bind but would prevent the binding of more than two molecules of S. As further S is added, it would tend to displace I in a normal competitive manner.

49. P. Reichard, *Biochemistry* **6,** 1016 (1967).

In that way I would act as an activator at very low concentrations of S and in inhibitor at high concentrations of S. This type of complex behavior has already been observed by Atkinson for isocitric dehydrogenase (*50*) and by Gerhart and Pardee for aspartyltranscarbamylase (*8*).

B. Protein Design

The calculation of a specific model depends on a knowledge of the arrangements of the subunits in the protein. Although many of the qualitative features such as negative and positive cooperativity will occur for any type of structure, the detailed quantitative evaluation involves a knowledge of the geometry of the subunits. In the early analysis, various illustrative geometrics (tetrahedral, square, etc.) and an averaged value for the subunit interaction energy were used (*14*). As the models become more sophisticated, it will be necessary to be more precise. It will also be necessary to consider that the amino acid contacts in one direction from a given subunit will be different from the interactions in a different direction (cf. Fig. 16). Although the average subunit interaction terms may be sufficient in many cases, the detailed consideration of these factors will be important in others.

The more refined treatment of subunit interactions has been considered in detail (*51*), and some of the possible subunit contacts in dimers, trimers, and tetramers of identical subunits are shown in Figs. 14 and 15. In these figures p, q, and r represent one or more amino acid residues in the binding domain of the subunit and adjacent pp domains indicate isologous binding in the sense of Monod *et al.* (*16*) whereas pq domains indicate heterlogous binding. The argument that such structures are thermodynamically possible and even probable is presented elsewhere (*51*). At this point we will merely examine how these structures affect our interpretation of subunit interactions. If we now extend the definition K_{AB} to $K_{AB_{pq}}$ meaning that we are considering the interaction between the p domain on one subunit in the A conformation with the q domain in the second subunit in the B conformation (cf. Section VII), then the procedures outlined above can again give detailed constants for these models. In Fig. 16, for example, the subunit interaction terms for some molecular species are shown. In Table III the intrinsic constants calculated for all-isologous (I_4) and all-heterologous (H_4) square tetramer and for the all-isologous tetrahedral tetramer (I_6) are shown.

The difference in the constants calculated considering the detailed binding domains and those using averaged subunit interaction terms can

50. D. E. Atkinson, J. A. Hathaway, and E. C. Smith, *JBC* **240**, 2682 (1965).
51. A. Cornish-Bowden and D. E. Koshland, Jr., *JBC* (1970) (in press).

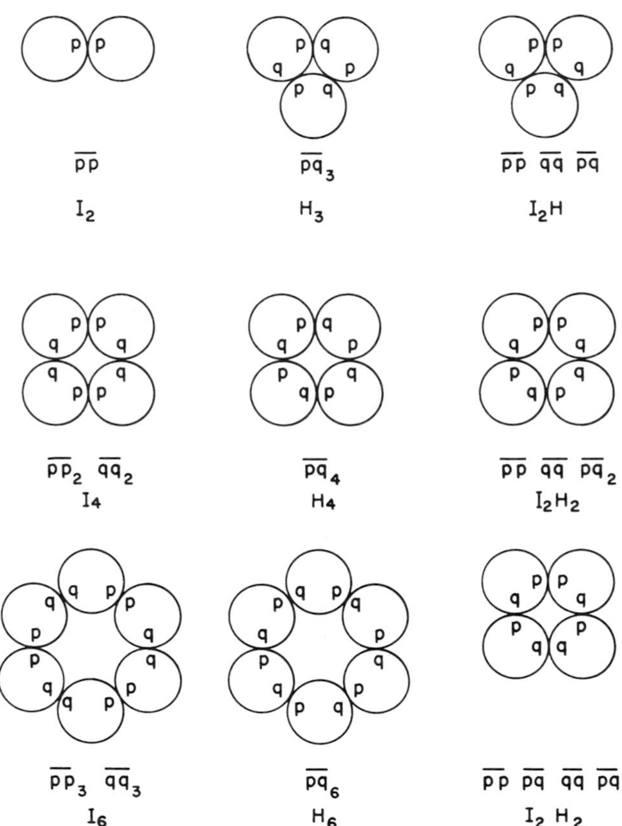

FIG. 14. Possible closed structures with identical subunits. Domains of bonding are indicated by p and q to suggest different amino acids or different arrangements of amino acids. The pp and qq bonds represent isologous binding and are also indicated by I, pq bonds indicate heterologous binding and are designated by H. All isologous structures are shown in column 1, all heterologous structures in column 2, and mixed isologous and heterologous in column 3. Structures can be specified by pp, pq, etc., or by I_4, H_4, etc.

be seen by comparing Table III with Table I. It is seen that the H_4 case gives the same constants as those derived with average constants although the detailed breakdown is slightly different. In the I_4 cases and I_6 cases however there are slight differences in the relations between the intrinsic constants, and these might be potentially available to experimental delineation when more sophisticated tools for conformations in solution becomes available. In many of these cases distinguishing between constants depends on highly accurate data. However, in other cases the complexity of the saturation curve already indicates that the simple

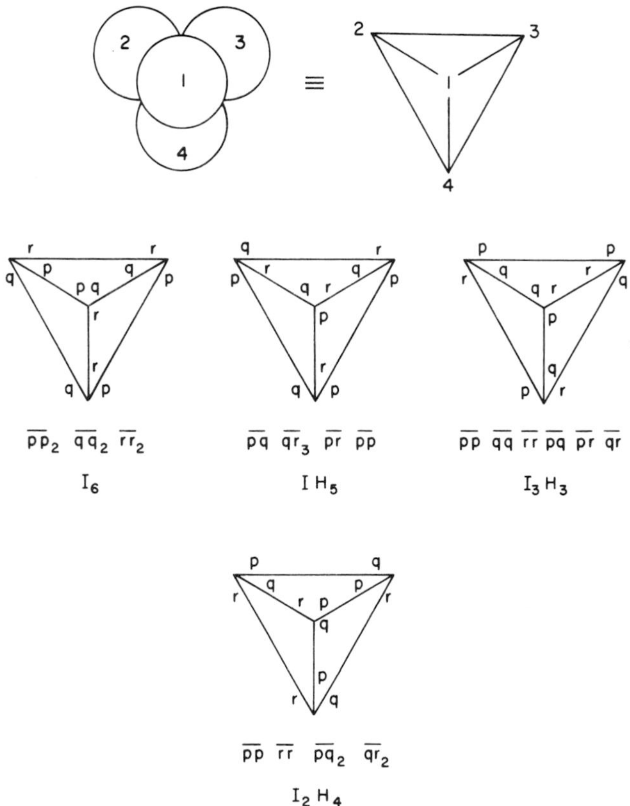

FIG. 15. Possible structures for tetrahedral arrangement of four subunits in which each subunit interacts with three other subunits. The schematic designation is illustrated below by placing the amino acid domains which interact with each other along the lines connecting the subunits. The same designations as in Fig. 14 are used for the domains of bonding.

models using averaged interactions between subunits will be insufficient to explain the data.

C. Status of Simple Models

Although the fundamental property of conformational changes appears to be accepted as a general basis for regulatory control, the quantitative complexity of the mathematics with many subunits leads to a desire for simplifying assumptions. Models based on such simplifying assumptions have been described briefly above and in detail elsewhere (*5*). Can it be

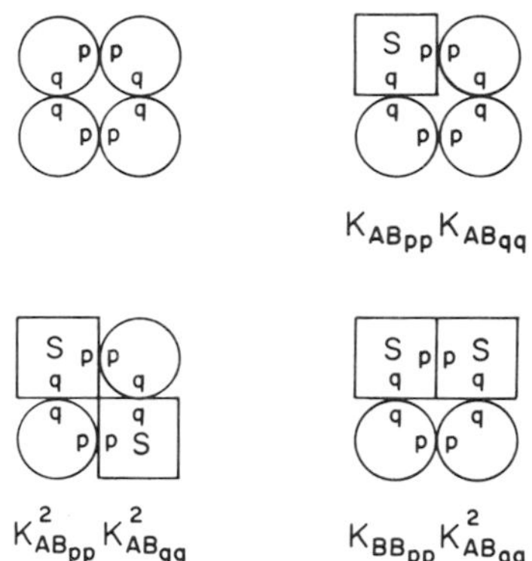

FIG. 16. Diagram to show ligand-induced changes in a tetramer containing identical subunits will not produce the same changes in subunit interactions at all interfaces. The subunit interaction terms for the different molecular species are indicated using the convention outlined in the glossary (Section VII). The corresponding terms using the averaged subunit interaction terms would be 1, K_{AB}^2, K_{AB}^4, and $K_{BB}K_{AB}^2$, respectively.

said that the present evidence allows us to say that a specific simple model can be used *a priori* in studying a new protein? The answer would appear to be "no."

The theory of Monod *et al.* (*16*) postulated two fundamental assumptions which would severely limit the possible molecular species in any regulatory protein. The first was the presence of a preexisting equilibrium between two or a few states and the second was that any conformational change affected all subunits identically, thus "conserving the symmetry of the protein." The model in this form is elegantly simple. Activators and substrates favor the R state. Inhibitors favor the T state. The shift from the T to the R state under the influence of substrate binding explains cooperativity. The model can be made somewhat more complex by allowing more than two states and by allowing nonexclusive binding, but the fundamental postulates of preexisting states and conservation of symmetry simplify mathematical and conceptual complexities. The model leads to predictions which can be tested experimentally, e.g., only positive cooperativity in homotropic interactions and the absence of hybrid conformational states. A number of proteins, e.g., phosphorylase and phosphofructokinase, give saturation curves which are qualitatively compatible with this theory (*51a*). Yeast glyceraldehyde-3-phosphate dehydrogenase at 40° has been found to give temperature jump data which are compatible with the symmetry model and incompatible with the simplest sequential model (*52*). However, data at 25° and in more com-

51a. D. Blangy, H. Buc, and J. Monod, *JMB* **31,** 13 (1968).

52. K. Kirschner, M. Eigen, R. Bittman, and B. Voigt, *Proc. Natl. Acad. Sci. U. S.* **56,** 1661 (1966).

TABLE III

INTRINSIC CONSTANTS FOR SOME TETRAMERIC MODELS IF THE DIFFERENT DOMAINS OF BINDING ON A SINGLE POLYPEPTIDE CHAIN ARE CONSIDERED

	Constants for polymeric structure indicated		
	I_4	H_4	H_6
$K_{\overline{AB}}$	$(K_{AB_{pp}} K_{AB_{qq}})^{1/2}$	$(K_{AB_{pq}} K_{AB_{qp}})^{1/2}$	$(K_{AB_{pp}} K_{AB_{qq}} K_{AB_{rr}})^{1/3}$
$K_{\overline{BB}}$	$(K_{BB_{pp}} K_{BB_{qq}})^{1/2}$	$K_{BB_{pq}}$	$(K_{BB_{pp}} K_{BB_{qq}} K_{BB_{rr}})^{1/3}$
$(K_1'/K_{S_B} K_{t_{AB}})$	$K^2_{\overline{AB}}$	$K^2_{\overline{AB}}$	$K^3_{\overline{AB}}$
$(K_2'/K_{S_B} K_{t_{AB}})$	$\dfrac{(K_{BB_{pp}} + K_{BB_{qq}} + K^2_{\overline{AB}})}{3}$	$\dfrac{K^2_{\overline{AB}} + 2K_{\overline{BB}}}{3}$	$\dfrac{K_{BB_{pp}} K^2_{AB_{qq}} K^2_{AB_{rr}} + K^2_{AB_{pp}} K_{BB_{qq}} K^2_{AB_{rr}} + K^2_{AB_{pp}} K^2_{AB_{qq}} K_{BB_{rr}}}{3K^3_{\overline{AB}}}$
$(K_3'/K_{S_B} K_{t_{AB}})$	$\dfrac{3K^2_{\overline{BB}}}{K_{BB_{pp}} + K_{BB_{qq}} + K^2_{\overline{AB}}}$	$\dfrac{3K^2_{\overline{BB}}}{K^2_{\overline{AB}} + 2K^2_{\overline{BB}}}$	$\dfrac{3K^3_{\overline{AB}} K^3_{\overline{BB}}}{K_{BB_{pp}} K^2_{AB_{qq}} K^2_{AB_{rr}} + K^2_{AB_{pp}} K_{BB_{qq}} K^2_{AB_{rr}} + K^2_{AB_{pp}} K^2_{AB_{qq}} K_{BB_{rr}}}$
$(K_4'/K_{S_B} K_{t_{AB}})$	$\dfrac{K^2_{\overline{BB}}}{K^2_{\overline{AB}}}$	$\dfrac{K^2_{\overline{BB}}}{K^2_{\overline{AB}}}$	$\dfrac{K^3_{\overline{BB}}}{K^3_{\overline{AB}}}$

plex circumstances suggest that sequential binding patterns are involved (*27, 53*) so that the final answer on this enzyme awaits further evidence.

Regardless of the ultimate role of symmetry, the contributions of this model in the field of allosterism are very great. Monod *et al.* provided a simple conceptual framework for many of the early ideas, emphasized the importance of symmetry features in protein design, provided penetrating analysis of the role of subunits in the cooperative process, and defined many of the terms which have added clarity and insight into this important area.

Atkinson proceeded from a different point of view and applied the reasoning of the induced-fit theory to presume that progressive deformation of the structure of a protein induced by ligand can explain regulatory properties. In this case the subunit to which the ligand was bound would be distorted and would transmit some of this distortion to neighboring subunits, thus altering their properties to produce cooperativity. Although this model did not provide a technique for quantitative calculation, its essence must in part be true in most cases and it has been strongly supported in a few cases. For example, the data for isocitric dehydrogenase (*13, 54*), rabbit muscle glyceraldehydephosphate dehydrogenase (*36*), and CTP synthetase (*22*) are compatible with this hypothesis. The basic idea of the Atkinson proposal is certainly true for many proteins.

Koshland, Nemethy, and Filmer explained allosteric effects also by assuming ligand-induced changes but emphasized several other aspects in their analysis. Their major postulates were (a) that the changes in conformation were ligand induced, (b) that subunit interactions were the key to the homotropic processes, and (c) that three molecular parameters describing the energy of the conformational change, the strength of the subunit interactions, and the affinity of ligand to defined conformational states could describe the complex behavior of the conformational states. This model basically suggested a general treatment of the type outlined in Sections II and III, but a much simpler model could be developed if the two postulates of the "simplest sequential model" described in Section III are made. It appears that this simplest sequential model may be a good first approximation for hemoglobin (see Section V,E below).

Like the symmetry model, the simplest sequential model appears to be too simple to fit many proteins but the mathematical tools developed for these simple equations, the curve fitting procedures, the concepts of subunit interactions, the features of protein design, and the diagnostic tools remain of key value in the more complete and general model.

A large number of proteins, however, fall in the category of the general

53. B. Chance and J. H. Park, *JBC* **242**, 5093 (1967).
54. J. A. Hathaway and D. E. Atkinson, *JBC* **238**, 2875 (1963).

sequential model if negative cooperativity is accepted as evidence for such a classification. Negative cooperativity is a strong argument (*22*) for ligand induced conformational effects if the protein is composed of identical subunits. Many of these proteins appear to fall in this category. Even in cases in which ultimate proof of subunit identity is absent, analogy to similar proteins provides strong presumptive support that ligand-induced changes are involved. More significantly in some cases a combination of negative and positive cooperativity is found in the same protein (*22*). Such a combination must result from ligand-induced changes. Finally, certain enzymes show "bumps" in their saturation curves, an indication, although not definitive proof, of ligand-induced changes (*54a*). The enzymes which apparently require ligand-induced sequential changes based on the first two critera are deoxythimidine kinase (*55*), homoserine dehydrogenase (*56*), human heart LDH (*57*), glutamic dehydrogenases (*58–62*), rabbit muscle GD (*32*), phosphoglucomutase (*63*), and purine nucleoside phosphorylase (*64*). Those which exhibit bumps and might be added to the list are CTP synthetase (*22*), PEP carboxylase (*48*), yeast TPD (*65*), threonine deaminase (*66*) and *E. coli* ADPG pyrophosphorylase (*67*).

D. Alternative Approaches

Two alternative approaches have been presented which involve quite different explanations for the unusual kinetic behavior of regulatory proteins. One of these postulates assumes rate determining steps in the interactions of species at a single site, and the other involves association–dissociation phenomena.

The rate determining interactions have been derived by Ferdinand and

54a. J. Teipel and D. E. Koshland, Jr., *Biochemistry* **8,** 4656 (1969).
55. R. Okazaki and A. Kornberg, *JBC* **239, 275** (1964).
56. P. Datta and H. Gest, *JBC* **240,** 3023 (1965).
57. J. S. Nisselbaum and O. Bodansky, *JBC* **236,** 323 (1961).
58. H. B. LéJohn and S. Jackson, *JBC* **243, 3447** (1968).
59. L. Corman and N. O. Kaplan, *Biochemistry* **4,** 2175 (1965).
60. L. Corman and N. O. Kaplan, *JBC* **242,** 2840 (1967).
61. B. O. Wiggert and P. P. Cohen, *JBC* **241,** 210 (1966).
62. C. Frieden, *JBC* **234,** 815 (1959).
63. O. H. Lowry and J. V. Pasonneau, *JBC* **244,** 910 (1969).
64. B. K. Kimm, S. Cha, and R. E. Parks, Jr., *JBC* **243, 1763** (1968).
65. D. E. Koshland, Jr., R. A. Cook, and A. Cornish-Bowden, *in* "Pyridine Nucleotide Dependent Dehydrogenases," p. 199. Springer, New York, 1970.
66. P. Datta, *JBC* **244,** 858 (1969).
67. N. Gentner and J. Preiss, *JBC* **243,** 5882 (1968).

Rabin and others (*68–71*). It is indeed possible to obtain higher powers of substrate concentration in the rate expression for unusual kinetic pathways with limited combinations of rate determining steps. One cannot simply conclude from a sigmoid kinetic curve that conformational changes are present. However, far more extensive studies are needed to show whether kinetic effects can provide the extended amount of cooperativity which has been observed in allosteric proteins. Moreover, no single enzyme has yet been shown to follow such a kinetic process. Therefore, this alternative remains a potential one rather than an established one.

Association–dissociation phenomena have also been postulated as an alternative hypothesis. It is well known that many proteins undergo association–dissociation behavior as has been pointed out by Frieden (*18*) and Nichol *et al.* (*72*). To explain cooperativity in the absence of conformational change involves special assumptions. One might, for example, imagine a dimeric protein in which the two active sites are blocked because of steric positioning as shown in Fig. 17a. In this case the monomers would be active and the dimer inactive. Or one could imagine an active site bridged between two subunits (Fig. 17b) in which case only the dimer would be active and the monomers inactive. In both of these cases addition of substrates would produce sigmoid curves and therefore provide an alternative explanation for sigmoid behavior.

In many other cases, however, association–dissociation phenomena are a natural consequence of ligand-induced conformation changes (*49*). As shown in Fig. 12 ligand-induced changes can affect subunit interactions and if this leads to weakening in the contacts of a protein which is on the verge of dissociation, the conformational change would be sufficient to induce dissociation of the protein. Conversely, a monomeric or dimeric protein might be induced into a new conformational state which would lead to association to form a tetramer. In the study of glutamic dehydrogenase by Frieden (*73*) it was demonstrated that the association–dissociation phenomena were a concomitant of ligand-induced conformational changes and that the primary regulatory effect was the conformational change not the association–dissociation phenomena. Since, however, the association of dimers to form tetramers or any other polymerization

68. W. Ferdinand, *BJ* **98,** 278 (1966).
69. B. R. Rabin, *BJ* **102,** 22C (1967).
70. J. R. Sweeny and J. R. Fisher, *Biochemistry* **7,** 571 (1968).
71. B. D. Sanwal and R. A. Cook, *Biochemistry* **5,** 886 (1966).
72. L. W. Nichol, W. T. H. Jackson, and D. J. Winzor, *Biochemistry* **6,** 2449 (1967).
73. C. Frieden, *JBC* **238,** 3286 (1963).

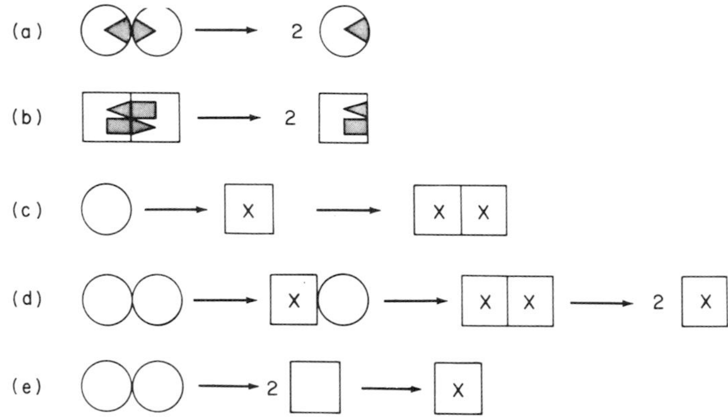

FIG. 17. Schematic illustration of a situation in associating–dissociating systems. (a) A possible situation in which a dimer could be inactive and a monomer active when there is no conformational change between the subunits. In this case the active site would be at or near the contact point between the two subunits so that it was blocked and not accessible to substrate in the dimer. (b) A situation in which the dimer is active and the monomer inactive because the active site overlaps the two subunits so that only when they are together is activity observed. (c) A dimerization induced by substrate in which the substrate first induces a conformational change which enhances the attraction between the two monomers. (d) The ligand induces a conformational change which decreases the stability of a dimer leading to dissociation. (e) A hypothetical situation in which dissociation of a dimer to a monomer is identified with a conformational change which is then stabilized by binding of substrate.

phenomena may well alter the conformation of subunits, the phenomena of association–dissociation can enhance or dampen ligand-induced changes and play an important role in the allosteric effect.

Since some proteins show cooperative effects which are not involved in the polymerization phenomena, the association–dissociation mechanism cannot be a universal explanation for allosteric effects. The association–dissociation phenomena as shown in Figs. 17a and b, which would be alternate mechanisms in the sense that no conformational change within the subunit is postulated, may have been observed in chymotrypsin (*74*), but they are not a normal feature of the behavior of that enzyme. Thus, the idea that association–dissociation provides an alternate explanation for conformational change is like the kinetic alternative postulated above: a potential alternate mechanism rather than one that has been established for any regulatory protein. The association–dissociation phenomena shown in Figs. 17c and d, where conformational changes are occurring either prior to or subsequent to the binding of ligand,

74. S. N. Timasheff, *ABB* **132,** 165 (1969).

probably do play an important role (*49, 73*) in regulatory proteins, but these types of association–dissociation represent a feature of ligand-generated changes and are not in essence alternative mechanisms.

E. Hemoglobin

Probably no protein has been studied more intensively than hemoglobin, and the interest in this protein has been heightened by the fact that the full three-dimensional structure has been elucidated by Perutz and co-workers (*75*). The phenomenon of cooperativity was discovered by Bohr (*21*) with this protein, and Hill (*24*) and Adair (*76*) derived their equations to describe this behavior. The extensive literature is impossible to cover here but fortunately excellent reviews have been published summarizing the history and recent developments (*77*). It may be worthwhile to examine this protein in a little more detail as an illustration of the application of the diagnostic techniques.

The sigmoid nature of the curve indicated a cooperative effect, and the early mathematical analyses of Pauling (*78*), Roughton and others (*79*) postulated theories with site–site interactions. Pauling was able to extend the empirical equations to a detailed consideration of the site–site interactions based on the geometrical arrangements of the subunits.

When conformational changes became a part of enzymology, models which incorporated these features were proposed and it was found that both the concerted model (*16*) and the sequential model (*14*) gave saturation curves which were in excellent agreement with the hemoglobin data. The fact that both the theories were able to fit the data did not mean that the constants calculated from the two theories were indistinguishable or that the curves were indistinguishable. Reference to Table I shows that the two models actually produce different theoretical curves. However, the theoretical curves derived from each equation were sufficiently close to the experimental data that no choice between them could be made. From these curves, however, deductions in regard to the conformational states at intermediate stages of saturation could be made, and Ogawa and McConnell (*34*) noted that the $\bar{B}$ vs. $\bar{Y}$ curves were

75. H. F. Perutz, H. Muirhead, J. M. Cox, and L. C. G. Goaman, *Nature* **219,** 131 (1968).

76. G. S. Adair, *JBC* **63,** 529 (1925).

77. A. Rossi-Fanelli, E. Antonini, and A. Caputo, *Advan. Protein Chem.* **19,** 74 (1964); E. Antonini, *Science* **158,** 1417 (1967).

78. L. Pauling, *Proc. Natl. Acad. Sci. U. S.* **21,** 186 (1935).

79. A. B. Otis, F. J. W. Roughton, and R. L. J. Lyster, *Proc. Roy. Soc.* **B144,** 29 (1955).

significantly different for the two models. They were able to test the two theories by introducing a spin label attached to residue 93 of the β chain of hemoglobin which could be used as a conformational probe. They were able to show a linear dependence of oxygen binding and conformation in agreement with the sequential model.

A possible argument against these results could maintain that the label modified the properties of hemoglobin so that the conclusions, although correct for the labeled protein, would not apply to native hemoglobin. Shulman and co-workers (*80*) took this study one step further using NMR techniques on the native hemoglobin and using the same techniques on hybrids in which oxidized subunits mimic the oxygen-liganded subunits. The NMR technique was found to be a highly sensitive conformational probe which, for example, revealed conformational changes not detected by X-ray crystallography and a conformational change during the binding of oxygen to myoglobin which had previously escaped detection. Applying this probe to hemoglobin, Shulman *et al.* were able to show that no direct heme–heme interactions occurred, i.e., the environment of the heme on a subunit without bound oxygen was identical to deoxy hemoglobin regardless of the subunits in the molecule. This confirmed conclusions of the simple sequential model. Apparently the binding of oxygen to heme induces a conformational change which affects the subunit to which the oxygen is bound and must extend at least to the interface between the subunits. Whether or not it distorts the neighboring subunit somewhat has not been determined, but whatever distortion occurs it does not proceed all the way to the unliganded heme of the neighboring subunit. It therefore is a ligand-induced distortion which affects the stability of the molecule and affects the binding of subsequent ligands. The same conclusion has been reached by Antonini and Brunori (*81*) using the reactivity of SH groups in the α and β chain. Since these studies were again on unmodified hemoglobin, they support and extend the conclusions of Ogawa and McConnell.

Although these studies established the sequential model, they did not provide the basis for evaluating all the subunit interactions. Extensive studies by many workers (*76*, *82–84*) indicated the amazing complexity of hemoglobin and theories which indicated the dimer as the functional unit, the monomer as the functional unit, the tetramer as the functional

80. R. G. Shulman, S. Ogawa, K. Wuthrigh, T. Yamane, J. Peisach, and W. E. Blumberg, *Science* **165,** 251 (1969).

81. E. Antonini and M. Brunori, *JBC* **244,** 3909 (1969).

82. G. Guidotti, *JBC* **242,** 3673 (1967).

83. R. E. Benesch, R. Benesch, and G. Macduff, *Proc. Natl. Acad. Sci. U. S.* **54,** 535 (1965).

84. M. Brunori and T. M. Schuster, *JBC* **244,** 4046 (1969).

unit, etc., could all be maintained with some credibility. Although it was apparent that the subunit interactions played a role, how many and which ones were yet to be resolved. These studies were then carried one step further using "frozen conformational states" as described above. These studies were able to place quantitative values on the β–β interactions and the α–α interactions showing that these crossed interactions had to be considered as well as the α–β interactions (*44, 45*); thus, a tetrahedral tetramer must be considered as the functional unit.

From the extent of the β–β interactions it could be calculated that K_{BB}/K_{AB} equaled 3.45 (*45*) and that the energy of the interactions was therefore approximately 700 cal (*44, 45*). Earlier, Wyman had estimated from plots of the type shown in Fig. 6 that the β–β interactions might be just this order of magnitude. Furthermore, this finding of the β–β interactions agreed qualitatively with the findings of Perutz that there were β–β contacts as well as α–β contacts. The studies by Brunori and Schuster (*84*) on this system and by Gibson and Parkhurst (*43*) using kinetic techniques are beginning to evaluate the rate constants for the conformational changes, which, superimposed on the equilibrium measurements described above, may give a fairly complete interpretation of the multisubunit interactions of this fascinating protein.

F. Evolutionary Considerations

The mathematical complexity of the ligand-induced model leads some biochemists to hope that the conclusion reached in Section V,C above, i.e., that simpler models are at best limiting cases, to be incorrect. It is hoped that somehow or other a simple picture will emerge from the many studies which will obviate the need for the many diagnostic tests outlined previously for the evaluation of an individual protein. The complexity of the mathematics should not obscure the fact that nature has devised a particularly simple mechanism for achieving a vast number of permutations and combinations. It may be worthwhile therefore to summarize in one section our understanding of the role of conformational changes in evolution.

Figure 18 shows the effect of ligand-induced changes both within a subunit and between subunits. As outlined in the previous figures in this chapter, it is seen that a ligand-induced conformational change at an allosteric site or within the active site can transmit its effects to neighboring peptide sequences by changes in the alignments of nonbonded amino acid residues. In fact many of these effects are transmitted in a manner which makes it irrelevant whether the peptide chains which move are connected ultimately by covalent bonds. Since the peptide chain winds

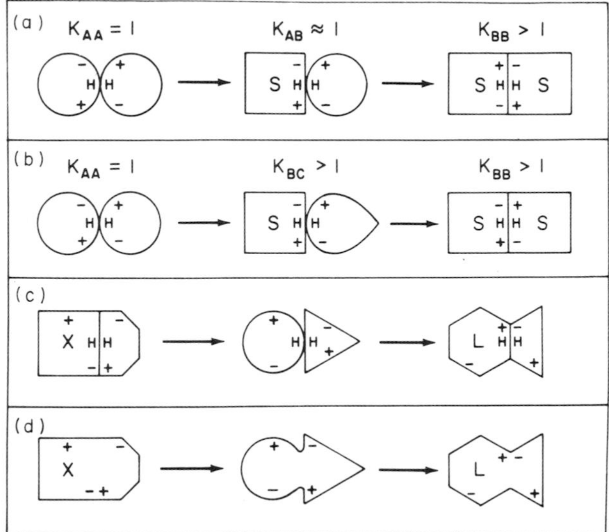

FIG. 18. A schematic illustration of conformational changes induced by ligands. (a) The simplest sequential model which is a limiting case assumes that only two conformations of an individual subunit can exist. (b) A more general sequential case in which the ligand-induced distortion of one subunit affects the stability and shape of the second subunit. (c) Two different ligands inducing different conformational changes in the subunit to which they are bound and causing different distortions of the neighboring subunit. Such might be the case of an activator or an inhibitor bound to a regulatory subunit influencing the conformation of the peptide chain containing the active site. (d) The same situation in which the regulatory site and active site are on the same peptide chain. In the latter case it is seen that the principles are precisely the same and the same mathematical treatment could be applied if the equations for subunit contacts were simply applied to the part of the peptide chain within the subunit.

like a piece of tangled rope, adjacent amino acid contacts may be very far from each other in the sequence and the splitting of an individual peptide bond as was found, for example, in the case of ribonuclease (*85*) may not make much difference to the interaction of residues which are held close together in three-dimensional space. It is therefore altogether logical that in some cases the regulatory site will be on a separate peptide chain as in the case of aspartyltranscarbamylase (*86*) and in other cases part of the same single peptide chain as the active site as in the cases of phosphorylase (*87*), glutamine synthetase (*87*), and CTP synthetase (*26*). If the ligand-induced conformational change alters immediately adja-

85. P. J. Vithayathil and F. M. Richards, *JBC* **235,** 2343 (1960).
86. J. Gerhart and H. K. Schachman, *Biochemistry* **4,** 1054 (1965).
87. E. R. Stadtman, B. M. Shapiro, H. S. Kingdon, C. A. Woolfolk, and J. S. Hubbard, *Adv. Enzyme Regulation* **6,** 257 (1968).

cent residues whether they are connected in the same peptide chain or not, this effect may be transmitted through nonbonded interactions over many portions of protein. It follows that distortion of neighboring subunits could occur as readily as distortion of distant portions of the same peptide chain. Thus, the allosteric effects resulting in homotropic interactions between identical sites on different subunits is a logical extension of ligand-induced changes altering the conformational state within a single peptide.

If such subunit interactions occur it might well be asked why they have survived and what advantage they provide for the cell. In the first calculations of the ligand-induced model the computer was given various values for subunit interactions, and it was found that the same values of K_S and K_t for a protein gave widely different saturation curves simply by varying subunit interactions (*14*). These terms could change the midpoint of the curve, i.e., the $S_{0.5}$, the steepness of the curve, and its shape. Both positive and negative cooperativity could be obtained. Subsequent studies have shown that almost all of these permutations can be identified in nature. This leads to the conclusion that in the process of evolution nature may well have operated as a sophisticated computer, varying subunit interactions in order to tailor the properties of an enzyme.

One might well ask why the same process could not be achieved by varying amino acid residues at the active site. It is clear that there would be substantial advantages in evolution to avoid that kind of mutation (although it must occur occasionally as well as mutations in subunit interactions). If an active site is constructed to carry out a specific reaction, mutation in a critical area may lead to widespread changes. The specificity, the catalytic coefficient, the very ability of the enzyme to catalyze a reaction depend on the intimate geometry at the active site. The active site was designed to carry out a link in a series of metabolic reactions, and alteration of this specificity may well change its entire relation to other enzymes in the cell. However, a microorganism might well find itself in a situation in which it desires to metabolize a given compound such as glucose in the same way as its ancestors but it must now cope with a supply of glucose which is 10^{-7} M in average concentration instead of 10^{-5} M. Varying the subunit contacts would shift the midpoint of the curve without changing the catalytic site and would be a mechanism which would produce an enzyme with a different apparent affinity but an identical active site. In another case an enzyme which operates as a step in a normal pathway might become a branch point in the metabolic process of a new living organism. In that case positive cooperativity, which would lead to greater sensitivity to metabolite fluctuations, might be needed by the organism. Mutations which would alter subunit interactions to produce this positive cooperativity would therefore be of advantage.

In the same way an enzyme that found itself in a fluctuating environment and needed mutations which would maintain the activity of the enzyme while reducing its sensitivity to fluctuations in the metabolite levels could obtain this desensitization by a negative cooperativity from subunit interactions. Finally, an enzyme subject to many controls might find it advantageous to exhibit both negative and positive cooperativity. Like CTP synthetase, such an individual enzyme would be quite sensitive to fluctuations in certain metabolites and rather insensitive to flucuations in others.

Once it is recognized that the capability for altering subunit interactions is an important feature of the cell, it must be further recognized that some of these subunit interactions may not be of great significance. Just as one can rationalize the need for twenty amino acids to produce a wide variety of enzymes, one may postulate the need for variety in subunit interactions. Nevertheless, some proteins with fewer than the total of twenty amino acids can operate quite effectively. Similarly, some proteins may need neither regulatory control nor cooperative interactions. When cooperative effects are observed but are extremely small, it would seem fruitless to attempt to rationalize in biological terms the necessity for these minor deviations from Michaelis–Menten kinetics. Rather it would seem that nature is preserving its capability for changing subunit interactions and that some enzymes may show minor deviations which are not of great biological significance.

VI. Conclusions

It is too early to say conclusively that a complete molecular understanding of regulatory control is now available. Yet the number of diverse phenomena which can now be explained in simple terms suggest that we may be closer than many people realize.

In the first place the qualitative features of regulatory control are probably known. The concept of conformational changes explain how a small molecule which is not part of the direct catalytic process can influence that process. It can do so in two ways: by affecting the affinity of the ligand and by affecting the velocity of catalysis. Both alternatives are used and both can operate by ligand-induced changes which affect K_m or V_M. If a ligand-induced change in one part of a peptide chain can alter a distant part of that chain and a protein is composed of multiple subunits, it is logical that distortions should affect neighboring subunits and that is observed.

The distortions induced by a nonreacting ligand may be either positive or negative. A ligand which causes the former is called an activator. One which leads to the latter is an inhibitor. It would seem logical that intersubunit effects would also be positive and negative and these have now been observed. Ligand-induced distortions therefore provide a consistent qualitative explanation for the patterns observed between regulatory sites on active sites in the same peptide chain and between like active sites or like regulatory sites on two different peptide chains. This important deduction that intrachain effects are qualitatively similar is illustrated schematically in Fig. 18.

Molecular parameters have been deduced which can express these qualitative effects in quantitative terms. The mathematics looks forbidding since the equations are long and contain many terms; in fact, the mathematics is amazingly simple, being a routine matter of accounting. It can be performed almost mechanically once the molecular species have been assigned. The resulting equations have accounted for regulatory behavior quantitatively as well as qualitatively and in a few cases have already led to predictions of molecular species which have been supported by independent physical probes. These examples should now increase rapidly in the years ahead since the mathematical method is available and the physical tools are proliferating rapidly.

The choice of molecular species likely to be present presents more difficulty than the mathematical manipulations. It would appear that very simple models—either the symmetry model or the simplest sequential model—are limiting cases which may be found in a few proteins and may be approximated in others, but which cannot be assumed *a priori* for any one protein. This conclusion does not in any sense minimize the importance of these simple models. They provided a basis for developing the principles on which the more complex model depends. However, simple models must be correct in all cases to be a general rule and their failure to explain the data in a reasonably large number of cases requires consideration from the outset of many molecular species.

If one cannot postulate *a priori* certain properties of a molecule, the molecular species must be defined by diagnostic tests. The number of these tests need to be enlarged, but many are already available which are extremely helpful. Hill plots, double reciprocal plots, equations of state, curve fitting techniques, etc., provide ways of presenting data which reveal molecular events occurring during the binding and kinetic processes. Physical tools such as NMR and spin labels and chemical tools such as susceptibility to proteolytic digestion and amino acid reactivity provide information about the conformational state of the protein. Together these mathematical analyses and conformational probes need to be improved and increased in resolving power, but already they give an indication that methods will be available to answer the essential questions in regard to molecular species.

The explanation of regulatory control in enzymes at the molecular level is of interest not only because of the important role of enzymes but also because so many other cellular organelles seem to follow a similar pattern. The repressors' interactions of inducers and corepressors with repressors, the nerve receptors' interactions with its stimuli, the permease interactions, the small molecules being transported, to name a few, all show properties similar to the regulatory properties of the allosteric proteins. It seems probable therefore that a molecular understanding of the more available enzymes can help in providing the diagnostic tools and molecular basis for these related and possibly more complex systems.

Despite the complexity of the mathematics, the reassuring feature is that simple fundamental concepts have made it possible to explain in qualitative terms the features of regulation and to provide the basis for a detailed quantitative determination. Moreover, these molecular events can be made consistent with evolutionary patterns and with the detailed structure of proteins. Much remains to be done but that it can proceed from a solid theoretical and experimental basis.

VII. Glossary

Conformations of individual subunits: A = ○, B = □, and C = △. The A conformation unless otherwise specified will be used as the conformation of the subunit when all subunits are devoid of ligands being studied. Obviously Cl^-, Br^-, or unknown ligands which are kept constant must be temporarily ignored.

Ligands will be specified as follows: X, general ligand which can be substrate, inhibitor, or activator; S, substrate; I, inhibitor; and J, activator. Since in allosteric systems a given compound can inhibit under some circumstances and activate under others, it is clear that the use of I and J is designed to be general but to indicate the predominant property of the ligand.

N_X is the number of moles of ligand X bound per mole of protein.

N_{X_J} is the number of moles of ligand X bound to an individual subunit which also contains J.

$\bar{Y}_X$ is the fraction of sites capable of binding X which contain X.

$X_{0.5}$, $X_{0.1}$, etc., are the concentration of ligand X in solution required to fill 0.5, 0.1, etc., of the sites which bind X.

n_m is the maximum number of molecules of ligand which can be bound to the protein when all the sites are saturated. This number is usually equal to the number of protomers in the protein.

n_H is the Hill coefficient.

K_{X_A} *is* the affinity constant of the subunit in conformation A for the ligand X, i.e., (AX)/(A) (X).

$K_{t_{AB}}$ is the conformational transition constant for the change from conformation A to conformation B, i.e., concentration (B)/(A).

K_{AB} is the averaged subunit interaction constant applying to interactions between subunits in the A conformation and the subunit in the B conformation defined in reference (*14*).

$K_{AB_{pq}}$ is the subunit interaction constant for the binding of domain p in the subunit with the A conformation interacting with the binding domain q of the subunit in the B conformation.

K_{AA}, $K_{AA_{pp}}$ are all defined as one and other interaction constants defined relative to K_{AA}. Since $K_{AA_{pp}}$ is not necessarily equal to $K_{AA_{qq}}$, this may seem incorrect. In fact this convention can be applied because the completely written out cooperative effect equations all actually involve ratios $(K_{AB_{pp}}/K_{AA_{pp}})$, $(K_{AB_{qq}}/K_{AA_{qq}})$, etc. Hence, the definition of K_{AA}'s as one involves no loss in generality and considerable convenience in mathematical manipulation. If extension to cases in which distinction between $K_{AA_{pp}}$ and $K_{AA_{qq}}$ is needed, that can easily be accommodated within this nomenclature (cf. reference *51*).

ψ_i is the coefficient of the term containing i molecules of ligand in the general binding equation $N_X = \Sigma_{i=0}^{i=n_m} i\psi_i X^i / \Sigma(1 + \Sigma\psi_i X^i)$.

K_i is the affinity constant for the binding of the ith molecule of ligand, i.e., $(EX_i)/(EX_{i-1})(X)$.

K_i' is the intrinsic constant for the binding of the ith molecule of ligand, i.e., K_i corrected for statistical factors [cf. Eq. (15)].

The terms *homotropic, heterotropic, allosteric,* etc., are as defined in references *4* and *16*.

K_m is the Michaelis constant.

V_M is the maximum velocity at enzyme saturation.

8

Mechanisms of Enzyme Regulation in Metabolism

E. R. STADTMAN

I. Introduction

Although the physiological significance of homeostasis has long been recognized, the magnitude of the problem it concerns became apparent only as the complexity of metabolism was disclosed through a detailed biochemical analysis of various metabolic pathways. These studies have demonstrated the overwhelming intricacy of metabolism which consists, on the one hand, of a maze of converging degradative pathways by which exogenously supplied food stuffs and endogenous reserves are converted to a relatively few simple products, and on the other hand, of a maze of diverging and interlocking biosynthetic pathways by which the catabolic end products are converted to simple metabolites, *viz.*, amino acids, sugars, nucleotides, and fatty acids. The latter, in turn, are utilized for the synthesis of macromolecules such as proteins, nucleic acids, complex lipids, and polysaccharides which are finally assembled into the structural elements of the cell. Regulation therefore involves the coordination of many highly interdependent biochemical functions including, in addition to those mentioned above, the transport of inorganic ions and metabolites across membrane barriers, the translocation of ions and metabolites to various organs, and other mechanical expressions of energy. The rates and direction of the catabolic processes must be governed in such a way as to provide an orderly supply of energy and catabolites, which together are required for the biosynthesis of all the end products (amino acids, nucleic acids, etc.) needed for the elaboration of macromolecules; moreover, the rate of each biosynthetic process must be controlled in order to achieve balanced partitioning of the common precursors among competing pathways in order to insure an adequate supply of each ultimate end product for final assemblage into the macromolecular components of the cell. In view of the diversity of functions that must be coordinated, it is evident that no single mechanism of control will suffice. Moreover, it seems likely that each biochemical function evolved independently of the others and that its regulation was developed at a later time as a consequence of selective pressures. It is, therefore, not surprising that organisms utilize many different principles in the regulation of their metabolism and, in addition, that the same principle is not always utilized for the regulation of comparable functions in a given organism nor for identical functions in different organisms.

Because of the vastness of the problem, and because of the infancy of its study, our knowledge of regulation is still very meager. Indeed, the knowledge is based mainly on *in vitro* studies of but a few isolated enzymes. Whereas these studies have led to the recognition of basic patterns and principles of regulation, they are limited in scope by the investigator's knowledge of metabolic pathways, of the interactions between various metabolic functions, and by his success in predicting which enzymes and metabolites are apt to have regulatory functions. Nevertheless, the field of study has developed rapidly during the past decade and though the knowledge gained appears superficial and limited in relation to the vastness of the problem, it is already too great to treat in a comprehensive way in a single chapter. Therefore, the present discussion will be restricted to a more general treatment whose aim is to summarize and illustrate some of the principles utilized by organisms in the regulation of their metabolism and to call attention to some of the unique problems involved.

II. Principles of Metabolic Regulation

The regulation of metabolism is concerned primarily with modulation of key reactions that determine the fluxes of metabolites throughout the various pathways. Although the contributions of thermodynamic and physical factors cannot be discounted, the modulation of reaction rates is achieved mainly by control of the activities and concentrations of the enzymes that catalyze key reactions.

A. Regulation of Enzyme Concentration

1. *Substrate Induction of Enzyme Synthesis*

The concentration of any given enzyme is determined by the rate of its synthesis and by the rate of its destruction. In exponentially growing bacteria, there is very little turnover of protein (*1–3*); the concentration of an enzyme is therefore determined primarily by the rate of specific gene expression. In the case of catabolic pathways, two classes of enzymes are formally recognized: (a) the so-called "constitutive" enzymes whose concentration is more or less independent of the presence of their substrates; and (b) the so-called inducible enzymes that are not normally produced unless their immediate substrates or suitable derivatives thereof are present. Often the presence of the first substrate in a catabolic path-

1. D. S. Hogness, M. Cohen, and J. Monod, *BBA* **16,** 99 (1955).
2. A. L. Koch and H R. Levy, *JBC* **217,** 947 (1955).
3. B. Rotman and S. Spiegelman, *J. Bacteriol.* **68,** 419 (1954).

way results in the simultaneous synthesis of all enzymes involved in its catabolism. This phenomenon has been referred to as *simultaneous induction,* and earlier studies suggested that it involves a sequential series of substrate-induced enzyme syntheses wherein the product of each consecutive reaction in the catabolic pathway serves as a specific substrate inducer for synthesis of the enzyme catalyzing the next step in the reaction sequence (*4–6*).

Subsequently, Stanier and his associates (*7–11*) showed that in addition to the mechanism of sequential induction, the simultaneous syntheses of functionally related enzymes may also result from coordinate induction of a block of enzymes whose synthesis is dictated by a specific operon that is subject to regulation by the first substrate or a later intermediate in the catabolic pathway. Studies on the bacterial oxidation of aromatic compounds have shown that the regulation of this complex metabolism involves a combination of the coordinate and sequential mechanisms of metabolite induction (*7–11*).

2. *Catabolite Repression*

Sometimes substrate-induced synthesis of catabolic enzymes is repressed when the energy and carbon requirements for growth are amply supplied by a different catabolic process (*12–16*). For example, when glucose is supplied in addition to a second substrate whose catabolism is under induced enzyme control the glucose will be utilized preferentially, and as long as it is present the enzymes required for the catabolism of the second substrate will not be formed. Various lines of evidence indicate that the so-called "glucose effect" (i.e., the repressive effect of glucose on induced enzyme synthesis) need not result from glucose per se but is attributable to one or more catabolites produced during the dis-

4. R. Y. Stanier, *J. Bacteriol.* **54,** 339 (1947).
5. M. Suda, O. Hayaishi, and Y. Oda, *Med. J. Osaka Univ.* **2,** 21 (1950).
6. B. P. Sleeper and R. Y. Stanier, *J. Bacteriol.* **59,** 117 (1950).
7. R. Y. Stanier, G. D. Hegeman, and L. N. Ornston, *Colloq. Intern. C.N.R.S. Symp. Mech. Regulation Activities Enzymatiques, Marseille, France, July 23–27, 1963.* Centre National de la Recherche Scientific, Paris, France, p. 228 (1965).
8. G. D. Hegeman, *J. Bacteriol.* **91,** 1161 (1966).
9. N. J. Palleroni and R. Y. Stanier, *J. Gen. Microbiol.* **35,** 319 (1964).
10. L. N. Ornston, *JBC* **241,** 3800 (1966).
11. J. L. Cánovas and R. Y. Stanier, *European J. Biochem.* **1,** 289 (1967).
12. H. M. R. Epps and E. F. Gale, *BJ* **36,** 619 (1942).
13. M. Cohen and J. Monod, *Symp. Soc. Gen. Microbiol.* **2,** 132 (1953).
14. J. Monod, *Growth* **11,** 223 (1947).
15. F. C. Neidhardt and B. Magasanik, *BBA* **21,** 324 (1956).
16. B. Magasanik, *Cold Spring Harbor Symp. Quant. Biol.* **26,** 249 (1961).

similation of glucose (*16*). Moreover, the effect is not restricted to glucose; any one of several substrates that can support rapid growth may yield catabolites that repress the synthesis of inducible enzymes (*16*). As yet the mechanism of this so-called catabolite repression (*16*) has not been clearly established.

3. *Feedback Repression of Enzyme Syntheses*

The important role of ultimate end products in the regulation of biosynthetic pathways was clearly established by Roberts *et al.* (*17*) who showed that the *de novo* biosynthesis of a given amino acid from isotopically labeled glucose is selectively discontinued when that amino acid is supplied exogenously in the growth medium. In the meantime, careful biochemical and genetic analysis of numerous biosynthetic pathways has demonstrated the generality of the basic principle that the concentration of the ultimate end product governs the rate of end product formation. Two kinds of end product control are definitely established; namely, *feedback inhibition* and *repression* (*18*). In the case of feedback inhibition (to be discussed later) the end product usually exerts a restraining action through inhibition of the enzyme that catalyzes the first step in the biosynthetic pathway. In the case of repression, however, accumulation of the end products leads to inhibition of the synthesis of one or more, indeed sometimes all, of the enzymes involved in its biosynthetic pathway. As in the case of enzyme induction, when several or all of the enzymes in a given biosynthetic pathway are products of closely linked genes comprising a single operon, they may be susceptible to coordinate repression (*19*). The regulation of histidine biosynthesis in *Salmonella typhimurium* is the classic example of this kind of control. In this case all ten of the structural genes concerned with the biosynthesis of histidine are contained in a single operon whose expression is subject to regulation by histidine (*20*). As a consequence, in normal cells the synthesis of all ten enzymes in the histidine pathway varies simultaneously and to an identical degree in response to fluctuations in the concentration of histidine (*20*).

Multivalent repression (*21*) represents still another unique mechanism

17. R. B. Roberts, P. H. Abelson, P. H. Cowie, D. V. Bolton, and R. J. Britten, *Carnegie Inst. Wash. Publ.* **607** (1955).

18. H. J. Vogel, *in* "The Chemical Basis of Heredity" (W. D. McElroy and B. Glass, eds.), p. **276**. Johns Hopkins Press, Baltimore, Maryland, 1957.

19. B. N. Ames and B. Garry, *Proc. Natl. Acad. Sci. U. S.* **45,** 1453 (1959).

20. B. N. Ames and R. G. Martin, *Ann. Rev. Biochem.* **33,** 235 (1964).

21. M. Freundlich, R. V. Burns, and H. E. Umbarger, *Proc. Natl. Acad. Sci. U. S.* **48,** 1804 (1962).

of feedback control which is important in the regulation of complex biosynthetic pathways. Its discussion is deferred until later (Section IV,G).

Protein synthesis involves both gene transcription and messenger RNA translation and evidence supporting a role of both functions in metabolite regulation of enzyme synthesis has been obtained. For further information the reader is referred to recent reviews (*22*, *23*) which are outside the scope of the present discussion.

4. *Balance between Enzyme Synthesis and Degradation*

In higher animals the problems of growth and development are significantly different from those in microorganisms, and the regulation of enzyme levels involves a fundamentally different mechanism (*24*). Animal proteins undergo continual synthesis and degradation; the half-lives for some proteins are only a few hours whereas for others they are several days (*24–28*). It is evident that the prevailing concentration of a given enzyme is determined by both the kinetics of its formation and of its degradation. A relative increase in enzyme concentration (apparent enzyme induction) can therefore be obtained by either increasing the rate of synthesis or by decreasing the rate of degradation. Similarly, a decrease in enzyme level (apparent derepression) may arise from either deceleration of synthesis or by acceleration of proteolytic degradation. It has been established for several enzymes that the fluctuations in the concentrations that occur in responses to nutritional and hormonal factors are related to specific effects on either the rate of synthesis or of degradation. For example, the increase of liver arginase concentration caused by feeding a high protein diet is solely the result of an increase in the rate of arginase synthesis (*26*). On the other hand, an increase in the arginase level that accompanies the abrupt change from a high protein diet to a starvation diet results from continued enzyme synthesis at a normal rate in the absence of enzyme degradation (*26*). Under still other conditions, rates of synthesis and degradation are both affected (*26*).

The unique contribution of hormones and substrates in the regulation

22. W. Epstein and J. R. Beckwith, *Ann. Rev. Biochem.* **37,** 411 (1968).
23. H. J. Vogel and R. H. Vogel, *Ann. Rev. Biochem.* **36,** 519 (1967).
24. R. T. Schimke, *in* "Current Topics in Cellular Regulation" (B. L. Horecker and E. R. Stadtman, eds.), Vol. 1, p. 77. Academic Press, New York, 1967.
25. D. L. Buchanan, *ABB* **94,** 501 (1961).
26. R. T. Schimke, *JBC* **239,** 3808 (1964).
27. R. W. Swick, *JBC* **231,** 751 (1958).
28. I. Arias, D. Doyle, and R. T. Schimke, *JBC* **244,** 3303 (1969).

of enzyme levels in animal cells is clearly evident from the studies with tryptophan oxygenase. The level of the enzyme is markedly increased by the administration of hydrocortisone and other glucocorticoid hormones, or, alternatively, by feeding tryptophan (*29–32*). However, Schimke and his colleagues have shown that the elevation of enzyme level induced by these two methods is entirely different. The hormone-induced change is related to an increase in the rate of enzyme synthesis whereas the tryptophan-induced increase mainly results from inhibition of enzyme degradation (*33, 34*).

B. Feedback Inhibition

The pioneering studies of Umbarger (*35*) and of Yates and Pardee (*36*) established the important principle that the first committed step in a biosynthetic pathway is sensitive to feedback control by the ultimate end product of that pathway. This is logically the most strategic point of control since regulation of the first step automatically governs the metabolic flux over the entire pathway and determines not only the rate of end product formation but also avoids unwanted accumulation of all intermediates in the pathway. It is significant that although "first" enzymes are generally inhibited by the ultimate end products of their respective biosynthetic pathways, the nature of the inhibitory effects is not always the same.

Kinetic analyses by classic procedures (*viz.*, double reciprocal plots of reaction velocity against substrate concentration in the presence and absence of inhibitors) suggest that the inhibitory effects of metabolites on regulatory enzymes may be either competitive, noncompetitive, partially competitive, uncoupled, or of the mixed type. When apparently noncompetitive kinetics are observed it seems probable that binding of the end product to an inhibitor specific site (not the catalytic site) on the enzyme results in the induction or stabilization of a conformational state that has a lower catalytic potential; i.e., inactive enzyme or enzyme with a lower turnover capacity. However, when an apparently competitive

29. W. E. Knox and A Mehler, *Science* **113,** 237 (1951).
30. W. E. Knox and V. H. Auerbach, *JBC* **214,** 307 (1955).
31. P. Feigelson, M. Feigelson, and O. Greengard, *Recent Progr. Hormone Res.* **18,** 491 (1962).
32. W. E. Knox and M. M. Piras, *JBC* **242,** 2959 (1967).
33. R. T. Schimke, E. W. Sweeney, and C. M. Berlin, *JBC* **240,** 322 (1965).
34. R. T. Schimke, E. W. Sweeney, and C. M. Berlin, *JBC* **240,** 4609 (1965).
35. H. E. Umbarger, *Science* **123,** 848 (1956).
36. R. A. Yates and A. B. Pardee, *JBC* **221,** 757 (1956).

type of kinetics is observed, the mechanism is less certain since metabolite effectors of regulatory enzymes generally bear no structural resemblance to the normal substrate for the enzyme. This is particularly true in the case of feedback inhibition of the first enzyme in a biosynthetic pathway by the ultimate end product of that pathway and also in situations where appropriate balance between two different metabolic processes is maintained by the specific action of a metabolite of one process on a regulatory enzyme of the other.

1. *The Allosteric Concept*

Monod, Changeux, and Jacob (*37*) emphasized the fact that metabolite effectors of regulatory enzymes are not isosteric but rather allosteric (i.e., they have no structural resemblance) with respect to the normal substrates of the enzymes. This fact, and the further consideration that enzymes are generally characterized by a high degree of substrate specificity, led them to propose that metabolite effectors of regulatory enzymes exert their effects by reacting at specific allosteric sites on the enzyme and thereby induce or stabilize conformational states which have either a lower affinity for substrates at the catalytic sites or which have a lower intrinsic catalytic potential. The possibility that regulatory enzymes contain separate binding sites for substrate and effector substances was suggested earlier by Gerhardt and Pardee (*38*).

Various lines of evidence support the conclusion that there is an allosteric site that is distinctly different from the catalytic substrate binding site:

(1) Regulatory enzymes can often be made insensitive to metabolite inhibition or activation without alteration of their catalytic activity. Thus, selective desensitization to feedback inhibition has been obtained by treatment of enzymes with mercurials (*38–41*), urea (*38, 41*), X-rays (*42*), proteolytic enzymes (*43*), high ionic strengths (*40*) or by aging at 0°–5° (*39, 40, 44*), dialysis (*44*), freezing (*44*), heating (*38, 39*), or

37. J. Monod, J. P. Changeux, and F. Jacob, *JMB* **6,** 306 (1963).
38. J. C. Gerhardt and A. B. Pardee, *JBC* **237,** 891 (1962).
39. R. G. Martin, *JBC* **237,** 257 (1962).
40. T. A. Murphy and G. R. Wyatt, *JBC* **240,** 1500 (1965).
41. R. H. Bauerle, M. Freundlich, F. C. Stormer, and H. E. Umbarger, *BBA* **92,** 142 (1964).
42. K. Kleppl and U. Spaeren, *Biochemistry* **6,** 3497 (1967).
43. K. Taketa and B. M. Pogell, *JBC* **240,** 651 (1965).
44. C. T. Casky, D. M. Ashton, and J. B. Wyngarden, *JBC* **239,** 2570 (1964).

by changes in pH of the environment (*41, 45, 46*). This desensitization is attributed to selective denaturation of the allosteric site.

(2) Feedback inhibitors (negative effectors) or metabolite activators (positive effectors) often protect the catalytic activity of a regulatory enzyme against denaturants, whereas the normal substrates will not (*47–49*). It seems unlikely that direct binding of effectors at the catalytic site would offer a protection that could not be obtained by substrate binding to that same site.

(3) Gene mutations have been obtained that affect the susceptibility of certain regulatory enzymes to feedback inhibition without affecting the catalytic activity (*37, 38, 47, 50–53*).

(4) Direct binding studies have shown that the binding of feedback inhibitors may not be influenced by the presence of saturating concentrations of substrates and vice versa. For example, the binding of the feedback inhibitor AMP to certain forms of *E. coli* glutamine synthetase is not affected by the presence of saturating concentrations of the substrates NH_3 or glutamate (*54*).

(5) With a few enzymes it has been demonstrated that the effector binding site and the catalytic substrate binding site occur on separate, dissimilar subunits of the enzyme; *viz.*, aspartotranscarbamylase (*55*) and, possibly, ribonucleotide reductase (*56*).

In view of these considerations it is evident that those feedback inhibitors exhibiting kinetics that are classically defined as competitive in nature do not exert their effects by direct competition with substrates at the catalytic site of the enzyme. Instead they may react at an allosteric site and thereby stabilize or induce conformational changes of the en-

45. C. Cennamo, M. Boll, and H. Holzer, *Biochem. Z.* **340,** 125 (1964).

46. J. A. Hathaway and D. E. Atkinson, *JBC* **238,** 2875 (1963).

47. J. P. Changeux, *Bull. Soc. Chim. Biol.* **46,** 927 (1964).

48. E. R. Stadtman, G. N. Cohen, G. LeBras, and H. de Robichon-Szulmajster, *JBC* **236,** 2033 (1961).

49. G. N. Cohen, J. C. Patte, P. Truffa-Bachi, C. Sawas, and M. Douderoff, *Colloq. Intern. C.N.R.S., Symp. Mech. Regulation Activities Enzymatiques, Marseille, France, July 23–27, 1963.* Centre National de la Recherche Scientific, Paris, France, p. 243 (1965).

50. G. N. Cohen, *Ann. Rev. Microbiol.* **19,** 419 (1965).

51. G. N. Cohen, *in* "Current Topics in Cellular Regulation" (B. L. Horecker and E. R. Stadtman, eds.), Vol. 1, p. 183. Academic Press, New York, 1969.

52. H. S. Moyed, *JBC* **236,** 2261 (1961).

53. H. S. Moyed, *JBC* **234,** 1098 (1960).

54. A. Ginsburg, *Biochemistry* **8,** 1726 (1969).

55. J. C. Gerhardt and H. K. Schachman, *Biochemistry* **4,** 1054 (1965).

56. N. C. Brown, A. Larsson, and P. Reichard, *JBC* **242,** 4272 (1967).

zyme that decrease the intrinsic affinity of the enzyme for substrates at the catalytic site.

a. K Systems and V Systems. Misleading mechanistic connotations arising from the use of "competitive" and "noncompetitive" inhibition to describe the effects of metabolites on allosteric enzymes are avoided by the use of the nomenclature introduced by Monod, Wyman, and Changeux (*57*). Monod *et al.* describe regulatory systems exhibiting competitive-type kinetic behavior as "K systems" since they involve changes in the apparent K_m for substrates but no changes in V_{max}. Similarly, noncompetitive systems are referred to as "V systems" since they involve changes in V_{max} but no change in apparent K_m.

2. *Cooperative Effects*

For most enzyme-catalyzed reactions, plots of initial reaction velocities against substrate concentrations yield hyperbolic curves as is predicted by the Michaelis–Menten equation. This is true also for some reactions catalyzed by regulatory enzymes. However, as noted first by Umbarger for *E. coli* threonine deaminase (*35*) and later for numerous other regulatory enzymes, the substrate saturation function is sigmoid rather than hyperbolic. Such complex kinetics are compatible with the existence of two or more interacting substrate binding sites on the enzyme such that the binding of one substrate molecule facilitates the binding of the next. Monod, Changeux, and Jacob (*37*) called attention to the fact that such abnormal kinetic behavior is a common feature of a large number of regulatory enzymes. Subsequently, Monod, Wyman, and Changeux (*57*) advanced a detailed hypothesis to explain these abnormal kinetics in terms of a specific enzyme model which takes into account the fact that enzymes are generally composed of two or more identical subunits arranged in a symmetrical manner. In essence they proposed that the enzyme may exist in two distinct conformational states, T and L, that are in dynamic equilibrium with each other, and that for each state all subunits are in an identical configuration. From this theory, it follows that the conversion of state L to state T, involves a concerted reaction in which all the subunits in the enzyme undergo simultaneous changes to a form characteristic of state T and vice versa. In other words, the conversion of one state to another involves a concerted transition in which symmetry is conserved. With appropriate assumptions as to the magnitude of the equlibrium constant that determines the distribution of enzyme molecules in the two states, and further assumptions regarding the affinities of substrates and other effectors for

57. J. Monod, J. Wyman, and J. P. Changeux, *JMB* **12,** 88 (1965).

subunits in each of the two configurations, it is possible to explain the kinetic behavior of most regulatory enzymes.

In the meantime various other theories have been advanced to explain the sigmoidal kinetics of regulatory enzymes. Among these, the nonconcerted transition models of Adair (*58*) and of Koshland and his associates (*59, 60*), the multisite hypotheses of Hill (*61*), Atkinson (*62*), and Freiden (*63*), the "relaxation" theory of Weber (*64, 65*), which was expanded by Rabin (*66*), and the "alternate pathway of two substrate reactions" theory of Ferdinand (*67*), which has been elaborated by others (*68, 69*), all deserve attention. Since these various hypotheses are considered in detail elsewhere in this volume (Chapters 7 and 9), they will not be dealt with further here. Suffice it to say that with appropriate assumptions each of these theories might explain the observed kinetics of some regulatory enzymes; however, on the basis of the available information it is unlikely that any one of the proposed mechanisms will account satisfactorily for the observed kinetics of all regulatory enzymes studied. But, as Atkinson and Walton (*70*) have pointed out, since each individual case must have evolved with a considerable degree of independence, there is no reason to assume that the mechanism by which sigmoid kinetics are obtained should be the same for all regulatory enzymes. Nevertheless, the sigmoidal response of enzymic activity to increasing substrate or effector concentration is probably of fundamental importance in cellular regulation. As a consequence there exists a threshold concentration below which the enzymic activity is relatively insensitive to variations in concentrations of substrates and effectors, and above which relatively slight changes in concentrations evoke marked effects in enzymic activity (*71*). Thus, the sigmoidal saturation function restricts the metabolite control to a very narrow, selected range of substrate and effector concentrations and should thereby facilitate a more

58. G. S. Adair, *JBC* **63,** 529 (1925).
59. M. E. Kirtley and D. E. Koshland, *JBC* **242,** 4192 (1967).
60. D. E. Koshland, G. Nemethy, and D. Filmer, *Biochemistry* **5,** 365 (1966).
61. A. J. Hill, *BJ* **7,** 471 (1913).
62. D. E. Atkinson, J. A. Hathaway, and E. C. Smith, *JBC* **240,** 2682 (1965).
63. C. Frieden, *JBC* **234,** 815 (1959).
64. G. Weber, *in* "Molecular Biophysics" (B. Pullman and M. Weissbluth, eds.), p. 269. Academic Press, New York, 1965.
65. G. Weber and S. R. Anderson, *Biochemistry* **4,** 1942 (1965).
66. B. R. Rabin, *BJ* **102,** 22c (1967).
67. W. Ferdinand, *BJ* **98,** 278 (1966).
68. J. R. Sweeney and J. B. Fisher, *Biochemistry* **7,** 561 (1968).
69. C. C. Griffin and L. Brand, *ABB* **126,** 856 (1968).
70. D. E. Atkinson and G. M. Walton, *JBC* **240,** 757 (1965).
71. J. P. Changeux, *Cold Spring Harbor Symp. Quant. Biol.* **26,** 497 (1963).

rigorous control of enzymic activity. If, in fact, any selective advantage is to be gained by the development of regulatory enzymes having cooperative saturation functions, then any mutational event that favors the development of cooperativeness in a regulatory enzyme might be favored irrespective of the mechanism that underlies this development. In other words the cooperative effect, rather than the mechanism by which it is achieved, could be the basis of natural selection. From this point of view, it follows that almost any mechanism that yields a sigmoid saturation function might be the basis of control of some regulatory enzyme. Or, stated otherwise, one would not expect any one mechanism to be responsible for the sigmoidal characteristics of all regulatory enzymes.

3. *A Note on Terminology*

In the development of their concerted transition hypothesis Monod *et al.* (*57*) took into account the fact that regulatory enzymes possess specific allosteric sites for negative and positive modifier compounds and also that many such enzymes exhibit cooperative-type kinetics. Unfortunately, this has led to a confusion in terminology. Many investigators now assume that any enzyme exhibiting sigmoidal kinetics is an allosteric enzyme and that the mechanism of activation of all allosteric enzymes involves a concerted transition as postulated by Monod *et al.* (*57*). As noted above the original definition of an allosteric enzyme was based upon the simple fact that the binding site for the effectors (the allosteric site) is distinctly different from the binding site for the substrate. This definition carries no connotation regarding the presence or absence of a cooperative kinetic behavior, much less so a connotation as to the mechanism of the cooperative effects when they do occur.

When properly used, the term *allosteric* is very useful since it describes a unique situation that is fundamental to many regulatory enzymes. However, when used in the broader sense to signify the existence of cooperative kinetic behavior or the mechanism thereof, the term becomes ambiguous and often misleading. Cooperative kinetics have not been demonstrated for all allosteric enzymes (as originally defined), and sigmoidal saturation functions could derive from any one of several mechanisms (see above) that are fundamentally different from that upon which the *concerted-transition* model is based. In the absence of information needed to discriminate between the various mechanisms of cooperative behavior, application of the term allosteric to a specific mechanism greatly limits its usefulness, especially since absolute proof of a given mechanism is almost impossible to obtain. For these reasons, in the present discussion, the term *allosteric* will be restricted to that intended in the original definition.

C. Energy-Dependent Covalent Modification of Regulatory Enzymes

The allosteric regulation of some key enzymes is reinforced and sometimes overridden by a second level of regulation that involves enzyme-catalyzed covalent modification of the enzyme. Thus the activities of phosphorylase (*72, 73*), glycogen synthetase (*74, 75*), and pyruvate dehydrogenase (*76*) are all modulated by the action of specific kinases that catalyze the ATP-dependent phosphorylation of certain seryl groups of the regulatory enzymes; the activity of *E. coli* glutamine synthetase is similarly modulated by adenylylation of a particular tyrosyl group of each subunit of the enzyme (*77–81*). These esterification reactions are accompanied by marked changes in catalytic parameters and in the responses to allosteric effectors. In each case the esterified enzyme is converted back to its unmodified state by the action of highly specific esterases that catalyze hydrolytic fission of the protein phosphate ester or diester linkages (*75, 76, 82, 83*). Since ATP is utilized as the source of high energy group potential, it is evident that the enzymes that catalyze the esterification and de-esterification reactions must themselves be under rigorous control; otherwise undirected coupling between the two processes would occur and the regulatory enzyme which is the common substrate in the reaction would serve simply as a catalytic vehicle for ATP hydrolysis.

1. *Adenylylation and Deadenylylation of E. coli Glutamine Synthetase*

Equations (1)–(3) in Table I show how the coupling of the adenylylation and deadenylylation of the *E. coli* glutamine synthetase would lead to the hydrolysis of ATP. Figure 1 shows how indiscriminate coupling is

72. E. G. Krebs, A. B. Kent, and E. H. Fischer, *JBC* **231,** 73 (1958).
73. E. G. Krebs, D. S. Love, G. E. Bratrold, K. A. Trayser, W. L. Meyer, and E. H. Fischer, *Biochemistry* **3,** 1022 (1964).
74. R. R. Trout and F. Lipmann, *JBC* **238,** 1213 (1963).
75. D. I. Friedman and J. Larner, *Biochemistry* **2,** 669 (1963).
76. L. C. Tracy, F. H. Pettit, and L. J. Reed, *Proc. Natl. Acad. Sci. U. S.* **62,** 234 (1969).
77. B. M. Shapiro, H. S. Kingdon, and E. R. Stadtman, *Proc. Natl. Acad. Sci. U. S.* **58,** 642 (1967).
78. H. S. Kingdon, B. M. Shapiro, and E. R. Stadtman, *Proc. Natl. Acad. Sci. U. S.* **58,** 1703 (1967).
79. K. Wulff, D. Mecke, and H. Holzer, *BBRC* **28,** 740 (1967).
80. B. M. Shapiro and E. R. Stadtman, *BBRC* **30,** 32 (1968).
81. B. M. Shapiro and E. R. Stadtman, *JBC* **242,** 5069 (1967).
82. E. G. Krebs and E. H. Fischer, *Advan. Enzymol.* **24,** 263 (1962).
83. B. M. Shapiro, *Biochemistry* **8,** 659 (1969).

TABLE I
ATP-DEPENDENT ENZYME MODIFICATION REACTIONS

System	Modifier reaction		Enzyme[b]	Activators	Inhibitors
A[d]	12 ATP + glutamine synthetase → glutamine synthetase $(AMP)_{12}$ + 12 PP_i	(1)	(1) Adenylyltransferase	Glutamine (*78, 84*)	α-Ketoglutarate (*85*) UTP (*85*)
	Glutamine synthetase $(AMP)_{12}$ + 12 H_2O → glutamine synthetase + 12 AMP	(2)			Glutamate (*78, 84*)
	Sum: 12 ATP + 12 H_2O → 12 AMP + 12 PP_i	(3)	(2) Deadenylylating enzyme system	α-Ketoglutarate (*83*) UTP (*83*)	Glutamine (*83*) Glutamate (*83*)
B	2 Phosphorylase b + 4 ATP → phosphorylase a + 4 ADP	(4)	(4) Phosphorylase kinase (*87, 90*)	3′,5′-AMP (*93, 100*) Glycogen (*100*) (*73*)	
	Phosphorylase a + 4 H_2O → 2 phosphorylase b + 4 P_i	(5)			
	Sum: 4 ATP + 4 H_2O → 4 ADP + 4 P_i	(6)	(5) Phosphorylase phosphatase (*82*)	Glucose (*106*)	AMP (*101*)
C	Glycogen synthetase I[c] + ATP → glycogen synthetase D[c] + ADP	(7)	(7) Glycogen synthetase kinase (*74, 75*)	ATP 3′,5′-AMP (*102*)	
	Glycogen synthetase D + H_2O → glycogen synthetase I + P_i	(8)			

	Sum: $ATP + H_2O \rightarrow ADP + P_i$	(9)	(8) Glycogen synthetase phosphatase (*75*)	Glucose (*106*)	ATP (*106*) Glycogen (*103, 104*)
D	ATP + phosphorylase b kinase $(I)^a \rightarrow$ phosphorylase b kinase (A) + ADP	(13)	(13) Phosphorylase kinase kinase (*92*)	3′,5′-AMP (*92, 93*) ATP (*92, 93*)	Glycogen (*93*)
	Phosphorylase b kinase $(A)^a + H_2O \rightarrow$ phosphorylase b kinase (I) + P_i	(14)			
	Sum: $ATP + H_2O \rightarrow ADP + P_i$	(15)	(14) Phosphorylase kinase phosphatase		
E	Pyruvate dehydrogenase $A^a + ATP \rightarrow$ pyruvate dehydrogenase I + ADP	(10)	(10) Pyruvate dehydrogenase kinase (*76*)		
	Pyruvate dehydrogenase $I^a + H_2O \rightarrow$ pyruvate dehydrogenase A + P_i	(11)	(11) Pyruvate dehydrogenase phosphatase (*76*)		
	Sum: $ATP + H_2O \rightarrow ADP + P_i$	(12)			

[a] Here A indicates activated form and I indicates inactive form.

[b] Numbers in parentheses refer to reaction catalyzed.

[c] Here I refers to Glucose-6-P independent form and D refers to the Glucose-6-P dependent form.

[d] Note added in proof: Since this review was written it was established (W. A. Anderson and E. R. Stadtman, unpublished data) that the deadenylylation of adenylylated glutamine synthetase involves phosphorolysis of the adenylyl–tyrosine bond to produce ADP as follows:

$$\text{Glutamine synthetase }(AMP)_{12} + 12\ P_i \rightarrow 12\ ADP + \text{glutamine synthetase}$$

The overall coupled adenylylation and deadenylylation reaction is therefore:

$$12\ ATP + 12\ P_i \rightarrow 12\ ADP + 12\ PP_i$$

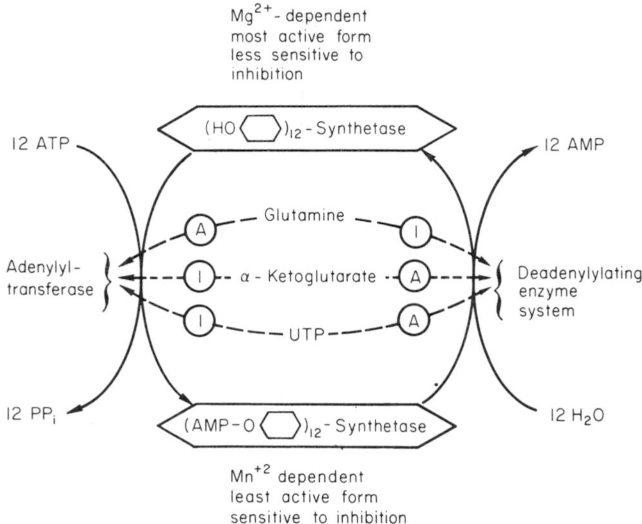

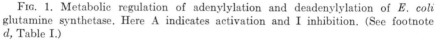

FIG. 1. Metabolic regulation of adenylylation and deadenylylation of *E. coli* glutamine synthetase. Here A indicates activation and I inhibition. (See footnote *d*, Table I.)

prevented by the reciprocal action of metabolites on the adenylyltransferase and the deadenylylating enzyme system. It can be seen that glutamine is required for the activity of the adenylylating enzyme (*78*, *84*), but it inhibits the activity of the deadenylylating enzyme (*83*). On the other hand, α-ketoglutarate and UTP are required for the activity of the deadenylylation system (*83*), but they inhibit the activity of the adenylyltransferase (*85*). A rational basis for these reciprocal effects is suggested by the fact that adenylylation of glutamine synthetase converts it to a form that is more susceptible to inhibition by various feedback inhibitors and which is catalytically less active under normal physiological conditions; adenylylation is also accompanied by a change in the divalent ion specificity of the enzyme from a Mg^{2+}-dependent form to a Mn^{2+}-dependent form (*78*). Since the intracellular concentrations of α-ketoglutarate and glutamine should vary in reciprocal manner with respect to each other in response to variations in the nitrogen supply, it follows that under conditions of nitrogen excess, when the level of α-ketoglutarate will be minimal and the level of glutamine should be high, adenylylation of glutamine synthetase will be favored and deadenylylation will be inhibited. As a result the activity of gluta-

84. K. Wulff, D. Mecke, and H. Holzer, *BBRC* **28**, 740 (1967).
85. B. Hennig and A. Ginsburg, unpublished data.

mine synthetase will be reduced, and it will become more susceptible to feedback control. On the other hand, when the supply of nitrogen becomes limiting, the level of glutamine will diminish and the α-ketoglutarate concentration will increase; then adenylylation will be inhibited and deadenylylation will be stimulated. Thus the activity of glutamine synthetase is determined by the relative activities of the two modifying enzymes, which are exquisitely regulated by the state of nitrogen nutrition of the cell. Moreover, metabolite regulation of the two modifier enzymes prevents indiscriminate coupling of the adenylylation and deadenylylation reactions and concomitant ATPase activity. (See footnote *d,* Table I.)

The physiological significance of this regulatory system is supported by the observation that the glutamine synthetase isolated from *E. coli* cells grown under conditions of nitrogen excess is almost fully adenylylated, whereas the enzyme isolated from nitrogen-starved cells is almost fully deadenylylated (*77, 86*).

2. *Phosphorylation of Glycogen Phosphorylase*

A detailed study of the properties of phoshorylase in mammalian tissues has disclosed the existence of a highly sophisticated control system for this enzyme (*82, 87–101*). Phosphorylase exists in two forms: phos-

86. H. Holzer, H. Schutt, Z. Masek, and D. Mecke, *Proc. Natl. Acad. Sci. U. S.* **60,** 721 (1968).
87. E. H. Fischer and E. G. Krebs, *Federation Proc.* **25,** 1511 (1966).
88. D. Brown and C. F. Cori, "The Enzymes," 2nd ed., Vol. 5, p. 207, 1961.
89. E. G. Krebs, D. J. Graves, and E. H. Fischer, *JBC* **234,** 2867 (1959).
90. W. L. Meyer, E. H. Fischer, and E. G. Krebs, *Biochemistry* **3,** 1033 (1964).
91. R. J. DeLange, R. G. Kemp, W. D. Riley, R. A. Cooper, and E. G. Krebs, *JBC* **243,** 2200 (1968).
92. D. A. Walsh, J. P. Perkins, and E. G. Krebs, *JBC* **243,** 3763 (1968).
93. E. G. Krebs, R. B. Huston, and F. L. Hunkeller, *Advan. Enzyme Regulation* **6,** 245 (1968).
94. J. B. Posner, R. Stern, and E. G. Krebs, *JBC* **240,** 982 (1965).
95. G. A. Robinson, R. W. Butcher, and E. W. Sutherland, *Ann. N. Y. Acad. Sci.* **139,** 703 (1967).
96. E. W. Sutherland and T. W. Rall, *Pharmacol. Rev.* **12,** 265 (1960).
97. I. Øye and E. W. Sutherland, *BBA* **127,** 347 (1966).
98. E. W. Sutherland, T. W. Rall, and T. Menon, *JBC* **237,** 1220 (1962).
99. E. G. Krebs and E. H. Fischer, *Ann. N. Y. Acad. Sci.* **88,** 378 (1960).
100. E. H. Fischer, D J. Graves, and E. G. Krebs, *Federation Proc.* **16,** 180 (1957).
101. E. H. Fischer, S. S. Hurd, P. Koh, V. L. Seery, and D. C. Teller, *in* "Control of Glycon Metabolism" (W. J. Whelan, ed.), p. 19. Academic Press, New York, 1968.

phorylase b and phosphorylase a (*88*). Phosphorylase b is intrinsically the less active form; it is a dimer and is catalytically inactive in the absence of the allosteric effector AMP (*88*). Phosphorylase a exists normally as a tetramer; it is intrinsically more active than phosphorylase b and does not require AMP for activity (*88*). Conversion of phosphorylase b to phosphorylase a occurs through phosphorylation of the protein catalyzed by a specific kinase called phosphorylase kinase (*72, 73, 82*) which utilizes ATP as the phosphoryl group donor. In the overall reaction, two seryl residues of phosphorylase b (one per subunit) are phosphorylated and two molecules of the phosphorylated derivative associate to form the tetrameric derivative, phosphorylase a (*72, 82*).

Phosphorylase a is converted back to b [reaction (4), Table I] by a phosphatase (*82*). As noted in Table I coupling of the phosphorylation and dephosphorylation reactions would result in ATPase activity [reaction (6)]. Although the metabolite regulation of these two processes is not well understood, phosphorylase kinase itself exists in active and inactive forms that are interconverted by phosphorylation and dephosphorylation, respectively.

a. The Cascade Phenomenon. In the presence of ATP, phosphorylase kinase can catalyze slow phosphorylation of itself in an autocatalytic manner (*91*). In addition, there is a separate protein kinase (phosphorylase kinase kinase) (*92*) that requires 3′,5′-cyclic-AMP for activity (*92, 93*). The absolute requirement for 3′,5′-cyclic-AMP for activity of the phosphorylase kinase kinase is thought to be the link between phosphorylase activation and the stimulating effect of epinephrine, a hormone known to enhance adenyl cyclase activity (*94–97*). We see then as illustrated in Fig. 2 that regulation of phosphorylase activity is achieved by a cascade of activation reactions that begins with hormonal induced activation of adenyl cyclase to form 3′,5′-cyclic-AMP (*98*). The latter is needed for the activation of phosphorylase kinase kinase which then catalyzes the activation (phosphorylation) of phosphorylase kinase which finally catalyzes the conversion (phosphorylation) of phosphorylase b to phosphorylase a (*99*). The net result of this series of reactions is the activation of phosphorylase and its conversion to a form whose activity does not require the presence of the allosteric effector, AMP. As noted above the two ATP-dependent activation reactions involve phosphorylation of the enzyme, and in each case this activation is reversed by the action of a phosphatase that catalyzes hydrolysis of the phosphate ester. It is essential that activities of the three protein kinases and the respective phosphatases be regulated in order to establish temporal polarization of function and to avoid useless ATPase activity that would result from unhindered coupling of the phosphorylation and dephosphorylation

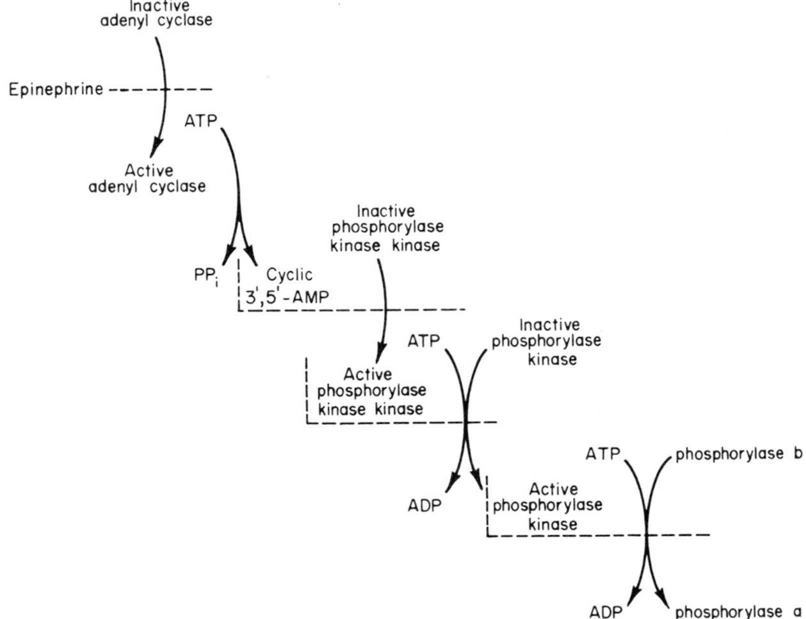

FIG. 2. Cascade effect in the conversion of phosphorylase b to phosphorylase a. The dashed lines indicate either an activation or a catalytic function for the indicated process.

reactions. This is illustrated in Table I which shows, also, that a similar situation exists in the phosphorylation of glycogen synthetase and pyruvate dehydrogenase. Although metabolite effects have been demonstrated for some of these enzymes (*102–106*), reciprocal effects of these on the coupled enzymes generally have not been observed. It seems likely, however, that controls comparable to those found in the regulation of *E. coli* glutamine synthetase exist also for ATP-dependent phosphorylation of key enzymes. Particularly noteworthy is the recent discovery by Krebs (*105*) that the phosphorylase kinase kinase and the glycogen synthetase kinase are one and the same enzyme. The significance of this will be discussed later (Section III,C).

102. F. Huijing and J. Larner, *BBRC* **23,** 259 (1966).
103. J. Larner, C. Villar-Palasi, N. D. Goldberg, J. S. Bishop, F. Huijing, J. I. Wenger, H. Sasko, and N. B. Brown, *Advan. Enzyme Regulation* **6,** 153 (1968).
104. J. Larner, *Ann. N. Y. Acad. Sci.* **29,** 192 (1967).
105. E. G. Krebs, *Advan. Enzyme Regulation* **8,** (1970) (in press).
106. H. G. Hers, *Advan. Enzyme Regulation* **8,** (1970) (in press).

D. Role of Proteolysis in Regulation

1. *Activation of Proteolytic Enzymes*

In the foregoing illustrations, covalent modification of regulatory enzymes was achieved by enzyme-catalyzed esterification of specific functional groups of the enzymes. Another important mechanism by which the activity of key enzymes is controlled involves activation by limited proteolysis. Perhaps the best known examples are the conversions of pepsinogen to pepsin (*107*) and of trypsinogen to trypsin (*108*). In each case the activation reaction is an autocatalytic process involving limited proteolysis of the inactive precursors and is catalyzed by their respective active derivatives. The conversion of trypsinogen is additionally catalyzed by enteropeptidase (*109*). It is of interest that proteolysis of several typical regulatory enzymes *in vitro* can produce changes in their catalytic and kinetic behavior that mimic effects achieved by other forms of regulation. For example, the activation of phosphorylase b kinase and the conversion of the glucose 6-phosphate-dependent glycogen synthetase to the independent form are both accomplished either by limited tryptic proteolysis (*73*) or by a Ca^{2+}-activated proteolytic enzyme (*93*, *110*) whose physiological role is obscure. The susceptibility of fructose-1,6-diphosphatase to inhibition by AMP is destroyed by limited digestion with papain (*43*).

2. *Blood Coagulation*

Enzyme activation by limited proteolysis is a part of the complex regulatory system that governs the coagulation of blood. As shown in Fig. 3 the conversion of fibrinogen to fibrin is an essential event in blood clotting. This conversion involves a limited hydrolysis of fibrinogen and is catalyzed by thrombin. Under normal physiological conditions thrombin exists in an inactive form, prothrombin. The conversion of inactive prothrombin to active thrombin is a highly complicated process that involves the interaction of at least six different proteins. The available evidence indicates that normally these proteins all exist in plasma in inactive proenzyme states. Conversion of the proenzymes to their active

107. R. M. Herriott, *J. Gen. Physiol.* **21**, 501 (1938).
108. J. H. Northrop, M. Kunitz, and R. Herriott, "Crystalline Enzymes," 2nd ed. Columbia Univ. Press, New York, 1948.
109. M. Kunitz, *J. Gen. Physiol.* **22**, 447 (1939).
110. M. M. Appleman, E. Belocopitow, and H. N. Torres, *BBRC* **14**, 550 (1964).

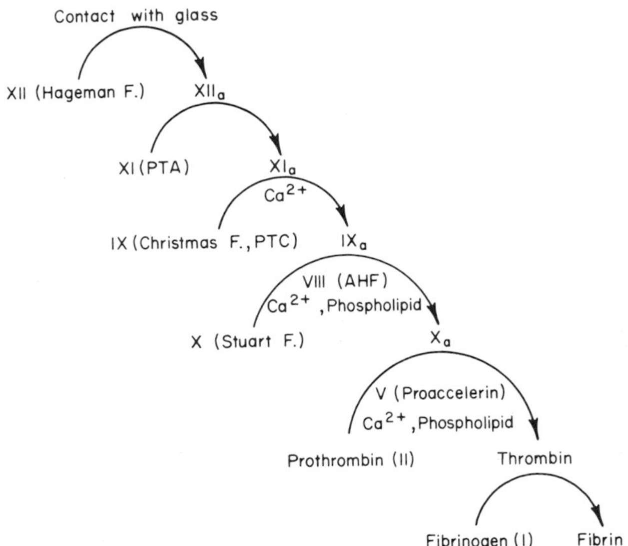

FIG. 3. Postulated cascade mechanism for blood coagulation, reproduced with permission of Davie (*111*).

enzyme forms is visualized (*111*, *112*) as a cascade of activation reactions analogous to that demonstrated for the conversion of phosphorylase b to phosphorylase a (Fig. 2), except that highly specific limited proteolysis and possibly other, as-yet undetermined mechanisms are the basis of the covalent modification involved rather than phosphorylation. As illustrated in Fig. 3, the cascade begins with surface contact, which causes activation of factor XII (Hageman factor) (*113*). Activated factor XII then catalyses the conversion of proenzyme factor XI to its active derivative; this probably involves a limited proteolysis (*111*, *114*). Activated factor XI then catalyzes the activation of factor IX, and so on. For the present discussion, it is important only to note that the regulation of blood clotting entails a cascade of activation reactions in which proteolysis forms the basis of at least one mechanism. For further details of this amazing system the reader is referred to several excellent reviews (*111*, *112*).

111. O. E. W. Davie, C. Hougie, and R. L. Lundblad, *in* "Recent Advances in Blood Coagulation" (L. Pollen, ed.), p. 13. Churchill, London, 1969.

112. H. S. Kingdon, *J. Biomed. Mater. Res.* **3,** 25 (1969).

113. O. D. Ratnoff and J. M. Rosenblum, *Am. J. Med.* **25,** 160 (1958).

114. O. D. Ratnoff, E. W. Davie, and D. L. Mallett, *J. Clin. Invest.* **40,** 803 (1961).

E. Significance of Chemical Equilibria in Regulation of Metabolism

Among other mechanisms involved in regulation of metabolism, Krebs (*115*) has emphasized that reaction equilibria may be a controlling force. He points out that some metabolic steps are so rapid, because the high activities of the enzymes involved, that near equilibrium always exists between the starting materials and products of the reactions, whereas other steps are never in a state of true equilibrium because of the low activity of the enzymes involved. The latter so-called "nonequilibrium" enzymes are therefore potentially rate limiting and are the most likely targets for cellular regulation. Neverthelss, "equilibrium" enzymes may play an important role in cellular regulation because the equilibria that they establish between their reactants and products determine the intracellular concentrations of these compounds, some of which are substrates for nonequilibrium, regulatory enzymes. As emphasized above, the concentrations of substrates for regulatory enzymes can be important in modulating their activity. Krebs (*115*) has cited several situations in metabolism wherein equilibrium control could be of considerable significance. For example, in liver and kidney, pyruvate carboxylase is a nonequilibrium enzyme that catalyzes a rate limiting step in the synthesis of glucose from lactate. However, the concentration of its substrate, pyruvate, is determined by the relative amounts of pyruvate, lactate, DPN, and DPNH which are always maintained nearly at equilibrium by the high catalytic potential of the equilibrium enzyme, lactic dehydrogenase. The absolute concentration of pyruvate is thus a function of the level of pyruvate plus lactate and the redox level of the pyridine nucleotide, i.e., the ratio of DPNH to DPN. An increase in the DPNH/DPN ratio, brought about, for example, by the availability of a suitable electron donor such as alcohol, will tend to inhibit gluconeogenesis by virtue of the fact that this results in a decrease in the pyruvate concentration (*115*). Similarly, a rise in lactate concentration, as occurs during physical exercise, will favor gluconeogenesis when the redox level of pyridine nucleotides is held constant.

The above example illustrates the importance of the DPNH/DPN ratio as a controlling factor in regulation; since pyridine nucleotides constitute common links between diverse dehydrogenase-catalyzed reactions, their relative concentrations will determine the equilibrium between the pools of certain oxidized and reduced metabolites. In addition these oxidized and reduced coenzymes can have specific allosteric affects on

115. H. A. Krebs, *in* "Current Topics in Cellular Regulation" (B. L. Horecker and E. R. Stadtman, eds.), Vol. 1, p. 49. Academic Press, New York, 1969.

certain regulatory enzymes. In a similar manner various other dissociable coenzymes such as CoA, FAD, pyridoxal phosphate, B_{12} coenzyme, and AMP, form common links in other metabolic processes, consequently, the ratios of free and group-charged forms of those coenzymes can affect the equilibria of the reactions in which they are involved as well as the metabolic pools which they interconnect. Atkinson (*116, 117*) has emphasized the importance in regulation of varying ratios of components whose sums make up more or less fixed pools. This is the basis of his "adenylate control" or "energy charge" hypothesis which is supported by *in vitro* studies (*118–120*). Since Atkinson discusses the concept elsewhere in this volume (Chapter 9) it will not be considered further here.

F. Role of Inorganic Ions in Regulation

1. *Monovalent Cations*

Many regulatory enzymes are either dependent upon or are markedly affected by specific inorganic ions. Variations in the kinds and concentrations of specific ions are therefore factors of tremendous potential importance in the regulation of metabolism. It has long been recognized that among the monovalent cations, K^+ has a unique capacity to activate certain regulatory enzymes, *viz.*, pyruvate kinase (*121*), aspartokinase (*122–124*), L-homoserine dehydrogenase I of *E. coli* (*125*), pyruvate carboxylase from baker's yeast (*126, 127*), δ-aminolevulinic acid dehydratase of *R. spheroides,* and muscle adenosine deaminase (*128, 129*). Whereas NH_4^+ is sometimes able to replace K^+ as an activator, Na^+ is generally an antagonist and sometimes quite inhibitory.

On the contrary, intestinal sucrase is specifically activated by Na^+ and

116. D. E. Atkinson, *Biochem. Soc. Symp.* **27,** 23 (1968).
117. D. E. Atkinson, *in* "Current Topics in Cellular Regulation" (B. L. Horecker and E. R. Stadtman, eds.), Vol. 1, p. 29. Academic Press, New York, 1969.
118. D. E. Atkinson and G. M. Walton, *JBC* **242,** 3239 (1967).
119. D. E. Atkinson, *Biochemistry* **7,** 4030 (1968).
120. L. Shen, L. Fall, G. M. Walton, and D. E. Atkinson, *Biochemistry* **7,** 7041 (1968).
121. P. D. Boyer, H. A. Lardy, and P. H. Phillips, *JBC* **146,** 673 (1942).
122. D. E. Wampler and E. W. Westhead, *Biochemistry* **7,** 1661 (1968).
123. H. Paulus and E. Grey, *JBC* **239,** PC4008 (1964).
124. P. Datta and L. Prakash, *JBC* **241,** 5827 (1966).
125. J. C. Patte, G. LeBras, T. Loving, and G. N. Cohen, *BBA* **67,** 16 (1963).
126. J. J. Cazzulo and A. O. M. Stoppani, *ABB* **121,** 596 (1967).
127. D. L. Nandi, K. F. Baker-Cohen, and D. Shermin, *JBC* **243,** 1224 (1968).
128. B. Setlow and J. M. Lowenstein, *JBC* **242,** 607 (1967).
129. K. L. Smiley and C. H. Suelter, *JBC* **242,** 1980 (1967).

is inhibited by K^+ and other monovalent cations (*130, 131*). Whereas Na^+ is an activator of all intestinal sucrases studied thus far, the nature of its effect varies from species to species; *viz.*, in the hamster and rat Na^+ decreases the apparent K_m for sucrose and has no effect on V_{max} (*132*), but in man and the rabbit (*132*) Na^+ increases V_{max} and has little or no effect on K_m. Noteworthy, is the fact that similar relationships exist in the effects of Na^+ and K^+ on the intestinal transport of sugar (*132–134*).

O'Brien and Stern (*135*) have shown that Na^+ has a unique role in determining alternative pathways of citrate catabolism in *Aerobacter aerogenes*. Under aerobic conditions and in the absence of Na^+ the catabolism of citrate occurs exclusively via the citric acid cycle; however, upon the addition of 10 mM Na^+, oxalacetate decarboxylase (Na^+- dependent) is activated and α-ketoglutarate dehydrogenase is repressed, thereby diverting citrate catabolism along the fermentative, citritase-dependent pathway; the further addition of 2% potassium acetate causes repression of citritase and derepression of α-ketoglutarate dehydrogenase, switching citrate catabolism back into the citric acid cycle. A somewhat different but nevertheless similar role of K^+ and Na^+ in the aerobic metabolism of citrate by *S. typhimurium* has also been observed (*136*).

It is evident from these studies and those mentioned above that the ratio of Na^+ to K^+ can be an important factor in the regulation of certain cellular activities.

2. *Divalent Cations*

The requirement for divalent ions in the activation of enzymes is more general. Numerous enzymes, especially those concerned with reactions of purine and pyrimidine nucleotides and orthophosphate, are found to require a divalent cation for activity. Disregarding those ions with electron carrier functions, the roles of others (*viz.*, Ca^{2+}, Mg^{2+}, Mn^{2+}, and Zn^{2+}) are less well understood. They are probably sometimes involved in the binding of the substrate to the enzyme (*137–143*), but in addition they

130. D. Miller and R. K. Crane, *BBA* **52,** 281 (1961).
131. D. Miller and R. K. Crane, *BBA* **52,** 293 (1961).
132. G. Semenza, *Protides Biol. Fluids, Proc. Colloq.* **15,** 201 (1967).
133. R. K. Crane, *Federation Proc.* **21,** 981 (1962).
134. R. K. Crane, G. Forstner, and A. Eichholz, *BBA* **109,** 467 (1965).
135. R. W. O'Brien and J. R. Stern, *J. Bacteriol.* **99,** 389 (1969).
136. R. W. O'Brien, G. M. Frost, and J. R. Stern, *J. Bacteriol.* **99,** 395 (1969).
137. L. Noda, T. Nihei, and M. F. Morales, *JBC* **235,** 2830 (1960).
138. S. A. Kuby, T. A. Mahowald, and E. A. Noltmann, *Biochemistry* **1,** 748 (1962).
139. L. Noda, "The Enzymes," 2nd ed., Vol. 6, p. 139, 1962.
140. M. Marshall and P. P. Cohen, *JBC* **241,** 4199 (1966).

play an important role in determining the secondary, tertiary, and quaternary structure of regulatory enzymes (*144–146*). Whereas many enzymes are relatively nonspecific with respect to the nature of the divalent cation required, others show a remarkably high degree of specificity.

The regulatory function of divalent ions is especially notable for those reactions in which ATP is a substrate. In many such cases the Me^{2+}–ATP complex appears to be the true substrate for the enzyme, and often maximal activity of the enzyme is obtained when the concentrations of ATP and Me^{2+} are nearly equal (*144*); relative excesses of either ATP or Me^{2+} are inhibitory. In such cases it follows that the ratio of ATP/Me^{2+} is an important factor in governing the activity of the enzyme. Factors that tend to decrease the concentration of Me^{2+} will decrease the concentration of ATP required for maximal activity and vice versa. Among the factors that might influence the intracellular concentration of Me^{2+}, the concentrations of various nucleoside di- and triphosphates probably are of greatest importance since nucleoside polyphosphates have unusually high affinities for divalent cations. Therefore, they may serve as either activators or inhibitors depending on the prevailing ratio of ATP to Me^{2+}. When the ATP/Me^{2+} ratio is equal to or more than 1.0, other nucleotides will act as inhibitors of the enzyme by reducing the effective concentration of Me^{2+} and thereby increasing the ATP/Me^{2+} ratio; similarly, when the ATP/Me^{2+} ratio is less than 1.0, other nucleotides will serve as activators of the enzyme as long as the concentration of the total nucleotide pool does not exceed the Me^{2+} concentration.

The intracellular concentrations of ATP and other nucleoside polyphosphates are determined by the balance between endergonic and exergonic metabolism; i.e., between biosynthetic processes on the one hand and electron transport linked phosphorylation on the other. The regulation of these functions is therefore geared to the regulation of certain divalent cation-dependent enzymes via the nucleotide/Me^{2+} ratios. In the more general sense, the above considerations are only an extension of Atkinson's energy charge hypothesis (*116, 117*) and adenylate pool systems (*114*) to encompass the regulation of certain divalent cation-dependent enzymes.

---

141. A. W. Murray and P. C. L. Wong, *BBRC* **29,** 582 (1967).
142. B. Kiech and G. J. Barritt, *JBC* **242,** 1983 (1967).
143. J. McD. Blair, *FEBS Letters* **2,** 245 (1969).
144. H. S. Kingdon, J. S. Hubbard, and E. R. Stadtman, *Biochemistry* **7,** 2136 (1968).
145. B. M. Shapiro and A. Ginsburg, *Biochemistry* **7,** 2153 (1968).
146. N. C. Brown, A. Larsson, and P. Reichard, *JBC* **242,** 4272 (1967).

a. Relationship between Divalent Cations and Nucleoside Polyphosphates. An extraordinary example of divalent metal ion control is exhibited by the glutamine synthetases of *E. coli* (*144, 145*) and *B. lichiniformis* (*139*). In the absence of divalent cations the *E. coli* enzyme assumes a so-called "relaxed" configuration that is catalytically inactive and is converted to a "tightened," active form when either Mn^{2+} or Mg^{2+} is supplied. But in addition to these divalent ion-dependent conformational changes, a second level of divalent ion control is afforded by enzyme-catalyzed adenylylation and deadenylylation of a particular tyrosine residue of each subunit of the enzyme. Adenylylation is accompanied by complete change in divalent ion specificity (*77, 78, 147*). Biosynthetic activity of the unadenylylated enzyme is almost completely dependent upon the presence of Mg^{2+}, whereas the adenylylated enzyme has a nearly absolute requirement for Mn^{2+}. Moreover, the activity of the unadenylylated enzyme with Mg^{2+} is not appreciably influenced by the ratio of ATP to Mg^{2+}; but optimal activity of the adenylylated enzyme is rigorously dependent upon the ATP/Mn^{2+} ratio (*139*). The latter form of the enzyme is therefore susceptible to activation or inhibition by the free nucleoside polyphosphate pool in the manner discussed above. Although adenylylation of the glutamine synthetase from *B. lichiniformis* has not been observed, the Mn^{2+}-dependent biosynthetic activity of this enzyme is governed by the ATP/Mn^{2+} ratio and has been shown *in vitro* to respond to nucleotide inhibition or activation according to the above predictions (*139*).

Another prime example of divalent ion control has been disclosed by Gentner and Priess (*148*) in their studies on the ADP-glucose pyrophosphorylase in *E. coli.* The kinetic characteristics of this enzyme are markedly influenced by the nature of the divalent cation utilized for activation. With Mg^{2+} the concentration of ATP required for half-maximal activity is unrelated to the Mg^{2+} concentration; but if Mn^{2+} is the activating cation, maximal activity is obtained when the ratio of ATP to Mn^{2+} is 1.0. Moreover, the sensitivity to inhibition by various metabolites and the saturation kinetics of the specific activator, fructose 1,6-diphosphate, are different for the Mn^{2+}- and Mg^{2+}-activated enzymes.

As yet, the physiological significance of divalent cations in regulation has not been systematically investigated; however, in view of the above considerations it seems likely that this could be the basis of an extremely important regulatory mechanism. Certainly this possibility has not received the attention it deserves.

Several other instances of specific ion effects have been observed. Of

147. H. S. Kingdon and E. R. Stadtman, *J. Bacteriol.* **94,** 949 (1967).
148. N. Gentner and J. Priess, *JBC* **243,** 5882 (1968).

these, the specific inhibition of pyruvate carboxylase by physiologically significant concentrations of sulfate ions (*149*) and the activation of yeast aspartate semialdehyde dehydrogenase by bicarbonate ion (*150*) are noteworthy.

G. Compartmentalization

It has long been recognized that certain metabolic processes are separated physically from each other in the cell. For instance, various oxidative activities including the oxidation of pyruvate to acetyl-CoA, the oxidation of acetyl-CoA to CO_2 via the Krebs cycle, the oxidation of fatty acids, and the oxidation of reduced pyridine nucleotides with coupled phosphorylation are all associated with the mitochondrion. Enzymes involved in gluconeogenesis, glycolysis, fatty acid synthesis, and the degradation of pyrimidines are found in the cytosol. A coupling of intra- and extramitochondrial functions must therefore be mediated by the transport of common metabolites across the mitochondrial membrane barrier. Control of the enzymes having transport functions and those uniquely concerned with conversion of common metabolites into transportable derivatives provides the basis of another level of cellular control.

Lehninger (*151*) has emphasized the importance of such cellular control in maintaining balance between intra- and extramitochondrial electron transport functions. DPNH produced in glycolysis is incapable of direct interaction with the respiratory chain because it cannot penetrate the mitochondrial membrane. Coupling between dehydrogenase activity in the cytosol and electron transport in the mitochondrion is facilitated by so-called shuttle systems. One shuttle is the α-glycerolphosphate–dihydroxyacetone phosphate system. In this system, illustrated in Fig. 4, DPNH generated by metabolism in the cytosol is oxidized by dihydroxyacetone phosphate through the action of an α-glycerolphosphate dehydrogenase in the extramitochondrial cytoplasm. The α-glycerolphosphate thus produced readily crosses the mitochondrial membrane where it is oxidized to dihydroxyacetone phosphate by a flavin-linked α-glycerolphosphate dehydrogenase which in turn passes the electrons to cytochrome b in the respiratory chain. The dihydroxyacetone phosphate resulting from the intramitochondrial oxidation of α-glycerolphosphate then passes back out of the mitochondrion to the cytosol, where it can pick up more electrons from DPNH. In essence then dihydroxyacetone

149. C. H. Fung and M. F. Utter, unpublished observations, cited in (*223*).
150. Y. Surdin, *European J. Biochem.* **2**, 341 (1967).
151. A. L. Lehninger, "The Metochondrion." Benjamin, New York, 1964.

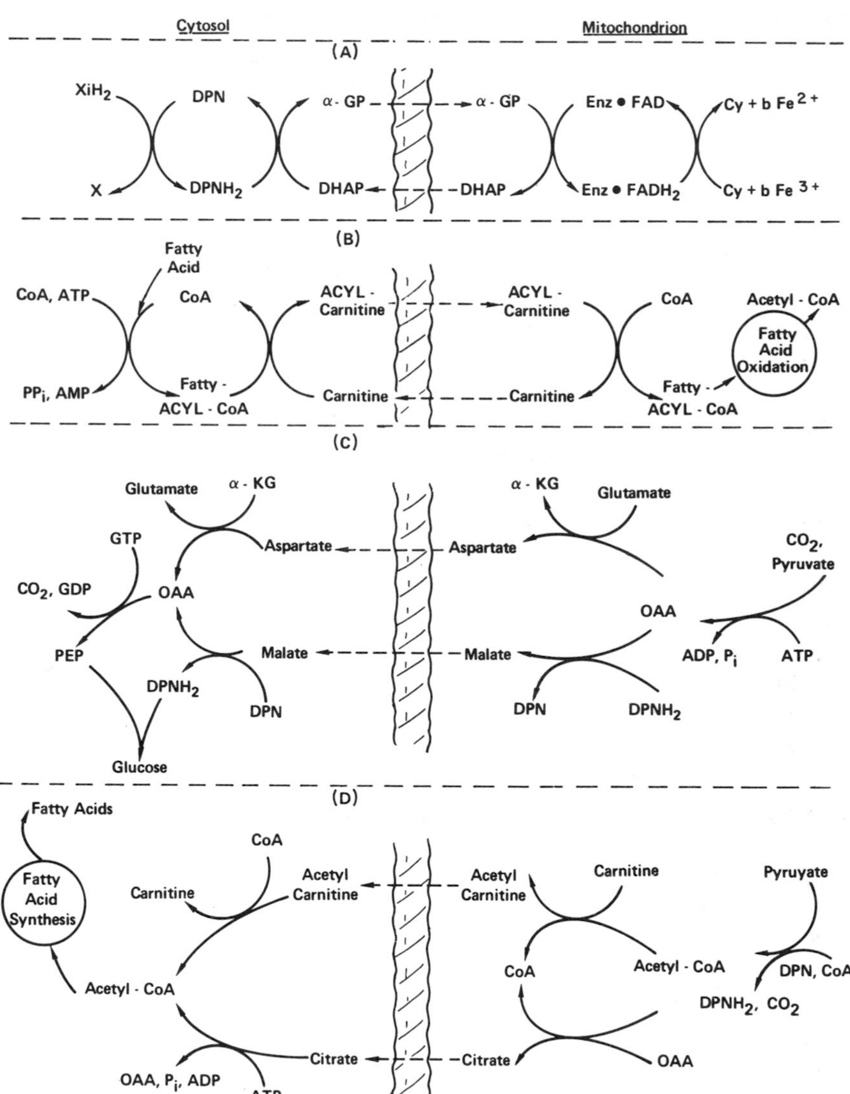

FIG. 4. Shuttle systems for the transfer of metabolites across the mitochondrial membrane. (A) Electron shuttle—based on the scheme suggested by Lehninger (*151*). (B) Fatty acid shuttle—based on the proposal by Fritz (*152, 153*). (C) Oxalacetate shuttle—based on that described by Lardy *et al.* (*158*). (D) Acetyl shuttle—based on schemes suggested by Fritz (*153*) and Lowenstein (*155*). Abbreviations are as follows: $X.H_2$ and X, reduced and oxidized substrates, respectively, for DPN-linked dehydrogenases; α-GP, α-glycerolphosphate; DHAP, dihydroxyacetonephosphate; OAA, oxalacetate; α-KG, α-ketoglutarate.

phosphate and α-glycerolphosphate represent "electron empty" and "electron filled" vehicles, respectively, for the shuttling of electrons across mitochondrial membranes. In addition to the α-glycerolphosphate dehydrogenase system, other dehydrogenases such as lactic dehydrogenase and malate dehydrogenase may be assigned electron shuttle functions.

Shuttle systems are not restricted to electron transport functions however. Among many other processes, shuttles are required to affect oxidation of extramitochondrial fatty acid by the fatty acid oxidizing enzymes that are contained in the mitochondrion. Since neither free fatty acids nor their acyl-CoA derivatives are able to penetrate the mitochondrial membrane, they can enter the mitochondrion only by a shuttle system and in this instance carnitine serves as the vehicle of transport (*152–154*). Free fatty acids from exogenous sources, or those produced in the cytosol by the fatty acid synthetase complex contained therein, are converted to their acyl-CoA derivatives by the action of the fatty acid thiokinase of microsomes. The fatty acyl-CoA carnitine transacylase in the cytosol then catalyzes transfer of the acyl group of the CoA derivatives to carnitine to produce the *O*-acylcarnitine derivatives which can pass through the mitochondrial membrane. Once inside the mitochondrion, fatty acyl-CoA carnitinetransferase catalyzes a second transesterification reaction with CoA to regenerate free carnitine and the fatty acyl-CoA derivatives. The latter are then oxidized to acetyl-CoA by the fatty acid oxidizing enzymes.

A similar problem is encountered in acetyl-CoA metabolism. Acetyl-CoA cannot penetrate the mitochondrial membrane; therefore, acetyl-CoA generated in the mitochondrion by the oxidation of pyruvate or fatty acids is not readily available to the extramitochondrial enzyme systems that utilize acetyl-CoA for biosynthesis of other metabolites including fatty acids.

A carnitine system as described above can mediate the translocation of acetyl groups from inside the mitochondrion to the cytosol (*152*, *153*), but the physiological significance of that shuttle has been questioned (*155*). In another shuttle system that is probably of physiological significance, citrate serves as the carrier of potential acetyl groups (Fig. 4) (*155–157*). In this shuttle the citrate synthetase in the mitochondrion

152. I. B. Fritz, *Advan. Lipid Res.* **1**, 285 (1963).
153. I. B. Fritz, *in* "Cellular Compartmentalisation and Control of Fatty Acid Metabolism" (F. C. Gran, ed.), p. 40. Academic Press, New York, 1968.
154. N. R. Marquis, R. P. Francesconi, and C. A. Valee, *Advan. Enzyme Regulation* **6**, 31 (1967).
155. J. M. Lowenstein, *Biochem. Soc. Symp.* **27**, 61 (1967).
156. P. A. Sere, *Biochem. Soc. Symp.* **27**, 11 (1967).
157. J. B. Chappell and B. H. Robinson, *Biochem. Soc. Symp.* **27**, 123 (1967).

catalyzes the conversion of acetyl-CoA and oxalacetate to citrate. The latter passes through the mitochondrial membrane and then is cleaved to oxalacetate and acetyl-CoA in an ATP-dependent reaction catalyzed by citrate lyase.

In some organisms, another shuttle system is required to couple oxalacetate synthesis inside the mitochondrion with gluconeogenesis in the cytosol (*158, 159*). Oxalacetate produced inside the mitochondrion by the action of pyruvate carboxylase must be translocated into the cytosol where it is converted to *P*-enolpyruvate by the action of *P*-enolpyruvate carboxykinase. As shown in Fig. 4 the translocation of oxalacetate is mediated through interconversion to either malate or aspartate on both sides of the membrane.

In gluconeogenesis, the malate shuttle may have a dual function: It mediates the translocation of both oxalacetate and electrons which are needed for glucose synthesis.

Because of the strategic importance in the coupling of intra- and extramitochondrial functions, it is obvious that the shuttle enzymes are particularly important sites of metabolic control. Although their regulatory characteristics have not been exhaustively studied, there is considerable *in vitro* evidence that they are susceptible to metabolitic control. For example, palmityl-CoA inhibits citrate lyase (*154*), citrate synthetase (*160, 161*), and carnitine acetyltransferase (*154*); moreover, inhibition of the latter two enzymes is antagonized by palmityl carnitine (*154*). Elsewhere in this volume, Atkinson discusses the susceptibility of ATP:citrate lyase to inhibition by ADP and the dependence of both citrate lyase and citrate synthetase activities on the energy charge of the cell (Chapter 9).

H. Macromolecular Complexes

The organization into a single complex of enzymes concerned with a particular metabolic function is of obvious advantage in the coordination of the individual catalytic activities and affords a mechanism for the channeling of metabolic intermediates along a given metabolic pathway (see Chapter 4 in this volume by Reed and Cox). With proper positioning of the individual enzymes the product of one enzymic reaction can be passed on directly to the enzyme catalyzing the next step in the

158. H. A. Lardy, V. Paetkau, and P. Walter, *Proc. Natl. Acad. Sci. U. S.* **53,** 1410 (1965).
159. M. C. Scrutton and M. F. Utter, *Ann. Rev. Biochem.* **37,** 249 (1968).
160. O. Wieland and L. Weiss, *BBRC* **13,** 26 (1963).
161. P. K. Tubbs, *BBA* **70,** 608 (1963).

metabolic sequence without equilibrating with the medium. The local concentrations of metabolites may thus be considerably different from those of the surrounding medium and may afford a finer system of metabolic control than is predicted from *in vitro* studies with isolated components. Moreover, conformational changes accompanying the binding of an allosteric effector to one enzyme component of the complex may, through protein–protein interactions, also affect the activities of other enzymes in the complex which by themselves are unaffected by the effector. Such a theoretical possibility is conceptually only an extension of the principle established for the regulation of aspartate transcarbamylase, in which case metabolite regulation of a catalytic subunit of the enzyme is mediated through its interaction with a dissimilar regulatory subunit. The structure and general properties of several multienzyme complexes are described elsewhere in this volume (see Chapter 4 by Reed and Cox); therefore they will not be discussed here in detail. However, the pyruvate dehydrogenase complex deserves special comment because of the unique regulatory function associated with it. The pyruvate dehydrogenase from *E. coli* exists as a multienzyme complex with a molecular weight of about 4.8 million (*162, 163*). It contains three enzyme components (pyruvate dehydrogenase, dihydrolipoyl transacetylase, and dihydrolipoyl dehydrogenase) that catalyze consecutive reactions involved in the DPN-linked oxidation of pyruvate to acetyl-CoA and CO_2 (*162, 163*). The overall activity of the complex is inhibited by the reaction products, acetyl-CoA and DPNH; these inhibitions are reversed by CoA and DPN, respectively (*164, 165*). DPNH exerts its effect on the lipoyl dehydrogenase component of the complex. The pyruvate dehydrogenase component, which catalyzes the first step in the oxidation of pyruvate, is inhibited by acetyl-CoA and is stimulated by *P*-enolpyruvate and by various nucleotides (*166, 167*). In addition, the overall activity of the complex is subject to regulation by the energy charge of the cell (*120*). The mammalian pyruvate dehydrogenase complex has been isolated from kidney, heart, and liver mitochondria and exhibits properties similar to the bacterial enzyme. For example, the activity is inhibited by acetyl-CoA and DPNH. However, in addition, the mammalian enzyme complex contains a specific kinase and a phosphatase that catalyze the phosphorylation and dephosphorylation, re-

162. L. J. Reed and D. J. Cox, *Ann. Rev. Biochem.* **35**, 57 (1966).
163. L. J. Reed and R. M. Oliver, *Brookhaven Symp.* **21**, 397 (1969).
164. P. B. Garland and P. J. Randle, *Biochem. J.* **91**, 6C (1964).
165. R. G. Hansen and U. Henning, *BBA* **122**, 355 (1966).
166. E. R. Schwartz and L. J. Reed, *Federation Proc.* **27**, 389 (1968).
167. E. R. Schwartz, L. O. Old, and L. J. Reed, *BBRC* **31**, 495 (1968).

spectively, of the pyruvate dehydrogenase component (*168, 169*). Phosphorylation results in inactivation and dephosphorylation in reactivation of the dehydrogenase activity. ATP is a specific phosphoryl group donor for the kinase reaction. The kinase and probably also the phosphatase are bound to the transacetylase fraction of the complex.

It is evident that the phosphorylation and dephosphorylation reactions represent an effective regulatory mechanism that is formally similar to that observed in the interconversions of phosphorylase a and phosphorylase b and of glycogen synthetase I and D. But as noted previously, the regulation of the kinase and phosphatase activities is of extreme importance; otherwise, they, together with their common enzyme substrate, would constitute a useless ATPase activity. Although the regulation of the kinase and phosphatase activities is not yet thoroughly understood, it may be significant that the kinase is specifically inhibited by pyruvate and that the level of Mg^{2+} required for activation of the kinase is much lower than that required for activation of the phosphatase. Reed *et al.* (*168*) have pointed out that variations in the concentration of Mg^{2+} as conditioned by the ratio of ATP to ADP and AMP might be a controlling force in the regulation of the two activities.

I. Unidirection of Reversible Reactions

It is commonly stated that nearly irreversible reactions are the most likely targets for cellular regulation. This is in fact often the case and is most certainly true for those irreversible reactions for which the catalytic potential is very high in relation to that for other reactions in the metabolic sequence. However, it is by no means a rule that target enzymes catalyze essentially irreversible reactions. As noted earlier, enzymes at branch points in metabolism and those catalyzing first steps in metabolic pathways as well as multifunctional enzymes are often subject to control irrespective of the reversibility of the reactions they catalyze. Moreover, there are some situations where it is the object of regulation to render a thermodynamically freely reversible reaction operative in one direction only. For example, in *Blastocladia emersonii* (*170*) the highly reversible DPN-linked oxidative deamination of glutamate [reaction (16)] proceeds only in the direction of glutamate syn-

$$DPN^+ + H_2O + \text{glutamate} \leftrightarrows NH_4^+ + DPNH + \alpha\text{-ketoglutarate} \qquad (16)$$

168. T. C. Linn, F. H Pettit, and L. J. Reed, *Proc. Natl. Acad. Sci. U. S.* **62,** 234 (1969).

169. T. C. Linn, F. H. Pettit, F. Hucho, and L. J. Reed, *Proc. Natl. Acad. Sci. U. S.* **64,** 227 (1969).

170. H. R. LeJohn, *JBC* **243,** 5126 (1968).

thesis in the absence of divalent cations or when any one of various metabolites including citrate, isocitrate, fructose-1,6-diphosphate, and fumarate or ATP is present. These ligands specifically inhibit the oxidative deamination of glutamate but have no effect on the initial rate of the reverse reaction, i.e., the reductive amination of α-ketoglutarate. On the other hand, divalent cations, or AMP or ADP, could reestablish reversibility of the reaction. The apparent irreversibility is probably achieved by allosteric effects of the various ligands which lead to a decreased affinity of the enzyme for DPN and increase the sigmoidicity of the saturation function for this substrate (*170*).

Several other examples of apparent irreversibility of thermodynamically reversible reactions have been noted. Particularly noteworthy is a situation encountered in the metabolism of ornithine in pseudomonads. The conversion of ornithine and carbamyl phosphate to citrulline [reaction (17)] is a key step in the biosynthesis of arginine; however, the reverse reaction is also an important step in the catabolism of arginine when the latter is supplied as a sole source of carbon and nitrogen for growth (*171, 172*). Studies in Wiame's laboratory (*171, 172*)

$$\text{Carbamyl phosphate} + \text{ornithine} \rightleftharpoons \text{citrulline} + \text{P}_i \tag{17}$$

have shown that difficulties which could arise in the regulation of the common step in these two processes are avoided by the elaboration of two separate ornithine carbamyltransferases that catalyze reaction (17). A biosynthetic function for one of these enzymes is apparent from the fact that its formation is selectively repressed by high concentrations of arginine, whereas a catabolic function for the other enzyme is indicated by the fact that its formation is induced coordinately with carbamate kinase and L-arginine iminohydrolyase, i.e., with enzymes concerned with the catabolism of arginine. Of interest to the present discussion is the fact that the biosynthetic enzyme catalyzes the reaction only in the direction of citrulline synthesis (*172*).

An analogous situation may exist in *Clostridium cylindrosporum*. This organism produces two separable formate kinases: One catalyzes the reversible phosphorylation of formate by ATP, but the other catalyzes the same reaction only in the direction of formyl phosphate synthesis, which is the thermodynamically less favored direction (*173*). In these latter situations it seems probable that the products of the unidirectional reactions are allosteric effectors of the enzyme. It is assumed that interaction of a product at a specific allosteric site on the enzyme

171. F. Ramos, V. Stalon, A. Pierard, and J. M. Wiame, *BBA* **139**, 91 (1967).
172. F. Ramos, V. Stalon, A. Pierard, and J. M. Wiame, *BBA* **139**, 98 (1967).
173. W. S. Sly and E. R. Stadtman, *JBC* **238**, 2639 (1963).

destroys its catalytic potential for the reverse reaction, possibly by decreasing its affinity for products at the catalytic site. This interpretation is consistent with the fact that allosteric roles of immediate reaction products or substrates have been established for certain other enzymes, *viz.*, the inhibition of phosphofructokinase by ATP (*174*) and of hexokinase by glucose 6-phosphate (*175, 176*).

III. Regulation of Balance between Metabolic Functions

The decision as to what constitutes the ultimate end product of a biosynthetic pathway is rather arbitrary. From the standpoint of cellular regulation it is generally assumed that simple metabolites such as amino acids, purine and pyrimidine nucleotides, fatty acids, and simple sugars are the end products of their respective biosynthetic pathways since all of the steps in each pathway are functionally concerned with the synthesis of such metabolites. However, these so-called end products are actually building blocks for the synthesis of more complex substances, *viz.*, proteins, nucleic acids, polysaccharides, and complex lipids. In its broader sense regulation of metabolism must take into account not only the biosynthesis of each type of building block but also the fact that a proper balance between the synthesis of all building blocks must be maintained in order to insure their uninterrupted assemblage into the more complex structures of the cell. The magnitude of this problem is overwhelming in light of the fact that a deficiency of any single component will limit overall production of macromolecules. Knowledge concerning the nature and kinds of control mechanisms that achieve balance between the multiplicity of individual metabolic functions is still meager. Nevertheless, some novel mechanisms as well as basic principles that underlie this control have been recognized and are considered in the following sections.

A. Metabolite Interconversion Systems

Balance between two metabolite pools is sometimes facilitated by interconversion of the metabolites, which can be rigorously controlled. An elegant example of such control is found in purine nucleotide metabolism. Inosinic acid is the primary end product of the pathway involved

174. T. E. Mansour, *Advan. Enzyme Regulation* **8**, (1970) (in press).

175. R. K. Crane, "The Enzymes," 2nd ed., Vol. 6, p. 47, 1962.

176. G. De la Fuente and A. Sols, *Abstr. Proc. 6th Intern. Congr. Biochem., New York, 1964*. Vol. 6, p. 506 (1964).

in the *de novo* synthesis of purine nucleotides. As shown in Fig. 5 it is a common intermediate in two divergent pathways of metabolism that lead to the biosynthesis of either guanine nucleotides or to adenine nucleotides; in addition, it forms a common link in pathways by which these two classes of nucleotides are interconverted. Since both kinds of nucleotides are required simultaneously for the synthesis of nucleic acids, it is essential that their production be balanced so that each is always available in adequate supply relative to the other. Figure 5 illustrates how this balance is achieved by strategically directed feedback controls. It is noteworthy that for each nucleotide, separate, oppositely directed pathways are utilized for the conversion of IMP to the product (AMP or GMP) and for the conversion of the product back to IMP. Also, in each case, synthesis of the product requires the utilization of one energy-rich phosphate bond. Furthermore, it appears appropriate that GTP is the source of high energy phosphate bond energy for the synthesis of the adenine nucleotides, whereas ATP is the source of energy for the synthesis of the guanine nucleotides.

Here, then, are two other examples of oppositely directed exergonic reactions that would couple in the absence of appropriate controls and lead to futile cyclic activities with a net loss of energy (see Table II).

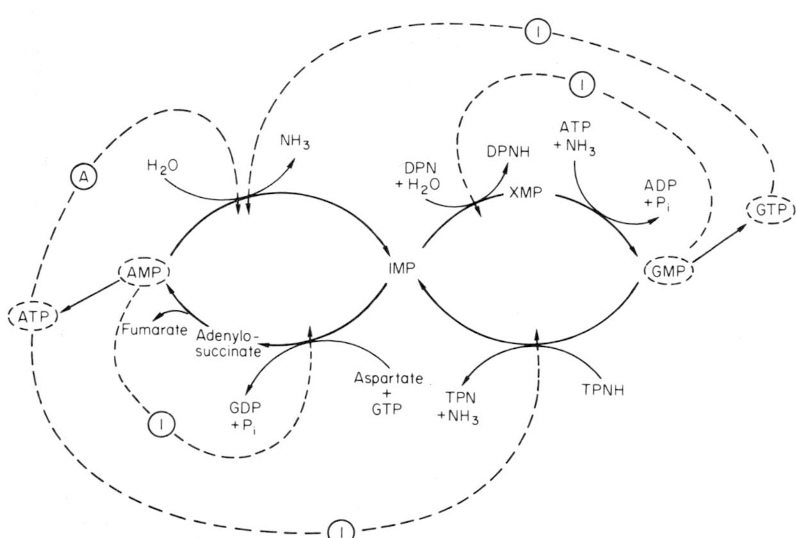

FIG. 5. Regulation of the purine nucleotide interconversion system. The dashed lines connect the steps that are sensitive to control with the metabolites that affect them (A and I as defined in Fig. 1). The scheme is modified after that originally presented by Magasanik (*180*).

TABLE II
FUTILE CYCLES

System	Opposing unidirectional reactions		Reaction	Enzyme	Activators	Inhibitors
(A)	ATP + fructose-6-P → ADP + FDP	(18)	(18)	Phosphofructokinase	Fructose-6-P (*199, 212*)	FDP (*3*)
	$ADP + H_2O \rightarrow P_i$ + fructose-6-P	(19)			AMP (*199, 212*) ADP (*199*) 3′,5′-AMP (*199*) FDP (*199*) Pi (*199*)	ATP (*12, 199, 200*) Citrate (*201–203, 212*)
	Sum: $ATP + H_2O \rightarrow P_i + ADP$	(20)				
			(19)	Fructose-1,6-diphosphatase	ATP (*204*)	AMP (*43*) FDP (*43*)
(B)	ATP + glucose → glucose-6-P + ADP	(21)	(21)	Hexokinase	Pi (*213*) ATP (*214*) Glucose (*204*)	Glucose-6-P (*215*) FFA (*3*)
	Glucose-6-P + H_2O → glucose + P_i	(22)				
	Sum: $ATP + H_2O \rightarrow P_i + ADP$	(23)	(21)	Glucokinase		FFA Acetyl-CoA (*216*) PEP
			(22)	Glucose-6-phosphatase	Glucose-6-P (*216*)	P_i (*213, 220*) Glucose (*220, 228*) PP (*22, 220, 221*)
(C)	$ATP + H_2O + CO_2$ + pyruvate → $OAA + P_i + ADP$	(24)	(24)	Pyruvate carboxylase	Acetyl-CoA (*222*)	$SO_4^{=}$ (*223*) PEP (*223*) ADP (*222*) L-Aspartate (*224, 225*) Malonyl-CoA (*226*)
	$OAA + GTP \rightarrow PEP + CO_2 + GDP$	(25)				
	PEP + ADP → ATP + pyruvate	(26)				
	Sum: $GTP + H_2O \rightarrow GDP + P_i$	(27)	(25)	*P*-Enolpyruvate carboxykinase		

			(26)	Pyruvate kinase	Glucose-6-P (*216*) Glycerol-3-P (*216*) FDP (*216*) PEP (*216*)	FFA (*216*), NADH (*216*) GTP chicken Acetyl-CoA (*216*) ITP liver ATP (*216, 217*) L-Alanine (*216, 219*) 3-Phosphoglycerate (*218*) Proline (*218*)
(D)	Glycogen + P_{i_n} → n(glucose-1-P)	(28)	(28)	Phosphorylase a	PEP (*207*), glycogen (*205*)	UDPG (*207*)
	Glucose-1-P + UTP → UDPG → PP_i	(29)			AMP (*206*) Glucose (*208*)	Malate (*207*) ATP (*209*)
	UDPG + glycogen → UDP	(30)				Glucose-6-P (*209*)
			(28)	Phosphorylase b	AMP (*88, 211*) G-1-P (*210*)	UDPG (*207*) Malate (*207*) GTP, 6-P-glycerate (*207*) DPNH (*207*) ATP (*209*)
			(30)	Glycogen synthetase I		
	UTP → P_i → UDP → PP_i		(30)	Glycogen synthetase D	Glucose-6-P (*74, 75*)	
(E)	IMP + DPN + H_2O → XMP + DPNH	(31)	(31)	IMP dehydrogenase		GMP (*177*)
	XMP + ATP + NH_3 → ADP + P_i + GMP	(32)	(32)	Xanthosine-5′-phosphate aminase		
	GMP + TPNH → IMP + TPN + NH_3	(33)				
	ATP + H_2O → ADP + P_i[a]	(34)	(33)	GMP reductase		AMP, IMP ATP (*179*)
	IMP + aspartate + GTP → adenylosuccinate + GDP + P_i	(35)		I Adenylosuccinate[b] synthetase		ADP, AMP, IMP (*177*)
	Adenylosuccinate → fumarate + AMP	(36)		III Adenylyl acid[c] deaminase	ATP (*178*)	
	AMP + H_2O → IMP + NH_3	(37)			ADP (*129*) AMP (*181*)	GTP (*179*)[c]
	Sum: Aspartate + GTP + H_2O = GDP + P_i + fumarate + NH_3	(38)				

[a] Reduced and oxidized form of the pyridine nucleotides respectively are regarded as equivalent.
[b] *E. coli.*
[c] Ox brain.

Effective regulation of this complex situation is achieved by two kinds of feedback control:

(a) The first step involved in the conversion of IMP to either AMP or GMP is specifically inhibited by the ultimate products; thus, IMP dehydrogenase that catalyzes the oxidation of IMP to XMP is specifically inhibited by GMP, and the enzyme that catalyzes the conversion of IMP to adenylosuccinate is specifically inhibited by ADP (*177*).

(b) The conversions of AMP and of GMP back to IMP are inhibited by GTP (*178*) and ATP (*179*), respectively. At first glance the latter effects appear to represent situations in which there is compensatory control of one cycle on another (*180*); however, in reality they are merely cases of a simple feedback inhibition by end products since the conversions of GMP and AMP to IMP are the first steps in the interconversion of these nucleotides to each other.

In any case, the concerted action of these two types of end product control prevents futile cycling of the various nucleotides and provides a mechanism by which the relative concentrations of the two kinds of nucleotides can be kept in proper balance to meet the changing demands for nucleic acid synthesis.

B. Compensatory Control Mechanisms

To facilitate balanced synthesis of various metabolites, relative excesses of one end product may sometimes stimulate the production of a second end product, either by antagonizing the feedback controls that limit synthesis of the second or by direct allosteric activation of the enzyme that catalyzes the rate determining step in its synthesis. This kind of pace keeping regulation has been aptly referred to as *compensatory control* (*182*).

1. *Metabolite Activation*

Sanwal and Maeba (*182*) have discovered that a fine system of compensatory control is involved in the regulation of *P*-enolpyruvate carboxylase activity in *Salmonella typhimurium*. As shown in Fig. 6 car-

177. I. Lieberman, *JBC* **223,** 327 (1956).

178. B. Setlow and J. M. Lowenstein, *JBC* **243,** 6216 (1968).

179. J. Mager and B. Magasanik, *JBC* **235,** 1474 (1960).

180. B. Magasanik, *in* "The Bacteria" (I. C. Gunsalus and R. Y. Stanier, eds.), Vol. 3, p. 295. Academic Press, New York, 1962.

181. C. H. Suelter, A. L. Kovacs, and E. Antonini, *FEBS Letters* **2,** 65 (1968).

182. B. D. Sanwal and P. Maeba, *JBC* **241,** 4557 (1966).

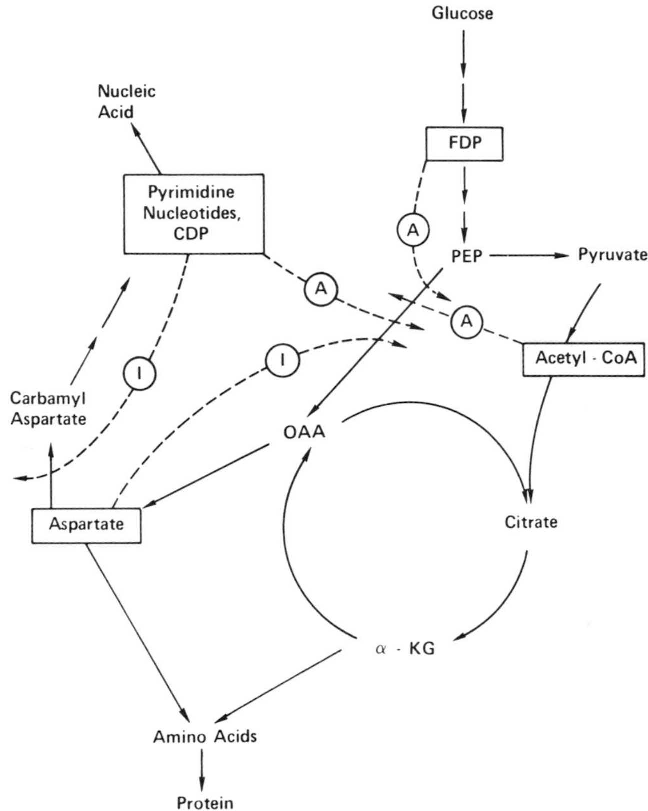

FIG. 6. Carboxylation of *P*-enolpyruvate.

boxylation of *P*-enolpyruvate represents the first common step in a highly branched pathway that leads ultimately to the formation of pyrimidine nucleotides, various amino acids, and oxalacetate, the latter has a catalytic role in the oxidation of acetyl-CoA via the tricarboxylic acid cycle. Coordination of these various functions is achieved by a basic pattern of sequential feedback inhibition which is tempered by a system of compensatory modulation. Sequential feedback inhibition begins with an accumulation of pyrimidine nucleotides which inhibit aspartate transcarbamylase (*36*), catalyzing the first committed step in pyrimidine nucleotide synthesis. This leads to an accumulation of aspartate which in turn inhibits *P*-enolpyruvate carboxylase (*183*). This could lead to a shortage of oxalacetate which is needed in the tricarboxylic acid cycle for citrate synthesis. A deficiency of oxalacetate is avoided however

183. P. Maeba and B. D. Sanwal, *BBRC* **21,** 503 (1965).

through compensatory activation of *P*-enolpyruvate carboxylase by three different basic mechanisms: (a) feedback activation by pyrimidine nucleotides (*182*); (b) precursor activation by fructose diphosphate (*184*); and (c) activation by acetyl-CoA (*183, 185*), a compound which would accumulate for a lack of oxalacetate. The influence of this compensatory control is augmented by the fact that acetyl-CoA and pyrimidine nucleotides are synergistic in their effects; acetyl-CoA increases markedly the affinity of the *P*-enolpyruvate carboxylase for pyrimidine nucleotides. Each of the three patterns of compensatory control utilized by *S. typhimurium* in the regulation of *P*-enolpyruvate carboxylase activity is found to occur in the regulation of other complex metabolic situations. A few examples are discussed in the following sections.

2. *Antagonism of End Product Inhibition*

A striking example of this kind of compensatory control is found in the regulation of carbamyl phosphate synthetase by *E. coli* as shown in Fig. 7. Carbamyl phosphate is a common precursor for the biosynthesis of pyrimidine nucleotides on the one hand and for arginine on the other. In *E. coli* the carbamyl phosphate synthetase is subject to strong feedback inhibition by pyrimidine nucleotides but not by arginine (*186*). Consequently, when pyrimidine nucleotides become present in excess, the synthesis of carbamyl phosphate is curtailed and may become too low to support adequate synthesis of arginine. However, imposed deficiency of carbamyl phosphate will lead to an accumulation of ornithine, which, at high concentrations, is able to antagonize the inhibitory effects of pyrimidine nucleotides and thereby to restore activity of carbamyl-

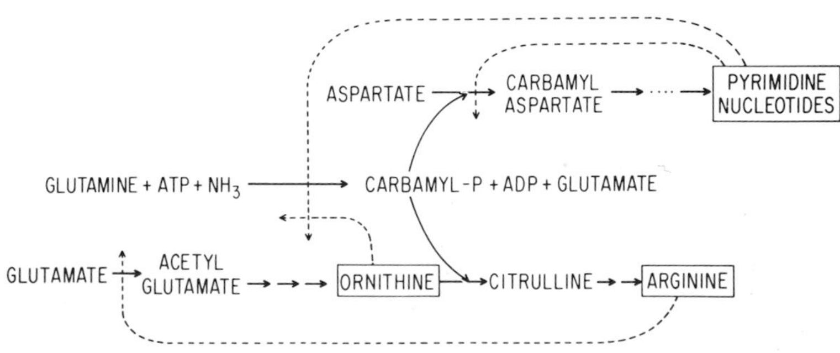

Fig. 7. Compensatory control of carbamyl phosphate synthetase activity in *E. coli*. The scheme is similar to that published by A. Pierard (*186*).

184. P. Maeba and B. D. Sanwal, *BBRC* **22,** 194 (1966).
185. J. L. Canovas and H. L. Kornberg, *BBA* **96,** 169 (1965).

phosphate synthetase (*186*). It is significant from the standpoint of cellular regulation that the carbamyl phosphate derived from this renewed synthesis will be available exclusively for arginine biosynthesis since feedback inhibition of aspartate transcarbamylase by CTP (*36*) will prevent its use for continued synthesis of unwanted pyrimidine nucleotides.

Several other situations have been noted wherein feedback inhibition by one end product is antagonized by another metabolite. To mention a few:

(a) The inhibition of deoxycytidylate deaminase by (*d*)TTP is reversed by *d*CTP (*187–189*). This facilitates balanced synthesis of these two nucleic acid precursors.

(b) In *Bacillus subtilis* feedback inhibition of threonine deaminase by isoleucine is antagonized by valine (*190*).

(c) In *E. coli* the inhibition of ribonucleotide reductase by *d*ATP is antagonized by ATP (*191, 192*). This, together with the complex cascade system that regulates deoxyribonucleotide synthesis (Section III,D), could serve to keep the synthesis of ribonucleotides and deoxyribonucleotides in balance for nucleic acid synthesis.

(d) In *Rhodospirillum rubrum* the feedback inhibition of aspartokinase by threonine or the concerted feedback inhibition of this enzyme by a combination of threonine plus lysine is antagonized by isoleucine (*193*). As a result of this compensatory control, continued synthesis of lysine and methionine can occur in the presence of an excess of threonine and isoleucine thus keeping the production of all four amino acids in balance.

(e) The deoxythymidine kinase of *E. coli* is activated by *d*CDP and other deoxynucleotide diphosphates all of which would accumulate owing to a lack of (deoxy)thymidine nucleotides needed for DNA synthesis (*194*).

186. A. Pierard, *Science* **154,** 1572 (1966).
187. G. F. Maley and F. Maley, *JBC* **239,** 1168 (1964).
188. G. F. Maley and F. Maley, *JBC* **237,** PC3311 (1962).
189. E. Scarano, G. Geraci, A. Polzella, and E. Campanile, *JBC* **238,** PC1556 (1963).
190. W. G. Hatfield, Ph.D. Thesis, Purdue University, Lafayette, Indiana, 1968; quoted by H. E. Umbarger, *Ann. Rev. Biochem.* **38,** 323 (1969).
191. A. Holmgren, P. Reichard, and L. Thelander, *Proc. Natl. Acad. Sci. U. S.* **54,** 830 (1965).
192. A. Larsson and P. Reichard, *JBC* **241,** 2533 (1966).
193. P. Datta and H. Gest, *JBC* **240,** 3023 (1965).
194. R. Okazaki and A. Kornberg, *BBA* **81,** 150 (1964).

3. *Precursor Substrate Activation*

In all of the foregoing examples of compensatory control it is significant that one end product that would accumulate for want of a second end product is able to stimulate the production of the second either by direct activation or by antagonism of negative feedback controls of the enzyme catalyzing the rate determining step in the synthesis of the second end product.

The activation of *P*-enolpyruvate carboxylase by fructose diphosphate (Fig. 6) (*184*) represents a situation in which an enzyme catalyzing a rate limiting step in a metabolic sequence is activated by a precursor. This might be regarded as a compensatory type of control by a metabolic intermediate that would tend to accumulate when activity of the sensitive enzyme becomes too low. Several other instances of precursor activation have been described. Thus, in several organisms, citrate is an activator of the DPN-dependent isocitrate dehydrogenase (*195–197*) and glucose 6-phosphate is an activator of glycogen synthetase (*198*). In each of these instances the precursor is an intermediate at least one step removed from the immediate substrate for the target enzyme. However, in a broader sense immediate substrates of key enzymes catalyzing the rate determining steps in metabolism could be regarded as compensatory activators of their respective enzymes. This follows from the fact that the overall rate of metabolism is determined by the activities of these key enzymes, and this in turn is a function of the substrate concentration. In other words, a tendency of the critical substrates to accumulate is partially offset by mass action effects that serve a compensatory role in acceleration of the rate limiting steps. This kind of compensatory control reaches an even greater degree of sophistication when there are homotropic (*57*) interactions of the substrates with the regulatory enzymes, giving rise to sigmoidal saturation functions. The significance of these "cooperative" phenomena in modulating substrate effects, by establishing thresholds beyond which the regulatory enzymes are hypersensitive to substrate activation, has already been discussed (Section II,B).

It seems likely that compensatory control systems are much more widespread than present knowledge indicates. This follows from the fact that compensatory effectors are not readily recognized since they often bear no obvious relationship to the sensitive enzyme or the end products of the pathway under their control. Until the complex interrelationships

195. B. D. Sanwal, C. S. Stachow, and R. A. Cock, *Biochemistry* **4,** 410 (1965).
196. D. E. Atkinson, J. A. Hathaway, and E. C. Smith, *JBC* **240,** 2682 (1965).
197. J. A. Hathaway and D. E. Atkinson, *JBC* **238,** 2875 (1963).
198. L. F. Leloir, J. M. Olavarria, S. H. Goldenberg, and H. Carminatte, *ABB* **81,** 508 (1959).

of various metabolic pathways are better understood, the disclosure of compensatory controls is apt to be the result of chance observations rather than the result of logically designed experimental efforts.

C. Regulation of Oppositely Directed Exergonic Reactions

Often in metabolism situations exist in which an enzyme in one metabolic pathway catalyzes the conversion of a given substrate to a product which can then be converted back to the given substrate by an enzyme whose function is concerned with a different metabolic process. Situations of this kind are often encountered when biosynthetic and degradative pathways of metabolism share common intermediates. In such cases both processes may utilize identical steps for those reactions that proceed with little or no change in free energy; however, when a reaction is highly exergonic in one direction of metabolism, it is often replaced by another reaction for metabolism in the opposite direction. For example, in carbohydrate metabolism, where pathways of glycolysis and gluconeogenesis share several reactions in common, the three highly exergonic reactions of glycolysis (i.e., the conversion of glucose to glucose 6-phosphate, the conversion of fructose 6-phosphate to fructose 1,6-diphosphate and the conversion of phosphoenolpyruvate to pyruvate) are all replaced by different reactions in the pathway of gluconeogenesis.

1. *The Phosphofructokinase–Fructose-1,6-Diphosphatase Couple*

In glycolysis, phosphofructokinase (PFK) catalyzes a reaction between fructose 6-phosphate and ATP to produce fructose 1,6-diphosphate [reaction (18)]. This reaction is highly exergonic ($\Delta F_0' = -4.2$ kcal), and its reversal constitutes an appreciable energy barrier. This energy barrier is circumvented in gluconeogenesis by the action of fructose-diphosphatase (FDPase), which catalyzes the hydrolysis of fructose diphosphate to fructose 6-phosphate [reaction (19)]. Since reactions (18) and (19) are both exergonic, it is evident that when fructose-1,6-diphosphatase and phosphofructokinase are both present coupling between reactions (18) and (19) will occur and the net result would be hydrolysis of ATP [reaction (20)]. In effect reactions (19) and (18) would be uncoupled from their respective metabolic functions and a useless recycling of fructose-6-phosphate and fructose-1,6-diphosphate

$$\text{ATP} + \text{fructose-6-P} \xrightarrow{\text{PFK}} \text{ADP} + \text{fructose 1,6-diphosphate} \qquad -4.2 \text{ kcal} \quad (18)$$

$$\text{Fructose 1,6-diphosphate} + H_2O \xrightarrow{\text{FDPase}} \text{fructose-6-P} + P_i \qquad -3.8 \text{ kcal} \quad (19)$$

$$\text{ATP} + H_2O \rightarrow \text{ADP} + P_i \qquad -8.0 \text{ kcal} \quad (20)$$

would occur with the concomitant hydrolysis of ATP. It is evident that the activities of the two enzymes must be controlled in order to maintain their normal function and to prevent a futile cycling (*159*) of intermediates with attendant loss of energy accompanying ATP hydrolysis. In this and other analogous situations, coupling of the oppositely directed reactions is prevented by the reciprocal action of various metabolites on the two enzymes involved (*201–221*). Thus effectors that activate one enzyme inhibit the other and vice versa.

As shown in Fig. 8 and Table II, phosphofructokinase is activated by AMP and FDP (*199, 212*) and is inhibited by ATP (*199, 200, 212*); whereas these metabolites have opposite effects on fructosediphosphatase activity (*43, 204*). It follows that high concentrations of AMP and

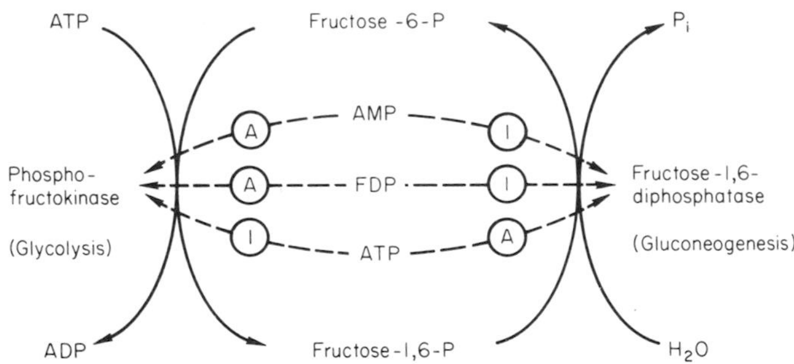

FIG. 8. Metabolite regulation of phosphofructokinase and fructosediphosphatase activities. (A and I as defined in Fig. 1.)

199. O. H. Lowry and J. V. Passonneau, *Arch. Exptl. Pathol. Pharmakol.* **248,** 185 (1964).

200. H. A. Lardy and R. E. Parks, Jr., *in* "Enzymes: Units of Biological Structure and Function" (O. H. Gaebler, ed.), p. 584. Academic Press, New York, 1956.

201. J. V. Passonneau and O. H. Lowry, *BBRC* **13,** 372 (1963).

202. A. Parmeggiani and R. H. Bowman, *BBRC* **12,** 268 (1963).

203. P. B. Garland, P. J. Randle, and E. A. Newsholme, *Nature* **200,** 169 (1963).

204. E. A. Newsholme, *BJ* **89,** 38P (1963).

205. J. H. Wang, M. L. Shanka, and D. J. Graves, *Biochemistry* **4,** 2296 (1965).

206. O. H. Lowry, D. W. Schultz, and J. V. Parsanneau, *JBC* **239,** 1947 (1964).

207. H. E. Morgan and A. Parmeggiani, *JBC* **239,** 2440 (1964).

208. J. H. Wang, M. L Shonka, and D. J. Graves, *BBRC* **18,** 131 (1965).

209. A. Parmeggiani and H. E. Morgan, *BBRC* **9,** 252 (1962).

210. N. B. Madsen, *BBRC* **15,** 390 (1964).

211. E. Helmreich and C. F. Cori, *Proc. Natl. Acad. Sci. U. S.* **51,** 131 (1964).

212. J. V. Passonneau and O. H. Lowry, *Advan. Enzyme Regulation* **2,** 265 (1964).

213. I. A. Rose, J. V. B. Warms, and E. L. O'Connell, *BBRC* **15,** 33 (1964).

214. H. J. Fromm and V. Zeive, *JBC* **237,** 1662 (1962).

215. R. K. Crane and A. Sols, *JBC* **203,** 273 (1953).

FDP will stimulate glycolysis and inhibit gluconeogenesis, whereas a high concentration of ATP will stimulate gluconeogenesis and retard glycolysis. The effectiveness of this control is further augmented by the fact that the relative concentrations of AMP and ATP must vary inversely.

2. *The Glycogen Synthetase–Glycogen Phosphorylase Couple*

Table II, system D, shows how the coupling of glycogen synthetase and glycogen phosphorylase could result in a futile cycle (*159*). However, this coupling is prevented because of the unique interrelationship between the two enzymes. For example, both enzymes exist in phosphorylated and nonphosphorylated forms, but a reciprocal relationship exists with respect to the effects of phosphorylation on catalytic activity; thus, phosphorylation of phosphorylase b leads to an active form (phosphorylase a) (*89, 90*), whereas phosphorylation of glycogen synthetase I leads to an inactive form (synthetase D) (*74, 75*). The opposite effects of phosphorylation on catalytic activity are of particular significance in mammals since phosphorylase kinase kinase and glycogen synthetase kinase appear to be one and the same enzyme (*105*). Therefore, the activation of this enzyme by 3′,5′-cyclic-AMP can lead to the simultaneous activation of phosphorylase and inactivation of glycogen synthetase. Furthermore, as can be seen from Table I, glycogen and glucose have opposite effects on the phosphatase that catalyzes the dephosphorylation of phosphorylase and glycogen synthetase. Thus, glycogen inhibits the conversion of inactive glycogen synthetase D to the active I (*103, 104*) form, but it stimulates the conversion of inactive phosphorylase to active phosphorylase a (*73, 100*). On the other hand, glucose stimulates the

---

216. G. Weber, M. A. Lea, and N. B. Stamm, *Advan. Enzyme Regulation* **6,** 101 (1968).

217. W. Seubert and W. Huth, *Biochem. Z.* **343,** 176 (1965).

218. W. Seubert, H. V. Henning, W. Schoner, and M. L'Age, *Advan. Enzyme Regulation* **6,** 153 (1968).

219. T. E. Mansour and C. E. Ahlfors, *JBC* **243,** 2523 (1968).

220. R. C. Nordlie, *in* "Control of Glycogen Metabolism" (W. J. Whelan, ed.), p. 153. Academic Press, New York, 1967.

221. R. C. Nordlie and W. J. Arion, *JBC* **240,** 2155 (1965).

222. D. B. Keech and M. F. Utter, *JBC* **238,** 2609 (1963).

223. D. B. Keech and M. F. Utter, *in* "Current Topics in Cellular Regulation" (B. L. Horecker and E. R. Stadtman, eds.), Vol. 1, p. 253. Academic Press, New York, 1969.

224. J. J. Cazzulo and A. O. Stopani, *ABB* **121,** 596 (1967).

225. J. J. Cazzulo and A. O. Stopani, *ABB* **127,** 563 (1968).

226. M. C. Scrutton and M. F. Utter, *JBC* **242,** 1732 (1967).

activation of glycogen synthetase and the inactivation of phosphorylase (*106*). These oppositely directed metabolite effects, together with complicity of the common kinase, provide an unusually effective means of regulation of glycogen metabolism.

The importance of regulating situations of this kind is suggested by Table II, which shows how coupling of several oppositely directed steps in metabolism would result in futile cycles (*159*) and a loss of energy. The table also shows that reciprocal responses to various metabolic effectors have been observed *in vitro*. Whereas these reciprocal effects could be the basis of a rational regulation system, the physiological significance of such regulation has not been definitely established. In addition to those effectors that exert opposite effects on the coupled enzymes, it is noted that other metabolites are also able to influence one or another of the coupled enzymes. The physiological basis for the activities of some of these effectors is not yet apparent but may become evident as our knowledge of the interrelationships between various metabolic pathways becomes more complex.

D. Allosteric Determination of Catalytic Function

1. *Ribonucleotide Reduction*

Reichert and his associates (*56, 191, 192, 227, 228, 229*) have disclosed the existence of a highly sophisticated system for the regulation of ribonucleotide reductase activity in *E. coli*. The remarkable feature of this system is the fact that a single enzyme catalyzes the reduction of all purine and pyrimidine nucleoside diphosphates to their corresponding deoxyribonucleotide derivatives; yet under various conditions the enzyme exhibits a degree of substrate specificity. Available data support the view that variations in substrate specificity and catalytic potential are related to the existence of multiple conformational states of the enzyme which are induced or stabilized by the binding of different allosteric effectors (*191, 229*). Thus the reduction of CDP or UDP is facilitated by the presence of ATP, the reduction of ADP and GDP is facilitated by *d*GTP, and the reduction of all nucleoside diphosphates is stimulated by (*d*)TTP and is inhibited by *d*ATP. The possibility that the enzyme consists of a regulatory subunit as well as a catalytic subunit is suggested by the fact that it is composed of two kinds of subunits, B_1 and B_2, both of which are required for catalytic activity and whose association is

227. A. Larsson and P. Reichard, *Progr. Nucleic Acid Res.* **7,** 303 (1967).
228. A. Larsson and P. Reichard, *BBA* **113,** 407 (1966).
229. A. Larsson and P. Reichard, *JBC* **241,** 2540 (1966).

mediated by the presence of Mg^{2+} (*56*). The allosteric effectors *d*ATP, *d*GTP, and (*d*)TTP have been shown to bind only to the B_1 subunit, which is presumed to be the regulatory subunit (*56*). The B_2 subunit contains two equivalents of iron and is assumed to contain the catalytic site.

The positive and negative effects of various nucleotides on ribonucleotide reduction are compatible with a cascadelike control of deoxyribonucleotide synthesis which begins with the activation of CDP and UDP reduction by ATP; *d*UTP thus formed is partly converted to (*d*)TTP which serves as an activator for the reduction of all nucleoside diphosphates. Since *d*GDP is a specific activator of purine nucleoside diphosphate reduction, its accumulation leads to accelerated synthesis of *d*GDP and *d*ADP. In this cascade of events *d*ADP would be the last deoxynucleoside diphosphate to be formed, and *d*ATP derived from it could be regarded as the terminal end product of ribonucleotide reduction. Viewed in this way it is perhaps reasonable that *d*ATP serves as a feedback inhibitor of the reduction of all ribonucleoside diphosphates. As discussed earlier (Section III,B), an antagonistic action of ATP in overriding the inhibition of ribonucleotide reduction by *d*ATP may constitute a compensatory action that helps to keep the synthesis of deoxyribonucleotides in balance with the synthesis of ribonucleotides.

2. *The Glutamic-Alanine Dehydrogenase System*

The ability of allosteric effectors to alter the substrate specificity of an enzyme is clearly illustrated by the studies of Tomkins and his associates (*230, 231*) with bovine liver glutamate dehydrogenase. The enzyme can exist in at least three different interconvertible conformations.

$$\text{Polymer} \overset{\text{I}}{\leftrightarrow} \text{monomer } x \overset{\text{II}}{\leftrightarrow} \text{monomer } y$$

Equilibrium I is highly dependent upon the protein concentration; whereas equilibrium II is markedly influenced by specific ligands, pH and ionic strength. At low protein concentrations equilibrium I is shifted almost completely to the right and the enzyme exists mainly as a monomer. For the present discussion it is important to note that the monomer x conformation is stabilized by the binding of ADP, whereas the conformation y is stabilized by the presence of various steroids and by CTP as well as other ligands (*232*). It is of further significance that the

230. G. M. Tomkins, K. L. Yielding, J. F. Curran, M. R. Summers, and M. W. Bitensky, *JBC* **240,** 3793 (1965).

231. M. W. Bitensky, K. L. Yielding, and G. M. Tompkins, *JBC* **240,** 663 (1965).

232. M. W. Bitensky, K. L. Yielding, and G. M. Tomkins, *JBC* **240,** 1077 (1965).

polymer and monomer x forms exhibit maximal glutamate dehydrogenase activity, whereas monomer y possesses little glutamate dehydrogenase activity but has considerable L-alanine dehydrogenase activity as well as a capacity to oxidize a number of other monocarboxylic amino acids (*230, 231*). It is evident, therefore, that in the presence of CTP the enzyme is converted from a glutamate dehydrogenase to a fairly nonspecific monocarboxylic amino acid dehydrogenase, and in the presence of ADP the enzyme is converted back to a form with glutamate dehydrogenase activity; this latter shift in equilibrium is further influenced by the protein concentration which, at high concentration, favors polymer formation and therefore glutamate dehydrogenase activity. Although the physiological significance of the ligand-induced changes in enzyme function has not been demonstrated, Kun *et al.* (*233*) have suggested that such effects may be important in the regulation of glutamate and alanine metabolism in kidney.

IV. Feedback Regulation of Multifunctional Pathways

As noted earlier, the regulation of linear biosynthetic pathways such as that involved in the synthesis of histidine presents no serious problems. Inhibition of the first step in the pathway by the ultimate end product provides an elegant control system. On the other hand, regulation of a branched biosynthetic pathway such as that illustrated in Fig. 9 presents a special problem since a part of the biosynthetic sequence, i.e., the conversion of A to C, is involved in the synthesis of both end products E and G. If the first common step (step a) is subject to independent feedback control by either or both of the ultimate end products, a situation could arise in which an excess of one end product would inhibit the conversion of A to B and thereby cause a deficiency in the production

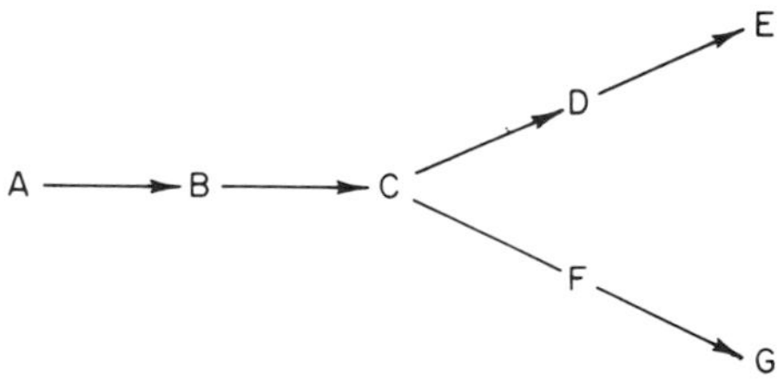

FIG. 9. Branched metabolic pathway.

233. E. Kun, J. E. Ayling, and B. G. Baltimore, *JBC* **237**, 2896 (1964).

of the other. Investigations of several branched biosynthetic pathways have disclosed that a number of fundamentally different mechanisms of regulation are utilized to avoid such difficulties. Some of these mechanisms are as follows.

A. Enzyme Multiplicity

In this mechanism, which is illustrated in Fig. 10, the first common step (step a) in the pathway is catalyzed by two different enzymes, one of which is under specific feedback control by the end product E and the other is specifically controlled by compound G. An excess of compound E can therefore inhibit only that fraction of step a activity that is catalyzed by the E-sensitive enzyme. The residual step a activity resulting from the G-sensitive enzyme would still be available for the production of compounds B and C needed for the continued synthesis of compound G. In order to prevent further utilization of B and C for the synthesis of E, there is a second feedback control of step c by the end product. This latter represents a simple case of feedback inhibition of the first committed step in the biosynthetic pathway by the ultimate end product. As illustrated by Fig. 10, a parallel set of feedback controls of the G-sensitive step a activity and of step d limits the synthesis of intermediates that are common to both end products but assures normal production of E in the presence of an excess of G. Finally, when both end products are in excess the entire pathway becomes inoperative.

Multiple enzymes form the basis of effective regulatory patterns for a number of biosynthetic pathways. Thus, in *E. coli* three different aspartokinases and two different homoserine dehydrogenases are involved in the feedback control of the highly branched pathway by which aspartate is utilized for the synthesis of lysine, threonine, methionine,

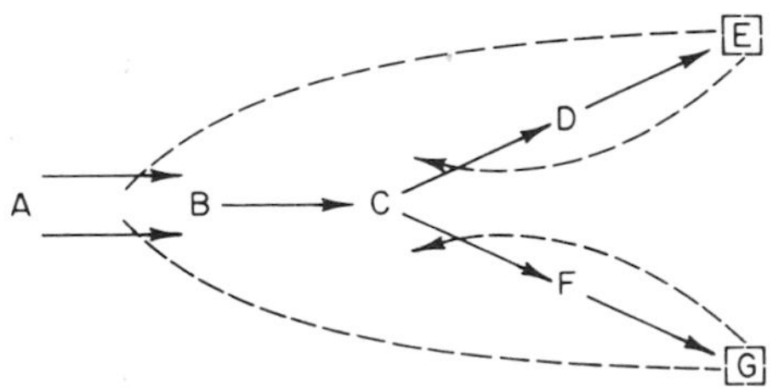

Fig. 10. Enzyme multiplicity in the regulation of branched pathways. Dashed lines interconnect the in product with the steps they inhibit.

and isoleucine (*234–236*). In *E. coli* (*237–240*) and *Aerobacter aerogenes* (*24*), the first step in the biosynthesis of the three aromatic amino acids, tryptophan, tyrosine, and phenylalanine is catalyzed by three separate enzymes, each of which is subject to feedback control by a different end product; moreover, there are two separate mutases that catalyze the conversion of chorismic acid to prephenic acid, one of which is regulated by tyrosine and the other by phenylalanine (*241*).

The synthesis of carbamoyl phosphate from ATP, glutamine and CO_2 is a common step in the biosynthesis of pyrimidine nucleotides and arginine. In *Saccharomyces cerevisiae,* there are two carbamoyl phosphate synthetases, one under feedback control by arginine and the other by uracil and UTP (*242*).

The elaboration of multiple enzymes to facilitate regulation is not restricted to branched biosynthetic pathways. In fact, whenever a given reaction is required for two or more functions, multiple enzymes may be involved in the differential control of these functions (*243*). For instances, when the same reaction is used in two different catabolic pathways or when it is used for both a catabolic and a biosynthetic function, multiple enzymes may form the basis of effective regulation. An example of the former situation is found in *Bacillus subtilis* where the oxidation of pyrrolidine-5-carboxylate to glutamic acid is a common step in the degradation of both arginine and proline. In this case it has been found that the common step is catalyzed by two different dehydrogenases (*244*). One form is produced when the organism is grown in the presence of proline, whereas the other is formed only in the presence of arginine. Both forms are produced when both arginine and proline are present. Whereas these enzymes catalyze the same reaction, they are structurally dissimilar and can be separated from one another by chromatography on DEAE-cellulose (*244*).

234. E. R. Stadtman, G. N. Cohen, G. LeBras, and H. de Robichon-Szulmajster, *JBS* **236,** 2033 (1961).

235. J. C. Patte, G. LeBras, and G. N. Cohen, *BBA* **136,** 245 (1967).

236. G. N. Cohen, *in* "Current Topics in Cellular Regulation" (B. L. Horecker and E. R. Stadtman, eds.), Vol. 1, p. 183. Academic Press, New York, 1969.

237. L. C. Smith, J. M. Ravel, S. R. Lox, and W. Shive, *JBC* **237,** 3566 (1962).

238. K. D. Brown and C. H. Doy, *BBA* **77,** 170 (1963).

239. C. H. Doy and K. D. Brown, *BBA* **104,** 377 (1965).

240. K. D. Brown and C. H. Doy, *BBA* **118,** 157 (1966).

241. R. G. H. Cotton and F. Gibson, *BBA* **100,** 76 (1965).

242. F. Lacroute, A. Pierard, M. Grenson, and J. M. Wiame, *J. Gen. Microbiol.* **40,** 127 (1965).

243. H. E. Umbarger, *in* "Control Mechanisms in Cellular Processes" (D. M. Bonner, ed.), p. 67. Ronald Press, New York, 1961.

244. G. DeHauwer, R. Lavalle, and J. M. Wiame, *BBA* **81,** 257 (1964).

A number of instances have been reported in which multiple enzymes are involved in the regulation of a step that is common to biosynthetic and degradative pathways of metabolism. Thus, in *E. coli* the conversion of threonine to α-ketobutyrate is the first committed step in the biosynthesis of isoleucine and is also an important step in the fermentative metabolism of threonine when it is provided as a substitute for growth under anaerobic conditions. Umbarger and Brown (*35, 245, 246*) showed that conflict between biosynthetic and catabolic functions is avoided by the production in *E. coli* of two threonine deaminases; one is subject to feedback control by isoleucine, the other is an inducible enzyme that is not susceptible to feedback control by isoleucine and is formed only when threonine is available as a substrate for anaerobic growth. Similar relationships between biosynthetic and catabolic functions have been demonstrated for two α-acetolactate synthetases in *Aerobacter aerogenes* (*246, 247*) and for two ornithine transcarbamylases in some pseudomonads (*171, 172*). The former are concerned with pathways of valine biosynthesis and the fermentative metabolism of pyruvate at low pH, whereas the latter are concerned with either the biosynthesis of arginine or with fermentation of arginine.

B. Sequential Feedback Controls

Figure 11 illustrates another pattern of feedback control that was discovered by Nester and Jensen (*248*) as a result of their studies on the

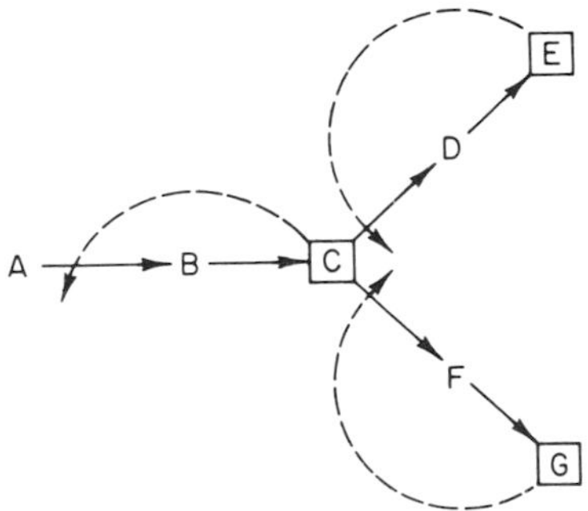

Fig. 11. Regulation of branched pathway by sequential feedback control. The dashed lines interconnect the metabolites with the steps they inhibit. Steps a, b, c, and d refer to conversions of A to B, B to C, C to D and C to F, respectively.

245. N. E. Umbarger and B. Brown, *J. Bacteriol.* **73,** 195 (1957).
246. H. E. Umbarger, *Ann. Rev. Biochem.* **38,** 323 (1969).
247. H. E. Umbarger and B. Brown, *JBC* **233,** 1156 (1958).
248. E. W. Nester and R. A. Jensen, *J. Bacteriol.* **91,** 1591 (1966).

biosynthesis of aromatic amino acids in *B. subtilis*. In this mechanism the first common step a is catalyzed by a single enzyme that is not susceptible to feedback control by either of the ultimate end products; however, step a is sensitive to feedback control by the last common intermediate (compound C) in the branched pathway. In addition, the first divergent steps c and d are under feedback control by the ultimate end products of their unique pathways.

The sequential nature of the control is evident if one considers the course of events that takes place when one of the end products becomes present in excess. For example, with an accumulation of compound E, step c will be inhibited thereby preventing further utilization of compound C for the synthesis of E. As a result more of compound C is diverted toward the synthesis of compound G. When the latter becomes present in excess also, it will inhibit step d, thus curtailing all utilization of compound C. As a consequence, compound C will accumulate and this will lead to feedback inhibition of step a, bringing the entire pathway to a halt.

As noted above sequential feedback control is the basis of regulating the biosynthesis of the aromatic amino acids in *B. subtilis*. As shown in Fig. 12 feedback inhibition of the first divergent steps in the synthesis of tryptophan, phenylalanine, and tyrosine by their respective end products will lead to an accumulation of equilibrium mixtures of

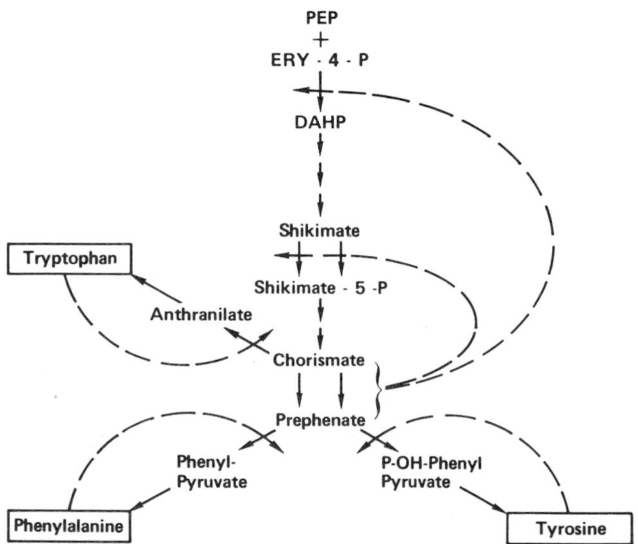

FIG. 12. Regulation aromatic amino acid biosynthesis in *B. subtilis*. The dashed lines interconnect the metabolites with the steps they inhibit.

chorismic and prephenic acids. These intermediates, in turn, are able to inhibit the first common step in the overall biosynthetic pathway. In *B. subtilis* this mechanism of sequential feedback control is supplemented by the elaboration of multiple forms of chorismic mutase and shikimate kinase (*249–251*). The mutases are insensitive to feedback inhibition but are susceptible to repression (*249–251*). In contrast, the shikimate kinase is feedback inhibited by chorismate and prephenate (*251*). In addition to these complicating influences, the regulation of aromatic amino acid biosynthesis in this organism is further facilitated by the organization of at least three enzymes (DAHP synthetase, one of the isozymes of chorismate mutase, and one isozyme of shikimate kinase) into a single multienzymes complex (*252*).

Another example of sequential feedback control is found in the regulation of threonine and isoleucine biosynthesis in *Rhodopseudomonas spheroides* (*253*). Here, the sequential control begins with an accumulation of threonine, which inhibits homoserine dehydrogenase and thereby causes aspartate-β-semialdehyde to accumulate; finally, the aspartate-β-semialdehyde inhibits aspartokinase which catalyzes the first step in the pathway (*254, 255*).

C. Concerted Feedback Inhibition

A different pattern of end product control, illustrated in Fig. 13, is referred to as *concerted feedback inhibition* or *multivalent feedback* inhibition because neither of the ultimate end products alone is able to affect the first common step (step a) in the pathway, but when both are in excess simultaneously they act in concert to inhibit the activity of that enzyme. When reinforced by specific controls of the divergent steps c and d by the end products of their respective branches, this mechanism provides an effective means of regulation. As in the case of the sequential mechanism discussed above, inhibition by one end product of the first divergent step leading to its synthesis facilitates accumulation of the second end product by making available more of the common precursor C for synthesis of the latter. Then together excesses of both end products shut down the entire pathway by inhibiting the common first step a.

249. J. H. Lorence and E. W. Nester, *Biochemistry* **6,** 1541 (1967).
250. E. W. Nester, J. H. Lorence, and D. S. Nasser, *Biochemistry* **6,** 1553 (1967).
251. D. S. Nasser, G. Henderson, and E. W. Nester, *Bacteriology* **98,** 44 (1969).
252. H. Paulus and E. Gray, *JBC* **239,** PC4008 (1964).
253. P. Datta, *Science* **165,** 556 (1969).
254. P. Datta and L. Prakash, *JBC* **241,** 5827 (1966).
255. P. Datta, *JBC* **241,** 5836 (1966).

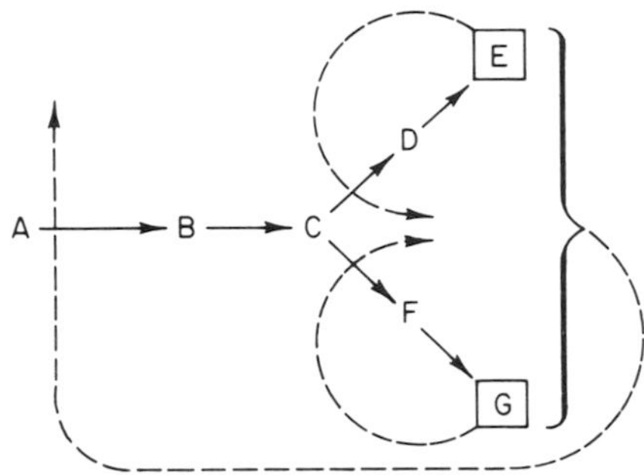

FIG. 13. Concerted feedback control of branched pathways. The dashed lines interconnect the end products with the steps they inhibit. Steps a, b, and d as defined in Fig. 12.

Examples of concerted feedback inhibition are found in *Rhodopseudomonas capsulata* (*256*) and *Bacillus polymxa* (*252*). Each organism contains only one aspartokinase that is insensitive to feedback inhibition by any one of the end products of aspartate metabolism; however, when the two end products, threonine and lysine, are both present in excess they inhibit aspartokinase activity.

D. SYNERGISTIC FEEDBACK INHIBITION

This pattern of feedback control differs from the concerted mechanism only in that each of the ultimate end products has a slight ability to act independently of the other; however, when excesses of both are present simultaneously the total inhibition is much greater than the sum of their independent activities. This is illustrated in Fig. 14, which shows that independently high concentrations of the end products E and G cause 5 and 10% inhibition, respectively, of step a enzymic activity, but together they can exhibit 90% of the enzymic activity.

This type of synergistic action was first observed in the regulation of purine nucleotide biosynthesis by 6-hydroxy and 6-amino-purine nucleotides. Both kinds of nucleotides are independently able to inhibit in an allosteric manner the enzyme glutamine phosphoribosylpyrophosphate amidotransferase which catalyzes the first step in purine nucleotide biosynthesis (*257*). Available data suggest that the allosteric enzyme contains separate binding sites for the two classes of purine nucleotides.

256. P. Datta and H. Gest, *Nature* **203,** 1259 (1964).
257. C. T. Casky, D.M. Ashton, and J. B. Wyngaarden, *JBC* **239,** 2570 (1964).

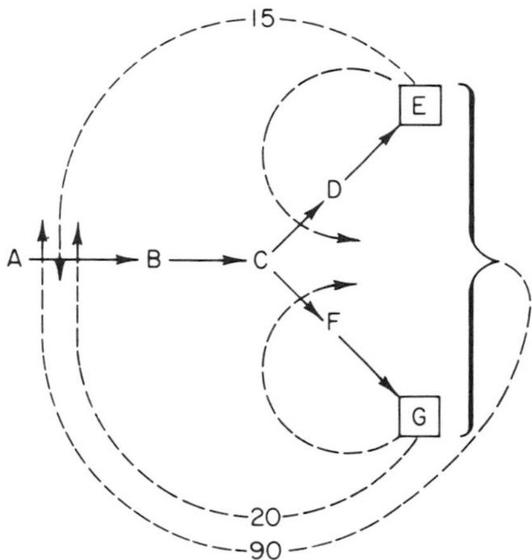

Fig. 14. Synergistic feedback control of branched pathways. The numbers **15** and **20** indicate the percent inhibition obtained by products E and G, respectively, when they are independently present in excess. The number **90** refers to the inhibition obtained when both are present in excess. Step a is conversion of A to B.

Mixtures of purine nucleotides containing a 6-hydroxy group (GMP + IMP, GMP + GDP) or those containing a 6-amino group (AMP + ADP) cause no more inhibition than the most effective inhibitor of each pair. However, the extent of inhibition exerted by mixtures containing both 6-amino purines and 6-hydroxy purines (*viz.*, GMP + AMP or AMP + IMP) is greater than the sum of the fractional inhibition obtained with each nucleotide independently (*257, 258*). This mode of synergistic control has been referred to as *cooperative feedback inhibition;* however, in view of the fact that allosteric enzymes often exhibit a different kind of cooperativity with respect to their kinetic behavior (see Section II,B), use of the term *cooperative* to describe the pattern of feedback inhibition illustrated in Fig. 14 may lead to confusion and should be avoided.

Another example of synergistic feedback inhibition is found in the regulation of glutamine synthetase activity in *Bacillus lichiniformis.* This enzyme is only slightly inhibited by low concentrations of either AMP, histidine, or glutamine; however, mixtures of AMP plus histidine or of glutamine plus histidine cause almost complete inhibition of glutamine synthetase activity (*259*). Synergistic effects have been observed also in

258. D. P. Nierlich and B. Magasanik, *JBC* **240,** 358 (1965).
259. J. S. Hubbard and E. R. Stadtman, *J. Bacteriol.* **94,** 1016 (1967).

the inhibition of aspartokinase activity in *E. coli* by a variety of amino acids (*260, 261*).

E. Cumulative Feedback Inhibition

In this kind of feedback control, high concentrations of each end product alone are able to cause only partial inhibition of the first common step in the branched pathway. Moreover, each end product acts independently of the others and the presence of one does not influence the activity of the others. Therefore, when they are both present simultaneously, their effects are cumulative. Cumulative feedback inhibition differs from the synergistic type discussed above in that the extent of inhibition caused by mixtures of end products is quantitatively determined by the capacity of each inhibitor by itself to inhibit the enzyme. For example, as illustrated in Fig. 15, if the two end products E and G alone cause 30 and 40% inhibition, respectively, then together they will cause 58% inhibition. This follows from the independent and noninteracting nature of their effects, from which it can be derived that the fraction of enzymic activity that remains when both end products are present simultaneously is equal to the product of the fractional activities ob-

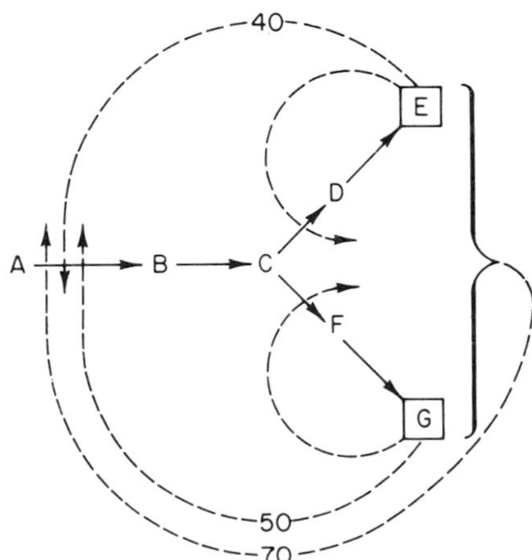

Fig. 15. Cumulative feedback regulation of branched biosynthetic pathways. The numbers 40 and 50 indicate the percent inhibition caused by end products E and G when they are independently present in excess. The number 70 indicates the percent inhibition obtained when both end products are in excess.

260. P. Truffa-Bachi and G. N. Cohen, *BBA* **113,** 531 (1966).
261. N. Lee, E. M. Lansford, Jr., and W. Shive, *BBRC* **24,** 315 (1966).

served for each independently. Thus, in the example given, the fractional activities with compounds E and G alone would be 0.7 and 0.6, respectively; hence, when both are present simultaneously the fraction of enzymic activity remaining would be $0.7 \times 0.6 = 0.42$, or the total inhibition would be $(1\text{–}0.42) \times 100 = 58\%$.

Cumulative feedback inhibition was first observed in the regulation of glutamine synthetase in *E. coli* (*262, 263*). Glutamine occupies a central role in the nitrogen metabolism of many bacteria since the amide group of glutamine is a source of nitrogen for the biosynthesis of a variety of end products including AMP, CTP, glucosamine 6-phosphate, histidine, and carbamoyl phosphate. In addition, the α-amino group can be used in transamination reactions for the synthesis of a variety of other amino acids. Glutamine synthetase can therefore be regarded as an enzyme catalyzing the first common step in the biosynthesis of a large number of different end products. Woolfolk and Stadtman (*262, 263*) showed that the activity of this enzyme is partially inhibited by each one of the above six end products of glutamine metabolism and also by L-alanine and glycine. Moreover, when two or more inhibitors are present at the same time, their effects are cumulative, and when all eight are present the glutamine synthetase activity is almost completely inhibited.

In the meantime, further investigations have disclosed that the apparently homogeneous enzyme preparation used by Woolfolk and Stadtman in their original studies was in fact a mixture of adenylylated and unadenylylated forms that differ markedly in their susceptibility to feedback inhibition by some of the end products and, also, in their requirement for divalent cations (*77, 78*). The original observation that each inhibitor can cause only partial inhibition at saturating concentrations may have been fortuitous. Further studies have disclosed that, although there are probably separate binding sites for each one of the different feedback inhibitors, under certain conditions of pH, ionic strength and substrate concentrations interaction between certain binding sites can occur (*264*). Nevertheless, at the concentrations of end products likely to occur under physiological conditions, each will cause only partial inhibition and in combination they should exhibit nearly cumulative behavior. Cumulative feedback inhibition provides the basis of an elegant control mechanism for branched pathways in which there are numerous end products. When reinforced by specific end product control of first divergent steps in each branch of the pathway, it modulates the activity of the first common step in a manner commensurate with the demands of

262. C. A. Woolfolk and E. R. Stadtman, *BBRC* **17,** 313 (1964).
263. C. A. Woolfolk and E. R. Stadtman, *J. Bacteriol.* **118,** 736 (1967).
264. A. Ginsburg, *Biochemistry* **8,** 1726 (1969).

the organism. As each end product becomes present in excess, it is able to curtail only a portion of the common activity and this is ideally only that portion needed for its own synthesis.

As noted earlier (Section II,C) in *E. coli,* cumulative feedback inhibition of glutamine synthetase is modulated by the interconversion of adenylylated and unadenylylated forms of the enzyme. Adenylylation converts the enzyme from a form that requires Mg^{2+} for activity and which is relatively insensitive to cumulative feedback inhibition to a form that requires Mn^{2+} and is very susceptible to cumulative feedback control. The physiological significance of the changes in divalent ion specificity is not fully understood. It might be noted, however, that in the absence of Mn^{2+}, which has not been shown to be required for the growth of *E. coli,* adenylylation of glutamine synthetase would be paramount to inactivation of the enzyme. This might be the desired result since adenylylation of the enzyme is induced by conditions (nitrogen excess) which favor the accumulation of the glutamine-dependent end products. However, when Mn^{2+} is available, as it could be in a natural environment, adenylylation would not lead to total inactivation of the enzyme since the adenylylated form is activated by Mn^{2+}. Under these conditions, inactivation of the enzyme could be obtained by cumulative feedback inhibition.

F. Heterogeneous Metabolic Pool Inhibition

An effect similar to cumulative feedback inhibition has been observed in the regulation of several other highly branched biosynthetic pathways. The mechanism of the effect differs from cumulative inhibition in one very important respect; namely, all of the various end products apparently react at the same regulatory site on the enzyme. Furthermore, whereas each one of the end products can by itself, at high concentration, cause nearly complete inhibition of the enzyme this potential is never realized under physiological conditions because the affinity of the regulatory enzyme for the feedback inhibitor is very low. In this situation, a physiological excess of each end product will lead to only partial inhibition of the regulatory enzyme. However, since all of the various end products react at the same effector site on the enzyme, the extent of inhibition will be determined by the sum of the concentrations of all end products. In other words, it is the concentration of the heterogeneous end product pool that determines the extent of inhibition of the regulatory enzyme. Heterogeneous pool control appear to be a major mechanism for the regulation of phosphoribosylpyrophosphate synthetase in *S. typhi-*

murium (*265*) and *E. coli* (*266*) and also for the regulation of purine mononucleotide pyrophosphorylase activity in *B. subtilis* (*267*).

G. Multivalent Repression

This novel mechanism for the regulation of complex pathways was first recognized by Freundlich *et al.* (*21*) in the course of their investigation on the repression of enzymes involved in the synthesis of the branched-chain amino acids in *S. typhimurium.* As shown in Fig. 16, valine and isoleucine are produced by parallel sequences of reactions that are catalyzed by a common set of enzymes. Thus the enzyme catalyzing a par-

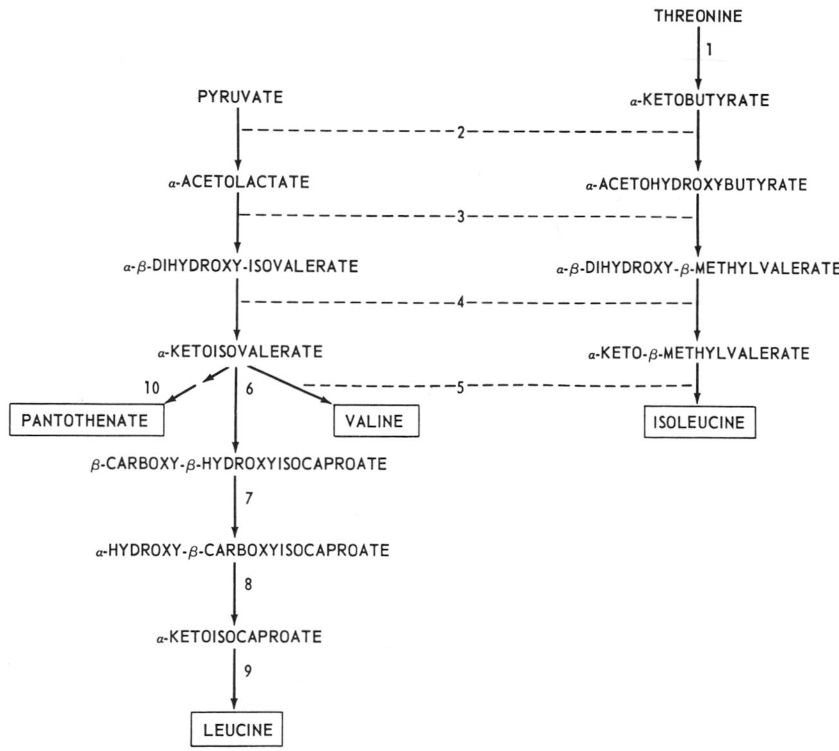

Fig. 16. The biosynthesis of branched-chain amino acids. The dashed lines show the steps in the parallel pathways that are catalyzed by a common enzyme.

265. R. L. Switzer, *Federation Proc.* **26,** 560 (1967).
266. D. E. Atkinson and L. Fall, *JBC* **242,** 3241 (1967).
267. R. D. Berlin and E. R. Stadtman, *JBC* **241,** 2679 (1966).

ticular step in the synthesis of valine is responsible also for catalysis of the analogous step in the synthesis of isoleucine. Moreover, enzymes catalyzing all steps up to and including that involved in the formation of the branched-chain α-ketoacids are required for the synthesis of leucine as well. Regulation of this complex pathway therefore poses a special problem and is still incompletely understood (*246, 268*). For the present discussion it is important to note that the end product, valine, is a specific feedback inhibitor of acetohydroxy acid synthetase, the first bifunctional enzyme in the pathway. In the absence of an appropriate compensatory control, an excess of valine should lead to a deficiency of isoleucine. This, in fact, appears to be the case in *E. coli* K12, where it is observed that an excess of valine inhibits growth and this inhibition can be overcome by the addition of isoleucine or any one of its six carbon precursors (*269, 270*). In *S. typhimurium,* however, an excess of valine does not cause inhibition of growth. The compensatory control that overrides the expected valine inhibition is not understood (*246, 268*). One mechanism which, in theory, could provide the necessary compensatory action is multivalent repression. In *S. typhimurium* the five structural genes for the isoleucine–valine biosynthetic enzymes are subject to repression by the concerted action of excesses of leucine, isoleucine, and valine (*21*). Neither of these alone, nor any combinations of two, will cause repression; adequate concentrations of all three must be present simultaneously.

Under normal growth conditions the endogenous levels of leucine, isoleucine, and valine are all sufficiently high to cause considerable repression of the branched-chain amino acid enzymes. Further accumulation of one or all of the end products causes only slight additional decrease in rate of enzyme synthesis (*21*). On the other hand, a decrease in the level of any one of the three amino acids below the level required for multivalent repression results in derepression of all bifunctional enzymes in the parallel pathway.

It is therefore evident that when acetohydroxy synthetase is inhibited by an excess of valine, the resultant deficiency of isoleucine production would lead to a derepression of enzyme synthesis and hence an increase in the level of acetohydroxy acid synthetase and all other bifunctional enzymes in the branched-amino acid pathway. Whereas this compensatory effect would tend to maintain a level of isoleucine required for growth, it would also lead to a further increase in the already over-

268. H. E. Umbarger, *in* "Current Topics in Cellular Regulation" (B. L. Horecker and E. R. Stadtman, eds.), Vol. 1, p. 57. Academic Press, New York, 1969.

269. D. M. Bonner, *JBC* **116**, 545 (1946).

270. R. I. Levitt and H. E. Umbarger, *J. Bacteriol.* **83**, 623 (1962).

abundant supply of valine (*219, 220*). It should be emphasized that whereas multivalent repression undoubtedly plays a dominant role in the regulation of the branched amino acid pathway it has not definitely been proven that the failure of excesses of valine to inhibit growth of *S. typhimurium* results from the compensatory action of multivalent repression (*270*).

A second example of multivalent repression of enzyme synthesis has been noted in the regulation of aspartate metabolism in *E. coli* and *S. typhimurium.* In these organisms the formation of the aspartokinase complex I is repressed by a combination of threonine plus isoleucine (*271, 272*).

H. Regulation of Aspartate Metabolism in *E. coli*

Few branched biosynthetic pathways have received as much attention from the standpoint of cellular regulation as has that concerned with the conversion of aspartate to threonine, lysine, methionine, and isoleucine. Mention has already been made of the fact that various organisms have developed fundamentally different patterns of control for the regulation of this pathway. Particularly noteworthy, however, are the studies in *E. coli* that have been carried out by Cohen and his associates in France. As a result of their investigation many of the regulatory enzymes in this pathway have been isolated in a relatively homogeneous state and their kinetics and feedback behavior have been studied in great detail. Since the body of knowledge obtained constitutes a comprehensive illustration of how various control mechanisms can be integrated into an overall effective regulatory system, a detailed description of this regulation pattern seems worthwhile. Figure 17 sketches the sequence of reactions involved in aspartate metabolism and shows the points of feedback controls that have been clearly established.

Note that the first divergent step in each branch of the pathway is subject to feedback control by the ultimate end product of that pathway. In addition to its inhibition by threonine, a step in the conversion of homoserine to threonine is subject to multivalent repression by a combination of threonine plus isoleucine (*271*). In a second level of control, the conversion of aspartate to aspartyl phosphate, which is the first common step in the synthesis of all four end products, is catalyzed by three separate aspartokinases: Aspartokinase I is subject to feedback inhibi-

271. M. Freundlich, *BBRC* **10,** 277 (1963).

272. G. N. Cohen and J. C. Patte, *Cold Spring Harbor Symp. Quant. Biol.* **28,** 513 (1963).

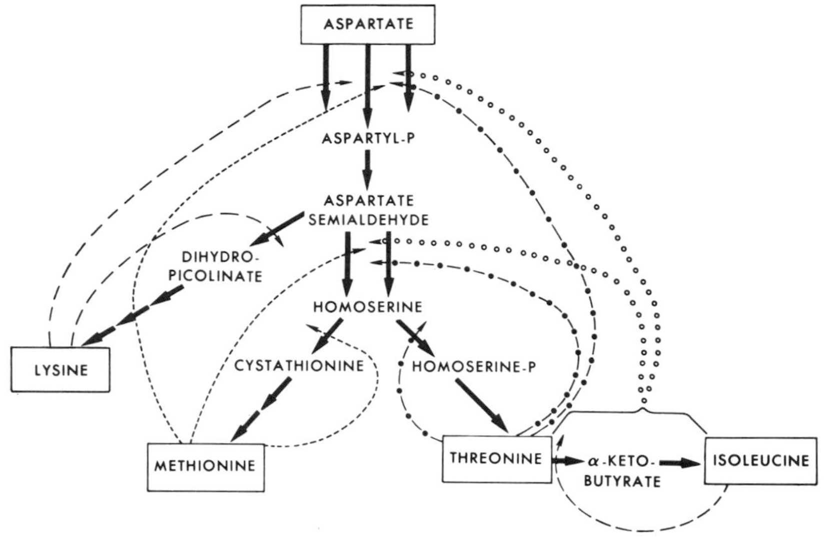

FIG. 17. The regulation of aspartate metabolism in *E. coli*. The broken lines interconnect the metabolites with the steps they control.

tion by threonine (*234*) and to multivalent repression by a mixture of threonine and isoleucine (*271, 272*); aspartokinase II is specifically repressed by methionine (*235*); and, aspartokinase III is subject to both feedback inhibition and repression by lysine (*234*) and to synergistic feedback inhibition by lysine plus L-phenylalanine and by lysine plus isoleucine (*210, 273*).

A third level of control involves the conversion of aspartate-β-semialdehyde to homoserine. This reaction is the first common step in the biosynthesis of only three amino acids (methionine, threonine, and isoleucine), and here again *E. coli* takes advantage of the multiple enzyme principle for its regulation. There are two homoserine dehydrogenases: One is repressed by methionine (*235*), whereas the other is subject to feedback inhibition by threonine (*124*) and to multivalent repression by threonine plus isoleucine (*242, 271*). A fourth level of control utilizes the sequential feedback principle. The sequence begins with an accumulation of isoleucine which inhibits threonine deaminase (*25*) and thereby causes accumulation of threonine; this is followed by threonine inhibition of homoserine kinase (*274*) and concomitant accumulation of homoserine; finally, homoserine inhibits the aspartokinase I activity (*275*). This sequential feedback mechanism is supplemented by multivalent repres-

273. J. C. Patte, T. Loving, and G. N. Cohen, *BBA* **99,** 523 (1965).
274. E. H. Wormser and A. B. Pardee, *ABB* **78,** 416 (1958).
275. J. C. Patte, P. Truffa-Bachi, and G. N. Cohen, *BBA* **128,** 426 (1966).

sion of homoserine synthetase homoserine dehydrogenase I and aspartokinase I activities by the joint action of an excess of isoleucine and the sequentially induced excess of threonine (*271, 272*). Finally, the action of aspartokinase III is subject to compensatory control via synergistic feedback inhibition by combinations of lysine and L-phenylalanine or leucine, or L-valine (*260, 273*).

Another potential level of control is derived from the fact that the aspartokinase I and homoserine dehydrogenase I activities are carried on a single bifunctional enzyme (*275, 276*) and the aspartokinase II and homoserine kinase II activities are carried on another single enzyme (*235*). Therefore, the concentration of these pairs of activities will vary in a coordinated fashion in response to multivalent repression and derepression. Furthermore the aspartokinase I and homoserine dehydrogenase I activities are negatively influenced by each other's substrates. Thus, the homoserine dehydrogenase I activity is inhibited by ATP and L-aspartate, whereas the aspartokinase I activity is inhibited by L-homoserine and TPNH (*275*). In a sense then the substrate sites for one activity are allosteric inhibitory sites for the other. The bicephalic nature of the enzyme affords a unique control system in which the two catalyzed activities will vary in a reciprocal manner in response to the relative substrate concentrations. Whereas inhibition of aspartokinase activity by homoserine could be regarded functionally as an example of sequential feedback control, a teleological explanation for the inhibitory effects of ATP and L-aspartate on homoserine dehydrogenase activities is not apparent. The latter would lead, in effect, to a relative accumulation of aspartate-β-semialdehyde and could be of significance as a compensatory control device designed to assure an adequate level of this metabolite for lysine synthesis. Finally, the lysine-sensitive aspartokinase III is sensitive to regulation by energy charge (*277*). Considered altogether, the regulation of aspartate metabolism utilizes many of the basic patterns of feedback control noted above. In addition to feedback inhibition of the first committed step in the linear portions of the pathway by the ultimate end products of each branch, *E. coli* makes use of multiple enzymes to control the first common steps in the main stream of metabolism and reinforces these controls with repression and multivalent repression of certain key enzymes and with sequential as well as synergistic and compensatory feedback inhibition of appropriate enzymes.

276. J. Janin, P. Truffa-Bachi, and G. N. Cohen, *BBRC* **26,** 429 (1967).

277. L. Klungsøyr, J. H. Hageman, L. Fall, and D. E. Atkinson, *Biochemistry* **7,** 4035 (1968).

9

Enzymes as Control Elements in Metabolic Regulation

DANIEL E. ATKINSON

I. Introduction

Enzymes are functional elements of the living cell and have been designed by selection to contribute to survival and reproduction of the cell or organism. The close correspondence between evolution through mutation and selection and the trial-and-error aspects of engineering design will not be discussed here. This correspondence will, however, be taken for granted. Components of biological systems and of man-made devices are both functionally designed elements, and analogies between them will be considered to be meaningful comparisons rather than mere verbalizations.

This chapter is not the place for a discussion of the origin of biological concepts, but it may be relevant to suggest that much progress in biology has rested on analogies with technological devices. The correlation between the spurting of blood from wounds and the beating of the heart was surely noticed by prehistoric hunters, but, as Adolf (*1*) pointed out, the existence of a sustained one-way flow of blood (rather than mere to-and-fro surging) was not recognized until after the development by mining engineers of pumps with internal valves. Other examples could be cited to support the generalization that real progress toward understanding biological function has usually come only after a technological device to perform an analogous function had been invented. The heart is a pump and the eye is a camera. This extensive flow of concepts from technology to fundamental biology has occurred in some cases by explicit analogy and in others by unconscious borrowing, but in either case knowledge of the technological device led to new insight regarding biological function. This debt to technology should be recognized, and we should draw on technological concepts consciously and explicitly in the study of living systems. An organism is a marvelously functional machine, exhibiting very great complexity of design, and we need all the help we can get in attempting to understand it. Individual enzymes are functional elements in this machine. In this chapter they will be considered to be, in many respects, analogous to functional elements of technological devices. Such an analogy led Umbarger to discover feedback control in isoleucine biosynthesis in 1956 (*2*). This was one of the key findings that initiated the study of metabolic regulation at the molecular level.

Because of the mass of information that has accumulated from the study of enzymes, it is not surprising that molecular and metabolic aspects of enzymology are often treated separately. This series of volumes, for example, concentrates on molecular aspects. But it should be remembered that this separation is made for human convenience and does not reflect a real distinction in nature. The molecular structure of an enzyme has evolved to subserve metabolic ends; hence, molecular and metabolic aspects of enzyme study must keep in close contact if they are to be realistic. A somewhat frivolous analogy may be illustrative. An interplanetary visitor to earth, landing briefly in an uninhabited region, might happen to pick up a discarded automobile distributor to take back home for study. Scientists there might emphasize the importance of the essential symmetry of the structure, especially the symmetrical arrangement of the points around a central axis. They would have an opportunity to apply the most advanced scientific approaches to the study of this strange object, and might soon come to feel that they had supplied a complete and satisfying description of it. But until they

1. E. F. Adolf, *Physiol. Rev.* **41,** 737 (1961).
2. H. E. Umbarger, *Science* **145,** 674 (1956).

learned of the existence of internal combustion engines, they could never, however intelligent and scientifically advanced, understand anything essential or really important about the distributor or its structure.

II. Analogy between Metabolic and Electronic Systems

Because of the importance of electronic devices in contemporary technology and the general acquaintance of scientists with elementary electronics, such devices probably supply the most useful analogy to metabolic systems. In the comparison between a metabolizing cell and an electronic device, metabolic fluxes correspond to currents, and concentrations (strictly speaking, chemical potentials) to voltages. An enzyme corresponds to a tube (valve) or transistor. The flux through the reaction catalyzed by the enzyme corresponds to tube current, substrate concentration corresponds to plate voltage, and the concentrations of modifiers are equivalent to voltages applied to the control grids of the tube. The solvent volume in which a given metabolite equilibrates corresponds to capacitance. Within this framework, a cell may contain, for example, both voltage amplifiers [in which a concentration (voltage) is regulated in response to fluctuations in another concentration (voltage)] or current amplifiers [in which a reaction flux (current) is regulated in response to fluctuations in a control concentration (voltage)]. Such analogies have been briefly discussed (*3*).

The first type of metabolic control to be demonstrated at the molecular level was end product negative feedback control of biosynthesis (*2, 4*). The corresponding electronic circuit is a current-controlling amplifier in which the stage that is controlled is in a circuit leading to the controlling voltage. When the phase relationships are such that an increase in output voltage leads to a decrease in current, the result is stabilization of the voltage. The analogy of such negative feedback voltage regulation with negative feedback regulation in biosynthetic pathways is obvious. More complex patterns, especially with regard to control inputs, will be discussed below.

No analogy, and certainly not this one, should be pushed too far. The physical principles underlying metabolic regulation are clearly entirely different from those involved in electronic devices. But if used with

3. D. E. Atkinson, *in* "The Neurosciences: A Study Program" (G. C. Quarton, T. Melnechuk, and F. O. Schmitt, eds.), p. 123. Rockefeller Univ. Press, New York, 1967.

4. R. A. Yates and A. B. Pardee, *JBC* **221,** 757 (1956).

restraint, the analogy may be valuable. Perhaps most importantly, it leads to the suggestion that one goal of studies on enzymes *in vitro* should be the construction of performance curves analogous to the "characteristic curves" by which electron tubes are routinely described—curves, for example, of tube current as a function of plate voltage when various voltages are applied to a control grid (Fig. 1B). An analogous family of curves for a typical regulatory enzyme is shown in Fig. 1A.

It is evidently impossible to predict, from its characteristic curves alone, how a tube will behave in an actual circuit. The performance of a circuit depends on the properties of all of its elements, not only on those of the tube. For example, in a simple voltage amplifier stage, an increase in control grid voltage causes an increase in tube current. Because a resistor is connected in series with the plate, the plate voltage falls. The amplitude of the fall will depend on the steepness of the curves in Fig. 1B and on their separation but also on parameters of the associated circuit, notably the size of the plate resistor. For voltage amplification, the circuit is designed so that the fall in plate voltage is larger than the rise in control grid voltage that produced it, and this change in plate voltage is the output of the amplifier stage. The same tube may, however, be used in other circuits with quite different functions. A similar reservation applies to all study of enzymes *in vitro*. Without detailed knowledge of its metabolic context, we cannot predict the detailed behavior of an enzyme *in vivo* from results of studies *in vitro*. There is in addition the troubling possibility that the enzyme may have been damaged in extraction or purification, so that its observed behavior does not faithfully reflect the properties of the native enzyme. Nevertheless,

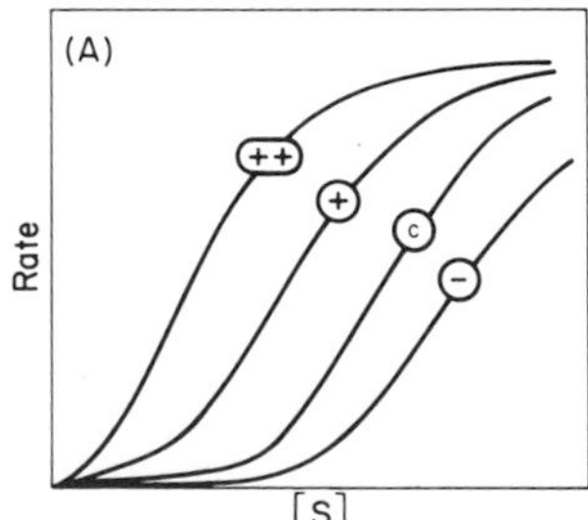

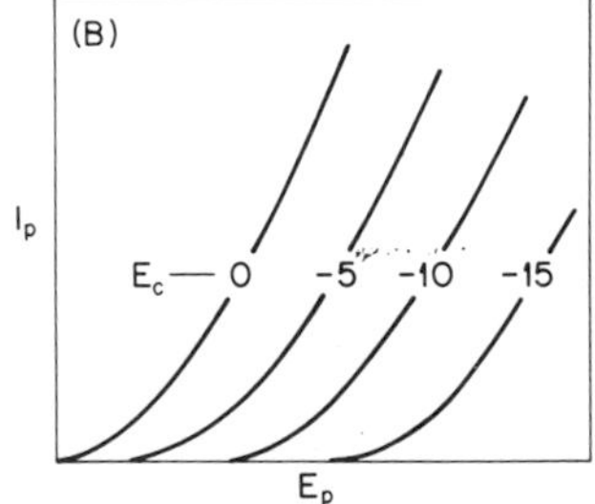

FIG. 1. Comparison of generalized performance curves for a regulatory enzyme and for a triode. (A) Rate of an enzyme-catalyzed reaction as a function of substrate concentration. Curve identifications: c, no modifier added; —, negative modifier (inhibitor) added; +, positive modifier added; ++, larger amount of positive modifier added. (B) Tube current (I_p) as a function of plate voltage (E_p) at several values of control grid voltage (E_c). From Atkinson (*3*).

study of its constituent parts is an appropriate approach in attempting to understand the operation of any mechanism. As applied to metabolism, this approach has led to much of our present understanding of cellular chemistry. As applied to metabolic regulation, it has led to generalizations that at least have a considerable degree of internal consistency and that, although it is not yet possible to test them rigorously *in vivo,* are in agreement with observations on living cells.

Enzymes are extremely complicated materials, and they regulate rates of reactions in complex ways. To define an enzyme as a protein catalyst is as inadequate as it would be to define an electron tube as a conductor of electricity. A tube with a short circuit between the plate and cathode is a better conductor than an intact tube, but it is useless. A cell in which the enzymes had been replaced by mere catalysts, no matter how effective, would resemble a living cell no more than a radio with each tube replaced by a piece of copper wire would resemble a functional radio.

III. Physiological Function and Design Requirements

Throughout the history of enzyme chemistry there has been a strange tendency to minimize the importance of the protein. This tendency reached its bizarre extreme in the long-standing and tenacious insistence by Willstätter, Waldschmidt-Leitz, and others that proteins have nothing to do with enzymic activity. In their view the enzyme, of unknown composition, merely tends to become adsorbed onto proteins. The enzyme might be stabilized as a consequence of adsorption, but the protein was otherwise viewed as a passive carrier. That episode has passed into history, and it has been universally recognized for forty years that the protein *is* the enzyme. However, during most of that time, the attention of enzymologists was directed exclusively to the catalytic site. Identification of that site and elucidation of its structure was considered to be the main, if not only, goal of structural work on enzymes. Except that the active site was recognized as being protein in nature, the prevailing attitude hardly differed basically from Willstätter's; the only feature of interest, the active site, was assumed to be carried on a nondescript and passive mass of protein. This attitude was strengthened by the observation that a few enzymes could be partially degraded without loss of catalytic activity. The protein was often dismissed, in seminars or in the classroom, with the casual comment that it is strange that enzymes are so large when the catalytically active site is so small. The prevailing attitude was somewhat like that of investigators who might seek an explanation of Michelangelo's genius by undertaking a detailed anatomical study of his hands.

In retrospect, Koshland's concept of induced fit at the active site (*5, 6*) will probably be seen to have been the first important step toward consideration of the role of the enzyme as a whole in the catalytic process, and thus toward modern enzyme chemistry. According to this view the naked enzyme must be in effect spring-loaded in a conformation in which functional groups are positioned for binding a substrate. As a consequence of binding (or in the course of binding) this substrate, the conformation must change in a precise way to allow for binding of a second substrate, or perhaps to facilitate reaction with a previously bound substrate. A high degree of specificity must be maintained throughout. This kind of action demands active participation of other parts of the protein in addition to the residues directly involved in the catalysis. Evidently the demands to be met in the evolutionary design of an enzyme that functions in this way are many times more stringent than would be the case for a passive enzyme bearing a simple rigid site.

The discovery of a class of regulatory enzymes in which catalytic behavior is altered on binding of a modifier not sterically related to the substrate (*2, 4*) led to a proposal that is essentially an extension of Koshland's hypothesis. Gerhart and Pardee (*7*) suggested that binding of the modifier at a specific site leads, through a conformational change, to altered properties (usually altered substrate affinity) at the active site. A protein that acts in this way must be designed not only to allow for highly specific local conformational change when a specific substrate is bound at the catalytic site but also to undergo an equally specific change when a specific modifier is bound at a site at some unknown distance from the catalytic region.

In the case of aspartate transcarbamylase, it appears that the modifier site is on a peptide chain distinct from that bearing the catalytic groups (*8*). The design requirements for such behavior are evidently much more stringent than for a single peptide chain bearing an induced-fit catalytic site. The subunits must recognize each other and join in a specific way; binding of modifier must alter the conformation of the regulatory subunit; and this alteration must be transmitted to the catalytic site on the other subunit, causing a specific change in catalytic behavior. Koshland

5. D. E. Koshland, Jr., *Proc. Natl. Acad. Sci. U. S.* **44,** 98 (1958).
6. D. E. Koshland, Jr., *in* "Horizons in Biochemistry" (M. Kasha and B. Pullman, eds.), p. 265. Academic Press, New York, 1962.
7. J. C. Gerhart and A. B. Pardee, *JBC* **237,** 891 (1962).
8. J. C. Gerhart and H. K. Schachman, *Biochemistry* **4,** 1054 (1965).

(*9*) has discussed conformational aspects of enzyme regulation with emphasis on interactions between subunits and on the transmission between subunits of the consequences of binding substrates or modifiers.

A. Kinetic Properties

Regulatory enzymes often exhibit so-called sigmoid kinetic behavior; that is, a plot of reaction velocity against substrate concentration is concave upward in the region of low velocity, indicating that more than one molecule of substrate directly or indirectly affects the rate-limiting step. Since none of the reactions that have been observed to give such behavior is bimolecular in the sense that two molecules of the same substrate could be expected to react at the same site, it seems evident that the presence of a substrate molecule at one site affects the properties of one or more other sites. Several enzymes give fourth-order kinetic curves with respect to substrate, which seems to require strong interactions between at least four sites. In principle the effect could be either to facilitate reaction at the other sites (increase the catalytic constant) or to enhance substrate binding (decrease the effective intrinsic dissociation constant). Perhaps because of the obvious resemblance of the kinetic response to the long-known cooperative binding of oxygen by hemoglobin, the effect was assumed from the first to be on binding, with little or no change in the catalytic constant. Observed curves of velocity as a function of substrate concentration have thus been assumed, usually tacitly, to be equivalent to substrate binding curves. As observations on regulatory enzymes have accumulated, this interpretation has become increasingly well established, especially by analogy with the action of modifiers, which, as discussed below, is undoubtedly of this type in most cases. However, cooperative binding of substrate has been directly demonstrated by nonkinetic means in very few cases.

As illustrated in Fig. 1A, modifiers usually change the apparent affinity of the enzyme for substrate but have little effect on the maximal velocity attained at saturating concentrations of substrate. Since it seems likely that the catalytic constant (or turnover rate) is a sensitive function of the steric relationships at the filled catalytic site, these results suggest that the conformation at all active sites bearing substrates is virtually identical, whether the modifier sites are occupied or empty. In other words, when modifier binds to the regulatory enzyme it facilitates (or hinders) adsorption of substrate at the catalytic sites, but it does not

9. D. E. Koshland, Jr., *in* "Current Concepts of Metabolic Regulation" (B. L. Horecker and E. R. Stadtman, eds.), p. 1. Academic Press, New York, 1969.

affect the conformation of the filled site. Most enzymes that respond to modifiers also exhibit sigmoid kinetic curves in response to variations in substrate concentration, so it seems likely, as noted above, that interaction between substrate binding sites is of this same type.

On the basis of kinetic properties that have been observed *in vitro,* it appears that regulatory enzyme behavior patterns cover a wide range from considerably simpler than those discussed in the preceding section to much more complex. Probably every enzyme has properties that could be termed regulatory in the sense that they tailor the enzyme's kinetic behavior to the precise needs of the species in which it occurs. Even in the narrower sense in which the term regulatory enzyme is usually applied, quite simple properties of a catalytic site may have important regulatory consequences. For example, the affinities at a catalytic site for substrate and product might not ordinarily be thought of as regulatory. Several enzymes have been found to exist in the same organism in two or more isozymic forms, however, with quite different affinities for substrate or product. It seems clear in such cases that these differences must fit the individual isozymes for different functional roles.

If two enzymes catalyzing the same reaction, $A \rightleftarrows B$, differ in relative affinities for A and B, it is metabolically reasonable to assume that the isozyme with the greater affinity for A is specialized to catalyze the reaction in the direction shown and that the isozyme with the greater affinity for B catalyzes the conversion of B to A *in vivo.* The isozymes of mammalian lactate dehydrogenase have been discussed in this way, especially by Kaplan [see Kaplan *et al.* (*10*) for recent discussion and references]. A similar interpretation has been applied to other pairs or sets of isozymes.

From the Haldane relationship,

$$K_{eq} = \frac{K_P}{K_S} \frac{V_1}{V_2} \tag{1}$$

where K_{eq} is the equilibrium constant for the overall reaction, K_P and K_S are dissociation constants for product and substrate, and V_1 and V_2 are maximal velocities as indicated in Eq. (2), Cleland (*11*) reached a conclusion opposite to that discussed in the preceding paragraph. He

$$E + S \overset{K_S}{\rightleftarrows} ES \underset{V_2}{\overset{V_1}{\rightleftarrows}} EP \overset{K_P}{\rightleftarrows} E + P \tag{2}$$

10. N. O. Kaplan, J. Everse, and J. Admiraal, *Ann. N. Y. Acad. Sci.* **151,** 400 (1968).
11. W. W. Cleland, *Ann. Rev. Biochem.* **36,** 77 (1967).

proposed that K_P should be smaller than K_S (that is, that the enzyme should have a higher affinity for product than for substrate) in order to maximize the ratio V_1/V_2, particularly when K_{eq} is small. This proposal seems unlikely on both metabolic and evolutionary grounds. A usefully high value of V_1 is evidently desirable, but this is not incompatible with a high value of V_2; it is difficult to understand why a high value of the ratio V_1/V_2 should confer selective advantage. Further, in such a case the product would necessarily be an effective inhibitor—a situation that is metabolically useful only in special cases.

Several very important metabolites, conventionally termed cofactors, participate as substrates in metabolic reactions but are regenerated in other reactions. The pyridine nucleotide pairs, DPN^+ and DPNH and TPN^+ and TPNH, are prominent examples. The total pool of each cofactor pair (for example, $[DPN^+] + [DPNH]$) will be essentially constant, at least on a short time scale, but the relative amounts of the individual compounds may vary. In such cases the relative affinities of an enzyme for the alternate forms of the cofactor may have a quite different significance from that discussed above for conventional substrates and products. Because such cofactors participate in many reactions, they must be involved in maintaining the chemical homeostasis of the cell. Wide fluctuations in such ratios as $DPNH/DPN^+$ would tend to disrupt metabolic balance. There appears to be no way to stabilize these ratios except by the properties of the enzymes that catalyze the reactions in which the cofactors participate. More complex regulatory interactions probably are also involved, but it seems likely that the simple ratios of affinities at the catalytic site for the two forms of the cofactor are an important control property of enzymes of this type. Figure 2 shows, for a constant cofactor pool $(A + B)$, the fraction of catalytic sites that would bind form A as a function of the mole fraction of A. Each curve is calculated for a different value of the affinity ratio K_B/K_A, where K_A and K_B are dissociation constants for A and B. It is clear that rather small changes in the affinity ratio cause large differences in the shapes of the response curves. Since the initial velocity of the reaction $S + A \rightleftharpoons P + B$ will depend on the number of catalytic sites at which A is bound, the curves of Fig. 2 can be taken as performance curves. The kinetic behavior of an enzyme catalyzing a reaction that involves a cofactor pair must depend to a large degree on the ratio of concentrations of the two forms that is maintained in the cell. Conversely, the metabolic steady-state ratio of these forms must be determined to a very considerable extent by the response curves of the enzymes that catalyze reactions in which the cofactors are used.

It is, of course, only in regions where response curves are steep that

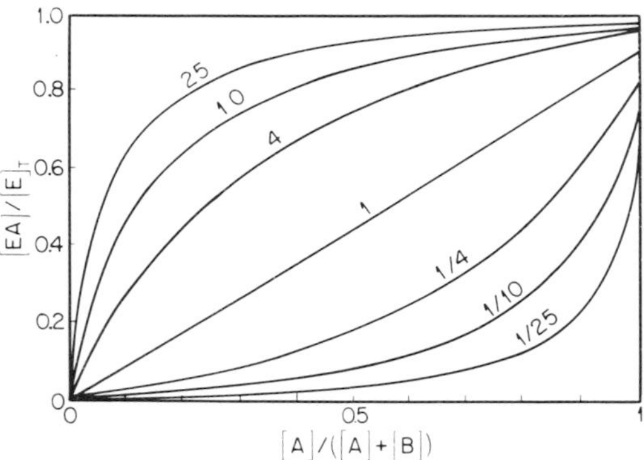

FIG. 2. Calculated fraction of catalytic sites binding A as a function of the mole fraction of A; effect of relative affinities for A and B. The ligands A and B are a structurally related reactant–product pair or the two forms of a cofactor, such as DPN^+ and DPNH. The combined pool (A + B) is assumed to be constant and is assigned the value 1. The numbers identifying the curves are the ratios of the dissociation constants K_B/K_A. The values of K_B and K_A used in the calculation were 25, 0.5 and 0.02; 10, 0.333 and 0.033; 4, 0.2 and 0.05; 1, 0.1 and 0.1; 1/4, 0.05 and 0.2; 1/10, 0.033 and 0.333; 1/25, 0.02 and 0.5. Detailed features of the figure depend on these assumptions; thus the spacing along the 1.0 charge axis results from the fact that both dissociation constants were changed in such a way as to hold their geometric mean constant. If, for example, K_A had been held constant and the desired ratios obtained by variation of K_B, the curves would have all intersected the 1.0 charge axis at the same point. The general shapes of the curves are, however, independent of such considerations.

sensitive regulatory interactions can be expected. The curves of Fig. 2 were calculated for a reaction that follows simple Michaelis kinetics, with no interactions between catalytic sites. For such a reaction, the steep parts of the curves necessarily occur near the two ends of the graph. It is evident that the steep portions of the curves could be moved toward the center of the mole fraction scale if the responses to A and B were sigmoid. Thus response curves of different shapes, with regions of steep response at different A/B ratios, might be evolved through changes in relative binding affinities and strengths of intersite interactions. Very little as yet has been done on regulation of the oxidation levels of the pyridine nucleotides; it will be interesting to see to what extent the response curves of the dehydrogenases can be recognized as having been shaped by evolutionary selection to aid in maintaining homeostasis.

B. ADENYLATE ENERGY CHARGE

The adenine nucleotides, AMP, ADP, and ATP, are the most widely used compounds in metabolism. In its role of accepting, storing, and transferring chemical potential energy, the adenylate system not only is involved in every extensive metabolic sequence but also serves as a stoichiometric link interconnecting all of the chemical activities of a living cell. From a design standpoint it would seem especially important that the ratio of AMP, ADP, and ATP concentrations should be buffered against wide fluctuations if the cell is to remain operational. On the other hand, a pool of compounds so intimately involved in all aspects of metabolism would, if not regulated, be expected to undergo wide changes in composition as the relative rates of various metabolic sequences change. Scattered observations indicate that for many types of cells the AMP/ADP/ATP balance does in fact remain relatively constant. Such constancy obviously must result from effective regulatory mechanisms.

Even if there were no observations on the concentrations of AMP, ADP, and ATP *in vivo*, it is clear that the existence of functioning metabolic systems implies that the rate of production of ATP must be somehow regulated by the need for its expenditure. A consequence of such regulation—the Pasteur effect—has been recognized for many years. This is the observation that various types of cells use substrate more rapidly under anaerobic conditions than when oxygen is available. It is evident that in the latter case the enzymes that degrade the substrate are regulated, since they are operating at less than maximal rate. The effect was initially described in terms of glucose as substrate, and the enzymes of interest are thus those of glycolysis. It has long been known that one of the major functions of glucose metabolism is regeneration of ATP and that far more ATP is produced per mole of glucose when the substrate is oxidized to CO_2 than when it is anaerobically fermented to ethanol and CO_2 or to lactate. A design engineer, given these facts and asked to devise a system to allow for adequate regeneration of ATP, while guarding against unnecessary expenditure of glucose, would almost certainly have suggested that the concentration of ATP should be sensed and used as a feedback signal to regulate the rate of consumption of glucose. The principle has been widely used in technology for nearly two hundred years, since Watt applied it in the form of a speed-regulating governor for a steam engine. A voltage regulator, controlling the generation of charging current in response to the charge of an automobile battery, is another example of this type of feedback control. Engineers were not consulted, however, and the very extensive literature on the Pasteur effect concerned itself with other approaches.

The discovery in 1962 and 1963 of AMP stimulation of two enzymes of glucose metabolism—phosphofructokinase (*12, 13*) and isocitrate de-

12. T. E. Mansour, *JBC* **238,** 2285 (1963).
13. J. V. Passonneau and O. H. Lowry, *BBRC* **7,** 10 (1962).

hydrogenase (*14*)—finally led to the realization that the concentrations of the adenine nucleotides may be directly involved in feedback systems that regulate the rate of ATP regeneration. The early development of this adenylate control hypothesis has been reviewed (*15*).

As observations accumulated, it was found that, at least under conditions of observation *in vitro,* some enzymes responded primarily to the concentration of AMP, some to ATP, at least one to ADP, some to the ATP/ADP concentration ratio, and some to the ATP/AMP ratio. All of the reported responses were in the right direction to be rationalized as participating in regulation of the regeneration of ATP, but some basis for comparing the responses in a way that would be both chemically and metabolically meaningful was needed. To meet this need, the parameter *energy charge* of the adenylate pool was proposed (*16*). This parameter is based on the chemical characteristics and metabolic function of the adenylate pool. It was defined in such a way as to be analogous with the charge of a storage battery—a linear stoichiometric expression of the amount of chemical change (current) that has been put into, and is available from, the system. The energy stored in the system is proportional to the mole fraction of ATP plus half the mole fraction of ADP. The energy charge is accordingly defined in this way [Eq. (3)]. Its value

$$\text{Energy charge} = \frac{(\text{ATP}) + \frac{1}{2}(\text{ADP})}{(\text{ATP}) + (\text{ADP}) + (\text{AMP})} \tag{3}$$

ranges from 0 at full discharge (when only AMP is present) to 1 at full charge (only ATP present). The development of the energy charge concept and its first applications to enzyme studies have been reviewed (*17*).

The three components of the adenylate pool are interrelated by the adenylate kinase reaction [Eq. (4)]. Adenylate kinase is present in all

$$\text{ATP} + \text{AMP} \rightleftarrows 2\ \text{ADP} \tag{4}$$

cells, and in the discussion and model calculations to follow it will be assumed that the concentrations of the adenine nucleotides are maintained near equilibrium of the adenylate kinase reaction. This assumption may not be valid for all cells under all conditions, and it is even possible that adenylate kinase may itself be a regulatory enzyme functioning on a more sophisticated level of control than those discussed here. But nonequilibration of the adenylate kinase reaction would affect the present discussion only in detail, not in essentials.

14. J. A. Hathaway and D. E. Atkinson, *JBC* **238,** 2875 (1963).
15. D. E. Atkinson, *Ann. Rev. Biochem.* **35,** 85 (1966).
16. D. E. Atkinson and G. M. Walton, *JBC* **242,** 3239 (1967).
17. D. E. Atkinson, *Ann. Rev. Microbiol.* **23,** 47 (1969).

It will be evident from the preceding discussion that the intended use of the energy charge is as a parameter in terms of which characteristic performance curves of enzymes can be constructed. For the most sensitive stabilization of energy charge, the response curves for enzymes involved in ATP regeneration and for those that catalyze reactions that use ATP should intersect in a region where both are steep, and all curves of both types should be steep in the same region. Both of these requirements are met by the behavior of the enzymes for which energy charge response curves have been obtained. Regulatory enzymes of ATP-regenerating sequences (phosphofructokinase, pyruvate kinase, pyruvate dehydrogenase, citrate synthase, and isocitrate dehydrogenase) respond to energy charge as illustrated by curve R of Fig. 3, and enzymes in sequences that use ATP (aspartokinase, citrate cleavage enzyme, phosphoribosylpyrophosphate synthase, and phosphoribosyl-ATP synthase) respond as shown by curve U. All of the observed curves are steepest at charge values above a value of about 0.8. These results have been reviewed (*17*). This consistency of response observed *in vitro* strongly suggests that the enzymes have been designed by selection to participate in maintaining the energy charge *in vivo* at values between 0.8 and 0.9 or slightly higher. Although review of the relevant literature would be outside the scope of this chapter, it may be noted that most of such analytical results as are available indicate that the energy charge in living cells is in fact between 0.8 and 0.95.

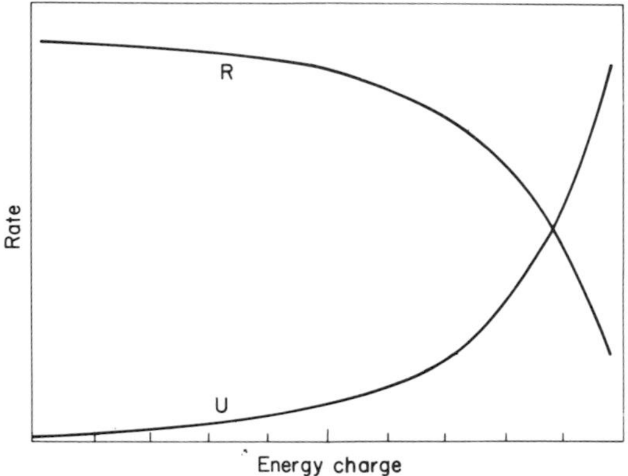

FIG. 3. Generalized response to the adenylate energy charge expected for enzymes involved in regulation of ATP-regenerating (R) and ATP-utilizing (U) sequences. From Atkinson (*18*). Reprinted from *Biochemistry* **7**, 4030 (1968). Copyright 1968 by the American Chemical Society. Reprinted by permission of the Copyright owner.

Any type of feedback control system functions circularly, and this fact must be kept in mind when such systems are considered. If the relationships illustrated in Fig. 3 are in fact general, they not only indicate that the rates of most or all metabolic processes are controlled by the energy charge of the adenylate pool but also show how the value of the charge may be maintained within a narrow range. The necessarily cyclic nature of regulatory systems of this type has not been appreciated by several authors, who have said, for example, that the concentration of ATP cannot be an important general control parameter because it varies so little. Such comments have a strong, though unintended, flavor of vitalism. They are equivalent to the statement that The Cell (or *élan vital,* or whatever) controls the concentration of ATP. In a rational world, there is no way to regulate the level of ATP except by controlling the rate at which it is regenerated, the rate at which it is used, or both. The control system must sense and respond to fluctuations in the concentration of ATP. If the control mechanism is sensitive, these fluctuations will be small. (It would obviously be incorrect to conclude that because it is so nearly constant the temperature of a thermostatic water bath can have no important effect on heater current.) Thus the observation that the concentration of ATP remains relatively constant, although this compound functions in every sequence of a very complex and dynamic chemical system, is itself strong *prima facie* evidence that the level of ATP is sensed by a system that regulates the rates of most or all of the metabolic sequences of the cell. Such reasoning involves nothing difficult or specifically biological. Circularity in cause-and-effect relationships is essential to regulation. It would obviously be meaningless to ask, in reference to a regulated water bath, whether the bath's temperature controls the current through the heating elements or the heater current controls the bath temperature. But a false dichotomy of this type is implied in many discussions of metabolic regulation.

Much ATP is used in simple phosphate transfer reactions of the type shown in Eq. (5). As indicated in Fig. 2, by simple adjustment of relative

$$\mathrm{X + ATP \rightleftarrows X\text{-}P + ADP} \tag{5}$$

affinities of the catalytic site for ATP and ADP, a U-type curve of any degree of steepness can be obtained. [Curves calculated specifically for kinases have been published (*18*); they differ from those of Fig. 2 because the adenylate pool contains three components rather than two, but the differences are slight.] This type of response, involving as it does only adjustment for an optimal ratio of binding energies for two compounds that are necessarily bound at the catalytic site in any case, is

18. D. E. Atkinson, *Biochemistry* **7**, 4030 (1968).

relatively simple, and it should be easily obtained by evolutionary selection (in comparison, for example, with the development of a regulatory site *de novo*).

Most kinases participate in biosynthetic or other energy-expending pathways. For such sequences, a U-type response like that given by strong binding of ADP at the catalytic site is obviously appropriate. There are a few kinases, however, for which this type of control would not be suitable. Phosphofructokinase is the most interesting enzyme of this type. The reaction catalyzed by phosphofructokinase [Eq. (6)] uses ATP, but this reaction participates in glycolysis, which leads to net

$$\text{ATP} + \text{fructose 6-phosphate} \rightleftarrows \text{ADP} + \text{fructose 1,6-diphosphate} \qquad (6)$$

ATP production. Thus, despite the chemistry of the reaction itself, it is metabolically essential that the response curve of phosphofructokinase be of the R, rather than the U, type. Phosphofructokinase clearly poses a much more difficult design problem than do kinases of the usual type. As noted in a previous section, phosphofructokinase in fact does exhibit an R-type response curve. It is very unlikely that such a response could be obtained by interactions at the catalytic site of a kinase, and the observations on phosphofructokinase clearly indicate the presence of a separate regulatory site. The properties of the yeast enzyme (*19*) will be briefly described; they are in most respects reasonably typical of the various phosphofructokinases that have been studied. A plot of reaction velocity as a function of ATP concentration rises rapidly, reaches a maximum, and then falls. In the physiological range of ATP concentrations, the reaction is of approximately negative first order with respect to ATP. This inhibition is specific for ATP; other nucleoside triphosphates serve as phosphate donors in the reaction but do not inhibit. Inhibition by ATP is specifically overcome by 5′-AMP. These results seem to require a catalytic site that is relatively nonspecific for phosphate donor and a highly specific regulatory site that binds only ATP or AMP. These nucleotides appear to compete for the regulatory site, and their effect is primarily on the affinity for fructose 6-phosphate at the catalytic site: ATP causes a decrease in strength of binding and AMP a slight increase.

A comparison of the properties of phosphofructokinase and phosphoribulokinase strikingly illustrates the design requirements dictated by physiological function. These enzymes catalyze reactions that are chemically nearly identical: Each is phosphorylation at C-1 of an ω-phosphoketose, and the substrates differ by only one CHOH unit. The responses of the two enzymes to adenine nucleotides, however, are oppositely directed.

19. A. Ramaiah, J. A. Hathaway, and D. E. Atkinson, *JBC* **239**, 3619 (1964).

Phosphofructokinase, as we have seen, is inhibited by ATP and stimulated by AMP, as is appropriate for an enzyme in an ATP-regenerating pathway. Phosphoribulokinase participates in autotrophic CO_2 fixation, an ATP-using process, and phosphoribulokinases from several organisms have been shown to be inhibited by AMP (*20–23*).

Conway and Koshland (*24*) have discussed cases of apparent negative cooperativity, defined as behavior that leads to Hill plots [see Atkinson (*15*) for discussion and references] with slopes less than one. Such reactions are formally of kinetic order less than one. An order (or Hill slope) less than zero, as seen for ATP in the phosphofructokinase reaction, may be considered an extreme example of apparent negative cooperativity; evidently if the effect of ATP at the regulatory site were weaker, the apparent order with respect to ATP might have been reduced merely to some value between one and zero, rather than all the way to —1. An effect on the interaction between enzyme and another substrate must thus be added to Conway and Koshland's list of ways in which kinetic behavior that might be termed negatively cooperative can be produced.

C. Interactions between Input Signals

1. *Amphibolic Pathways*

In a cell growing on glucose, glycolysis is an amphibolic (*25*) pathway; it not only leads to regeneration of ATP but also supplies the starting materials needed for biosynthesis. A process that serves more than one function must be regulated by more than one input. If glycolysis were regulated solely by the effect of energy charge on enzymes with R-type response curves, it would be severely inhibited whenever the energy charge was high. This would limit the cell's ability to produce biosynthetic starting materials and thus to grow. Such a limitation would clearly be an undesirable response to a ready availability of energy. The appropriate design solution is clear (*18*): Glycolysis should be regulated by both energy charge and the concentration of one or more metabolites that are indicative of the level of biosynthetic starting materials. Sensitivity to each parameter should be a function of the value of the other,

20. N. L. Gale and J. V. Beck, *BBRC* **24,** 792 (1966).
21. B. A. McFadden and C. L. Tu, *J. Bacteriol.* **93,** 886 (1967).
22. R. D. MacElroy, E. J. Johnson, and M. K. Johnson, *BBRC* **30,** 678 (1968).
23. J. V. Mayeux and E. J. Johnson, *J Bacteriol.* **94,** 409 (1967).
24. A. Conway and D. E. Koshland, *Biochemistry* **7,** 4011 (1968).
25. B. D. Davis, *Cold Spring Harbor Symp. Quant. Biol.* **26,** 1 (1961).

and glycolysis should not be severely inhibited unless both of its products—ATP and biosynthetic starting materials—are in good supply. Figure 4 illustrates the type of response curves to be expected. Just this type of response has been observed for phosphofructokinase (*26*). The indicator metabolite is citrate, which had been previously shown to inhibit phosphofructokinase from various sources [see Atkinson (*15*)]. When the concentration of citrate is low (Fig. 4, curve l) the enzyme is relatively insensitive to the inhibitory effect of high energy charge, and when energy charge is low citrate has very little effect. When both energy charge and citrate concentration are in their normal ranges, phosphofructokinase should respond sensitively to fluctuations in either, but the reaction will be strongly inhibited only when both energy charge and citrate concentration are high; that is, when neither of the major metabolic functions of glycolysis is needed. This behavior seems well designed to insure flexible control of glycolysis in response to changing metabolic needs. Some animal phosphofructokinases, but not those so far studied from microbial sources, respond to 3′,5′-cyclic AMP, as first observed by Mansour and Mansour (*27*). This nucleotide has been shown by Sutherland and colleagues [for a recent review, see Robison *et al.*

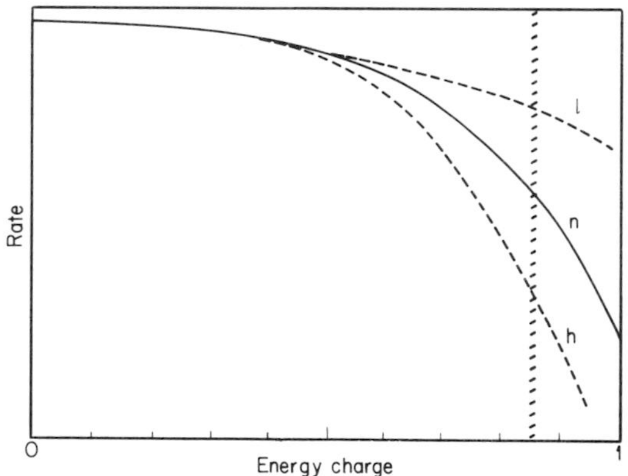

FIG. 4. Generalized interaction between adenylate energy charge and the concentration of a metabolite modifier expected in the control of a regulatory enzyme in an amphibolic sequence. The curves correspond to low (l), normal (n), and high (h) concentrations of the metabolite modifier. From Atkinson (*18*). Reprinted from *Biochemistry* **7**, 4030 (1968). Copyright 1968 by the American Chemical Society. Reprinted by permission of the copyright owner.

26. L. C. Shen, L. Fall, G. M. Walton, and D. E. Atkinson, *Biochemistry* **7**, 4041 (1968).
27. T. E. Mansour and J. M. Mansour, *JBC* **237**, 629 (1962).

(*28*)] to function as an intracellular "second messenger," which is elicited by a hormone, the "first messenger," and which is the actual effector of the cell's response to the hormone. In skeletal muscle, production of cyclic AMP is stimulated especially by epinephrine (adrenalin), the alarm hormone. The many effects of epinephrine all seem to be directed toward preparing the animal for violent physical activity. Among these effects is acceleration of carbohydrate breakdown in skeletal muscle. In the case of phosphofructokinase from rabbit muscle, the cyclic nucleotide has no effect at low values of energy charge but specifically prevents the normal reduction of activity at high energy charge (*26*). The picture that emerges is strikingly similar to a technological device that is automatically regulated but has provision for manual override. In the unstressed animal, glycolysis in muscle may be envisaged as being controlled on the basis of local instantaneous needs for ATP and citrate cycle intermediates. When the central nervous system perceives a menacing or otherwise exciting situation, epinephrine is released. In muscle, this hormone activates the enzyme that produces cyclic AMP, which in turn releases phosphofructokinase from local control. Phosphorylase b, which catalyzes the first step in utilization of glycogen, is also under local energy charge control (AMP stimulates its activity) in the unstressed animal. Cyclic AMP overrides local control of phosphorylase also, but in this case by a much more complex mechanism involving activation of several enzymes [see Krebs *et al.* (*29*)]. Acceleration of glycogen breakdown and of glycolysis presumably leads to higher than normal levels of all of the intermediates involved in ATP regeneration and thus provides for an intense initial level of energy release. Like other animal tissues, muscle contains a phosphodiesterase that hydrolyzes cyclic AMP; thus, the state of central override of local control will terminate when excitation of the central nervous system ends, leading to a fall in epinephrine level and consequently to slowing or cessation of cyclic AMP production.

Several other compounds have been shown to affect the behavior of phosphofructokinase *in vitro,* and in each case the observed effect is in the direction that would be expected to contribute to metabolic stability *in vivo*. These effects will not be described here. The purpose of this chapter is not to discuss metabolic regulation as such, but to illustrate some of the ways in which enzyme behavior is adapted by selection to

28. G. A. Robison, R. W. Butcher, and E. W. Sutherland, *Ann. Rev. Biochem.* **37,** 149 (1968).

29. E. G. Krebs, R. J. DeLange, R. G. Kemp, and W. D. Riley, *Pharmacol. Rev.* **18,** 163 (1966).

meet metabolic needs, and to emphasize the importance of considering functional design requirements in any study of enzyme properties.

2. *Biosynthetic Pathways*

The needs of the cell for such metabolites as amino acids and nucleotides, which serve as building blocks for the production of the macromolecules of which the cell is mainly composed, will vary widely with changes in external and internal conditions. It is obviously desirable that the concentrations of these and other metabolites should be held reasonably constant by regulating the rate of each biosynthetic pathway in response to the level of the end product. As noted above, this was the first type of metabolic control to be recognized at the molecular level. All biosynthetic processes require ATP; hence, it seems desirable that the biosynthetic activities of the cell should also be regulated in response to the availability of ATP (the energy charge of the adenylate pool). Biosynthesis is essential for growth and reproduction but generally not for immediate survival. Most substrates require preliminary phosphorylation or some other type of ATP-consuming activation before they can be metabolized to yield ATP; hence, even a cell starving for energy should conserve its ATP supply (maintain a high energy charge) to allow for such activation when a substrate becomes available. Thus ATP-utilizing processes such as biosynthesis should be sharply curtailed in response to only a small decrease in energy charge. It has been noted in an earlier section that several enzymes in biosynthetic pathways have been observed to respond appropriately to variation in energy charge as indicated by curve U of Fig. 3.

An engineer designing a biosynthetic control system would want the control signals to affect the enzyme that catalyzes the first step unique to the synthesis. There are two main reasons for this: (a) Direct competition between the first enzymes of alternative pathways for the same branchpoint metabolite (the common substrate for the two enzymes) permits very sensitive control, and (b) by controlling at this step the concentrations of all intermediates can be kept low; control at a later step would lead to increases in the concentrations of intermediates before the controlling point. Avoidance of high metabolite concentrations is probably among the most fundamental imperatives in the design of a living cell (*30*).

That product feedback controls do in fact operate on enzymes at metabolic branchpoints (enzymes catalyzing first committed steps in

30. D. E. Atkinson, *in* "Current Concepts in Metabolic Regulation" (B. L. Horecker and E. R. Stadtman, eds.), p. 29. Academic Press, New York, 1969.

sequences) is one of the major generalizations to have emerged from work on biosynthetic regulation. The arguments for control at branch-points apply to energy charge control as cogently as to end product control; hence, we should expect to find both types of control affecting the same enzymes. The type of interaction to be expected is illustrated in Fig. 5. Under normal conditions the rate of biosynthesis should be jointly controlled by energy charge and end product concentration, but when end product concentration is high synthesis should be strongly inhibited regardless of energy charge, and when energy charge is low synthesis should be inhibited regardless of end product concentration (*18*). The pattern is evidently similar to that of Fig. 4, but in a sense the control logic is reversed. For the amphibolic pathway (Fig. 4), two conditions—high energy charge and high indicator metabolite concentration—must be met if the sequence is to be inhibited. The biosynthetic pathway, on the other hand, requires two conditions—high energy charge and low end product concentration—for high activity. Both types of interaction clearly reflect the metabolic roles of the corresponding pathways.

Dual control of the type shown in Fig. 5 has been found in every case where appropriate experiments have been done (*31*). The lysine-sensitive aspartokinase of *Escherichia coli* responds in this way to energy charge and lysine concentration, whereas phosphoribosyl-ATP synthase, which

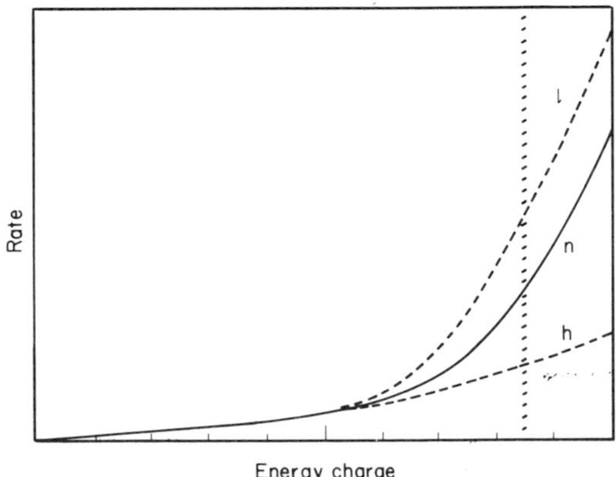

FIG. 5. Generalized interaction between adenylate energy charge and end product concentration expected in the control of a regulatory enzyme in a biosynthetic sequence. The curves correspond to low (l), normal (n), and high (h) concentrations of the end product modifier. From Atkinson (*18*). Reprinted from *Biochemistry* **7**, 4030 (1968). Copyright 1968 by the American Chemical Society. Reprinted by permission of the copyright owner.

31. L. Klungsøyr, J. H. Hageman, L. Fall, and D. E. Atkinson, *Biochemistry* **7**, 4033 (1968).

catalyzes the first step unique to histidine synthesis, responds to energy charge and histidine concentration. In the case of phosphoribosylpyrophosphate synthase, which supplies the phosphoribosyl group for synthesis of nucleotides, the end product inhibition is exerted cumulatively by ribo- and deoxyribonucleoside di- and triphosphates. The response is thus more complex than for the other enzymes described, but the interaction pattern again resembles that of Fig. 5.

3. *Interactions between Amphibolic and Biosynthetic Pathways*

It seems likely that families of response curves like those illustrated in Figs. 4 and 5 are characteristic of many enzymes that catalyze reactions at metabolic branchpoints, and that interaction between them is a primary factor in maintenance of the metabolic stability of the cell. The general type of interaction to be expected between many enzymes of the U type (Fig. 5) and the smaller number of R-type enzymes (Fig. 4) that regulate amphibolic sequences is obvious (*18*).

Several of the enzymes discussed above are known to have other modifiers in addition to energy charge and the end product or indicator metabolite feedback effectors mentioned, and it is to be expected that additional modifiers will be found for most of them. A regulatory enzyme is a multi-input device. Its response to a given modifier will necessarily be affected by the levels of other modifiers. In some cases [for example, phosphoribosyl-ATP synthase (*31*)] the curve showing response to energy charge assumes the typical shape only in the presence of a feedback modifier. Such an observation is not surprising. It must be remembered that enzymes have evolved in the context of a metabolizing cell, where all of the modifiers were present. In attempting to understand enzyme behavior, it is desirable that the parameters studied be as realistic as possible. A plot of reaction velocity as a function of the concentration of a single adenine nucleotide may serve to illustrate some property of interest, but it is not applicable to conditions *in vivo,* where a range of concentrations of one nucleotide in the absence of others never exists. Plots against the energy charge of an adenylate pool of constant size, as symbolized by Fig. 3, seem much more realistic in terms of enzyme function. In turn, plots showing interactions between responses to energy charge and to other modifiers, as illustrated by Figs. 4 and 5, should be more realistic than plots in terms of energy charge alone or modifier concentration alone. Incorporation of the effects of other modifiers would complicate the making and interpretation of response plots but would approach a meaningful functional description of the enzyme more closely.

In one important sense curves of the type illustrated here are still unrealistic. Enzymes compete for branchpoint metabolites, and the outcome of the competition will depend largely on the relative apparent affinities of the competing enzymes for their common substrate. The effects of energy charge and other modifiers are usually on the affinity of the enzyme for substrate rather than on maximal velocity or catalytic constant. The generalized curves shown here are all presented in terms of relative velocity at a fixed substrate concentration; plots showing the variation in $(S)_{0.5}$, the substrate concentration required for half-maximal velocity, would be more functionally realistic. Even such plots would not be adequately descriptive in some cases. For some regulatory enzymes, the apparent order, in terms of substrate, changes on the addition of modifiers. It is clear that in these cases where both the shape of the plot of velocity versus substrate concentration and its midpoint concentration, $(S)_{0.5}$, are changing, adequate graphical representation of the effects of several modifiers acting simultaneously will be very difficult.

4. *Operational Response Curves*

No serious effort has yet been made to characterize a regulatory enzyme thoroughly by means of operational response curves. Such an effort might be premature in view of the very unsatisfactory state of our knowledge of conditions inside the living cell, for example the effective ionic strength and the chemical potentials of water and solutes. It seems reasonable to suggest, however, that characterization of this type may prove to be both more attainable and more useful than is description in terms of the parameters and rate equations of classical kinetics. Description of functional elements in terms of performance curves and simple parameters derived from them is generally more useful to design engineers than would be a complete mathematical description of the element, even if it were available. It seems far from certain that the beneficial results would be commensurate with the effort of attempting to derive a classical rate equation that really comes to grips with the behavioral complexities of a typical regulatory enzyme. The reaction catalyzed by phosphofructokinase, for example, is second order with respect to fructose 6-phosphate. The concentration of this substrate required for half-maximal velocity depends on the concentration of the other substrate, ATP, in such a way that under some conditions the apparent kinetic order with respect to ATP is −1. The effect of ATP on affinity of the active site for fructose 6-phosphate is overcome by AMP. For the enzyme from some sources, the order with respect to fructose 6-phosphate is also changed by the relative concentrations of adenine nucleotides present. Magnesium ion is required at the active site, as for all kinases, and in addition appears to enhance fructose 6-phosphate binding. Citrate strongly reduces affinity for fructose 6-phosphate; it is not clear whether this effect is exerted directly or by alteration of affinities for ATP and AMP. (The affinity of an enzyme for one modifier is generally affected by the binding of other modifiers.) Fructose diphosphate, although a product of the reaction, is a positive modifier. All of these effects appear to be exerted primarily on binding affinities, but ammonium ion very strongly increases the maximal velocity of the reaction catalyzed by phosphofructokinases from some sources. A rate equation capable of quantitatively describing the behavior of this enzyme would be horrendous.

Similarly, it may be noted that the classic description of interactions between two ligands and an enzyme as "competitive," "noncompetitive," or "uncompetitive" may not be appropriate when applied to the behavior of regulatory enzymes. These adjectives are defined in terms of the relative positions of curves in a Lineweaver–Burk double reciprocal plot, and refer, respectively, to lines that intersect on the ordinate, that intersect on the abscissa, or that are parallel. These are evidently three special cases, and there is no reason why a given interaction need fit any of them. The imprecision and ambiguity of Lineweaver–Burk plots (*32*) have probably contributed to a general belief that most interactions fit one of the three arbitrary categories: The relative weight that should be given to various points in such a plot is ambiguous, and lines that are expected to meet at a common point will frequently do so. At least in the case of regulatory enzymes, the categories are so oversimplified as to be of very dubious value. Under certain conditions, for example, one of the substrates of the phosphofructokinase reaction, ATP, would appear by classical criteria to be a competitive inhibitor of the other, fructose 6-phosphate; in other concentration ranges, the interaction will be that typical for two substrates. As the concentrations of modifiers change, there is no reason why an interaction previously classified as competitive may not change to one of the other types or, more likely, to "mixed," the name sometimes given to the whole range of behavior that fits none of the three special cases. Regulatory enzymes, characterized by complex mutual interactions between proteins and a number of ligands, must be considered on their own terms, and their behavior cannot usefully be described by categories defined for application to much simpler situations.

32. J. E. Dowd and D. S. Riggs, *JBC* **240,** 863 (1965).

5. *Covalently Bonded Modifiers*

This discussion has dealt entirely with modification of the behavior of regulatory enzymes as a result of noncovalent binding of modifiers. At least four examples are known of a still more complex type: enzymes that change their response to regulatory parameters as a result of covalent attachment of a specific group to a specific site on the enzyme. Phosphorylation of phosphorylase b [for review, see Krebs *et al.* (*29*)], pyruvate dehydrogenase (*33*), and glycogen synthase (*34*), all from mammalian sources, leads to significant changes in response to metabolite modifiers. Glutamine synthase from several bacterial species exhibits the most complex behavior of this type yet described. This enzyme contains twelve subunits. Twelve adenyl groups may be covalently attached to the enzyme from *E. coli,* presumably one to each subunit. In this form the enzyme is inhibited cumulatively by at least eight feedback inhibitors, each of which contains a nitrogen atom derived from glutamine. When the adenyl groups are removed, the enzyme becomes catalytically more active, and most of its response to feedback inhibitors is lost. The same cells contain an enzyme capable of catalyzing transfer of adenyl groups from ATP to glutamine synthase and an enzyme capable of removing the adenyl groups. The adenylylating enzyme is activated by glutamine, and the deadenylylating enzyme is inhibited by glutamine but activated by α-ketoglutarate. Taken together, the properties of these three enzymes seem to insure that glutamine synthase will be active and relatively insensitive to inhibition when the glutamine/α-ketoglutarate ratio is low, but when glutamine is present at an adequate level the activity of the synthase will be regulated by the levels of the more remote end products. In addition to the other changes, the adenylylated enzyme requires Mn^{2+} ion for activity, while the deadenylylated form specifically requires Mg^{2+}. The properties of this fascinating enzyme and the others that behave somewhat similarly are discussed by Stadtman in another chapter of this volume (*35*).

D. Design Requirements

A typical regulatory enzyme is capable of sensing the values of several input parameters, combining them in a precise and nonlinear way, and using the result to control the response of the functional unit of the system, the active site. Integration of a complete regulatory system,

33. T. C. Linn, F. H. Pettit, and L. J. Reed, *Proc. Natl. Acad. Sci. U. S.* **62,** 234 (1969)

34. D. L. Friedman and J. Larner, *Biochemistry* **2,** 669 (1963).

35. E. R. Stadtman, Chapter 8, this volume.

including sensing, data processing, and responding elements, into a single molecule or molecular aggregate must approach the ultimate in miniaturization. The question often asked, explicitly or implicitly, a few years ago, "Why is an enzyme so large, since it does one simple thing?" may now be replaced by "How can a regulatory enzyme be so small, when it does so many complicated things?"

An assignment to design, merely by specification of one or a few sequences of amino acid residues, a control and catalytic system of the type just described would seem self-evidently impossible. Only our observation that evolutionary processes have produced many different catalytic and regulatory molecules of this type allows (and forces) us to believe that such systems are possible. Nevertheless, the magnitude of the design problem remains incomprehensible. It is perhaps of some comfort that the number of possible permutations of amino acid residues along a polypeptide chain of typical length is also incomprehensibly large. As has often been pointed out, the approximately 10^{200} possible sequences of a rather short chain of 150 residues is enormously larger than any physically meaningful number, including such estimates as the number of protons and neutrons in the known universe.

Because it seems well established that the three-dimensional structure of a protein is determined by its primary amino acid sequence, it might be thought that any random sequence would specify a unique and stable three-dimensional structure. Such an expectation would almost certainly be wrong. Perhaps the most important generalization that has yet emerged from X-ray diffraction studies on proteins is that globular proteins contain a dense hydrophobic interior from which water is excluded, with polar residues restricted largely to the periphery [for a recent discussion, see Perutz (*36*)]. The avoidance of spatial conflict between residues, while still packing the chains so densely that water is excluded, poses acute design problems. Flory (*37*) has pointed out the unlikelihood of such a conformation, and emphasized that in terms of random sequences the formation of a stable globular structure would be a rare and improbable event. It is possible to make a globular protein "only by strategically fashioning the sequence of residues according to their special conformational capacities" (*37*). Only a small fraction of the possible sequences can be arranged in space to form a stable globular protein. Since a unique and reasonably stable structure is essential for biological function, the great majority of possible sequences are probably already eliminated by this requirement. The ability of a protein to as-

36. M. F. Perutz, *European J. Biochem.* **8**, 455 (1969).

37. P. J. Flory, *in* "Conformation of Biopolymers" (G. N. Ramachandran, ed.), Vol. 1, p. 339. Academic Press, New York, 1967.

sume a unique globular form thus appears not to be a chemical or physical property of polypeptide chains in general but a biological consequence of selection for sequences with this capacity.

Probably a very small fraction of those peptide sequences that are capable of forming a stable globular conformation would possess significant catalytic activity. Not only do those that have been evolutionarily selected have catalytic activity of a very high order, but the design problem is further complicated by the requirement for extremely high specificity for substrates, extending, for example, to such metabolic coding symbols as the phosphate group, remote from the pyridine "business end" of the molecule, that distinguishes TPN from DPN. If, as seems likely, specific conformational changes at the catalytic site are involved in the action of most or all enzymes [for a recent review, see Koshland and Neet (*38*)], provision for a conformation that recognizes and binds substrate and then changes to a form that catalyzes the reaction introduces additional design complication. The amino acid sequence must specify not one, but at least two, highly specific conformations, and also provide that binding of a specific ligand will cause transformation from one to the other.

The preceding paragraph may indicate, in a loose verbal way, the major design requirements for an enzyme of the simplest type. It is evident that the actual requirements are much more stringent than those indicated. The binding energies for substrate and products, for example, must be precisely adjusted to fit the needs of the cell's metabolism and the properties of other enzymes. When the enzyme contains regulatory sites distinct from, but affecting, the catalytic site, or consists of interacting subunits (or both) the design problem becomes exponentially more difficult. Superposed on all of the requirements for a functional enzyme, adjusted to the cell's economy, are the needs for specific binding of one to a dozen modifiers. The binding energy of each must be precisely adjusted; for example, in biosynthetic regulation this interaction will to a large extent determine the steady-state pool level of the end product modifier. Binding of the modifier must change the conformation of the subunit, and probably in most cases also the interaction between subunits, in such a way that the properties of the catalytic site are altered in the right direction and to an appropriate extent. When there are two or more modifiers, their effects must be combined so as to yield an appropriate integrated response. In such cases as glutamine synthase, the whole system must be so constructed that removal of a specific small covalently attached group will inactivate most of the control interactions while not interfering with catalytic activity.

At present we have only the most primitive understanding of the evo-

38. D. E. Koshland, Jr. and K. E. Neet, *Ann. Rev. Biochem.* **37,** 359 (1968).

lutionary, structural, and functional relationships among polypeptide chains. It seems certain that the relatively new techniques of amino acid sequence determination and of three-dimensional structure assignment by X-ray diffraction will add immensely to our knowledge in these regards in the near future. These approaches are discussed in other chapters of this volume (*39, 40*). Both have already led to contributions of the most fundamental importance but are clearly entering an era of much wider application. Sequence determination seems capable of supplying evolutionary insight of a type hitherto totally inaccessible. Such frequently discussed questions as whether all or most of the enzymes that catalyze a given type of reaction but differ in specificity (for example, dehydrogenases or kinases) have evolved from a common precursor may soon be answered. The finding that a change of only a few amino acid residues has caused a dramatic difference in the specificities of two alcohol dehydrogenases (*39, 41*) is both an exciting example of the advances to be expected and a demonstration that enzyme behavior may be much more plastic, in an evolutionary sense, than has been expected.

X-ray diffraction studies have shown (*40*) that the catalytic sites of some enzymes are in definite clefts or depressions. Perutz (*36*) has suggested that a large part of the catalytic effectiveness of enzymes may result from the enhancement of activity of polar functional groups that results from their location in a cleft that extends into the nonpolar interior of the enzyme. It is still too early to know whether a cleft or pocket is a general, or even a common, feature of enzymes, but the suggestion is attractive. It will be interesting to see whether one consequence of the conformational change induced in an enzyme by binding of a substrate is translocation of the substrate and the functional groups of the catalytic site from the polar environment of the solvent to a relatively nonpolar environment—that is, whether a more open conformation closes into a cleft or pocket as a consequence of substrate binding.

Crystallographic studies on lactate dehydrogenase from dogfish muscle by Rossman and colleagues (*42*) indicate that DPN^+ bridges across the subunit junction. The nicotinamide end of the cofactor is necessarily bound at the catalytic site, but the results suggest that the adenylate end is adsorbed at a site on the adjacent subunit. A conformational change involving a decrease in symmetry results when the cofactor is bound; comparative studies with analogs show that the change is a consequence specificially of interaction between the protein and the adenylate end of

39. E. L. Smith, Chapter 6, this volume.
40. D. Eisenberg, Chapter 1, this volume.
41. H. Jörnvall, *BBRC* **35,** 542 (1969).
42. M. J. Adams, D. J. Haas, B. A. Jeffery, A. McPherson, Jr., H. L. Mermall, M. G. Rossmann, R. W. Schevitz, and A. J. Wonacott, *JMB* **41,** 159 (1969).

the cofactor. The enzyme apparently is able to bind pyruvate only after this conformational change has occurred; thus, the adenylate end of the cofactor, which has no chemical role in the reaction catalyzed by the enzyme, can be identified as the trigger for induced formation of a functional catalytic site. These results are important, in the context of this chapter, especially because they supply direct evidence for a substrate-induced conformational change in the Koshland sense, and because they demonstrate the importance of interactions at a boundary between subunits. Such interactions may be of general importance for both catalysis and regulation.

In considering the evolutionary, structural, and metabolic relationships among enzymes, it is important not to lose sight of the truism that mutation is expressed at the level of the protein, and often at the level of individual amino acid substitution, but selection can act only at the level of the whole organism. The amino acid sequence of any protein has necessarily evolved from the sequences of earlier proteins, and it may soon be possible to construct phylogenetic maps of evolution of enzymes that will superficially resemble the diagrams of descent of species. But the evolution of enzymes is fundamentally different from that of species. The "ancestors" of the enzyme have not necessarily been selected for excellence as enzymes—for greater catalytic effectiveness, higher specificity, increased affinity for substrate, etc.—but for selective advantage to the organism of which they were a part. It is evident that great catalytic effectiveness, high specificity, and strong affinity for substrate have in general been advantageous, but it is equally obvious that a lower degree of specificity has sometimes been selected for, and that too strong affinity for substrates or modifiers could be as harmful to the organism as too weak affinity. Compatibility among the properties of the enzymes of a given cell is obviously a prime essential for viable metabolism, and selection for a change in properties of one enzyme, perhaps in response to changed environment, must lead to gradual adjustment in the properties of several or many others. The marked difference in amino acid pool levels in different organisms, for example, must be accompanied by differences in ligand affinities at the catalytic and modifier sites of dozens of enzymes.

IV. Summary

Enzymes are functional units of complex chemical systems. Although present knowledge in this area is primitive, it is already apparent that the properties of enzymes generally, and of regulatory enzymes in particular, are very precisely tailored to the needs of the cells in which they occur. The kinetic behavior of an enzyme can be rationally interpreted only in terms of metabolic function, and the structural features of an enzyme molecule are only the means by which this behavior is attained. Each enzyme has evolved in the context of a metabolizing cell containing hundreds of other enzymes. Progress in understanding the kinetic and structural properties of any enzyme will be aided by progress toward understanding the interactions in which that enzyme participates in the living cell.

Author Index

Numbers in parentheses are reference numbers and indicate that an author's work is referred to although his name is not cited in the text.

A

B

D

E

G

H

K

L

M

N

O

P

T

X

Y

Z

Subject Index

C

D

G

I

K

N

O

P

R

T

U

Z